D1776390

The Fascination of
PHYSICS

Jacqueline D. Spears
Dean Zollman

Kansas State University

THE BENJAMIN/CUMMINGS PUBLISHING COMPANY, INC.
Menlo Park, California • Reading, Massachusetts
Don Mills, Ontario • Wokingham, U.K. • Amsterdam • Sydney
Singapore • Tokyo • Mexico City • Bogota • Santiago • San Juan

To Kevin and Kim
who are never afraid to ask another question.

Sponsoring Editor: Andrew Crowley
Production Supervisors: Susan Harrington and Mimi Hills
Cover and Interior Designer: John Edeen
Artists: Corbin Hillam (creative art) and House of Graphics, Palo Alto, California (technical art)

Cover Photo: Patterns of motion are revealed by color dyes as a delta wing shape is tested in a water tunnel near Paris. Courtesy of Office National D'Etudes et de Recherches Aerospatiales, Paris.

Copyright © 1985 by The Benjamin/Cummings Publishing Company, Inc.

All rights reserved. No part of this publication may be reproduced, stored in a retrieval system, or transmitted, in any form or by any means, electronic, mechanical, photocopying, recording, or otherwise, without the prior written permission of the publisher. Printed in the United States of America. Published simultaneously in Canada.

Library of Congress Cataloging in Publication Data

Spears, Jacqueline D.
   The fascination of physics.

   Includes index.
   1. Physics.   I. Zollman, Dean.   II. Title.
QC21.2.S64   1985   530   84-21616
ISBN 0-8053-6974-0

abcdefghij-HA-898765

The Benjamin/Cummings Publishing Company, Inc.
2727 Sand Hill Road
Menlo Park, California 94025

Preface
to the Instructor

The students for whom this book has been written typically have one exposure to physics—this course. Given the enormous breadth of interesting material, a major challenge facing any instructor is the selection of topics. As the twentieth century draws to a close, it seems appropriate that the enormous strides taken in modern physics be meaningfully integrated into the intellectual traditions of the past. Consequently, this book presents a somewhat different approach to the topics and concepts typically included in a course at this level.

ORGANIZATION

The book is divided into five major units. Each unit begins with a concept in classical physics and builds toward a topic that generally falls into the category of modern physics. The first, *Space and Time,* begins with a fairly traditional presentation of position and motion, continues with classical relativity, and concludes with an introduction to the special theory of relativity. The care we have taken to integrate the concept of reference frame into descriptions of position and motion lays a strong foundation for the results of special relativity. Unit 2, *Interaction and Force,* begins by describing interactions and how we know they occur, next looks carefully at forces and Newton's Laws, and closes with a discussion of the four fundamental interactions in nature and current attempts to build a unified theory. Students often find the concept of force to be one of the more elusive ideas in physics. We have taken special care in developing the concept of force carefully. Unit 3, *Energy,* uses the concepts of kinetic and gravitational potential energy to introduce energy conservation and the various forms of energy that have emerged from our commitment to the principle of energy conservation. The discussion then moves to thermal energy and the laws of thermodynamics. We conclude with a look at the atomic and molecular view of matter and how kinetic energy is related to thermal phenomena on the atomic scale. Wave motion, classical and quantum, is the topic of Unit 4, *Waves and Particles.* It begins with mechanical waves, presents diffraction and interference, and then shows how experimental results led to wave-particle duality and quantum mechanics. The final section, *From Electricity to the Nucleus,* starts with the ideas of a circuit and electrical energy, looks at electromagnetism and methods of generating electrical energy, and then concludes with discussions of converting nuclear energy into electricity. Throughout the book, in an attempt to show both the unity and continuity of physics, modern ideas are integrated with the ideas that preceded them in the logical development of physical concepts.

Philosophical, social, and historical discussions are included in the development of the concepts when it seems appropriate. In addition, four interludes and an epilogue treat these aspects of physics in some detail.

OUR AUDIENCE

In this text, we make a few assumptions. First, we assume that many of the students' prior experiences can help them learn physics. So, we employ an inductive approach, and try to start each major section with some common experiences and move to generalizations and concepts. Our second assumption is that the students in this course have limited proficiency with mathematics. However, we cannot completely separate mathematics from physics. (One cannot understand a conservation law unless one sees that a value does not change.) Thus we expect the students to use arithmetic to evaluate some algebraic expressions but not to perform algebraic manipulations. Our final assumption is that the students must do in order to learn. Three different levels of written exercises and a set of activities conclude every chapter. In addition, Self-Checks (with answers at the ends of the chapters) are scattered throughout the book. (See *Preface to the Student*.)

Throughout the book we have used metric units. SI units are used almost exclusively. The only exceptions are for pedagogical reasons. (For example, N/m^2 seems to emphasize the meaning of pressure better than pascals.) Our experience has been that essentially all 18–25-year-old students have been taught the metric system and have a reasonably good feeling for the sizes of units such as the meter and kilogram. (Few have a good idea of the size of a joule, but they have no better feeling for a calorie or BTU.) Older students may not have any formal training in the metric system, but they seem to have picked it up from interactions with younger people, usually their children. Thus we have used only metric units and presented conversions with traditional U.S. units only in an appendix.

ACKNOWLEDGMENTS

We received help from a variety of people as we wrote this book. First and foremost we owe a special thanks to all the students (approximately 1000 of them) who used preliminary editions of the manuscript and were always more than willing to tell us when something was not quite presented in the clearest possible way. In addition, we relied heavily on the reviews of Murray Alexander, Milo V. Anderson, Claire Chapin, Russell Coverdale, James R. Crawford, Dewey Dykstra, Jr., David J. Ernst, John Giles, John S. King, Bernard Kramer, James A. Lock, Bernard F. Long, John E. Maling, Kaye Martin, Allan Miller, Fernando B. Morinigo, Carl J. Naegele, Barton Palatnik, Paul Phillipson, D. L. Rutledge, Lawrence C. Shepley, Jack White, and S. J. Yarosewick. Our colleagues at Kansas State University and Marymount College were seemingly infinite sources of helpful suggestions. One of us (D. Z.) also enjoyed the hospitality of the University of Utah for a sabbatical leave during which time part of this text was written. At Benjamin/Cummings, Andy Crowley provided the editorial guidance to bring this book to a successful completion, and Robin Fox's careful reading of the entire manuscript helped us tighten the final version. Sue Harrington and Mimi Hills directed the manuscript through production. Finally, we have been greatly influenced by discussions with many of our colleagues on subjects ranging from how students think to the principles of physics. We thank all of them and especially acknowledge the influences of Bob Fuller, Paul Hewitt, and Bob Karplus.

No set of acknowledgments would be complete without paying special tribute to our spouses, who only occasionally complained about the amount of time we spent on this project.

Jacqueline Spears
Dean Zollman
Manhattan, Kansas

Preface
to the Student

As you begin your study of physics you probably have two questions.

    What do I need to know before I start?
    How can I learn physics most effectively?

Because students' learning styles vary, no unique answers can be given to these questions. However, our students have found some general guidelines to be useful.

    We assume that you have not taken a formal physics course before starting this one. However, we also believe that you already know a lot of physics. You have learned this physics just by living and doing your normal, everyday activities. What you probably do not know are the underlying concepts that connect and explain your many observations and experiences. So, the main items you need to bring to your study are your experiences and observations.

    Physicists use two languages—ordinary speech with some specialized vocabulary and mathematics. As you study physics, you will pick up the vocabulary. We have tried to keep new words, as well as familiar words with new meanings, to a minimum. Thus the vocabulary of physics should not hinder your learning.

    We have also kept the mathematics at a level with which you should feel comfortable. We do not expect you to solve algebraic equations. We do expect you to be able to use arithmetic to evaluate an algebraic expression. (For example, speed = distance/time; if distance = 10 meters and time = 5 seconds, what is the speed?) Most evaluations involve simple expressions, which you can do in your head. A few may require pencil and paper or an inexpensive calculator. Nothing fancy is required; anything that adds, subtracts, multiplies, and divides will do. Armed with your paper and pencil or your calculator, you will be more than adequately prepared for the mathematics in this book.

    As you study physics, your first introduction to new concepts will come from reading and listening. But, that is only the start. Understanding comes from using these ideas to explain observations and problems. An effective way to acquire this understanding involves several steps.

    Before each new concept is introduced, we present some common experiences or observations related to the concept. After you have read a description of the new concept, think back on the opening. Try to relate in your own words how the concept explains and ties together the experiences and observations. Also think about your own experiences. Can some of them be explained with your new knowledge? After the introduction of a concept we present some of its applications. Again, read these carefully to follow the reasoning in applying your new knowledge.

Within the text are Self-Checks. These short exercises give you a chance to see how well you have understood the material which you have just studied. Each Self-Check is also a warning. You will need to know this material to understand future concepts. So, complete the Self-Checks and compare your answers with those at the end of the chapter. Only after you have written out the answers to the Self-Checks completely and understood the correct answers are you ready to continue your study of physics.

Your learning becomes more active when you apply your newly acquired knowledge. Exercises at the end of each chapter are designed for this purpose. We have divided the exercises into four groups: Reviewing Chapter Material, Applying the Chapter Material, Extensions to New Situations, and Activities. You cannot learn physics effectively unless you complete at least some of these exercises.

The first exercises, Reviewing Chapter Material, are labeled with a prefix of A and are a series of questions taken directly from the material in the text. The questions ask you to state the ideas of the chapter in your own words. You should be able to complete all the questions for every chapter that you study.

Applying the Chapter Material, with a prefix of B, gives you the opportunity to practice using the concepts in situations that are very similar to the ones presented in the text. Sometimes these exercises involve using a general idea and some logic; other times they will require a calculation or two. In both cases your understanding will improve as you complete the exercises.

Extensions to New Situations, with a prefix of C, enables you to take an idea from the chapter, follow some logical steps, and reach a conclusion that is not stated in the chapter. We do not expect you to do this entirely on your own, so we provide some help. These exercises are where the fun can really begin. You know enough to start with one idea and see where it takes you.

Finally, the Activities present some things you can do—short experiments you can complete at home, books or articles you can read, essays you can write, and so on. They can frequently help more than anything else. Doing is one of the best ways of learning.

Whenever you are working on any of the exercises, you should remember that the underlying question is always: How do the concepts of physics explain this situation? You should not simply state the answer but give an explanation in terms of what you have learned. In physics, Yes is never a correct answer; Yes, because . . . may be correct.

In general, learning physics effectively involves not only reading and listening but also doing exercises and reflecting on your experiences. Next winter when you slip on some ice, do not think "Oh darn, I almost broke my leg," but "Let's see, what forces didn't I consider while walking on the ice?"

Jacqueline Spears
Dean Zollman
Manhattan, Kansas

Contents

Preface to the Instructor iii
Preface to the Student vii
Introduction xiii

I Space and Time 1

1 Position and Change 2
Describing Position 3
Coordinate Systems 6
Describing Changes in Position 10
What to Do When the Car Turns a Corner 14
Saving a Few Steps? 16

2 Describing Motion 22
Is It Moving? 23
Describing Motion 24
Answering Other Questions 25
Average and Instantaneous Descriptions of Motion 27
Describing Changes in Motion 30
Pedal Power Plus 34
Analyzing Motion with Photography 35

3 Relative Motion at Low Speeds 41
Moving Reference Frames 43
Rolling Right Along 45
Relative Speed and Velocity 47
The Principle of Relativity 51

4 Special Theory of Relativity 58
Why a New Theory? 59
Relative Velocity at High Speeds 62
Simultaneity at High Speeds 64
Time Intervals at High Speeds 66
Time Drags On 70
Twin Paradox 72
Length Contraction 72
Fisk's Disk 75
How Fast is Fastest? 77

Interlude: Past, Present, Future, and Elsewhere 83

II Interactions and Forces 85

5 Interaction and Momentum 86
Interactions 87
Factors Affecting the Interaction 88
Momentum 90
Conservation Logic 94
Conservation of Momentum 96
A Very Stubborn Earth 99
Interactions with Large Masses 100

6 Interaction and Force 107
Forces—A First Look 108
Force and Change in Momentum 110
Lengthening the Collision Time with Pedestrians 114
A Force Is a Force Is a Force 115

Kinds of Forces 118
More Than One Force 120
Force, Acceleration, and the Speed of Light 124
Off Tall Buildings in a Single, Safe Bound 125

7 Newton's Three Laws 132

Newton's First Law 133
Newton's Second Law 136
Accelerated Reference Frames 140
Helping "Mom" 143
Newton's Third Law 143
All Three Laws 146

8 The Fundamental Interactions 153

Gravitational Interactions 154
What Is It Like on Mars? 157
Electromagnetic Interactions 158
Nuclear Interactions 164
Thanks to Electrical Charge . . . 165
Interaction at a Distance 166

Interlude: Each Step Beyond 175

Energy 179

9 Energy 180

Interaction, Work, and Energy 181
Energy of Position 184
Energy and Motion 189
Energy Is Conserved 191
Throbbing Motion 193
Easier than Walking? 195
Energy Takes Many Forms 195
Changing Energy Forms: Canine to Kinetic 199

10 Thermal Energy in Matter 206

The Energy Stored in Matter 207
Temperature 209

Thermal Energy Transfer and Changes in Temperature 213
Blankets to Keep You Cool 215
Hot, Cold, and Lukewarm 218
You Cannot Sleep if the Flowers are Cold 220
Thermal Energy Transfer and Change of State 221
Really Cool Tea! 227

11 How Thermal Energy Is Transferred 233

Convection 234
Floating from Town to Town 237
Conduction 238
Closer is Warmer 243
Cost of Heating a Home 245
Radiation-Absorption 246
All Three Processes 249

12 Thermodynamics 256

The First Law of Thermodynamics: You Can't Get Something for Nothing 257
Something for Nothing—Almost 261
The Second Law of Thermodynamics: You Can't Even Break Even 263
How Well Can We do? 270
The Second Law Gets in the Way, Too 271
Entropy and the Universe 276

13 Atoms, Molecules, and Thermal Energy 281

Inside Matter 282
Inside the Molecule 284
Thermal Energy in Transit 287
Second Law of Thermodynamics 290
A Philosophical Conclusion—Or Is It a Beginning? 293

Interlude: Social Issues and the Energy Crisis 298

IV Waves and Particles 301

14 Making Waves 302
Wave Motion 303
Characteristics of Waves 306
Moving Sources and Receivers 311
Using Doppler Shifts to Measure Speed 313
Waves Meet Matter 314
Waves Meet Waves 318
Getting the Message Across 327

15 Waves: Sound and Electromagnetic 333
Wave Classification 334
Sound 340
Electromagnetic Waves 348
"Seeing" the Bait 354
Right Off the End of the Piano! 357

16 Interference and Diffraction 363
Interference Patterns 364
Describing Interference Patterns Quantitatively 370
Interference Phenomena 371
Interference and Acoustics 375
Diffraction Patterns 376
Diffraction and Resolution 379
Waves Versus Particles 381

17 Wave-Particle Duality 387
The Particle Nature of Waves 388
Seeing with Photons 394
Wave Nature of Matter 396
De Broglie Wavelengths 397
Wave-Particle Duality 400

18 Light, Quanta, and Atoms 406
Light Spectra and Atomic Structure 407
Matching the Emission Spectrum 413
Lasers 414
The Quantum-Mechanical Atom 418
Probability and Uncertainty 424
Heisenberg's Uncertainty Principle 427

Interlude: Newton to Heisenberg 434

V From Electricity to the Nucleus 437

19 Turning on the Lights 438
Sparks and Lightning 439
Electrical Circuits 440
Taking the Light With You 443
Circuits, Resistance, and Safety 446
Energy Transfer in Circuits 448
More Than One Receiver 450
Why Do My Christmas Lights Burn Out So Fast? 455
Calculating Resistance in Parallel Circuits 457

20 Electromagnetism 464
Magnetism 465
Storing Information with Magnetic Domains 469
Electric Currents and Magnetic Fields 470
Move, Cow, Move 473
Electromagnetic Induction 476
Electromagnetism 482

21 Radioactivity 488
The Nature of the Nucleus 489
Nuclear Transformations 492
Alpha Decay 496
Beta Decay 497
Beta Decay in the Study of Art 499
Gamma Decay 500
Half-Life 503
Half-Lives in Archeological Dating 505

22 Nuclear Energy 513

Binding Energy 514
Nuclear Fission 518
Nuclear Fusion 525
The Bomb 527
Nuclear Radiation and the Human Body 531

Epilogue: Science, Technology, and Risk 536

Appendix A: Systems of Measurement 538

Appendix B: Powers of Ten 541

Index 543

Photo and Illustration Credits 548

Introduction

Our children, Kevin and Kim, ask lots of questions:

    Why do things fall down?
    Why won't my sled work on the sidewalk?
    What is a star?
    Why did the cat scratch me?
    Why aren't all people treated fairly?
    How do we know about atoms?

    You no doubt asked many of the same questions. In a lifetime, we ask literally thousands of questions—questions to gain information about ourselves and our world. Each question is an attempt to understand, to organize better the experiences we have had. They become the basis for the knowledge that we hand down from generation to generation.

    Scientists ask only questions that have specific characteristics. First, the questions must deal with concepts upon which we can agree. Why aren't all people treated fairly? is not a scientific question because its answer depends on different definitions of *fair*—definitions that depend on the values and emotional response of the person asked. On the other hand, How do we know about atoms? is a scientific question. Knowledge accumulated during centuries of investigation has resulted in a definition of the atom upon which everyone agrees. A second characteristic of a scientific question is that it must be able to be answered by experimentation. Normally, Why did the cat scratch me? cannot be answered experimentally because we cannot duplicate the event. If we could, we could vary the circumstances and discover what combination motivates the cat to scratch. The requirement that we be able to answer a question by experimentation helps assure us that we can agree upon the answer, that the answers will become part of the knowledge base.

Asking questions that can be answered by experimentation allows scientists to agree upon the answers and describe their observations in common terms. However, the answers to many questions change over a span of centuries, sometimes even over a span of a few months. How, you might ask, can the answers change? Because the experiments we can perform change. As we develop more-sophisticated ways of observing nature, we gain new perspectives. Our concepts of space, time, and matter have changed as we have been able to explore the space beyond our earth, to observe objects moving at incredibly high speeds, and to "see" smaller and smaller pieces of matter. Each new perspective adds new answers, answers that modify our view of nature.

When Kevin asks, "Why doesn't my sled work on the sidewalk?" he is beginning to be a scientist. His question can be answered by experimentation. But he needs to ask and answer other questions—questions that enable him to see patterns: "Will it work on the lawn? On a wet lawn? On snow covered with dirt?"

You may have done puzzles in which you are given a sequence of numbers and asked to predict the next one: 2, 4, 6, ?. To solve such a puzzle, you look for a pattern. Once you find a pattern, you can use it to predict the next number. Kevin has to do much the same thing with his answers. He looks for a pattern as he tries the sled on various surfaces.

Were patterns all that concerned us, we could be content with a list of patterns—a list of observations, so to speak. But science tries to place a series of patterns within a broader pattern, several broader patterns within still broader patterns, and so forth. In short, scientists try to understand the patterns. In the process they build theories.

We can illustrate the increased power of theories with an everyday example. In a letter to Ann Landers, a woman complained about her mother-in-law, who would habitually arrive an hour ahead of the prescribed time for dinner. The daughter-in-law recognized the pattern and responded by inviting her an hour later than she really wished her to arrive. At this point she observed a single pattern. The immediate problem is solved but not really understood. What if the two women were to meet for a concert? Should the daughter-in-law add an hour to the starting time? Without understanding the overall problem—looking for broader patterns and developing a theory—the daughter-in-law cannot answer this question. Only when she looks at other situations can she build a theory of her mother-in-law's behavior. Is her mother-in-law habitually early for all invitations or only for the daughter-in-law's? Is she early for movie invitations as well as for dinner invitations? Broader and broader patterns lead to a theory that is ultimately more useful.

Asking questions, recognizing patterns, and building theories help define science. Physics is one of several disciplines that share this approach to knowing. As you study physics in the chapters that follow, you will share the knowledge accumulated by thousands of physicists over a span of the last 400 years. As you tackle the task of learning about the theories of intellectual giants like Galileo, Newton, Maxwell, Einstein, and Curie, please bear in mind the words of Didacus Stella, a Roman general:

Pygmies placed on the shoulders of giants can see more than the giants themselves.

Compared to Newton and Einstein, we are all intellectual pygmies. Yet their knowledge can become ours, granting us an even broader and richer understanding of our world. This gift is, after all, what makes us human.

Space and Time

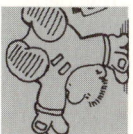
CHAPTER 1
Position and Change

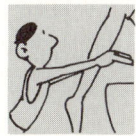
CHAPTER 2
Describing Motion

CHAPTER 3
Relative Motion at Low Speeds

CHAPTER 4
Special Theory of Relativity

We think nothing of our differing points of view as we rate the latest movie or last week's bestsellers. In many respects, each of us sees something different. So it is with the concepts of position and motion. A book is on my left but your right. We say we're standing still, yet the earth upon which we stand rotates once every twenty-four hours, carrying us past the sun at a rate of 1670 kilometers per hour. Descriptions of position and motion do indeed depend upon our point of view.

In 1905 Albert Einstein added two more quantities—length and time—to our list of relative concepts. Objects are shorter and time moves more slowly to observers moving past at speeds near the speed of light. These ideas seem absurd, partly because our thoughts are dominated by low-speed experiences. However, the relativity of length and time has been verified experimentally.

Position, motion, length and time—our descriptions of space and time depend on our separate points of view. Physics offers a process by which separate points of view can be merged—a common reference frame from which descriptions can agree.

Position and Change

The question of different points of view is a very basic one. Some people like modern art; others call it scribbling. Some like the latest movie; others find it mediocre. When we deal with opinions, it is rare that we ever share completely another's point of view. When we describe position, however, we can agree. All the Wizard (Figure 1-1) had to do to avoid confusion was turn around and face the two baskets in the same direction as the unfortunate citizen. Scientists consciously choose concepts that enable them to share points of view—to agree upon what they observe.

We begin studying motion by defining position and change in position. This chapter will show how *reference objects* combined with *reference directions* provide us with a *reference frame* with which to agree upon the position of an object. In order to be more precise, we will introduce *coordinate systems*, which incorporate the process of measurement into reference frames. *Distance* and *displacement* describe the change in position of a moving object. These last two concepts are the foundation for our later descriptions of motion.

By permission of Johnny Hart and News Group Chicago, Inc.

Figure 1-1

DESCRIBING POSITION

Every day we describe the position of many objects. It's on the table. She's at the office. Seaton Hall is north of the Student Union. All these statements describe the location of one object using another object and some reference direction.

Reference Objects and Reference Directions

To convince yourself of the need to describe one object's position in terms of another object, try a short exercise. Figure 1-2 shows a black dot inside a square. Without referring to any other object, try to describe the position of the dot. You probably find the task impossible. You might say that the dot is in the upper center of the square, but then you have used the square itself. If you now try the same exercise using objects inside the square, the task is much easier! You might say that the dot is directly above the person.

The exercise using Figure 1-2 shows that we need other objects to describe an object's location. A **reference object** is anything used to describe

Figure 1-2

We use the person as a reference object to describe the position of the dot.

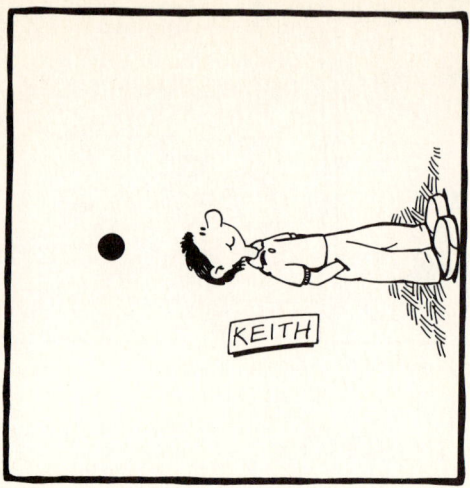

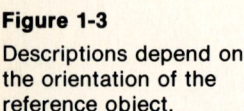

Figure 1-3
Descriptions depend on the orientation of the reference object.

the location of another person or thing. When you say that the dot is directly above the person, you are using the person as the reference object. If you tell a friend that the library is next to the Fine Arts Building, you are using the Fine Arts Building as your reference object. Whenever you use reference objects, you describe another object's position *relative to* the reference object. You describe the position of the dot relative to the person and the position of the Fine Arts Building relative to the library.

When two people choose the same reference object, they still may not agree in their description of the location of another object. You can readily see this if you consider Figure 1-3. First describe the location of the dot using Keith as the reference object. Then describe the dot's location as Keith would describe it using himself as the reference object.

    You: The dot is to the left of Keith.

    Keith: The dot is above me.

In both cases Keith was the reference object, yet the descriptions of the dot's location were different. Here the differences arise from the terms *above, below, left,* and *right.* Keith used himself to establish *above* and *below.* You used yourself to establish *left* and *right.* You and Keith each used Keith as the reference object but your own bodies to establish the **reference directions.** If you turn the book so that Keith is right side up, your use of reference directions will agree with Keith's. Then, both of you will report that the dot is above Keith.

In real life, as in Figure 1-3, we use our own bodies to establish the reference directions *right* and *left.* The confusion this causes is apparent in the Wizard of Id cartoon (Figure 1-1). In contrast to the subjectivity of right and left, the reference directions *up* and *down* seem more objective. Gravity defines our sense of up and down, so that the terms mean essentially the same to everyone. Down is the direction things fall. Up is the direction opposite to down. Thus, the situation in Figure 1-3 was rather contrived. Normally, one person's *up* and *down* would be the same as another's.

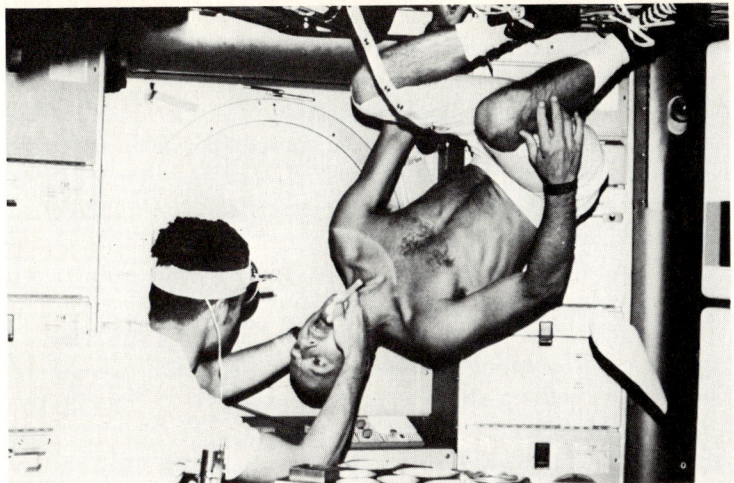

Figure 1-4

The space program has given us an opportunity to explore our sense of up and down in a weightless, or gravity-free, environment. The photograph of the Skylab astronauts (Figure 1-4) looks strange because the things in Skylab are not being pulled toward the earth. Thus we have no clues to establish "right side up." When asked about his sense of orientation in space, Astronaut Joseph Kerwin responded:

> *You do have a sense of up and down, and you can change it in two seconds, whenever it's convenient for you. If you go from one module to the next and you're upside down, you say to your brain "I want that way to be up" and your brain says "OK then that way is up." It's strictly eyeballs and brain.*

Once the physical sensation of gravity is removed, up and down are as subjective as left and right.

Reference Frames

Since we must be able to describe the positions of objects in weightless space as well as here on earth, we need to rely upon more than "eyeballs and brain." In order to agree upon the location of an object, we have to establish a common reference frame. A **reference frame** consists of the reference object and the reference directions used in our description. When we say that Seaton Hall is north of the Student Union, the reference frame consists of Seaton Hall and the compass direction north. Once we define our reference frame, others should be able to place themselves in that same reference frame and agree upon the location of Seaton Hall.

While the strict definition of a reference frame requires both a reference object and a set of reference directions, ordinarily we just mention the reference object. "It's on the table" and "she's at the office" both identify reference objects—the table and the office. The terms *on* and *at*, however, imply a set of commonly agreed-upon reference directions. This works fine as long as we are all in the same earth-based reference frame; but, as illustrated by the Wizard of Id cartoon, assumed reference directions can cause problems.

When we describe an object's location differently from someone else, we usually do so because we have chosen different reference frames. There is no end to the number of reference frames we can invent. Given the standard set of reference directions implicit in our vocabulary, we can invent as many reference frames as there are reference objects. "John is standing five meters west of the tree" and "John is standing two meters north of the house" could both be describing the location of the same person, but from two different reference frames. In the first, the tree defines the reference frame; in the second, the house defines the reference frame. Similarly, statements like "Ed is standing five meters west of the tree and Mary is standing two meters north of the house" tell us nothing about where Ed is relative to Mary. Each description uses a different reference frame.

Ordinarily we can resolve these differences by agreeing on one of the reference frames or by inventing a new reference frame common to the two descriptions. We might both describe John's location relative to the tree, agreeing to use the same reference frame. Or we might describe where the tree is relative to the house, allowing us to invent a common reference frame in which to describe John's location. Either solution would enable us to agree on our descriptions. The important characteristic of reference frames is that they allow people to describe position and change from a common point of view.

SELF-CHECK 1A

For each statement below, identify the reference frame used and whether the reference directions are stated or implied. Is the same reference frame used for each pair of statements?

a. The book is lying on the nightstand.
   The book is next to the window.

b. Venus is 20° north of the western horizon.
   The moon is about 25° north of the western horizon.

c. The Student Union is north of the Library.
   The Fine Arts Building is north of the Student Union.

COORDINATE SYSTEMS

Even with a common reference frame, our descriptions of position may still be imprecise. Suppose I tell you that the flowers are a little to the right of and slightly higher than the book. Missing here is a precise description of distance. How far to the right is *a little?* How much higher is *slightly?* Coordinate systems, which define distance as well as direction, allow us to be more precise.

Adding Measurement

Let's return to the dot, this time without the square. Figure 1-5 shows the dot amidst a series of lines. Use the lines to describe the location of the dot.

Using the convention of left and right and up and down while facing the page, you could say that the dot is at the intersection of the third line from the left and the fourth line from the bottom. Someone else might describe it as the intersection of the third line from the right and the third line from the top. Using the numbers associated with the lines allows us to establish immediately a common reference frame. Now we can say that the dot is at the intersection of the vertical line marked 3 meters and the horizontal line marked 4 meters.

The numbered lines combined with the reference directions vertical and horizontal provide a reference frame called a coordinate system. A **coordinate system** is a reference frame that shows units of measurement for each of the directions, or **dimensions.** Each number, or **coordinate,** describes the location of an object along one dimension. The reference object for the entire system is the point from which all others are measured, at the intersection of the lines marked 0. This point is known as the **origin** of the coordinate system.

A variety of coordinate systems can be invented. Typically each system is named by the number of dimensions it includes and the orientation among those dimensions. Because two coordinates, one for the vertical dimension and one for the horizontal dimension, are required to describe a position, this particular coordinate system is two-dimensional. Since the lines in Figure 1-5 form small rectangles, it is called a *rectangular coordinate system*. The complete name of our reference frame is a two-dimensional rectangular coordinate system.

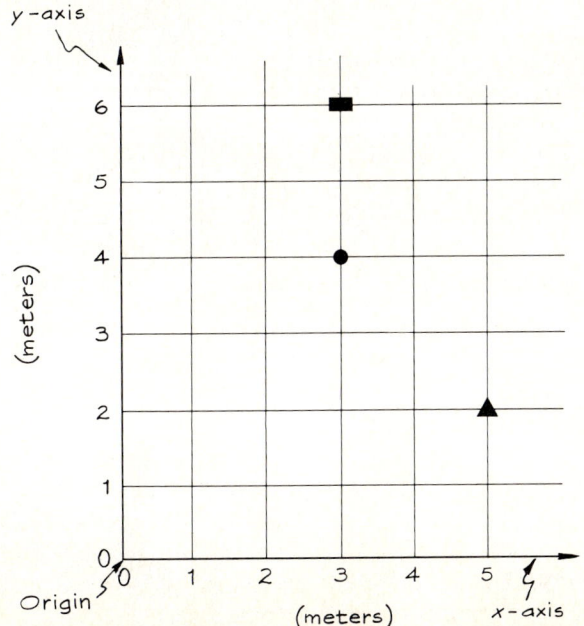

Figure 1-5

Numbered lines provide a more precise description of the dot's location. Distances along the coordinate axes are measured relative to the origin.

8 Chapter 1. Position and Change

Coordinate systems are not restricted to two dimensions nor to lines at right angles to one another. Football players care only about the distance to the goal line. Their coordinate system is one-dimensional. Hikers, however, need to know whether their trail goes over or around the mountain. Topographical maps add contour lines to give us a three-dimensional map of the terrain. A magnetic compass describes orientation about a reference circle. Its coordinates are expressed in degrees. Figure 1-6 shows a few of the many coordinate systems we invent to help us agree on the location of objects.

Figure 1-6
Everyday uses of coordinate systems.

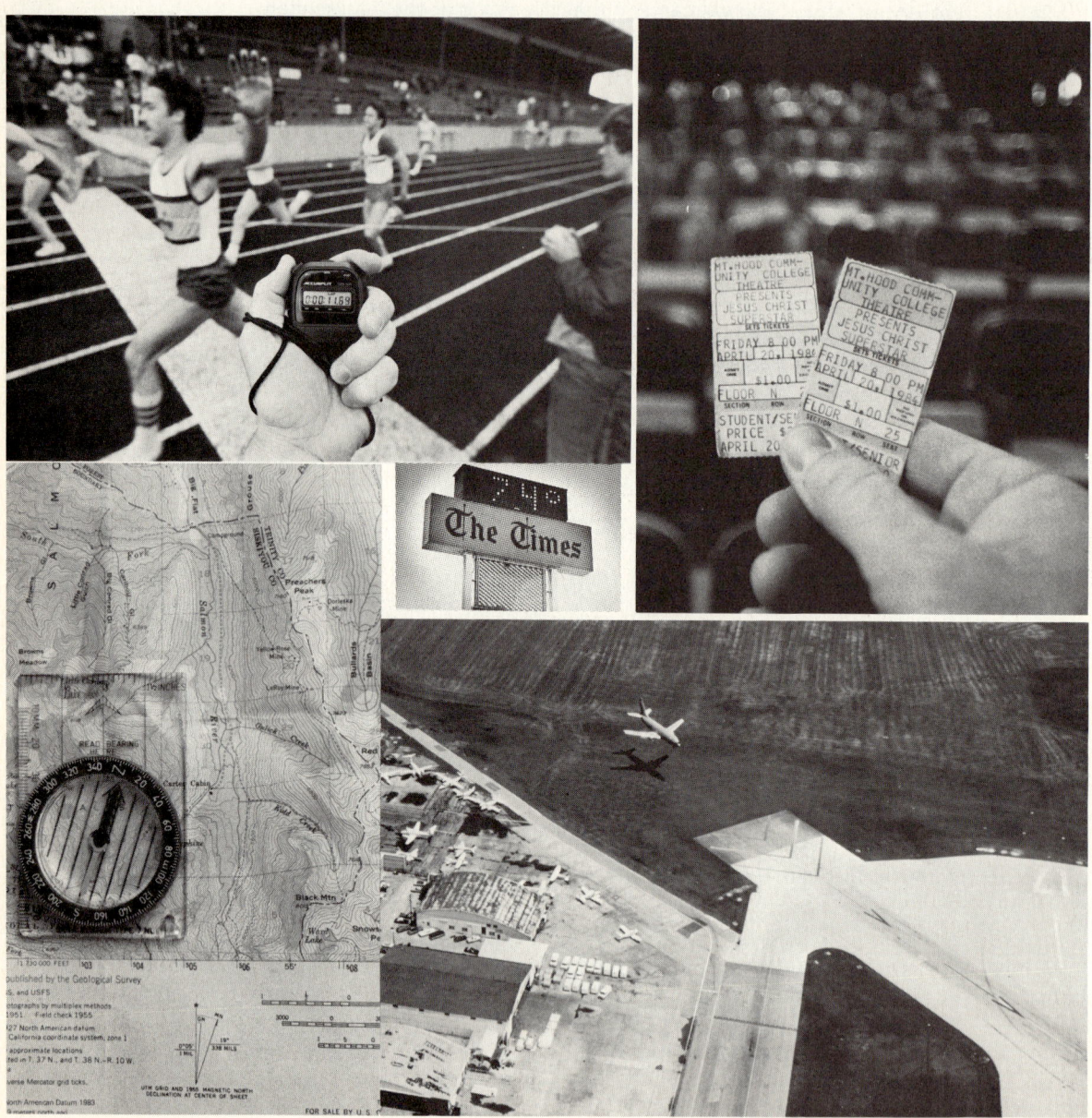

Using Rectangular Coordinate Systems

Physicists frequently use a two-dimensional rectangular coordinate system similar to that shown in Figure 1-5. The two lines that define the reference directions of the coordinate system and intersect at the origin are called the **coordinate axes.** Each coordinate axis has a scale along which distances from the origin can be measured. Scales are chosen for convenience and need not be the same for both axes. The orientation of coordinate systems can be chosen at will, but we generally use ourselves as the guide. If you hold a coordinate system so it faces you, up on the system is the same as up for you; right and left mean the same as your right and left. By convention, the horizontal and vertical dimensions are frequently called the *x*- and *y*-dimensions.

In using this coordinate system, we adopt a shorthand notation to describe position. The location of an object is given by two numbers with the appropriate units, separated by a comma: (10 meters, 5 meters) (meters are abbreviated m). The first number and unit gives the distance along the horizontal, or *x*-axis; the second gives the distance along the vertical, or *y*-axis. In Figure 1-5, the square is located at (3 m, 6 m).

The process of establishing a coordinate system with an origin as the reference object, coordinate axes to define reference directions, and measured distances to add precision is the physicists' way of defining the position of an object. To turn the statement around, the **position** of an object is defined by the *x*- and *y*-coordinates of the object in a coordinate system in which the origin and the orientation of the coordinate axes have been specified. Admittedly longer than a definition like "on my right," such a description is clearer because it allows different observers to agree upon the position of any object.

SELF-CHECK 1B

Use the shorthand notation to describe the position of the triangle in Figure 1-5.

BEETLE BAILEY

© King Features Syndicate, Inc.

Figure 1-7
What is Zero's coordinate system?

DESCRIBING CHANGES IN POSITION

If their purpose were only to describe fixed positions, coordinate systems would be useful to mapmakers and geographers but of comparatively little value to physicists and astronomers, who deal with things in motion. But motion involves a change in position, and coordinate systems help us describe these changes.

Consider an everyday example: moving about a city. Figure 1-8 shows a map of the northern section of Salt Lake City. Suppose you ride from P Street, First Avenue to H Street, Fifth Avenue along the route sketched. One way to describe your change in position would be to describe the route. "I started at First and P, rode to First and N, then to Fifth and N, and finally to Fifth and H." Another might be to describe the total distance you traveled along the route, 1.6 kilometers (km). A third way would be to describe the distance from P Street, First Avenue to H Street, Fifth Avenue "as the crow flies." As shown in Figure 1-8, this straight-line distance is 1.2 km. All three descriptions provide information about your motion, but the types of information are quite different. We will discuss the last two descriptions—distance and displacement—in more detail.

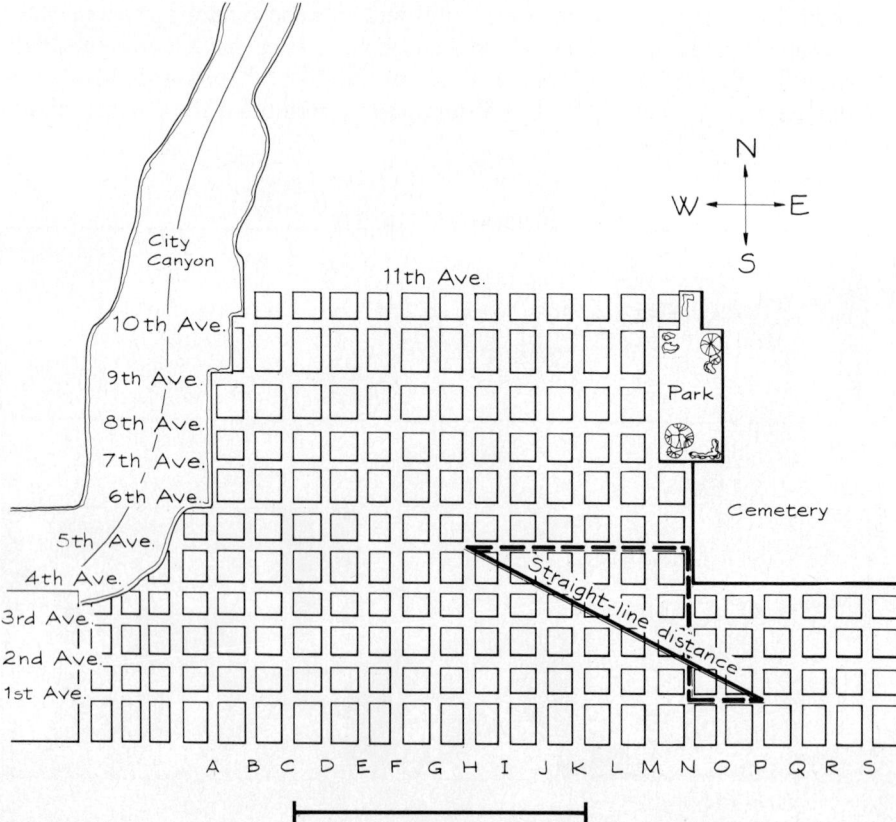

Figure 1-8

A route followed in traveling from P Street, First Avenue to H Street, Fifth Avenue in northern Salt Lake City.

Distance

We use the concept of distance to describe the length of the route you followed, 1.6 km. **Distance** is defined as the *length of path* traveled when an object changes position. We can measure distance in one of two ways. When traveling by car, we can use the beginning and ending odometer readings. The difference between these two numbers is the distance we have traveled. However, using the rectangular coordinate system provided by the streets and avenues, we can also measure the distance along your route by counting the number of blocks you traveled and multiplying this number by the length of each block. The first method involves continuous measurement of the path and does not depend on knowing where you started or stopped. The second involves calculating the length of the path from your known starting and stopping positions.

In a conventional rectangular coordinate system, distance is usually calculated from the coordinates that define your route. Figure 1-9 shows a route followed by a student in moving from A to B in a library. A coordinate system is superimposed on the route to allow us to measure distances. The student

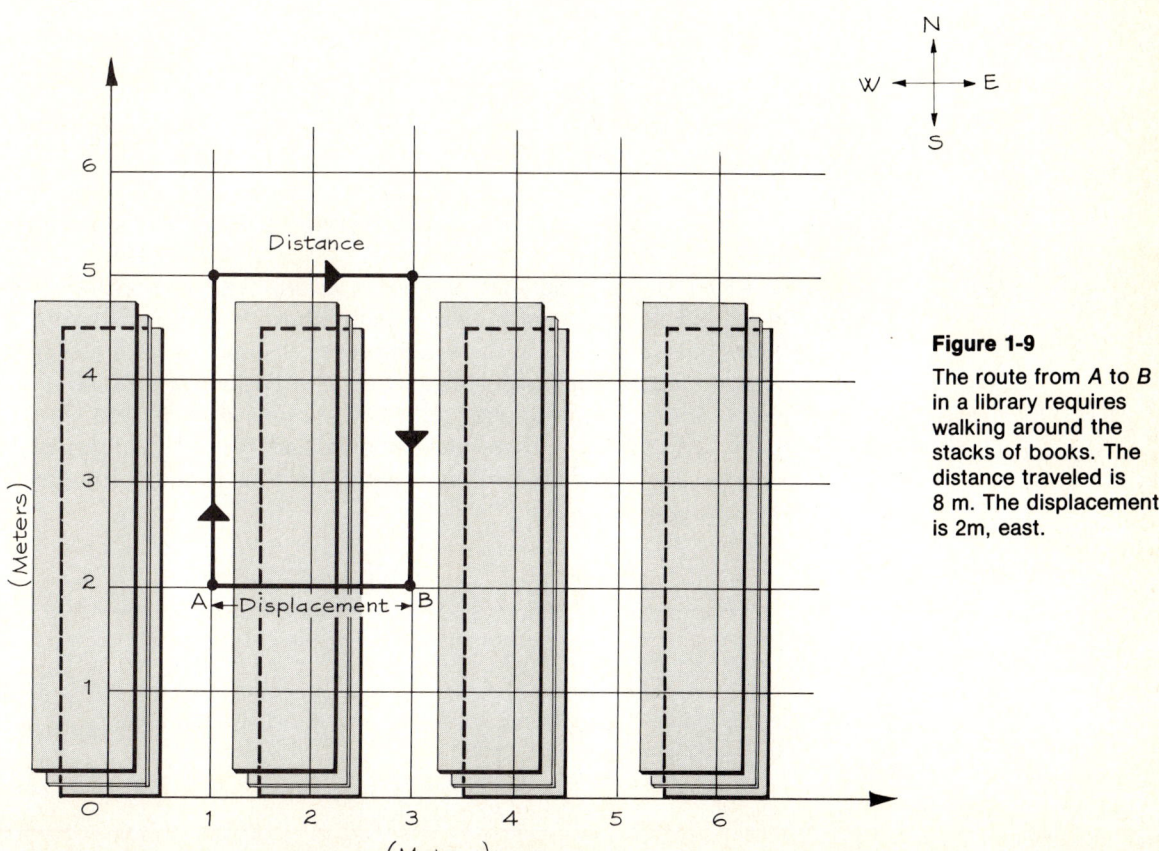

Figure 1-9

The route from A to B in a library requires walking around the stacks of books. The distance traveled is 8 m. The displacement is 2m, east.

moved from (1 m, 2 m) to (1 m, 5 m) to (3 m, 5 m) and finally to (3 m, 2 m). We can find the distance traveled by adding the distance of each leg of the trip. The distance traveled in the first leg is 5 m minus 2 m, which equals 3 m. Similarly, the distance traveled in the second leg is 3 m minus 1 m, which equals 2 m. For the third leg, it is 5 m minus 2 m, which equals 3 m. The distance traveled over the entire route is 3 m + 2 m + 3 m, which equals 8 m. We can calculate the distance traveled in this manner as long as the route is along the lines of a rectangular coordinate system.

Displacement

You are walking down the street when someone asks where the nearest post office is. "Oh, that's easy," you reply. "It's about three kilometers." If you walk on at this point, the stranger will probably ask someone else. "Three kilometers" is not very useful. "Three kilometers east" would have been much more helpful.

We use the concept of displacement to describe both the distance traveled as the crow flies and the direction in which the motion occurred. The **displacement** of an object is the distance and direction along the straight-line path from its initial position to its final position. A statement of displacement includes a straight-line distance and a direction, written as: straight-line distance, direction. In Figure 1-8, the displacement is 1.2 km, northwest. Quantities such as 30 km, north and 4 m, to your left are displacements. Thirty kilometers is not a displacement; neither is south.

The distinction between distance and displacement can be made clearer by returning to the library example in Figure 1-9. The straight-line distance between A and B is 2 m. When you arrive at B, you are 2 m east of A. The displacement is 2 m, east, even though you traveled a distance of 8 m to get there.

One way to distinguish between the terms *distance* and *displacement* is to think about the old stereotype of the expectant father. While his child is being born, he paces back and forth in the waiting room. He moves from one position to another and back again. With each change of position, he travels a distance of a few meters. Then he goes right back to where he started. By the time the baby is born, the father may have moved through a distance of several thousand meters. But his displacement is zero because he always comes back to where he began.

SELF-CHECK 1C

A car drove 3 km east and then 2 km in a different direction. Two possible routes are illustrated in Figure 1-10. Calculate the distance and displacement for each route.

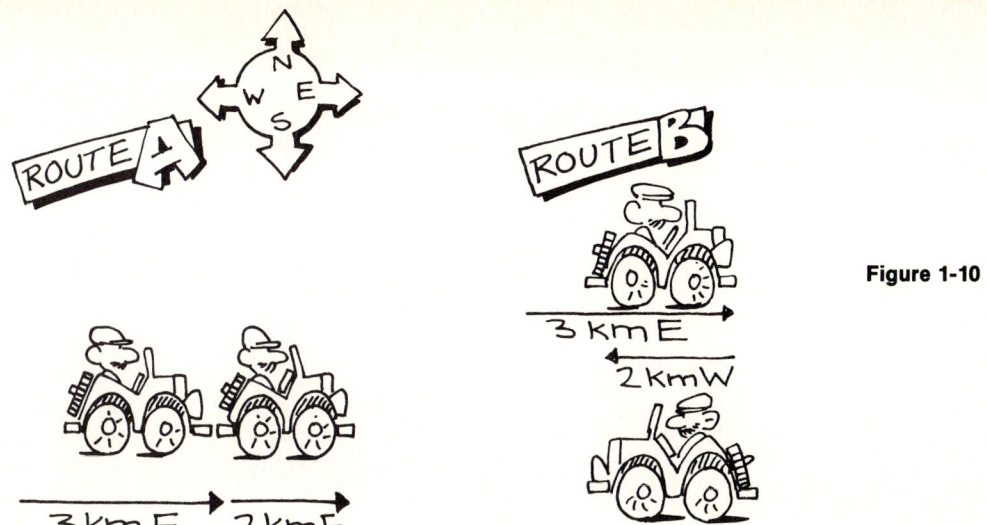

Figure 1-10

Vectors and Scalars

Distance and displacement are related concepts, but they have an important difference: direction. Distance is specified by a number with units, such as 30 km. Displacement requires a number, a unit and a direction, such as 30 km, east. This distinction occurs so frequently in physics that we define terms to distinguish the two types of quantities. A **scalar** is any quantity which can be completely described by a number and a unit. Distance is a scalar. A **vector** is any quantity which can be completely described by a number, a unit, and a direction. Displacement is a vector. The number and unit are often referred to as the **magnitude** of the vector. The magnitude of a displacement of 50 km, north, is 50 km. Quantities like speed, time, and temperature are scalars. Force and momentum, two quantities we will examine in more detail in later chapters, are both vectors. To help you recognize vector and scalar quantities in equations, we designate vectors by boldface type (such as **d** for displacement) and scalars by regular type (such as *l* for distance).

The addition of direction does more than simply change a scalar into a vector. Distance and displacement both describe a change in position. But, as illustrated by the stereotype of the expectant father who paces back and forth but never goes anywhere, these descriptions provide different types of information. Displacement uniquely locates the position of an object but tells us little about the path it took in reaching that position. A displacement of 30 km, north, means simply that the object is 30 km north of its starting location. By contrast, distance provides us information about the length of path followed but no unambiguous information about the object's final location. "The car traveled 30 km" tells us nothing about where to find the car, although we do know how far the car actually traveled. Vectors and scalars are clearly two different types of quantities.

A STEP FURTHER—MATH

WHAT TO DO WHEN THE CAR TURNS A CORNER

The two examples in Self-Check 1C are relatively simple since the car either continued in the same direction or turned around. But cars do turn corners. In the route shown in the figure, the driver traveled 3 km east and then 2 km north. Here the displacement is the straight-line distance from the tail of vector **A** to the tip of vector **B**. Since east and north are at right angles to one another, the displacement is actually the hypotenuse of a right triangle whose other two sides are 3 km, east, and 2 km, north. (The **hypotenuse** is the side of a right triangle that is opposite the 90° angle.) A bit of mathematics enables us to determine the length of the hypotenuse from the other two sides—the net displacement from the two legs of the trip.

As you may remember, the Pythagorean theorem describes the relationship among the three sides of a right triangle. The square of the hypotenuse is equal to the sum of the square of each of the other two sides. Applying this to the route shown:

$$(\text{Length of hypotenuse})^2 = (\text{length of side X})^2 + (\text{length of side Y})^2$$

$$\text{Length of hypotenuse} = \sqrt{(3 \text{ km})^2 + (2 \text{ km})^2}$$

$$= \sqrt{(9 \text{ km}^2 + 4 \text{ km}^2)}$$

$$= \sqrt{13} \text{ km}$$

$$= 3.6 \text{ km}$$

The displacement is 3.6 km, northeast.

While this procedure works for one corner, think about what would happen if we drove 3 km east, 2 km north, and then 2 km west—turning two corners instead of just one. Even a single corner can cause problems if it is not at a right angle. Determining the displacements in these situations becomes a little more complex. In the next section we will present a graphical technique for estimating displacements when the car turns lots of corners or moves along curves. If you find some of these discussions tough going, don't feel dismayed. While turning corners and going around curves is a daily occurrence, calculating displacements is not. What is important is that you realize what displacement describes and that there are mathematical techniques for calculating it.

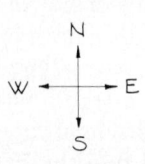
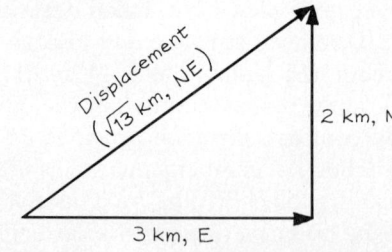

Working with Vectors

Since vectors have a direction associated with them, arithmetic operations like adding, subtracting, multiplying, and dividing are a little different. Three of these, adding vectors, subtracting vectors, and multiplying a vector by a scalar, crop up several times in later chapters. Let's illustrate the procedure we use with a few examples involving displacement.

In the process of adding vectors, we must keep track of the direction associated with each. A displacement of 1 km, east, added to a displacement of 2 km, west, does not lead to a total displacement whose magnitude is 3 km. The simplest way to add vectors is the **tail-to-tip method,** shown in Figure 1-11(a). In our example, an arrow is drawn to scale to represent each of the two displacements. Since the magnitude of vector **B** is twice as great as the magnitude of vector **A,** vector **B** is twice as long. The point at which each arrow begins is called the vector's **tail.** The point at which each arrow ends is called the vector's **tip.** To add the vectors, place them tail to tip so that the tail of vector **B** lies at the tip of vector **A.** The sum of the two vectors, called the **net displacement,** is a vector that goes from the tail of vector **A** to the tip of vector **B.**

Some situations in which we need to add vectors are relatively simple. When the two vectors are in the same direction, you can simply add the

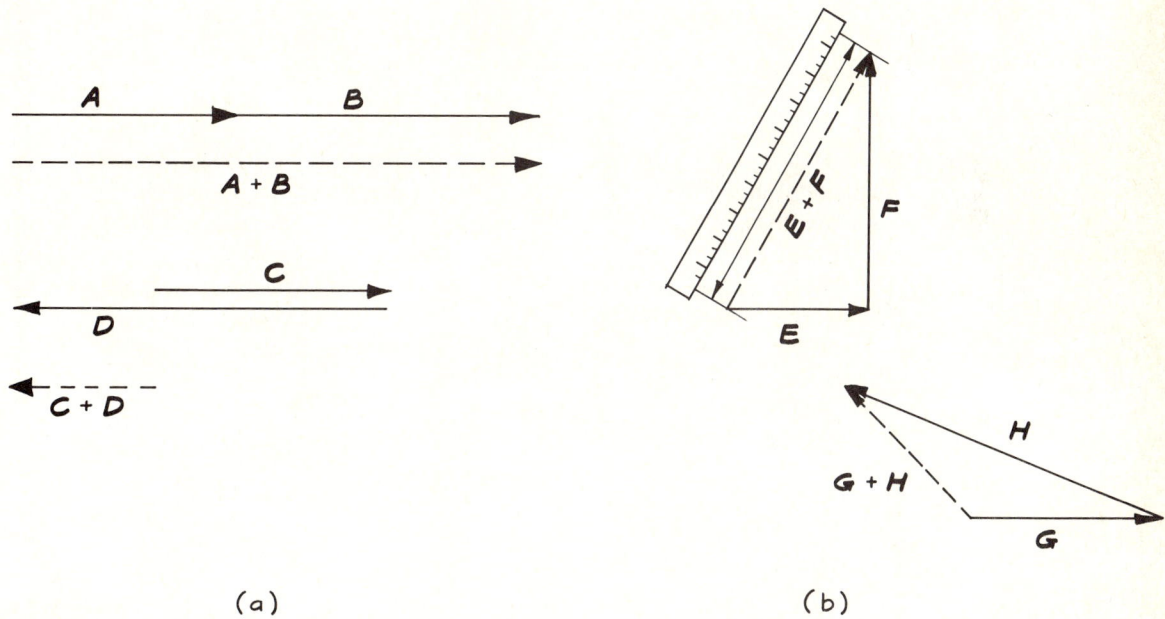

(a) (b)

Figure 1-11 **(a)** The tail of vector **B** is placed at the tip of vector **A**. The sum of the two vectors is **A** + **B**, the vector that goes from the tail of vector **A** to the tip of vector **B**. **(b)** Vector **E** + **F** is the vector that goes from the tail of vector **E** to the tip of vector **F**. Vector **G** + **H** describes the sum of vectors **G** and **H**.

SAVING A FEW STEPS?

Waiters and waitresses are very much aware of the difference between distance and displacement. They travel long distances but always end up at the same place that they started—back in the kitchen. So, while they may be very tired at the end of their working day, their displacement is zero. During the nineteenth century, William Lance thought of a way to decrease the distance traveled by the waiters and waitresses. His self-waiting table utilized a large mechanized circle of shelves. The servers (A) placed the food on shelves (B) which traveled past the diners (C, D, and E). When the customers saw something appealing, they picked it off the moving shelf. When finished, they placed their dirty dishes on the next empty shelf. The dishes returned to the server (A), who sent them back to the kitchen. Thus, the distance traveled by the dishes was huge, but the distance traveled by the servers was small. Needless to say, the waiters and waitresses were delighted!

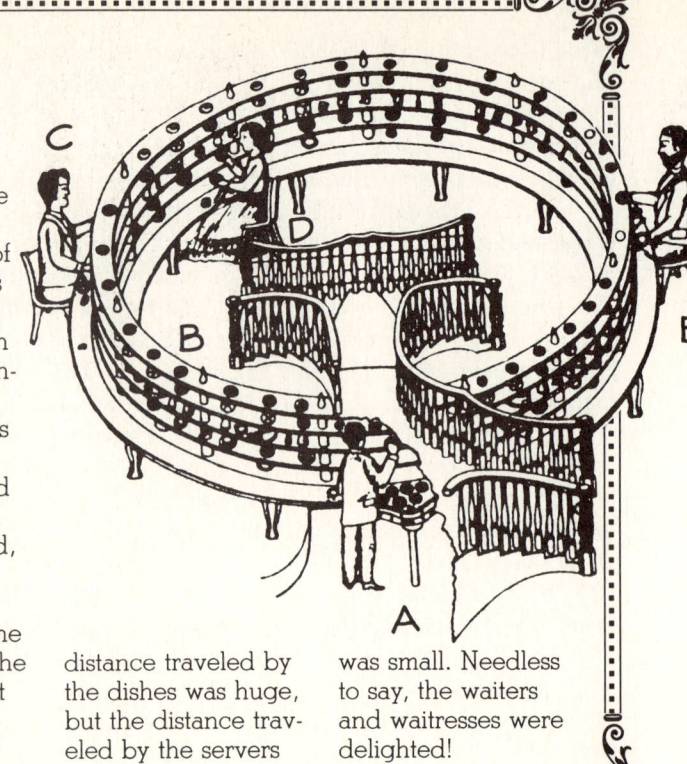

magnitudes of the two vectors. A displacement of 1 km, east, added to a displacement of 4 km, east, leads to a net displacement of 5 km, east. When the two vectors are in opposite directions, like **C** and **D,** you could get the result by subtracting the two magnitudes. When the vectors are at right angles to one another, the Pythagorean theorem (page 14) offers an algebraic way of determining the sum of the two vectors. At other angles, however, the tail-to-tip method offers the easiest way to determine the sum of the two vectors. The magnitude of the net displacement can be determined by measuring the length of its arrow to scale (Figure 1-11(b)).

Subtracting vectors takes advantage of the fact that subtraction is the reverse of addition. Suppose we need to subtract two displacements, 1 km, east, and 2 km, west. Vector **A** − vector **B** is the same as vector **A** + (−vector **B**). Consequently, we can subtract vector **B** from vector **A** by adding the negative of vector **B** to vector **A**. Figure 1-12(a) shows vector **B** and its negative, −**B**. To find the difference between the two vectors, we add vector **A** to the negative of vector **B** according to the tail-to-tip method. The resultant displacement is 3 km, east (Figure 1-12(b)).

Multiplying and dividing vectors by scalars is much simpler. Suppose your displacement is 2 km, east, and someone asks you to double your displacement. Doubling the displacement would affect the magnitude of the dis-

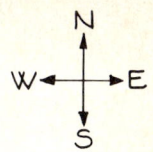

Figure 1-12 To subtract two vectors, you add the negative of the second vector to the first vector. **(a)** The negative of vector **B**, called −**B**, is a vector of the same length but in the opposite direction.
(b) **A** − **B** is the same as **A** + (−**B**). Here **A** and (−**B**) are added using the tail-to-tip method.

placement but not the direction. There is no such thing as "twice as east." A displacement that is twice as great is 4 km, east. A displacement that is half as great is 1 km, east. Multiplying or dividing a vector by a scalar affects only the magnitude of the vector. The direction remains the same.

Reference frames, coordinate systems, distance, and displacement define our concepts of position and change in position. In familiar examples, these concepts seem almost commonsense. In complete definitions, they may seem needlessly complex. As objects start moving, however, we find it increasingly difficult to establish common reference frames, or common points of view. The definitions become increasingly important to us.

CHAPTER SUMMARY

The position of an object must be described in terms of other objects, called *reference objects*. Reference objects and reference directions make up a *reference frame*. In order for two people to agree on the location of an object, they must describe the object's location in terms of the same reference frame.

Coordinate systems are standard reference frames. Rectangular coordinate systems are constructed from lines at right angles to one another. The location of an object in a rectangular coordinate system is specified in terms of distances from a single point, called the *origin*. By convention, we use a shorthand notation in which the position is given by two numbers and units separated by a comma. The first number and unit specifies the distance along the horizontal axis; the second specifies the distance along the vertical axis.

The change in position of an object (its motion) can be described in terms of distance or displacement. *Distance* is the length of the path traveled by an object in moving from its initial to its final position. *Displacement* is the straight-line distance traveled by the object in moving from its initial to its final position and the direction the object travels along that straight-line path. A *scalar* is defined by a number and a unit; a *vector* is characterized by a number, a unit, and a direction. The *magnitude* of a vector is its number and unit. Distance is a scalar quantity; displacement is a vector quantity. Vectors can be added using the *tail-to-tip method*. To subtract two vectors, we add the negative of the second vector to the first vector. Multiplying or dividing vectors by a scalar affects the magnitude of the vector but not its direction.

ANSWERS TO SELF-CHECKS

1A. a. Two different reference frames are used—one defined by the nightstand and the second by the window. Both descriptions are based on implied reference directions.
b. The two statements use the same reference frame—the western horizon. The reference directions have been stated.
c. Two different reference frames are used—one defined by the Library and the second by the Student Union. The reference directions have been stated.

1B. The triangle is located at (5 m, 2 m).

1C. Route A: Distance is 5 km. Displacement is 5 km, east.
Route B: Distance is 5 km. Displacement is 1 km, east.

PROBLEMS AND QUESTIONS

A. Review of Chapter Material

A1. Define the terms listed below:
Reference object
Reference direction
Reference frame
Coordinate system
Rectangular coordinate system
Scalar
Origin
Position
Distance
Displacement
Vector
Tail-to-tip method

A2. How do scientists avoid the "different point of view" problem illustrated in the Wizard of Id cartoon?

A3. In outer space astronauts do not feel the effects of gravity. Which of our common terms for directions lose their meanings in outer space?

A4. How do coordinate systems improve our ability to describe the location of an object?

A5. List three ways you could use to describe the change of position of an object.

A6. How are displacement and distance similar? How do they differ?

A7. How do scalars and vectors differ? Give an example of each.

A8. Why do we add vectors differently than we add scalars?

B. Using the Chapter Material

B1. In each sentence below, identify the reference object(s) used to describe the position of the object.
a. You will find the diary in the upper right-hand drawer of the chest of drawers.
b. Sandwiched between Manhattan and Long Island, Queens supports a substantial, largely commuter, population.
c. The fifth car from the corner has Oregon plates.
d. The North Star is the first bright star you reach as you trace the Big Dipper and extend the line upward from the bowl.

B2. You tell a friend that a book is to the right. She moves to your left to pick up the book. Where is she relative to you? (There is more than one possibility!)

B3. A football player starts halfway between two side boundaries on the 50-yard (yd) line. He drags several tacklers and is finally stopped halfway between the boundaries on the 5-yd line located at the north end of the field. With this information you can calculate only one of the quantities: distance or displacement. Which one can you calculate? Calculate the one you can and explain why you cannot calculate the other.

B4. Three astronauts, labeled A, B, and C, are oriented differently in Skylab, as illustrated in Figure 1-B4. Their descriptions of the location of the toolbox can differ because of their different orientations. For any two astronauts, write observations that differ because of:
  a. reference object chosen
  b. reference directions chosen
  c. both reference object and reference directions chosen

B5. A friend gives you these instructions: To get to my house from the bank, go two blocks north, then three blocks east and two blocks south. What distance, in blocks, do you travel to get to your friend's house? What is your displacement? (If you have trouble, draw a picture of the route.)

B6. Describe the location of the five dots in Figure 1-B6.

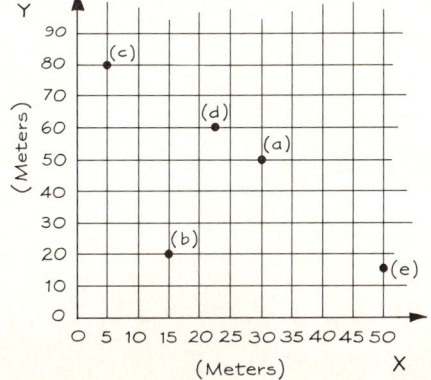

B7. Figure 1-B7 shows two different models of the universe that sparked an intense controversy between science and religion during the time of Galileo. Use the concept of reference frame to describe how the two models are different.

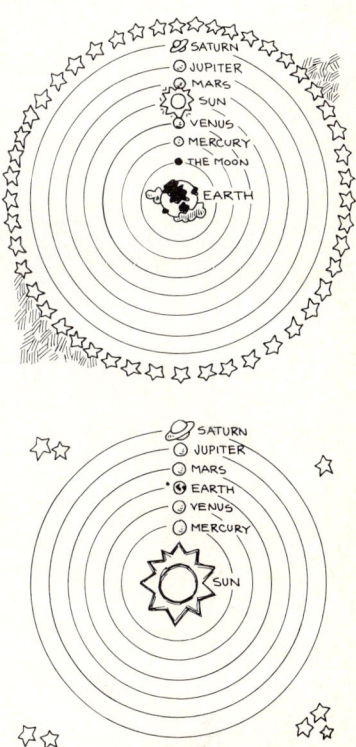

B8. What are the distances moved for each set of starting and ending coordinates listed below?
  a. (0 m, 10 m) to (0 m, 5 m)
  b. (2 m, 15 m) to (2 m, 60 m)
  c. (3 m, 5 m) to (1 m, 5 m)
  d. (20 m, 15 m) to (15 m, 15 m)

B9. How many dimensions are required of a coordinate system used in each of the following situations?
  a. highway mileage signs
  b. crossword puzzle
  c. football field
  d. location of airplane in flight

B10. The distance between New York and London is 5500 km. What are the distance and displacement of the Concorde supersonic airplane when it completes a New York-London round trip?

C. Extensions to New Situations

C1. Read the cartoon in Figure 1-7. Draw a picture of Zero's coordinate system. Show the location of the origin and orientation of the coordinate axes.

C2. One design for space stations is a large ring as sketched in Figure 1-C2. As the ring rotates, artificial gravity is created, allowing people to walk on the outer edges of the space station. How do each of the people in the diagram define the direction down?

C3. From *Through the Looking Glass*:

>And they went so fast that at last they seemed to skim through the air, hardly touching the ground, till suddenly, just as Alice was getting quite exhausted they stopped . . . .
>
>Alice looked round her in great surprise. "Why, I do believe we've been under this tree the whole time! Everything's just as it was!"
>
>"Of course it is," said the Queen . . . . "Now, here, you see, it takes all the running you can do to keep in the same place. . . ."

a. What reference object is Alice using to describe her motion?
b. What is her position relative to this object at the beginning and at the end of her run?
c. How far did she move relative to her reference object?
d. Can you explain Alice's lack of change of position? (Remember, this story is a fantasy, so anything is possible.)

C4. A construction very similar to a two-dimensional coordinate system combines one dimension in space with time. For example, in describing our location while traveling we could say, "I will leave Springfield now. In 30 minutes I will be 50 kilometers away; in 60 minutes, 100 kilometers away; and in 90 minutes, 150 kilometers away."

a. Draw a (space, time) system with 0 km, 0 minutes (min) as the origin.
b. Locate the given (space, time) coordinates on the system drawn in (a).
c. By looking at the information on this coordinate system can you predict the distance away from Springfield at 120 minutes?

C5. Draw a coordinate system in which a dot's location is (0 m, 10 m). Draw a second coordinate system with the same scale in which the same dot's location is (0 m, 20 m). What is different about the two coordinate systems?

C6. Look at the map shown in Figure 1-C6. Are there regions where rectangular coordinate systems have been used? Regions where there is no rectangular system? Can you invent another type of coordinate system to fit the maps?

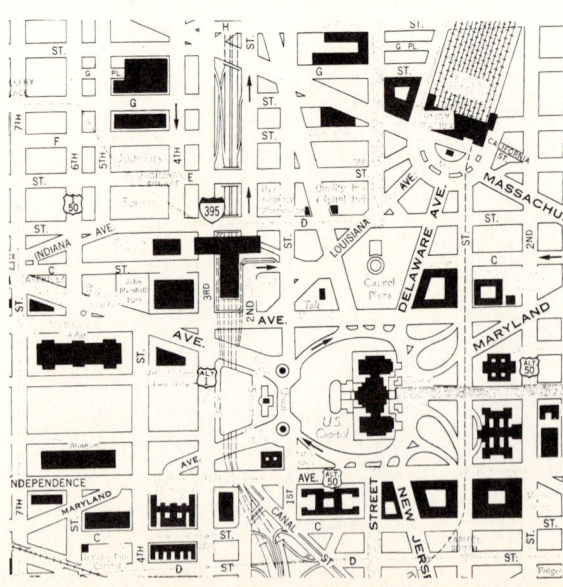

C7. We can measure the length of a piece of paper in two different reference frames. In reference frame A, one corner of the paper is located at 0.00 m and the other at 0.29 m. In reference frame B, the first corner is located at 0.10 m and the second at 0.39 m.
   a. What is the length of the paper in reference frame A?
   b. What is the length of the paper in reference frame B?
   c. Compare your answers in (a) and (b). Does the length of an object depend on the reference frame chosen?
   d. How is length different from position?

C8. Planets were first distinguished from stars because they appeared to wander relative to the constellations. Figure 1-C8 shows the motion of Mars relative to the constellations of Virgo, Libra, and Scorpius. As measured from Earth, will the distance Mars travels be different from its displacement?

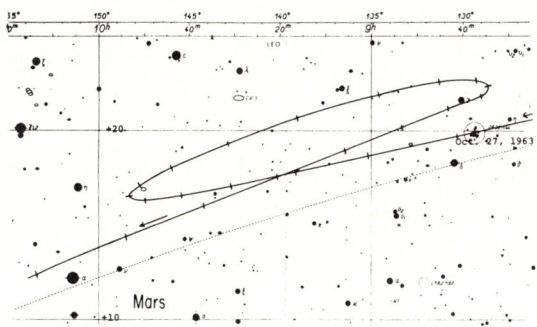

C9. In each of the situations below, will the distance and the magnitude of the displacement of the object be the same or different?
   a. A batter hits a fly ball to center field.
   b. An apple falls from the tree.
   c. The moon orbits the earth.
   d. A golfer tees off, sending the ball to the green.
   Use the examples to describe the circumstances under which the magnitude of the displacement will equal the distance moved.

C10. In Self-Check 1C you calculated the distance and displacement for the two routes shown in Figure 1-10. Use vector addition to measure several other routes. Convince yourself that the two routes in Figure 1-10 are the minimum and maximum displacements possible. Any other choice of direction for the 2-km displacement leads to a net displacement between 1 km, east, and 5 km, east.

D. Activities

D1. Select one building on your campus. Describe the coordinate system used to locate that building on a campus map. What reference objects are needed?

D2. Obtain a globe or map of the earth.
   a. What coordinate system is used to locate positions on the earth's surface?
   b. How many dimensions does the system have?
   c. Where is its origin?
   d. What are the locations of Sydney, Australia; Moscow, USSR; the South Pole; and your present position?

D3. Visual illusions such as that shown in Figure 1-D3 are often created because a reference object is used to trick our eyes. Consult a book dealing with illusions, such as R. L. Gregory, *The Intelligent Eye* (McGraw-Hill, 1970) and describe how a reference object was used to help create the illusion.

© The Exploratorium. Photo by Nancy Roger.

D4. List common situations in which:
   a. Distance would be more useful than displacement.
   b. Displacement would be more useful than distance.

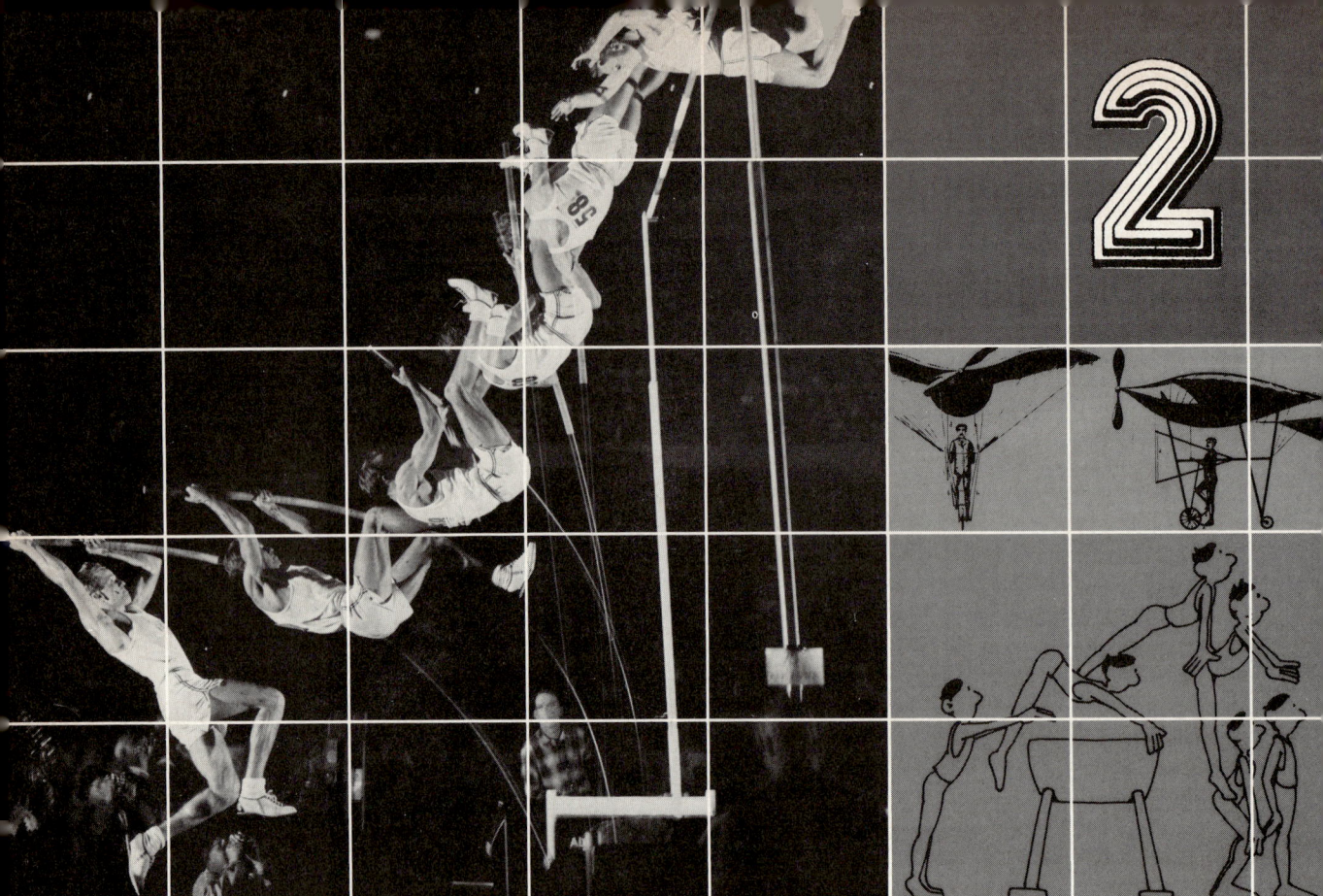

Describing Motion

Motion is so much a part of our lives that we can scarcely communicate without using words that refer to it—fast, slow, stop, go, swift, unhurried, hustle, leisurely. This abundant vocabulary reflects the amount and diversity of motion we experience. For survival purposes, many species of animals detect moving rather than stationary objects, a fact every hunter learns. Motion attracts the attention of infants, a fact every parent learns. Because we sense motion directly, many psychologists now suggest that we understand motion before we understand the related concepts of space and time. We experience space by moving about in it. We observe time with the motion of the hands of the clock or the periodic rising and setting of the sun.

Motion permeates all aspects of physics. Atoms and molecules remain in constant motion. The temperature of an object depends on the motion of its molecules. Our model of the internal structure of the atom is based largely on the motion of incredibly small particles. Electrical current arises from the almost imperceptible motion of tiny electrons. Even the apparently fixed stars that fill the night sky are in motion—often at astonishingly high speeds.

In this chapter we introduce the concepts used to describe motion. Motion is defined as a change of position that occurs during a time interval. The concepts of *speed* and *velocity,* used to describe motion, are derived from the concepts of distance and displacement introduced in Chapter 1. The terms

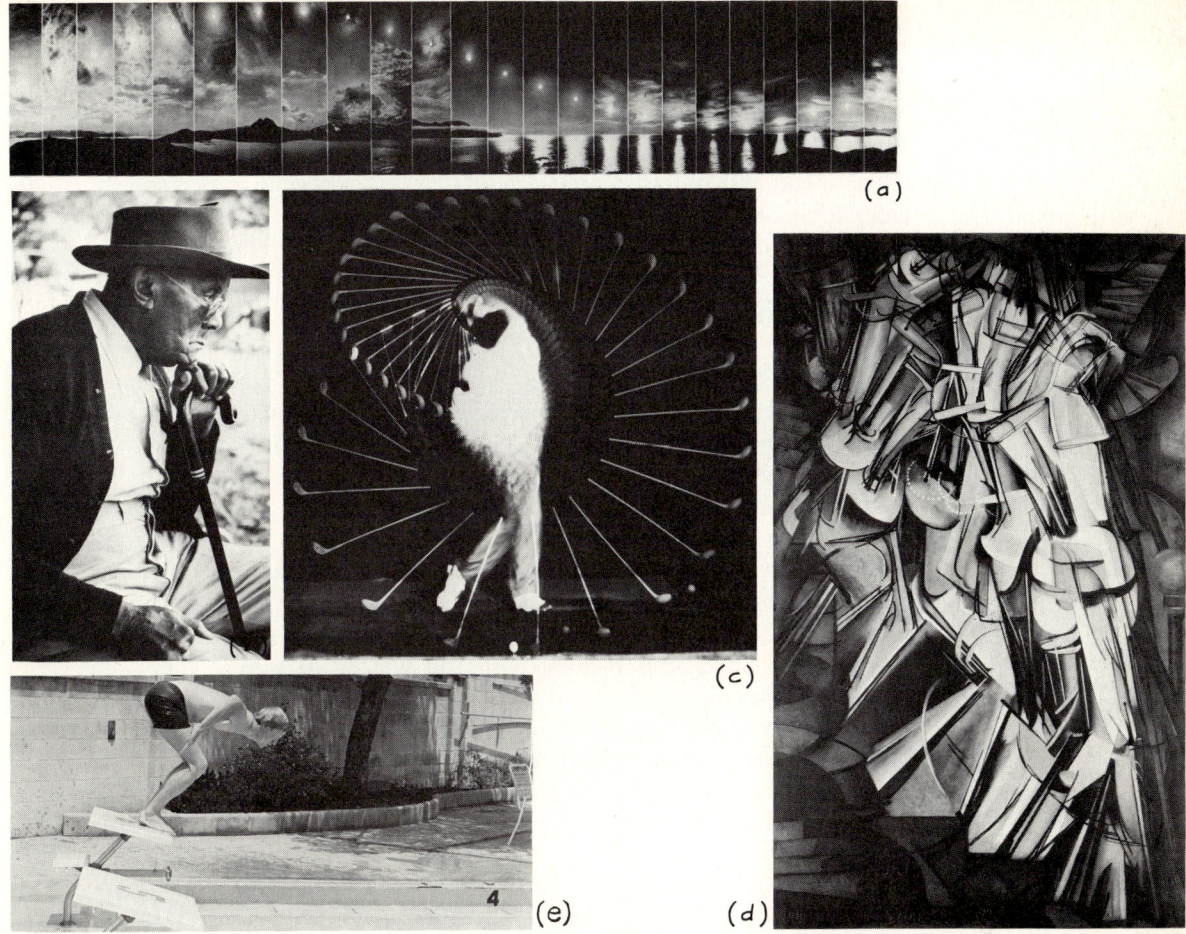

Figure 2-1
Which depict motion?

average and *instantaneous* are added to distinguish speed or velocity over a long time interval from speed or velocity at an instant in time. We use the concept of *acceleration* to describe motion in which the velocity is changing, as when a car starts to move away from a stop sign. Finally, we use the concepts of speed, velocity, and acceleration to analyze several common examples of motion.

IS IT MOVING?

Cars moving along in traffic, children riding bicycles, airplanes flying overhead—you can easily tell that motion occurs. Relative to their surroundings, the cars, children, and airplanes change position. In a textbook we cannot actually watch objects move, but we can capture and analyze motion in photographs, drawings and paintings.

Look at the pictures in Figure 2-1. Which depict motion? What visual clues can you use to support your choice? You probably said that the people or things in A, C, and D were moving, while those in B and E were still. Photographs A and C suggest motion because of the succession of images on the film. Marcel Duchamp used this same idea in his painting, *Nude Descending a Staircase,* shown in D. By contrast, the single image in photographs B and E suggests that the man and the swimmer were stationary, at least during the interval of time in which the photographs were taken. Perhaps they moved before or after the photograph. To know we would have to see another photograph taken at a different time. Only by seeing a *change of position* over *time* can we be certain that motion occurred.

DESCRIBING MOTION

Position and time are both needed to describe motion. When we want to compare the motion of different objects, the simplest strategy is to hold one of them constant and compare results on the second. Take, for example, a traditional race where all runners complete the same route. Distance is held constant and elapsed times are compared. The runner with the shortest elapsed time is the fastest. An alternative approach, used in some auto races, is to race for a fixed time interval (usually 24 hours) and compare distances. The car that traveled the greatest distance at the end of the fixed time interval is the fastest. Either strategy—shortest time with a fixed distance or longest distance over a fixed time interval—is a fair way to determine the winner. Frequently, however, we are faced with comparing two motions when both quantities (distance and elapsed time) are different. The concepts of speed and velocity allow us to make such comparisons. Speed describes the rate of change of position, while velocity describes both the rate and the direction in which position changes.

Speed

Suppose that one runner covers 80 meters (m) in 10 seconds (s) and a second runner completes 108 m in 12 s. Who is faster? To make such a comparison we take the ratio of the distance to the elapsed time, called the **speed** of each runner.

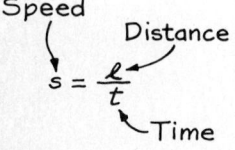

$$\text{Speed} = \frac{\text{distance traveled}}{\text{time taken}}$$

In our example, the speed of the first runner is 80 m divided by 10 s, which equals 8 meters/second (m/s). The speed of the second runner is 108 m divided by 12 s or 9 m/s. The second runner was faster than the first.

SELF-CHECK 2A

In a speed trial, a bicycle traveled 200 m in 10 s. What was its speed? How did its speed compare to the two runners?

A STEP FURTHER—MATH

ANSWERING OTHER QUESTIONS

So far we have used the definition of speed to answer the question: How fast? Sometimes we need to answer different, but related, questions. How long will it take to get to Milwaukee? How far can we travel before it gets dark? Fortunately, the definition of speed is versatile.

Suppose you are traveling from Minneapolis to Milwaukee, a distance of about 450 km, and your speed is 90 km/h. How long will it take to make the trip? Since speed and distance are both known, we can rearrange the definition of speed (in equation form) to isolate time, the unknown.

$$\text{Speed} = \frac{\text{distance}}{\text{time}} \quad \text{becomes} \quad \text{time} = \frac{\text{distance}}{\text{speed}}$$

Substituting the values for distance and speed, we find that

$$\text{Time} = \frac{450 \text{ km}}{90 \text{ km/h}} = 5 \text{ h}$$

The trip will take five hours.

Suppose you have six more hours until dark and want to figure out how much farther you will get before stopping for the night. You can figure on averaging 70 kilometers/hour during the six hours. Here the question we're asking is: How far? Again, we can manipulate the definition of speed to isolate distance.

$$\text{Distance} = (\text{speed}) \times (\text{time})$$
$$= (70 \text{ km/h}) \times (6 \text{ h}) = 420 \text{ km}$$

You would travel 420 kilometers before dark.

A single definition enables us to answer three categories of questions, most of which crop up rather frequently. We can relate each category of questions to one of the three variables in the definition. How fast? describes speed. How long? describes time. How far? describes distance. You will find that this same procedure applies to nearly every definition we will introduce in later chapters. Definitions that allow us to summarize relationships among several variables can be very useful!

Velocity

In the 1929 Rose Bowl, a Georgia Tech halfback fumbled. Roy Riegels, the California center, scooped up the loose ball and dashed toward the goal line. Faster than the other players, he neared the goal line virtually untouched. Suddenly, his teammate tackled him—it was the wrong goal! "Wrong Way" Riegels had the right speed but the wrong velocity.

Velocity is defined as the ratio of the displacement of an object to the time interval required for the displacement.

$$\text{Velocity} = \frac{\text{displacement}}{\text{time}}$$

Velocity (a vector)

Displacement (a vector)

$$v = \frac{d}{t}$$

Time

Figure 2-2

Roy Riegels picked up the fumble on about the 44-yard (yd) line, ran 2 yd toward the sidelines and then 5 yd in the right direction. In the process of dodging tacklers, he became disoriented, ran 60 yd in the wrong direction, and was tackled at the Georgia Tech 1-yd line. He traveled a distance of 67 yd and a displacement of 55 yd, wrong way.

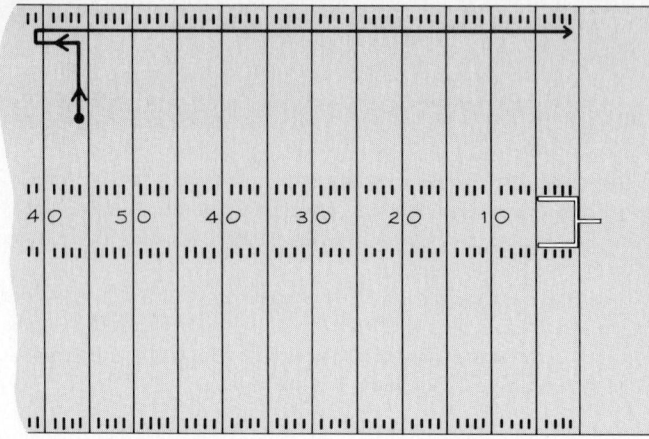

Displacement is a vector quantity while time is a scalar. A displacement of 50 kilometers (km), north, divided by a time interval of 2 h gives us a velocity of 25 kilometers/hour (km/h), north. The direction of the velocity is the same as the direction of the displacement. Velocity is a vector—we need a number, a unit, and a direction in order to describe it completely. Thus 50 km/h, north, is a velocity; 90 km/h is not.

Speed and velocity are related to each other in the same way that distance and displacement are related. As an example, consider "Wrong Way" Riegels' infamous run. His path is shown in Figure 2-2. The distance he traveled was 67 yards (yd), while his displacement was 55 yd, wrong way. We estimate that his run took about 15 s. Using this information, we can compare his speed and velocity. Riegels' speed was 67 yd divided by 15 s, which equals 4.47 yd/s. By contrast, his velocity was 55 yd, wrong way, divided by 15 s, or 3.67 yd/s, wrong way. Riegel probably had good speed for a center in 1929. However, it was his velocity during the play that made him famous.

---- **SMALL WARNING** ----

In everyday life we frequently use the terms *speed* and *velocity* interchangeably. In physics each term has a different meaning. Speed is a scalar quantity, related to the distance and time involved in the motion. Velocity is a vector quantity, derived from displacement and time.

---- **SELF-CHECK 2B** ----

Because of limited air service, you have to fly from Chicago to Albany via New York City. The straight-line distance between Chicago and New York City is 1294 km; between New York City and Albany, 250 km; and between Chicago and Albany, 1306 km. The total time for the trip was 4 h. What was your velocity? (Albany is roughly east of Chicago.)

AVERAGE AND INSTANTANEOUS DESCRIPTIONS OF MOTION

When we say that a runner's speed is 8 m/s, someone might imagine the runner being motionless one instant and running down the track at 8 m/s an instant later, much in the style that cartoon characters take off. From experience, we realize that such an image is absurd. Sprinters begin moving slowly, increasing their speed continuously until they reach their maximum speed. They may run only briefly at a speed of 8 m/s.

Figure 2-3 illustrates this situation more clearly. The photograph recorded images of a ball at 0.033-s intervals as it was being dropped. A meterstick was placed beside the ball's path to provide a coordinate system against which we could measure the ball's distance or displacement. If someone were to ask you the ball's speed, what would you answer? Using the definitions provided in the last section, the ball's speed from B to C is 0.07 m/0.033 s = 2.12 m/s. Yet, moving during an equal time interval from D to E, the ball's speed is 0.19 m/0.033 s = 5.76 m/s. If we made similar calculations between any two images, we would have many different speeds. To eliminate this ambiguity, we distinguish between average and instantaneous motion.

Average Motion

When we say that the sprinter runs at a speed of 8 m/s, we mean that the average speed during the 10-s time interval is 8 m/s. **Average speed** is defined to be the total distance traveled during a time interval divided by that time interval. Analogously, the **average velocity** is the displacement during a given time interval divided by that time interval. You may choose any time interval you wish, but you must specify it so others know which average speed or velocity you mean.

To illustrate the process, we return to the example in Figure 2-3. If we want to describe the speed of the ball during its entire fall, then we calculate its average speed during the time interval taken to go from A to E (0.561 s). The average speed over the entire trip is 1.587 m divided by 0.561 s, which is 2.83 m/s. Earlier, we calculated the average speed over smaller time intervals—from B to C and from D to E. The same process would apply in calculating average velocity, once displacements were substituted for distances. In this particular example, the magnitude of the displacements are equal to the distances traveled; consequently, we need only to add a direction. The average velocity in moving from A to E is 2.83 m/s, down.

When people describe speeds or velocities, they are usually talking about average speeds or average velocities. Generally, we know what time interval they mean—like the time required for a sprinter to complete a race. While these "averages" do not describe the detailed motion of the runner during the entire race, they resolve the issue important to the runner: Who wins.

Instantaneous Motion

To describe a runner's motion in detail, we need to know his speed at each instant of time. Such a speed is called his **instantaneous speed.** Analogously, an **instantaneous velocity** is the velocity of an object at a particular

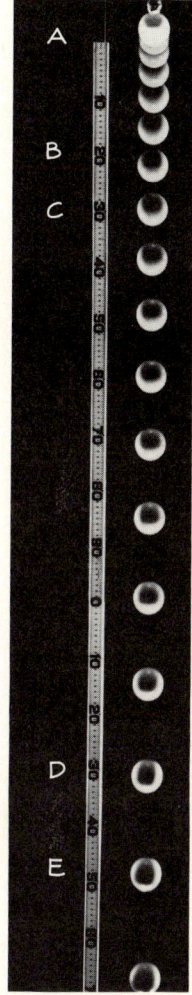

Figure 2-3

The ball's speed can be calculated from its distance and time measurements. The speed from A to E, however, is not the same as the speed from D to E.

instant of time. Practically speaking, we cannot measure speed or velocity at an instant of time. Measurements over smaller and smaller time intervals, however, allow us to estimate instantaneous speeds and velocities. Some instruments, like speedometers and police radar, measure average speeds over extremely short time intervals, so we often think of them as measuring instantaneous speeds. When we have indirect measurements like the strobe photo in Figure 2-3, then the best we can do is simply to estimate the instantaneous speed based on the smallest time interval available.

The procedure used to estimate an instantaneous speed is similar to that used to calculate average speeds. The only difference is that we choose the smallest time interval possible. In Figure 2-3, the best we can do is to calculate the speed between successive images, like from B to C. Earlier we calculated the average speed of the ball in moving from B to C to be 2.12 m/s. This represents our best estimate of the instantaneous speed of the ball midway through the time interval. Consequently, we say that the ball's instantaneous speed at 0.214 s (halfway between 0.198 s and 0.231 s) is 2.12 m/s. The only way that we could make a better estimate of the instantaneous speed at 0.214 seconds is to take another strobe photograph in which the time interval between flashes was even shorter. This would produce a photograph with additional images of the ball between B and C, and we could calculate an average speed over a smaller time interval. Theoretically, our best measurement of the instantaneous speed occurs when the elapsed time between images is zero. Practically, however, we cannot make such measurements, so we make the best approximation from the measurements available.

Instantaneous Motion Describes Change

The distinction between average and instantaneous motion is important only when the object's motion is changing. A car moving down the highway at a constant velocity of 90 km/h, east, has the same average and instantaneous velocity—90 km/h, east. Only when the car slows down, speeds up, or changes direction do the values of average and instantaneous velocity differ.

Figure 2-4 illustrates this distinction. The displacements of a Corvette, a Volkswagen, and a Model T are recorded at 1-s intervals in a series of strobe-like drawings. For the moment we will contrast the motions of just the Volkswagen and the Model T. The average velocity of each is simply the total displacement divided by the total time: 15 m/s, east, for the Model T and 11.5 m/s, east, for the Volkswagen. Successive images allow us to estimate instantaneous velocities midway through each time interval, at 0.5 s, 1.5 s, 2.5 s, and 3.5 s. These values for both the Model T and the Volkswagen are listed in Table 2-1.

Let's compare the average velocity over the entire 4 seconds with the instantaneous velocity during each 1-second interval. The Model T moved at a constant velocity throughout. Its instantaneous velocity always equaled its average velocity, 15 m/s, east. We really did not need to bother with instantaneous velocities. By contrast, the Volkswagen changed velocity. Its instantaneous velocity increased from 10 m/s, east, to 11 m/s, east, to 12 m/s, east, and finally to 13 m/s, east. Because the Volkswagen's motion changed, its

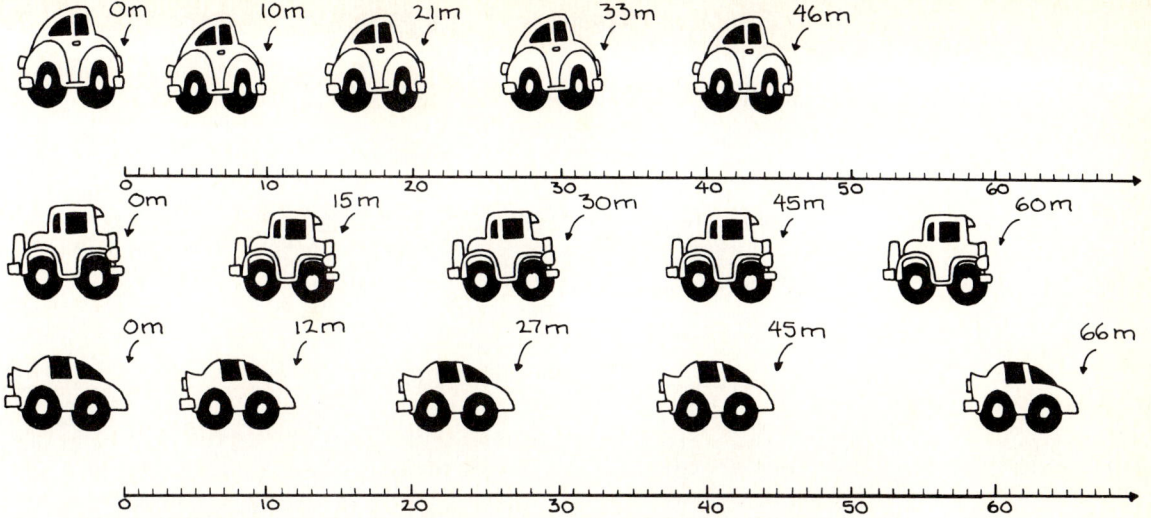

Figure 2-4
The Model T moves at a constant velocity—15 m/s, E. The Volkswagen changes speed—increasing its speed during each 1-s time interval. The Corvette increases its speed at a faster rate.

Table 2-1 Instantaneous Velocities for Figure 2-4

Time Interval	Time	Instantaneous Velocity (Model T)	Instantaneous Velocity (Volkswagen)
0–1 s	0.5 s	$\frac{15 \text{ m}}{1 \text{ s}} = 15$ m/s, E	$\frac{10 \text{ m}}{1 \text{ s}} = 10$ m/s, E
1–2 s	1.5 s	$\frac{15 \text{ m}}{1 \text{ s}} = 15$ m/s, E	$\frac{11 \text{ m}}{1 \text{ s}} = 11$ m/s, E
2–3 s	2.5 s	$\frac{15 \text{ m}}{1 \text{ s}} = 15$ m/s, E	$\frac{12 \text{ m}}{1 \text{ s}} = 12$ m/s, E
3–4 s	3.5 s	$\frac{15 \text{ m}}{1 \text{ s}} = 15$ m/s, E	$\frac{13 \text{ m}}{1 \text{ s}} = 13$ m/s, E

average and instantaneous velocities are no longer the same. While an average velocity of 11.5 m/s, east, provides a general description of the Volkswagen's motion, it does not tell us whether the motion ever changed or not. The various instantaneous velocities provide a more complete picture.

Changes in motion are generally obvious in strobelike drawings or photographs. Since the object's position has been recorded at the end of equal time intervals, we need only compare displacements to see if the motion has changed. Figure 2-5 shows an exaggerated comparison of the motion of two cars. Successive images of the Model T are equally spaced, since the displacements during each 1-s time interval are equal. Successive images of the Volkswagen are not equally spaced. Displacements during each time interval increase, reflecting the Volkswagen's change in motion.

Figure 2-5

In this exaggerated drawing, the Volkswagen is accelerating, but the Model T is not. The spacing between images increases when an object accelerates.

SELF-CHECK 2C

Turnpike authorities often monitor speed by checking the travel time of cars. A car travels from Kansas City to Topeka, a distance of 100 km. If the time of entry was 9:06 A.M. and the time of exit was 10:36 A.M., the travel time would be 1.5 h. Calculate the car's speed. Did it exceed the posted speed limit, 90 km/h? Does this procedure measure instantaneous or average speeds? Could the car still have exceeded the speed limit at some time during the trip?

DESCRIBING CHANGES IN MOTION

Now let's turn to the motion of the Corvette in Figure 2-4. Like the Volkswagen, successive images of the Corvette are further and further apart. Its motion is changing. However, the Corvette's motion is changing more rapidly than the Volkswagen's. If we compare instantaneous velocities during each time interval, as in Table 2-2, our qualitative observations are confirmed. The Volkswagen's instantaneous velocity increased from 10 m/s, east, to 11 m/s, east, to 12 m/s, east, and—finally—to 13 m/s, east. During each 1-s time interval its instantaneous velocity increased by 1 m/s, east. The Corvette's instantaneous velocity increased from 12 m/s, east, to 15 m/s, east, to 18 m/s, east, and—finally—to 21 m/s, east. Its instantaneous velocity increased by 3 m/s, east, during each time interval—three times as much as the Volkswagen. We say that the Corvette had a greater acceleration than the Volkswagen.

Acceleration

The **acceleration** of an object is the ratio of the change of velocity of an object to the time interval during which that change occurs.

$$a = \frac{v_f - v_i}{t}$$

(Final velocity, Initial velocity, Time, Acceleration)

$$\text{Acceleration} = \frac{\text{change in \textbf{velocity}}}{\text{time interval}}$$

Table 2-2 Instantaneous Velocities for Figure 2-4

Time Interval	Time	Instantaneous Velocity (Corvette)	Instantaneous Velocity (Volkswagen)
0–1 s	0.5 s	$\frac{12 \text{ m}}{1 \text{ s}} = 12$ m/s, E	$\frac{10 \text{ m}}{1 \text{ s}} = 10$ m/s, E
1–2 s	1.5 s	$\frac{15 \text{ m}}{1 \text{ s}} = 15$ m/s, E	$\frac{11 \text{ m}}{1 \text{ s}} = 11$ m/s, E
2–3 s	2.5 s	$\frac{18 \text{ m}}{1 \text{ s}} = 18$ m/s, E	$\frac{12 \text{ m}}{1 \text{ s}} = 12$ m/s, E
3–4 s	3.5 s	$\frac{21 \text{ m}}{1 \text{ s}} = 21$ m/s, E	$\frac{13 \text{ m}}{1 \text{ s}} = 13$ m/s, E

The change in velocity is the difference between the final instantaneous velocity and the initial instantaneous velocity during the time interval over which they were measured. Let's use the motions of the Volkswagen and the Corvette to illustrate how to apply this definition.

Compare the accelerations of the two cars during the time interval from 0.5 s to 2.5 s. The velocity of the Volkswagen changed from 10 m/s, east, to 12 m/s, east. Its change of velocity was 2 m/s, east. The time interval during which the velocity changed was 2.5 s minus 0.5 s, which equals 2.0 s. Applying the definition, the acceleration of the Volkswagen was 2 m/s, east, divided by 2.0 s, equals 1 (m/s)/s, east, which we read as "one meter per second per second, east." By contrast the Corvette changed its velocity from 12 m/s, east, to 18 m/s, east, during the same time interval. Its acceleration was 6 m/s, east, divided by 2.0 s, equals 3 (m/s)/s, east. While the unit (meters per second) per second looks strange, it is descriptive of what is occurring physically. The velocity, measured in meters per second, is changing during each second. (For convenience, these units are often written as m/s^2, read "meters per second squared.")

Acceleration is defined in terms of velocity; consequently, acceleration is a vector. It requires a number, a unit, and a direction. Any change in velocity results in an acceleration. Since velocity involves both a magnitude (number and unit) and a direction, a change in either magnitude or direction means that an acceleration occurred. In Figure 2-6 the Volkswagen changed the magnitude of its velocity but not its direction. On a merry-go-round that is rotating at a constant speed, we change direction but not magnitude. Other motions, such as a car traveling along a winding road, can involve both a change in magnitude and a change in direction. In each case, acceleration occurs.

In conversations we often distinguish between speeding up and slowing down by saying that vehicles *accelerate* and *decelerate*. In physics the single

Figure 2-6
Acceleration occurs if an object's velocity changes in direction, changes in magnitude, or changes in both magnitude and direction.

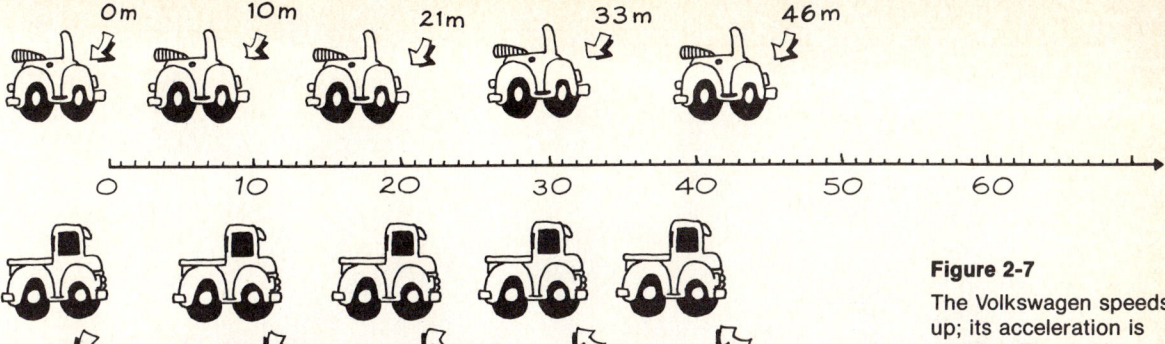

Figure 2-7
The Volkswagen speeds up; its acceleration is positive. The truck slows down; its acceleration is negative.

concept—acceleration—is used to describe both an increase and decrease in speed. For example, compare the motion of the Volkswagen and truck in Figure 2-7. The Volkswagen is accelerating, while the truck is decelerating. The Volkswagen increases its velocity from 10 m/s, east, to 12 m/s, east, during the 2 s. Its acceleration is 1 (m/s)/s, east. The truck decreases its velocity from 12 m/s, east, to 10 m/s, east, during the same time interval. Its acceleration is −1 (m/s)/s, east. The accelerations of the two vehicles differ in sign. A positive sign—positive acceleration—indicates that the vehicle is speeding up. A negative sign—negative acceleration—indicates that the vehicle is slowing down. A negative acceleration is a deceleration.

SELF-CHECK 2D

What is the acceleration of a bicycle whose velocity is initially 4 m/s, north, if 10 s later it has slowed to 1 m/s, north?

Acceleration Due to Gravity

Our environment provides us plenty of experience with acceleration. Pencils, books, apples—all objects fall to the ground when not supported by another object. If we examine their motions carefully, we find that they accelerate at a common rate—the **acceleration due to gravity.**

Figure 2-3 shows a ball being dropped near the surface of the earth. The images of the ball become more and more separated—our visual clue for acceleration. Careful measurements would show a constant acceleration of 9.8 (m/s)/s, down. The same is true for all other objects dropped near the surface of the earth, except for objects that are measurably affected by air resistance. Air resistance makes the motion of very light objects, like feathers, slower than that of heavier objects, like balls. But when we perform experiments in containers from which the air has been evacuated, as illustrated in Figure 2-8, light and heavy objects fall with the same acceleration.

In the now-famous, but perhaps fictional, Leaning Tower of Pisa experiment, Galileo first demonstrated that objects fall with the same acceleration.

PEDAL POWER PLUS

We are always in a hurry. We run when we could walk; drive when bicycling would do; fly when a train would suffice. Sometimes we get the impression that some people would fly everywhere if it were convenient. And, flying would be convenient if only we could do it under our own power.

The idea of increasing speed and acceleration by flying under human power (no engines) has fascinated inventors for centuries. Leonardo da Vinci designed but never built a device that would enable humans to imitate birds by flapping wings. Some nineteenth-century inventors attempted similar devices, while others created bicyclelike flying machines. Unfortunately, none of them worked. However, in the late 1970s a group of engineers successfully built human-powered aircraft. One of them, called the Gossamer Albatross, crossed the English Channel. It traveled the 36 kilometers from England to France in 2 hours 49 minutes—an average speed of 12.8 km/h, or 3.5 m/s. While the speed of the Gossamer Albatross does not yet rival that of a Boeing 747, the dream of human-powered flight is coming true.

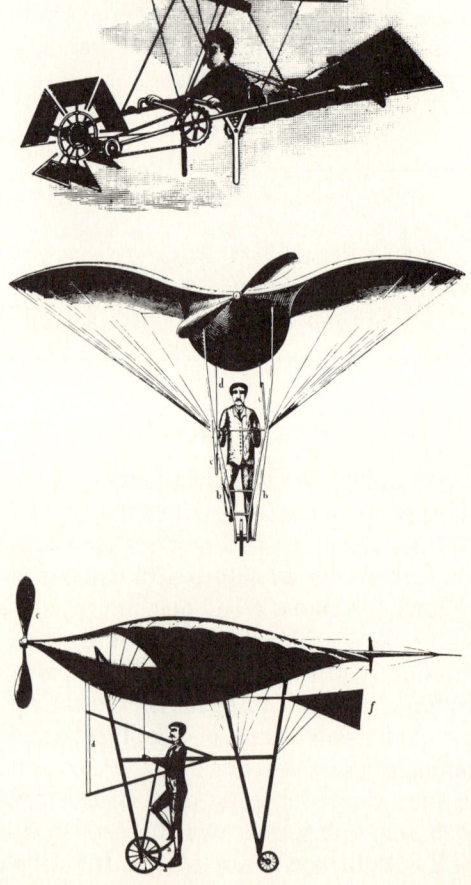

Since air resistance posed a problem, he cleverly chose objects heavy enough that the effect of air resistance was minimal. Today, we simply go to the moon. Since the moon has no atmosphere, air resistance is conveniently absent. Apollo Astronaut David Scott dropped a hammer and feather at the surface of the moon. Both fell with the same acceleration, 1.6 (m/s)/s, down. While the value of the acceleration due to gravity differs on the earth and moon, acceleration toward the ground is a natural part of all planetary environments.

ANALYZING MOTION WITH PHOTOGRAPHY

Photographs such as Figure 2-1A and C are often used to study motion. These photos are created by two processes. With very slow motion, such as the motion of the sun across the sky in Figure 2-1A, the camera shutter is opened and closed at several equal time intervals, such as once an hour. Each time the shutter opens, an image of the sun is recorded at a different location on the photographic film. In the case of very rapid motion, like the motion of the golf club in Figure 2-1C, the camera shutter is left open and a strobe light is used to illuminate the golfer. Each time the strobe light flashes, the film records an image of the golfer. Successive strobe flashes are always separated by equal time intervals.

These photographic techniques allow us to study motion both qualitatively and quantitatively. Qualitatively, we can use the spacing between images to compare both speed and acceleration. In Figure 2-1A the distance between images seems rather constant, suggesting that the sun moves across the sky at a constant speed. By contrast the distance between successive images of the golf club varies. The golf club changes speed. The spacing between images also allows us to compare speeds. Slower speeds result in images more closely spaced than higher speeds. For example, the golf club is moving faster at the bottom of the swing than at the top. Quantitatively, we can superimpose coordinate systems to calculate speeds, velocities, and accelerations. The procedures are exactly the same as those used throughout the chapter.

Athletes have been particularly interested in using stroboscopic photographs to analyze the various techniques of their given sport. Golfers were amazed to discover that the golf club and ball (Figure 2-9A) are actually in contact for less than a centimeter of the swing. The golfer is essentially unable to guide the ball with follow-through contact. Stroboscopic pictures were used extensively in studying the introduction of the fiberglass pole in pole vaulting competition (Figure 2-9B). Strobelike drawings are frequently used to teach correct body position and posture in a variety of sports (Figure 2-9C). Because the goal in any competitive sport is continued improvement, this detailed analysis of motion has become an important part of an athlete's training.

Stroboscopic analysis of motion has led us into fascinating new worlds, where things move either too quickly or too slowly for the naked eye to follow. Yet these worlds can be described with the same vocabulary as our everyday perceptions. Speed, velocity, and acceleration describe motion—for dancers, athletes, car designers, naturalists, and physicists alike.

Figure 2-8

In an evacuated tube, a rock and feather fall together at an acceleration equal to the acceleration due to gravity.

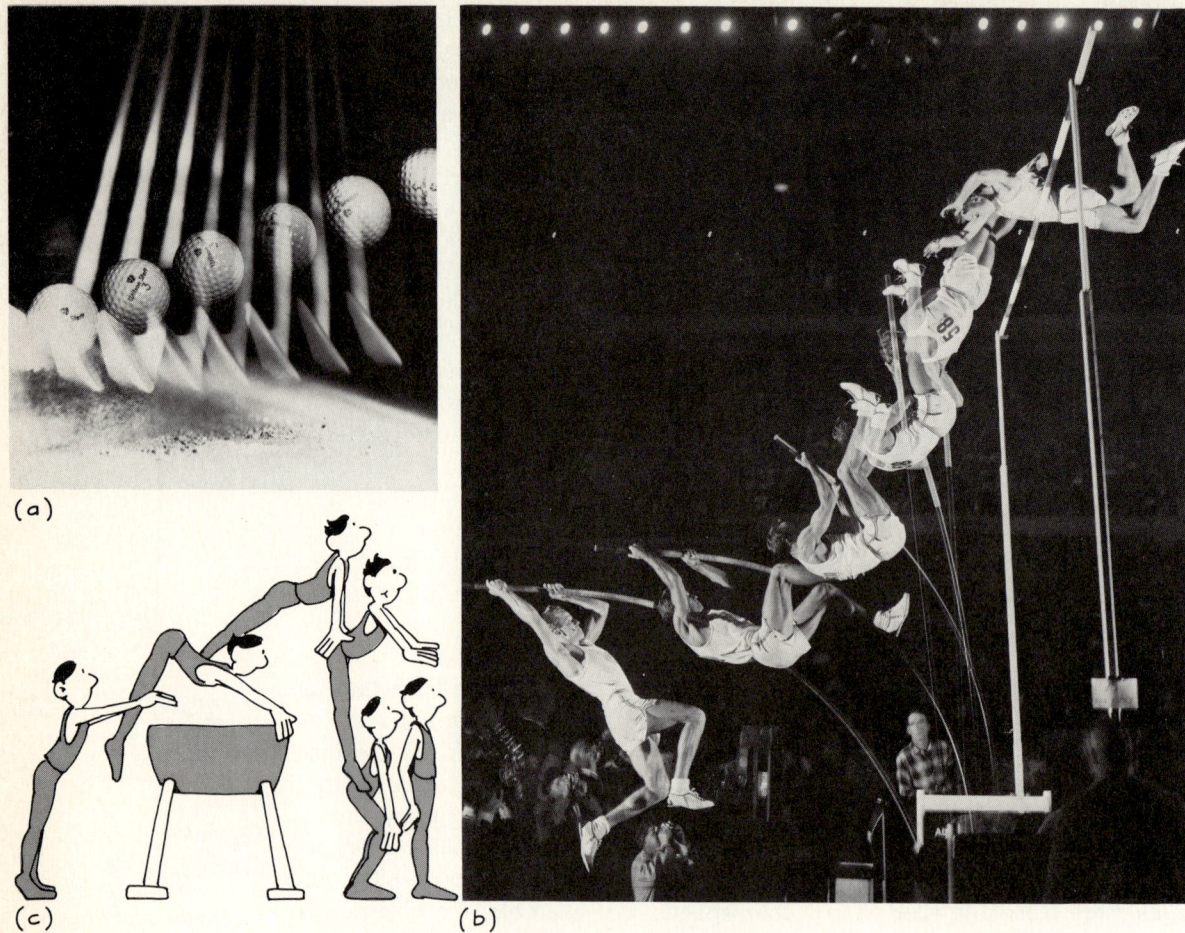

Figure 2-9 Stroboscopic pictures and drawings aid athletes in learning and perfecting various sports.

CHAPTER SUMMARY

Motion occurs when an object changes its position. Two concepts, speed and velocity, describe the rate at which a change in position occurs. *Speed* is the ratio of the distance traveled to the time interval in which the motion occurs. *Velocity* is the ratio of the displacement to the time interval. Speed is a scalar; velocity is a vector.

In describing motion, we distinguish between average and instantaneous speeds or velocities. *Average speed* is the total distance traveled during a time interval divided by that time interval. *Instantaneous speed* is the speed at one instant of time. We estimate instantaneous speed by calculating an average speed over as short a time interval as possible. *Average* and *instantaneous velocities* are defined in the same fashion, except that displacements are used instead of distances.

Acceleration is the ratio of the change of velocity of an object to the time interval during which the change occurs. An acceleration occurs when an object changes the magnitude or direction of its velocity. Like velocity, acceleration is a vector. A positive acceleration means that the object is speeding up; a negative acceleration means that the object is slowing down. Objects falling freely near the surface of the earth experience a common acceleration—the *acceleration due to gravity*. Measurements show the acceleration due to gravity at the earth's surface to be 9.8 (m/s)/s, down.

ANSWERS TO SELF-CHECKS

2A. Speed = distance/time = 200 m/10 s = 20 m/s. The bicyclist travels more than twice as fast as the runners.

2B. Velocity = displacement/time = (1306 km, east)/4 h = 326.5 km/h, east.

2C. Speed = distance/time = 100 km/1.5 h = 66.7 km/h. The car's speed did not exceed the posted speed limit. This procedure measures the average speed of the car. The driver could have exceeded the speed limit during the trip and then compensated by stopping for lunch. The instantaneous speed could have exceeded the speed limit, while the average speed did not.

2D. Acceleration = (change in **velocity**)/(time interval) = (1 m/s, north − 4 m/s, north)/10 s = −0.3 (m/s)/s, north. The bicycle slowed down.

PROBLEMS AND QUESTIONS

A. Review of Chapter Material

A1. Define the terms listed below:
Speed
Average speed
Instantaneous speed
Acceleration
Velocity
Average velocity
Instantaneous velocity
Acceleration due to gravity

A2. Which of the terms in Question A1 are vectors? Which are scalars?

A3. What properties of a photograph indicate that motion occurred while the photograph was being taken?

A4. Why do we use the concepts of speed and velocity rather than simply comparing travel times or distances?

A5. Explain the difference between average and instantaneous speeds. When do instantaneous speeds provide a better picture of an object's motion than average speeds?

A6. What two possibilities for change in an object's motion result in an acceleration?

A7. Explain the meaning of the units (m/s)/s.

A8. Describe the motion of an object experiencing positive acceleration; experiencing negative acceleration.

A9. When is the acceleration due to gravity a constant? When does it change?

A10. Describe the motion shown by a strobe photograph when:
   a. Images are a constant distance apart.
   b. Images are becoming further and further apart.
   c. Images are becoming closer and closer together.
   In which situation(s) is there acceleration?

B. Using the Chapter Material

B1. The average speed of travel has changed enormously:
   a. The longest foot race was in 1929 from New York to Los Angeles, a distance of 6000 km. The winning time was slightly less than 526 h. What was the average speed?
   b. In 1934 a cross-country train ride from New York to Los Angeles took 57 h. What was the average speed?
   c. Today, jets travel from New York to Los Angeles in 5.5 h. What is their average speed?

B2. You walk 100 m in 50 s. What is your average speed? Why can you not give the average velocity for this situation?

B3. You wish to travel from St. Louis to Chicago (465 km). If you average 80 km/h, how long will the trip take?

B4. Describe the direction of the instantaneous velocity of the skater in Figure 2-B4 for each of the four images. What is the direction of the average velocity from the first to the last image?

B5. Calculate the best estimate of the instantaneous velocity of the ball between points A and B in Figure 2-3. At what time, relative to the time at which the ball was released, does this instantaneous velocity occur?

B6. In a 100-m dash, a sprinter is timed at 10-m, 50-m, and 100-m distances from the starting block. It takes 2 s to reach the 10-m mark, 6.5 s to reach the 50-m mark, and 14 s to complete the entire 100-m race.
   a. What is the runner's average speed over the entire race?
   b. What is our best estimate of the runner's instantaneous speed at the beginning of the race? At what time, relative to the beginning of the race, does this instantaneous speed occur?
   c. How does this instantaneous speed compare to the average speed?

B7. Calculate the accelerations for the motions described below. Then compare each to the magnitude of the acceleration of gravity, 9.8 (m/s)/s.
   a. A good sprinter can change speed from 0 m/s to 12 m/s in 6 s.
   b. An average automobile can increase its speed from 0 m/s to 30 m/s in 20 s.
   c. At liftoff a rocket will increase its speed from 0 m/s to 56 m/s in 4 s.

B8. A baby slides on a floor and changes speed from 2 m/s to 1 m/s in 5 s. The direction of motion does not change. What is the baby's acceleration?

B9. On a merry-go-round you make five complete revolutions in 250 s. Each revolution covers a circular distance of 100 m.
   a. What is your average speed?
   b. What is your average velocity?
   c. Does acceleration occur? Why or why not?

B10. In Figure 2-9(a) you can see the motion of both the golf ball and the golf club after impact. Compare the velocity of the golf ball to the velocity of the golf club.

C. Extensions to New Situations

C1. When variables like speed, time, and distance are related by an equation, we find it useful to see how one variable changes with another one. We can do this by chang-

ing one variable and determining how the others change.
a. A person moves 10 m. What is the average speed if the trip takes each of the following times: 1 s, 2 s, 4 s, 8 s?
b. Moving from one place to another takes 2 s. What is the average speed if the places are separated by: 1 m, 2 m, 4 m, 8 m?
c. You walk at a speed of 1.5 m/s. How far will you travel in: 0.5 s, 2 s, 6 s, 12 s?

Use the results from parts (a), (b), and (c) to answer parts (d), (e), and (f).

d. If the distance traveled stays the same and the average speed doubles, how does the time change? What happens if the average speed triples?
e. If the time stays the same and the average speed doubles, what happens to the distance? What happens if again the time stays the same and the average speed becomes 12 times what it was before?
f. If the average speed stays the same and the time doubles, what happens to the distance? What happens to the distance if the average speed again stays the same and the time becomes five times as great as it was?

C2. You are taking a trip of 400 km and wish to complete the trip in 5 h.
a. What average speed must you maintain?
b. After 2 h you find that you have averaged only 50 km/h. How far do you have left to go?
c. Before calculating, estimate the average speed required for the rest of the trip in order for you to arrive on time. Explain how you made your estimate.
d. What is the distance and time you have yet to travel?
e. Calculate the average speed for the rest of the trip.
f. If your estimate was different from your calculation, can you explain why?

C3. We all know what happens to a ball when it is thrown straight up. It rises to some height, reverses direction, and returns to the earth. During its flight, the only acceleration is that due to gravity.

a. On the way up, how does the ball's velocity change?
b. On the way down, how does the ball's velocity change?
c. Does the acceleration of the ball ever change during the flight?
d. Is there any time during the flight when the ball's velocity is zero? (If you have trouble with this part, throw an object up and watch it carefully.)
e. Use the answers to parts (a) through (d) to describe the changes in velocity and the acceleration of an object going up and coming down.

C4. In which of the situations below do you expect the object's instantaneous speed at some point to be very different from its average speed?
a. A car is moving in heavy traffic.
b. A baseball has been released by the pitcher and is moving toward the batter.
c. A runner is competing in the Boston Marathon.
d. The sun moves across the sky.
e. An airplane takes you from the terminal in Chicago to the terminal in Kansas City.

C5. We can rearrange the definition of acceleration to answer a variety of interesting questions about motion. Algebraically the procedure is the same as that used to rearrange the definition of speed. In each of the situations below, rearrange the definition of acceleration to answer the questions posed.
a. Initially at rest, a ball is dropped from a rooftop. How long will the ball fall before its velocity reaches 39.2 m/s, down? (Since the ball falls freely, its acceleration is the acceleration due to gravity.)
b. What will be the ball's velocity after it has fallen for 10 s?

C6. Speed, velocity and acceleration introduce the use of the ratio as a convenient mathematical tool. Describe why ratios are useful and try to think of other ratios you might have encountered.

C7. An astronaut floating in space has no acceleration. Suppose the astronaut gently

releases a screwdriver so that it also has no acceleration.
  a. Will the screwdriver have a velocity? If yes, how does the screwdriver's velocity compare to the astronaut's?
  b. How far will the screwdriver move from the astronaut's hand?

C8. Objects thrown both upward and outward near the surface of the earth follow the path shown in Figure 2-C8. The path looks more complex than the ones we have studied, partially because motion occurs in two directions. We can analyze this motion by looking at the horizontal and vertical motions separately.
  a. Does the object move with constant velocity or accelerate along the horizontal dimension?
  b. Does the object move with constant velocity or accelerate along the vertical dimension?
  c. Sketch the horizontal and vertical motions separately. (Draw the horizontal motion as though there were no vertical motion and vice versa.)
  d. Show that the path illustrated in Figure 2-C8 is the sum of the two motions sketched in (c).
  e. Will the average speed and the average velocity of the object be the same?
  f. How would you expect this motion to be different on the surface of the moon? Sketch the motion you would expect to see.

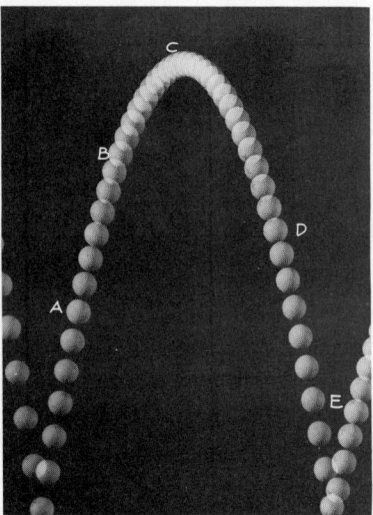

C9. Sprinters often begin a race from a crouched position, while long-distance runners start from a normal standing position. Some research suggests that a crouched position produces a greater acceleration than the standing position. Why would a large acceleration be important to the sprinter but not to the long-distance runner?

C10. Figure 2-1C shows the complete swing of a golfer as he tees off. If you are not a golfer, swing once or twice so you can relate the experience to the photograph.
  a. Where does the speed of the golf club increase? Decrease? Remain relatively constant?
  b. Where does the velocity of the golf club increase? Decrease? Remain relatively constant?
  c. Describe the acceleration of the golf club.

D. Activities

D1. As a scientist investigating Mars you are assigned the task of measuring the speeds of small pebbles that are blown across the ground. The measurements will be made by an unmanned space vehicle, which will land on the Martian surface and send photos back to Earth. What measurements would you need to take and how would you obtain them from the photographs?

D2. The San Andreas fault in California is one of many places where one part of the earth's surface is moving in a direction different from another part. Suppose you are trying to measure the speed with which the land at the edge of the fault is moving. How would you do it? (Include all measurements you would take and how you would use those measurements.)

D3. If you are active in a sport, consult some textbooks to learn about the use of stroboscopic photographs or drawings in describing the motions involved. Describe the use of the concepts introduced in this chapter in analyzing the various techniques used in the sport.

D4. If you are not interested in sports but enjoy photography, you might look at *Moments of Vision* by Harold E. Edgerton and James R. Killian, Jr. (MIT Press, 1979). It reveals a fascinating world!

Relative Motion at Low Speeds

Ever walked on walls—or even on ceilings? Taken from *2001: A Space Odyssey,* the scene in Figure 3-1 depicts a flight attendant walking in the weightless environment of space. It was, of course, filmed right here on earth. This scene and a variety of other special effects are created by changing the reference frame from which we view motion.

The last chapter dealt with motion only within stationary reference frames. However, motion also occurs in moving reference frames. You can walk around on an airplane as it transports you from one city to another. This chapter examines motion in coordinate systems that are moving at a constant velocity relative to each other. We will restrict ourselves to low speeds, that is, speeds that are far less than the speed of light. Reference frames moving at speeds near the speed of light will be discussed in the next chapter. We will see that the motion of the reference frame changes our description of the velocity of an individual object but not the *relative velocity* between two objects. This tells us that some quantities remain constant regardless of the

Chapter 3. Relative Motion at Low Speeds

Figure 3-1

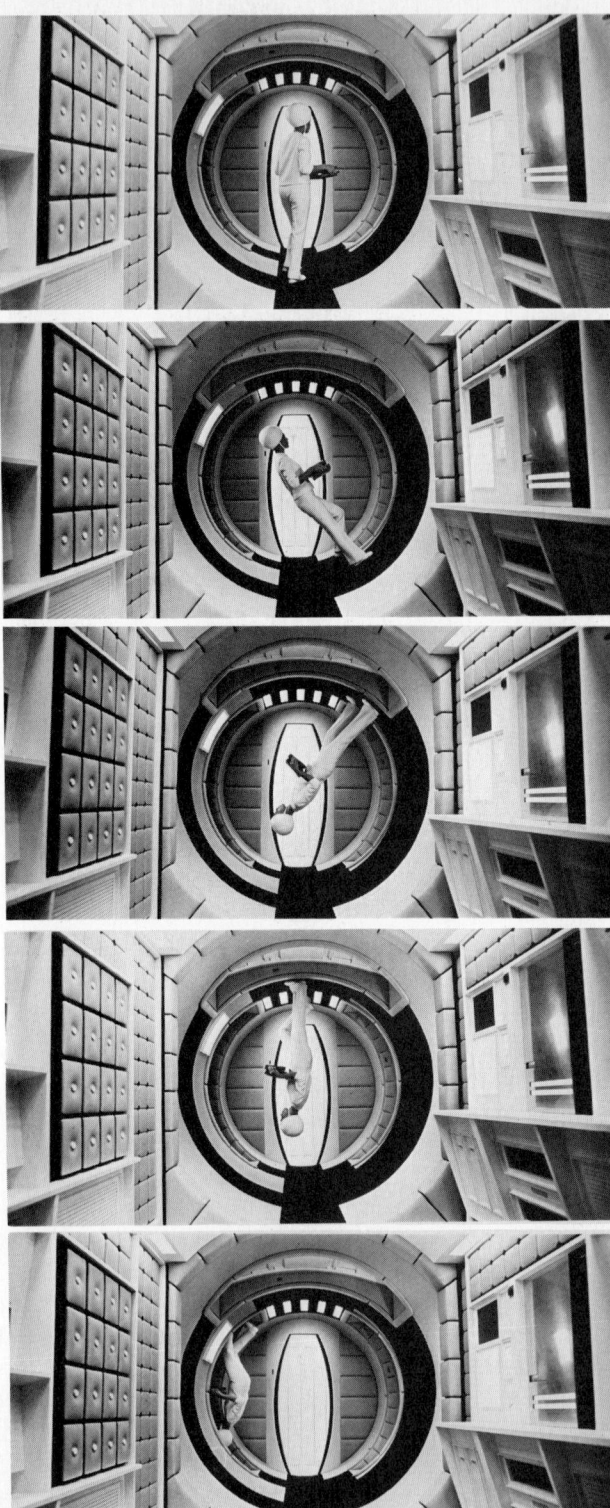

From the MGM release "2001: A SPACE ODYSSEY" © 1968 Metro-Goldwyn-Mayer Inc.

motion of the reference frame. For reference frames moving at constant velocity relative to each other, the *principle of relativity* states that events in any reference frame are the same.

MOVING REFERENCE FRAMES

Until as late as the eighteenth century, controversy raged over the earth-centered and sun-centered views of the universe. Religious teaching caught up in the conflict made the controversy extraordinarily complex. At face value, however, it was simply a disagreement about how to interpret the daily motion of the sun across the sky. Does the earth stand still and the sun move across the sky? Or, does the sun stand still and the earth rotate beneath it, creating the apparent motion of the sun? While mountains of evidence have piled up to convince us that the sun-centered model is the better one, our naked-eye observations alone do not enable us to decide between the two points of view.

Different points of view can arise when we use the same reference frame but make different statements about the motion of that reference frame. In saying that the sun moves, we imagine a stationary reference frame, the earth. In saying the earth moves, we imagine a moving reference frame, the rotating earth. In order to agree on descriptions of motion, we must specify both the reference frame and its state of motion.

Motion of the Reference Frame is Often Ignored

Our language reveals an assumption that the earth is stationary. We say that the sun rises and sets, that it moves across the sky, and that it is lower in winter than in summer. These statements arise not from ignorance about the motion of the earth but from the fact that the earth's motion is irrelevant in our everyday conversations about the sun.

To pursue this idea further, suppose you travel by bicycle from Denver, Colorado, to Washington, D.C.—a distance of 2500 kilometers (km). Such a trip would take about 35 days. During that time interval, the earth moves in its orbit around the sun, traveling a distance of about 94,000,000 km. Would you tell your friends that you traveled 94,002,500 km? If you did, you would receive rather strange looks! You ignore the motion of the earth because it is not relevant to your trip. In fact, you ignore a lot of motion—the earth rotating on its axis, the earth revolving around the sun, the solar system moving through the galaxy, and the galaxy moving through the universe. None of these motions, even though they occur continuously, are needed to describe a trip from Denver to Washington.

All Reference Frames are Valid

Although it seems absurd to include the motion of the earth in describing a trip from Denver to San Francisco, it would still be valid, particularly in a

Figure 3-2
Your description of what moves depends on the reference frame.
(a) In a reference frame attached to the tree, the truck moves past at a velocity of 10 m/s, east.
(b) In a reference frame attached to the truck, the tree moves past at a velocity of 10 m/s, west.

reference frame outside our solar system. In fact, in such a reference frame the motion from Denver to San Francisco would seem minute compared to the motion of the earth itself.

Motion can be described from a variety of reference frames. To illustrate this, let's examine a simple, everyday event like a truck moving past a tree. We can describe this event from two different reference frames—one attached to the tree and a second attached to the truck itself (Figure 3-2). In the reference frame attached to the tree, Figure 3-2(a), the truck is moving relative to a stationary tree. We could even estimate its velocity—10 meters/second (m/s), east. In the reference frame attached to the truck, Figure 3-2(b), the tree moves relative to the truck. Its velocity is 10 m/s, west.

ROLLING RIGHT ALONG

James William Knell took advantage of the ideas of relative motion when, in 1882, he invented and patented the "Apparatus for Producing Illusory Dramatic Effects." Mr. Knell wished to enable stage directors to present the illusion of motion during a play. His apparatus consists of a large painted canvas (*a*) which moves from left to right on two rollers (*b* and *c*). At the center of the stage in front of this canvas is a horse and carriage (*i*), which are placed on a treadmill (*d*). As the horse trots, but does not move relative to the stage, the canvas is rolled past. The relative motion between the horse and buggy and the background presents the illusion of motion.

Your immediate response might well be: "What do you mean, the tree moves? Trees don't move. Trucks move!" Your reaction reflects our tendency to choose the earth (including its hills, trees, and general land features) to be a motionless reference frame. However, descriptions of motion depend on the reference frame. When we change to a reference frame attached to the truck, all parts of the truck are stationary relative to itself. Objects not attached to the truck can move; consequently, we say that the tree moved relative to the truck.

Both reference frames, the one attached to the tree and the one attached to the truck, are equally valid in describing the event. "Move the truck 10 meters farther from the tree" seems more direct than "Drive until the tree has moved 10 meters farther behind the truck." Likewise, "The sun rises at 6:30 A.M." is easier to say than "At 6:30 A.M. the earth will have rotated to a position in which the sun will become visible." However, both statements describe the same event. The truck and the tree become separated by a larger distance. Dawn occurs. The difference is not in the event itself but in our description of it based upon our choice of reference frames.

46 Chapter 3. Relative Motion at Low Speeds

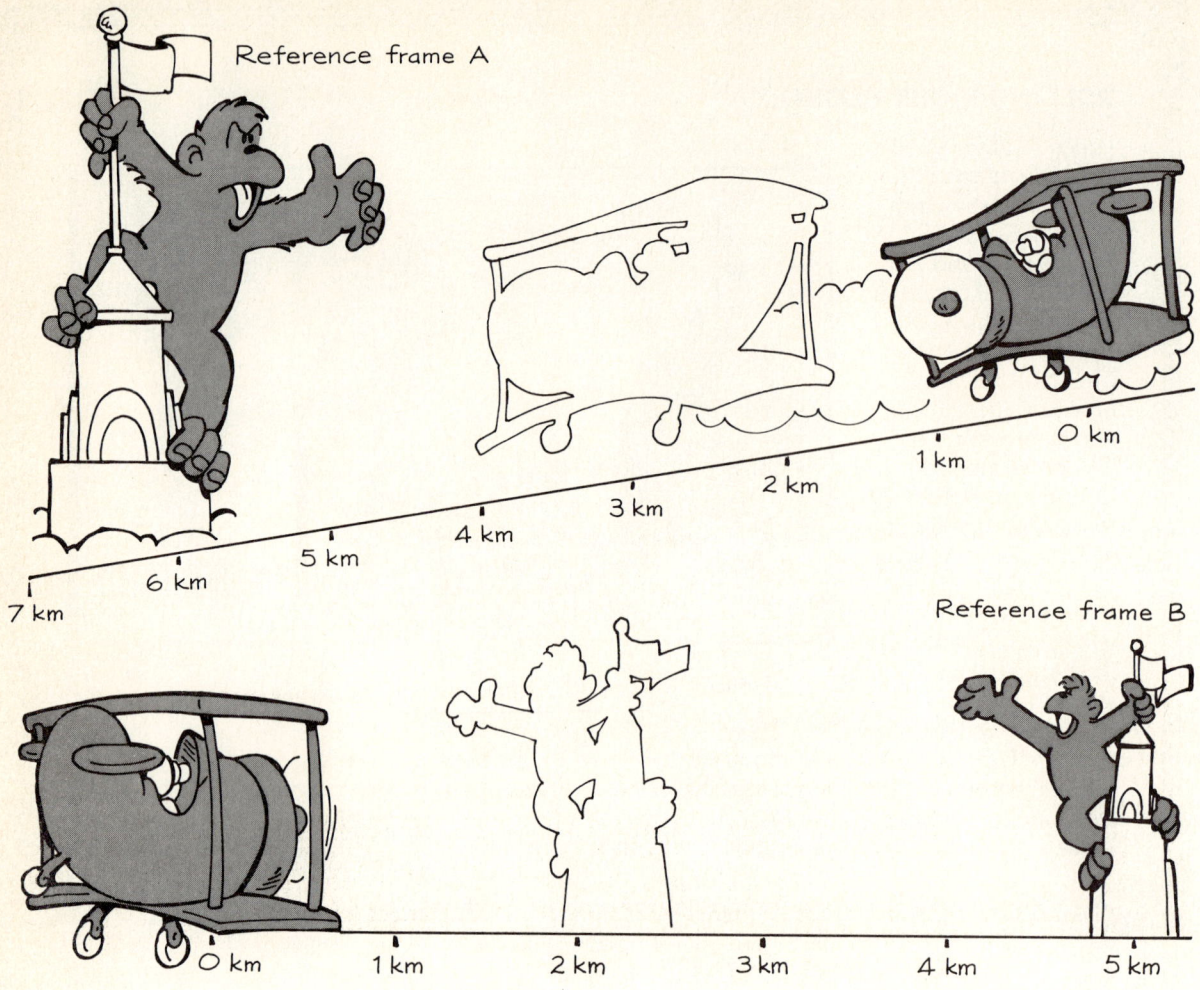

Figure 3-3

SELF-CHECK 3A

Figure 3-3 shows an event in two different reference frames. In each, identify the reference frame and describe the event in that reference frame.

Illusions of Motion

Many illusions in motion pictures take advantage of our natural tendency to assume that objects attached to the earth do not move. Animators cause their characters to move across a landscape, even though the body of the character

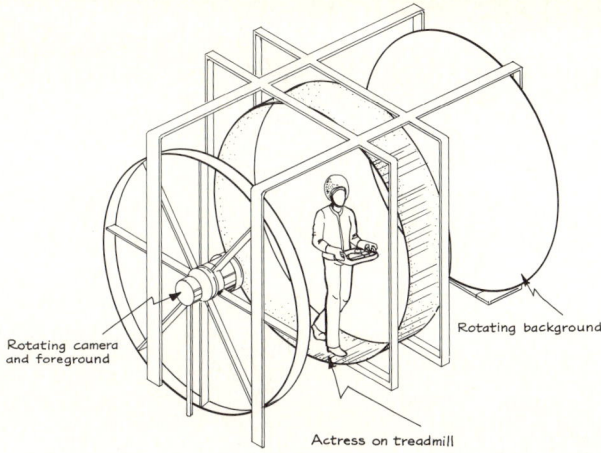

Figure 3-4
The motion in Figure 3-1 was created by rotating a camera, background, and foreground. The actress walked on a treadmill in a separate room.

and the camera are motionless in the studio. The character remains stationary relative to the camera and the background is pulled along behind it. Because we naturally assume that the earth and its scenery are not moving, we interpret the movement as the character walking. This technique is extended to nonanimated films, from the famous old-time cowboy "chase scenes" to spaceships moving through a background of stars. In *Return of the Jedi,* Luke Skywalker and Princess Leia were filmed while sitting on Imperial speeders that were essentially stationary. A second film was recorded by a camera mounted on a truck that drove through the forest. The two films were then superimposed to provide a breathtaking chase scene.

The more complex motion of the flight attendant in space that opened the chapter was filmed by a rotating camera. As shown in Figure 3-4, the actress remained fixed on the floor of the film studio, walking along a treadmill. As she walked, the camera and background were rotated. When the camera was upside down relative to the earth, it photographed the actress as being on the ceiling. This technique is effective photographically because we automatically use the room as a motionless reference frame.

Another common illusion occurs when you are stopped at a traffic light and a car on your right or left begins to move forward. With no other reference object to check your perception, you may think you are rolling backward and instinctively step on the brake. The impression created by our assumption that the other car is stationary can be so strong that many actually feel physical sensations of motion.

RELATIVE SPEED AND VELOCITY

All reference frames are equally valid. In one reference frame, trucks move; in another, trees move. If we calculated velocities, they would be different in the different reference frames. Most of us are uneasy with all these possibilities. We might wish we could find some absolute reference frame against which "true" motion could be described. Since we do not believe that such a

48   Chapter 3.   Relative Motion at Low Speeds

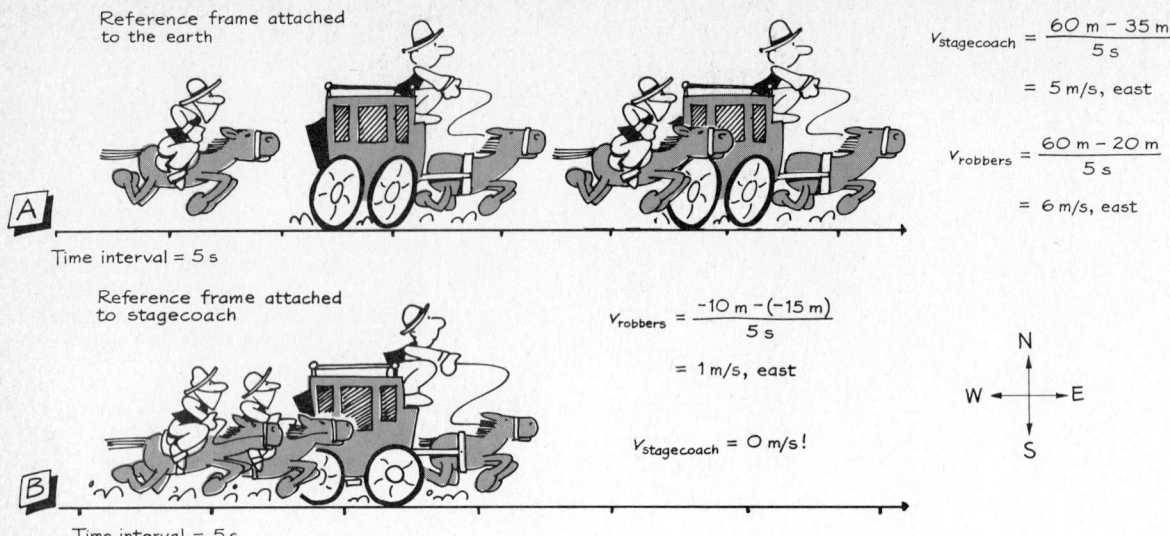

Figure 3-5
(a) In a reference frame attached to the earth, the stagecoach moves at a velocity of 5 m/s, E, and the robbers at 6 m/s, E.
(b) In a reference frame attached to the stagecoach, the stagecoach is stationary and the robbers move at a velocity of 1 m/s, E.

reference frame exists, we look instead for some quantity that does not change from one reference frame to another. For reference frames moving at a constant velocity relative to one another, relative speeds and relative velocities provide us with such a concept.

Velocity of One Object Relative to Another

Let's consider a familiar scene from old western movies. The stagecoach driver, who is being chased by the bad guys, looks behind him and says, "They're gaining on us!" The driver chose a reference frame attached to himself. The bad guys would still be gaining on him if he had chosen another reference frame—for instance, one attached to the earth. The rate at which the robbers catch up to the stagecoach remains constant regardless of the reference frame we choose.

We can convince ourselves of this by comparing the velocity of the robbers with the velocity of the stagecoach in two different reference frames. In Figure 3-5(a) the reference frame chosen is one attached to the earth. The velocity of the stagecoach is 5 m/s, east. The robbers are moving faster, at a velocity of 6 m/s, east. Using a coordinate system attached to the moving stagecoach, Figure 3-5(b), the robbers are moving at a velocity of 1 m/s, east. Relative to itself, however, the stagecoach is stationary. Its speed is 0 m/s.

As expected, the velocities of the stagecoach and robbers are different in the two reference frames. The stagecoach moves at a velocity of 5 m/s, east, in a reference frame attached to the earth and 0 m/s in a reference frame attached to itself. The robbers move at a velocity of 6 m/s, east, in a reference frame attached to the earth and 1 m/s, east, in a reference frame attached to the stagecoach. As we have already seen, the velocity of an object depends upon the motion of the reference frame chosen.

Now, compare the motion of the robbers to the motion of the stagecoach. In a reference frame attached to the earth, the stagecoach is moving at a velocity of 5 m/s, east, while the robbers are moving faster, at a velocity of 6 m/s, east. The robbers are gaining on the stagecoach at a rate of 1 m/s. When we change to a reference frame attached to the stagecoach, we find the same result: The robbers are moving at a velocity of 1 m/s, east, relative to the stationary stagecoach. If we were to choose yet another reference frame, we would find the same result. The robbers are always gaining on the stagecoach. The motion of one object relative to another remains constant when we change reference frames.

Relative Velocity Remains Constant

The realization that the velocity of the robbers relative to the stagecoach is the same regardless of the reference frame we choose makes this quantity more useful to us in describing the motion among objects. Consequently, we introduce the concept of relative velocity to describe the motion of one object relative to another. The **relative velocity** between two objects is the velocity of one object in a reference frame in which the second object is stationary. **Relative speed** is defined in the same manner, except that speeds are substituted for velocities. When we chose a reference frame attached to the stagecoach, Figure 3-5(b), the stagecoach became the reference object. Its velocity was 0 m/s. Consequently, the velocity of the robbers in this reference frame was their velocity relative to the stagecoach. When we chose the reference frame attached to the earth, we introduced a third object, the earth. The relative velocity must be determined from the velocity of each object relative to the earth.

A general rule for determining the relative velocity between two objects, A and B, given their velocities relative to the earth is:

$$\begin{matrix}\textbf{Velocity of A} \\ \text{relative to B}\end{matrix} = \begin{matrix}\textbf{Velocity of A} \\ \text{relative to Earth}\end{matrix} - \begin{matrix}\textbf{Velocity of B} \\ \text{relative to Earth}\end{matrix}$$

$$V_{AB} = V_A - V_B$$

where V_{AB} is the velocity of A relative to B, V_A is the velocity of B relative to Earth, and V_B is the velocity of A relative to Earth.

Velocity is a vector quantity, so we must keep track of direction as well as magnitude. Let's apply this general rule for determining relative velocities between two objects to the robbers and the stagecoach.

As shown in Figure 3-6(a), the velocity of the robbers (A) is 6 m/s, east. The velocity of the stagecoach (B) relative to the earth is 5 m/s, east. According to our rule, the velocity of the robbers (A) relative to the stagecoach (B) is the difference between the two velocities relative to the earth. Since the stagecoach and robbers are both moving in the same direction, we can subtract the two magnitudes. (You could also apply vector subtraction, as shown in Figure 3-6(a).) The relative velocity between the robbers and the stagecoach is 6 m/s, east, minus 5 m/s, east, which equals 1 m/s, east. The robbers are gaining on the stagecoach at a speed of 1 m/s.

Suppose the movie script requires that the robbers come toward the stagecoach rather than chasing it from behind. Let's assume that the velocity of the robbers (A) relative to the earth is 6 m/s, west, and the velocity of the

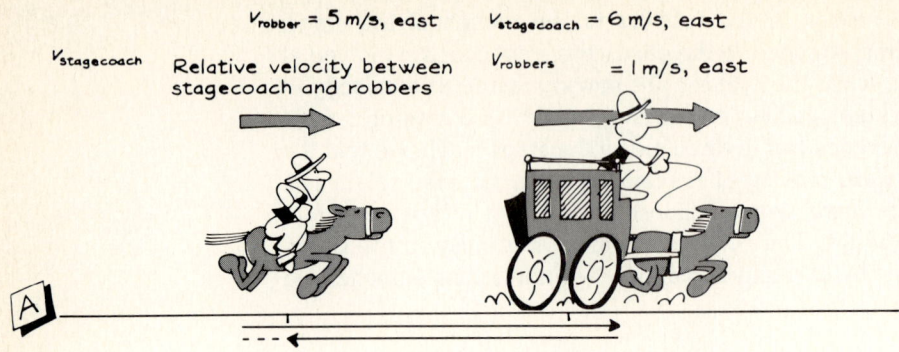

Figure 3-6 **(a)** In a reference frame attached to the earth, the stagecoach moves at a velocity of 5 m/s, E, and the robbers at 6 m/s, E. The relative velocity between the robbers and the stagecoach is 1 m/s, E.
(b) In a reference frame attached to the earth, the stagecoach moves at a velocity of 5 m/s, E, while the robbers move at a velocity of 6 m/s, W. The relative velocity between the robbers and the stagecoach is 11 m/s, W.

stagecoach (B) relative to the earth is still 5 m/s, east. Now the stagecoach and robbers are moving in opposite directions. Figure 3-6(b) illustrates the tail-to-tip method for subtracting the two vectors. A velocity of 6 m/s, west, minus a velocity of 5 m/s, east, equals a velocity of 11 m/s, west. The robbers are approaching the stagecoach at a speed of 11 m/s.

The fact that we are adding vectors makes the rule a little more complex, since the direction in which the two objects move must be taken into account. Many people find it easier to remember the general rule in terms of simply adding or subtracting the magnitudes of the velocities. When two objects are moving in the same direction, the magnitude of their relative velocity is the difference between the magnitudes of the velocities relative to the earth. When the two objects are moving in opposite directions, the magnitude of their relative velocity is simply the sum of the magnitudes of their velocities relative to the earth. That is,

$$v_{AB} = v_A - v_B \qquad \text{A and B in same direction}$$
$$v_{AB} = v_A + v_B \qquad \text{A and B in opposite directions}$$

The direction of the relative velocity is the direction in which object A is moving relative to the earth. These two expressions can be applied only to those cases where the two objects are moving along the same line, in the same direction or in opposite directions. If the two objects move toward one another at an angle, then the tail-to-tip method for subtracting vectors provides the simplest way of determining the relative velocity between the two objects.

We have described the process by which we can determine the relative velocity between two objects, given their velocities relative to the earth. Our rule can be extended to reference frames other than the earth. If we know the velocity of both objects (A and B) relative to some third object, then our rule can be applied to these situations as well. We substitute the third object, whatever it might be, for the earth in our expressions.

SELF-CHECK 3B

Traveling along the interstate highway, you pass a truck. Your velocity is 80 kilometers/hour (km/h), east, relative to the ground. The truck's velocity is 60 km/h, east, relative to the ground. What is your velocity relative to the truck? What would your velocity relative to the truck be if its velocity were 60 km/h, west, relative to the ground and yours were 80 km/h, east, relative to the ground?

When the robbers were chasing the stagecoach, the only velocity that mattered to the stagecoach driver was how fast the robbers were gaining on him. While the velocities of the stagecoach and robbers changed as we changed reference frames, the relative velocity between them remained constant. Relative velocity is, in fact, the velocity we perceive in everyday situations. When you see a friend walking away, you run to catch up. If you walked at the same velocity relative to the earth as your friend, your velocity relative to the friend would be zero. You would not catch up. You use the relative velocity between you and your friend to judge how fast you must run. The relative velocity between two objects is the same in any coordinate system moving at a constant velocity relative to any other coordinate system.

THE PRINCIPLE OF RELATIVITY

Careful observation over several centuries has confirmed the idea that relative velocity is the same in different reference frames, as long as the reference

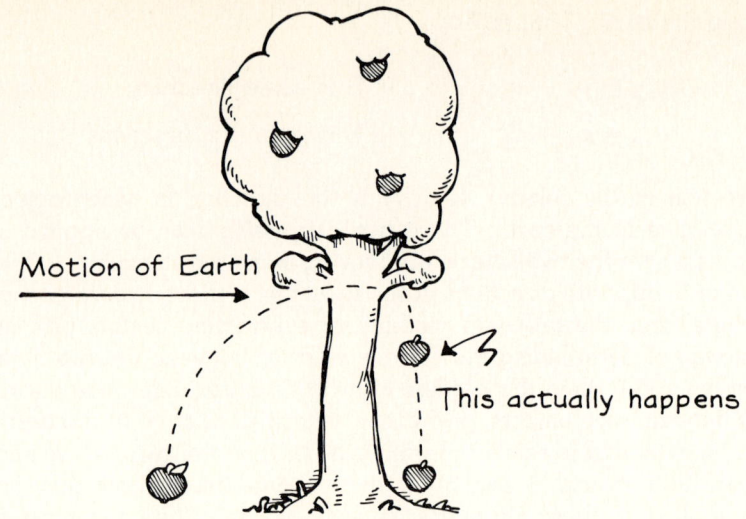

Figure 3-7
Those who believed that the earth was stationary argued that a moving earth would cause apples to land west of, rather than beneath, the tree. In truth, apples land beneath the tree in either a stationary earth or an earth moving at a constant velocity.

frames move at constant velocity relative to one another. Though this may have seemed quite obvious in our examples, it is not always obvious in everyday life. If you stand still and drop a rock, you know that the rock will fall in a straight line and land at your feet. Using yourself as a stationary reference frame, you expect objects to fall straight down. Many people suppose, however, that if they drop a rock while walking at a constant speed, the rock will land behind them. They expect the motion of the reference frame to affect the way in which the rock falls.

At the time of the controversy between the earth- and sun-centered views of the universe, some who held the earth-centered view argued that an apple that fell from a moving tree would land west of the tree rather than beneath it (Figure 3-7). Since the apple does in fact land beneath the tree, they reasoned, the earth does not move. In actuality, the apple always falls from the tree in exactly the same manner on a stationary earth or on an earth moving at a constant velocity, just as the rock always lands at your feet regardless of whether you are walking or stationary.

The observation that events occur in the same manner independently of the velocity of the reference frame led Galileo and others to suspect that the fundamental principles of physics do not depend upon the constant motion of the reference frame from which they are observed. Called the **principle of relativity,** this conclusion is: The laws of physics are the same in all reference frames moving at constant velocities relative to one another. The principle of relativity has been important as we have stepped beyond our own earth-centered reference frame into space. First stated by Galileo more than 300 years ago, this principle correctly guided our exploration into space—first into earth-centered orbits and later to the moon and back. It governs all events—those at everyday speeds and those at speeds near the speed of light. At speeds near the speed of light, however, our fundamental concepts of space and time have had to be altered in order to preserve our belief in the principle of relativity. These modifications are the subject of the next chapter.

CHAPTER SUMMARY

Motion, like position, must be described in terms of a reference frame. The velocity of an object is usually described in terms of a reference frame that is assumed to be stationary. The earth is rotating on its axis and revolving around the sun, yet we describe motion on the surface of the earth relative to a stationary earth. Two people who use reference frames that are moving at constant velocities relative to one another will describe the motion of an object differently. Both descriptions are equally valid, but one may be more useful than another in a given situation.

The velocity of an object will be described differently in reference frames moving at different—but constant—velocities relative to one another. But, the relative velocity between any two objects remains constant. The *relative velocity* between objects A and B is defined as the velocity of one object in a reference frame in which the second object is stationary. When the velocity of both objects is given relative to the earth, the relative velocity between the two objects is the vector difference between the velocity of each object relative to the earth. The relative velocity between two automobiles, for example, is the difference between their velocities relative to the ground. Relative velocity remains constant in all reference frames moving at a constant velocity relative to one another. More generally, the principles of physics in any reference frame are the same as those in any other reference frame moving at constant velocity relative to it. This is called the *principle of relativity*.

ANSWERS TO SELF-CHECKS

3A. In reference frame A, the airplane moves from (1 km) to (4 km). The coordinate system is stationary relative to the earth. In reference frame B, the top of the building moves from (5 km) to (2 km). The coordinate system is stationary relative to the airplane.

3B. 20 km/h, east; 140 km/h, east.

PROBLEMS AND QUESTIONS

A. Review of Chapter Material

A1. Under what circumstances will two people describe the motion of the same object differently?

A2. Give two examples that show how we ignore the motion of reference frames in describing events in everyday life.

A3. Is one reference frame better than another in describing motion? How do we choose which reference frame to use?

A4. What velocity remains unchanged as we change from one reference frame to another which is moving at a constant velocity relative to the first?

A5. Suppose you are given the velocities of objects X and Y relative to the earth. Describe how you could determine the velocity of X relative to Y.

A6. State the principle of relativity.

B. Using the Chapter Material

B1. While riding on a bus, you drop a book.
  a. How would you normally describe the motion of the book?
  b. What reference frame did you use in your description? Was the reference frame moving or stationary relative to the earth?
  c. What motion have you ignored in your description in (a)?
  d. Describe the motion of the book from a reference frame that is stationary relative to the earth.
  e. Do your descriptions in (a) and (d) agree? Why or why not?

B2. The photograph of the pole vaulter in Figure 2-9 was taken by a camera stationary relative to the earth. In this reference frame, the cross bar is stationary and the pole vaulter is moving.
  a. Describe how a photograph would look had the camera been attached to the pole vaulter.
  b. Which reference frame, the one attached to the earth or the one attached to the pole vaulter, would be most helpful to the coach in examining the pole vaulter's technique. Why?

B3. Figure 3-B3 is a strobelike drawing of a bicyclist moving in a reference frame attached to the earth.
  a. Describe the motion of the tree and the bicycle in a reference frame attached to the tree.
  b. Describe the same motion in a reference frame attached to the bicycle.
  c. Which reference frame is more valid?

B4. The time interval between the two images in Figure 3-B3 is 8 s.
  a. What is the velocity of the bicyclist relative to the tree?
  b. What is the velocity of the tree relative to the bicyclist?
  c. Why are the magnitudes of these two velocities equal?

B5. An airplane which has a velocity of 400 km/h, east, relative to the ground is flying in a wind with a velocity relative to the ground of 50 km/h, east. What is the speed of the airplane relative to the wind? (This speed is called the *airspeed* of the airplane.)

B6. A police officer, motionless relative to the earth, measures the relative speed between the police car and another car as 100 km/h. What would the relative speed be if the police officer measured the speed while driving at a constant speed of 50 km/h in a direction opposite to that of the other car?

B7. Maria is jogging at a velocity of 4 m/s, east. David is 12 m behind her, running at a velocity of 3 m/s, east. (Both velocities are relative to the earth.) What is David's velocity relative to Maria? Can David ever catch up with Maria? Why or why not?

B8. Kim and Kevin are rushing through an airport. Their velocities relative to the earth are 6 m/s, east, and 6 m/s, west, respectively.
  a. What is Kim's speed relative to Kevin?
  b. Kim and Kevin eventually bump into each other. Is it possible to change to a reference frame in which they do not bump into each other? If yes, describe the origin and motion of the reference frame. If no, explain why not.

B9. A police car is pursuing an automobile that has violated the speed limit. What velocity does the police officer use to judge how fast he must travel to overtake the automobile? Under what circumstances will the police officer radio ahead to another officer rather than pursue the automobile himself?

C. Extensions to New Situations

C1. The photograph in Figure 3-C1 was taken by aiming a camera toward the North Star and opening the shutter. The streaks are made by the light emitted from stars.
   a. How would you describe the motion of the church tower?
   b. How would you describe the motion of the stars?
   c. Describe the reference frame used in (a) and (b). What have you assumed about that reference frame?
   d. Someone else reports that the earth rotates beneath the stars. The stars are, in fact, stationary. What reference frame did this person use?
   e. What is different about the two reference frames used?
   f. Which reference frame, (c) or (d), is more valid?

C2. In coordinate system A, a ball moved from (0 m, 2 m) to (0 m, 10 m) in 4 s. Measured in a different coordinate system, system B, the ball moved from (2 m, 10 m) to (2 m, 18 m) in the same 4 s.
   a. What was the ball's velocity in coordinate system A?
   b. What was the ball's velocity in coordinate system B?
   c. Compare your answers in (a) and (b). What can you say about coordinate systems A and B? If they are moving relative to one another, find their relative velocity.

C3. Problem 2-C8 asked you to analyze the horizontal and vertical motions of an object thrown upward and outward near the surface of the earth. (See Figure 2-C8.)
   a. Describe the reference frame from which the photograph was taken.
   b. Sketch the motion of the object had the camera been moving horizontally at the same velocity as the object.
   c. In a reference frame fixed relative to the earth, how would you throw the object to make its motion appear as your description in (b)?
   d. Use the results in (a), (b), and (c) to explain why an object dropped from the top of a building lands immediately below where it was dropped rather than behind where it was dropped. (Since the building is attached to the rotating earth, it moves eastward during the time that the object falls to the ground.)

C4. Police radar systems determine a car's speed by measuring its speed relative to the police car. From this relative speed and the speed of the police car relative to the ground, a small computer determines your speed relative to the highway.
   a. Suppose you are traveling at 80 km/h, north, relative to the road and a police car is traveling at 70 km/h, south, relative to the road. Radar measures the speed of your car relative to the police car. What is the speed measured by the radar?
   b. A radar unit is moving at 50 km/h relative to the road. It measures the relative speed of an oncoming car to be 150

km/h. What is the speed of the oncoming car relative to the road?
  c. A radar unit traveling at 90 km/h relative to the road measures a relative speed of 0 km/h for a car traveling in the same direction. What is the speed of the car relative to the road?
C5. The earth rotates from west to east. If relative speeds were the only factor, which would be faster: a trip by air from New York to San Francisco or one from San Francisco to New York?
C6. A bicyclist is traveling at a constant velocity of 10 km/h, east, relative to the road.
  a. What is her velocity relative to a wind blowing 10 km/h, east, relative to the earth?
  b. What is her velocity relative to a wind blowing 10 km/h, west, relative to the earth?
  c. In which situation, (a) or (b), will the bicyclist feel a wind moving past her?
  d. Evaporation of sweat by the wind is an important factor in making a bicyclist feel cool on a hot day. Why will riding with the wind make the bicyclist less comfortable than riding against the wind?
C7. In this chapter we discussed the relative motion of two objects moving toward or away from one another. Another situation in which relative motion is important is when one object moves inside of another moving object. For example, you are on a bus moving forward 30 m/s relative to the earth and are walking toward the back of the bus at a speed of 2 m/s relative to the bus.
  a. In 1 s how far does the bus move forward?
  b. In 1 s how far do you move backward?
  c. What is your total displacement relative to the earth in 1 s?
  d. What is your velocity relative to the earth?
  e. Can you state a general rule for calculating relative velocities when one object is moving inside another?
C8. Test the rule you stated in Problem C7 on a new example. In still water you can paddle a canoe 2 m/s. That is your speed relative to water that is not moving. Suppose you continue to paddle in exactly the same manner; however, now you are floating on a river that flows at a rate of 1 m/s.
  a. If you paddle in the same direction as the current, what is your velocity relative to the earth?
  b. If you paddle against the current, what is your velocity relative to the earth?
C9. A lion pursues an antelope. Relative to the earth, the lion moves at a velocity of 23 m/s, west, while the antelope runs at a velocity of 15 m/s, west.
  a. What is the relative velocity between the lion and the antelope? Will the lion catch the antelope?
  b. In reference frame A, the lion's velocity is measured to be 100 m/s, west, and the antelope's velocity is 88 m/s, west. What do we know about reference frame A relative to the reference frame attached to the earth?
  c. If we compared our value of the relative velocity between the lion and the antelope with that measured in reference frame A, would there be any way that we could determine that reference frame A is moving relative to the earth?
C10. Police radar must accurately measure two relative speeds: the speed of the suspect speeder relative to the police car and the speed of the police car relative to the highway. With these two speeds, the radar system's computer can compute the speed of the suspect relative to the highway. To determine these two relative speeds, moving police radar uses two radar signals. One bounces off the road and measures the speed of the highway relative to the police car. The second bounces off the suspect's car and measures its speed relative to the police car. Recently some of the radar measurements have been questioned because of possible error. If the radar used to determine the police car's speed hits a large object on the side of the road rather than the road itself, the speed of the car relative to the road will be measured to be less than it really was. (This malfunction is named after a mathematical function and is called *cosine error*.) However, the speed

of the alleged speeder's car relative to the police car will be accurate. To investigate this situation, suppose a police car is actually traveling at 70 km/h, but its radar reports its speed to be 60 km/h.

a. The radar measures the speed of car A moving in the same direction as the police car as 30 km/h. What is the actual speed of car A relative to the highway?
b. What speed relative to the highway does the radar calculate for car A?
c. Car B is traveling toward the police car. Its speed relative to the police car is measured to be 160 km/h. What is car B's actual speed relative to the highway?
d. What speed relative to the highway does the radar calculate for car B?
e. The speed limit on the interstate is 90 km/h. Which car—A or B—is actually speeding? Which driver will receive a speeding ticket?
f. If you wish to use the cosine error as a defense in court, which way relative to the police car must you have been traveling?

D. Activities

D1. Have you ever been in a situation where you were confused about what was moving? If so, describe the situation.

D2. Select special effects from current movies and analyze how they might have been filmed using the concepts of relative motion.

D3. As a fair-weather jogger, you enjoy getting your exercise indoors. But, you want to jog at least 2 km per day. Design a device which will move 2 km relative to your feet while your body is fixed relative to your house.

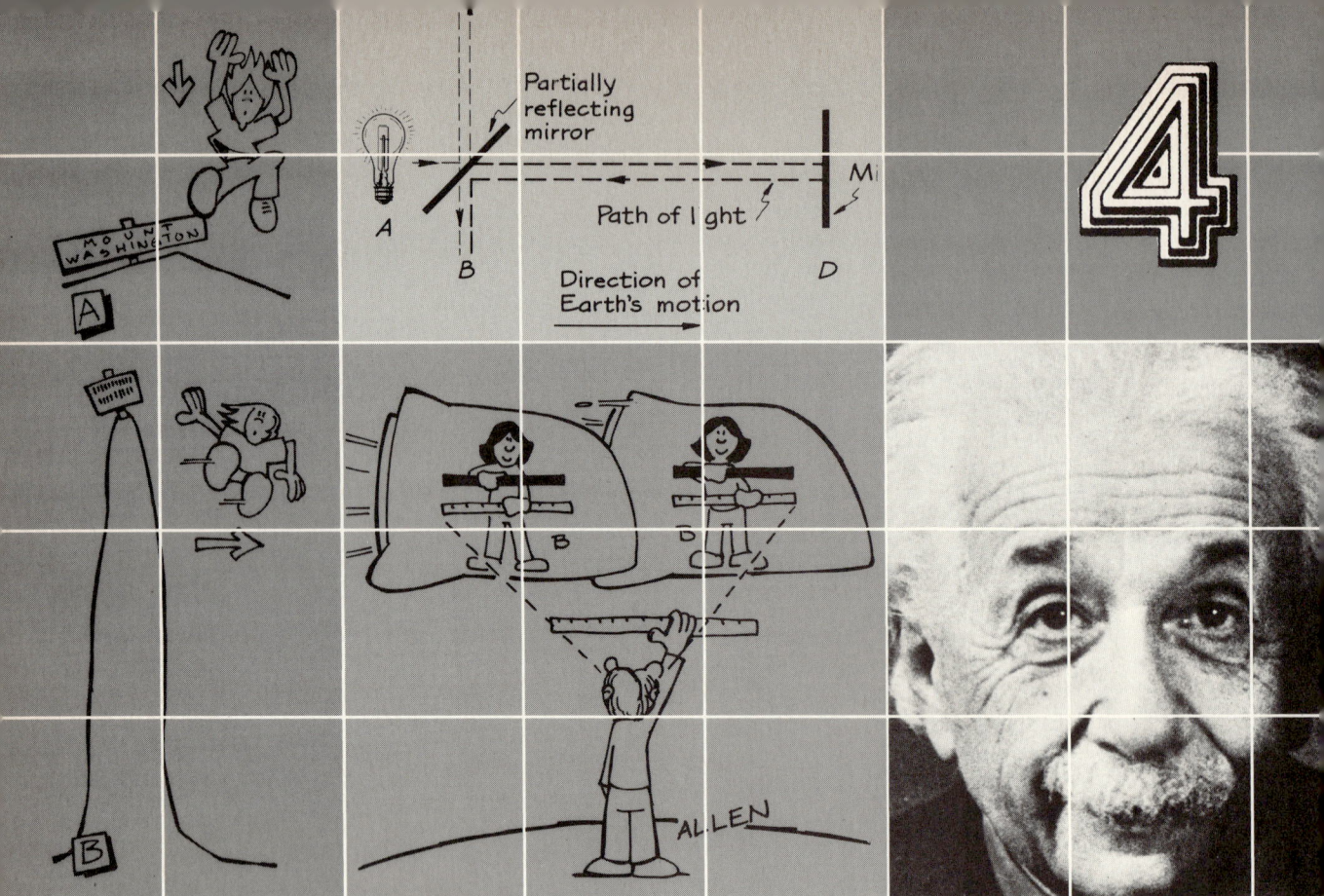

Special Theory of Relativity

What difference does it make anyway, whether we move faster or whether the object becomes shorter? I have to go ten blocks to get to the post office, and if I step harder on the pedals the blocks become shorter and I get there quicker.

George Gamow, *Mr. Tompkins in Paperback*

Faster speeds? Shorter blocks? Logically it makes no difference. A greater speed or a shorter distance will get you to the post office more quickly. However, never having seen a city block become shorter, we rely upon greater speeds. The excerpt from Gamow's book seems absurd—not because it violates our logic, but because it violates our common sense.

Our common sense includes an absolute concept of space and time. While position and time clearly are relative concepts, other concepts that are derived from measurements of position and time seem absolute. Take the concept of duration, for example. If you leave New York at 8:00 A.M. EST and arrive in Kansas City at 11:00 A.M. EST (Eastern Standard Time), the duration of your trip is 3 hours. Measured in a different reference frame, such as Central Standard Time, you would leave New York City at 7:00 A.M. CST and arrive in Kansas City at 10:00 A.M. CST. The duration of your trip is still 3 hours. While the departure and arrival times will be different in different reference frames, the *duration* of the event seems to be constant. The same can be said about the length of an object. The position of each end of a city block is described differently in coordinate systems that use different origins, but the *length* of the city block—found by subtracting the coordinates of each end—seems to be constant. In the last chapter we saw that measurements of the velocity of the stagecoach or the velocity of the robbers would be different in different reference frames. But the *relative velocity* between the stagecoach and robbers—found by subtracting their velocities relative to the earth—seems to be constant. This experience of constant length, constant duration, and constant relative velocities allows us to imagine an absolute reference frame against which we really can know the position or velocity of an object or the time of an event.

At the turn of this century, many physicists imagined, either consciously or unconsciously, an absolute reference frame for light. They expected the speed of light measured here on earth to reflect the motion of the earth through this absolute reference frame. An experiment performed by Albert Michelson and Edward Morley in 1887, called the *Michelson-Morley experiment,* shattered those beliefs when it showed that the speed of light is the same in all reference frames. In 1905 Albert Einstein introduced the *special theory of relativity,* which assumes that the speed of light is constant in all reference frames. In so doing, Einstein did away with all notions of an absolute reference frame—forcing us to rethink our concepts of position, time, and motion. Length, duration, and relative speeds vary with the motion of the reference frame, though imperceptibly at ordinary speeds. At speeds near the speed of light, city blocks do get shorter.

WHY A NEW THEORY?

Most people expect physical theories to develop in a stepwise fashion, almost in the manner that a computer executes a program. In fact, such development seldom occurs; the special theory of relativity is one such example.

Measurements by Michelson and Morley raised questions about the nature of light. The speed of light seemed to be constant in all reference frames. Apparently unaware of those measurements, Albert Einstein proposed a remarkable theory based upon the assumption that the speed of light was constant in all reference frames. In a sense, he had answered a question without knowing that it had been asked. We begin with this question.

Speed of Light and Relative Speed

By 1887 the speed of light had been measured rather accurately. Light travels at about 300,000,000 meters/second (3×10^8 m/s) in a vacuum. In materials like glass or water, light slows down a bit but always in predictable ways. The surprise comes when we measure the speed of light emitted from a source which is moving relative to us.

Figure 4-1
Albert Einstein.

Suppose you are driving past a sign advertising hamburgers. Light travels from the sign to you as you move along in much the same way that the robbers traveled toward the moving stagecoach (Chapter 3). Suppose your speed relative to the earth is 20 m/s. The speed of light relative to the earth is about 300,000,000 m/s. Using the concept of relative speeds and velocities, we would *expect* the speed of light relative to you to be 300,000,000 m/s + 20 m/s, or 300,000,020 m/s, as you approach the sign. If you pass the sign and move away from it, then we would *expect* the speed of light relative to you to be 300,000,000 m/s − 20 m/s, or 299,999,980 m/s. If you decrease your speed to 10 m/s, the *expected* speed of light relative to you would be 300,000,010 m/s as you move toward the sign and 299,999,990 m/s as you move away from it. We *expect* the speed of light relative to ourselves to depend upon our motion. As we shall see, these *expectations* are contradicted by experiments.

Michelson-Morley Experiment

When we talk about the speed of light relative to ourselves, we mean the speed of light that we would measure as we move along. Admittedly, a speed of 10 m/s or 20 m/s seems insignificant compared to the speed of light, but experiments can be performed in which the observer is moving fast enough to produce measurable differences. Albert Michelson and Edward Morley com-

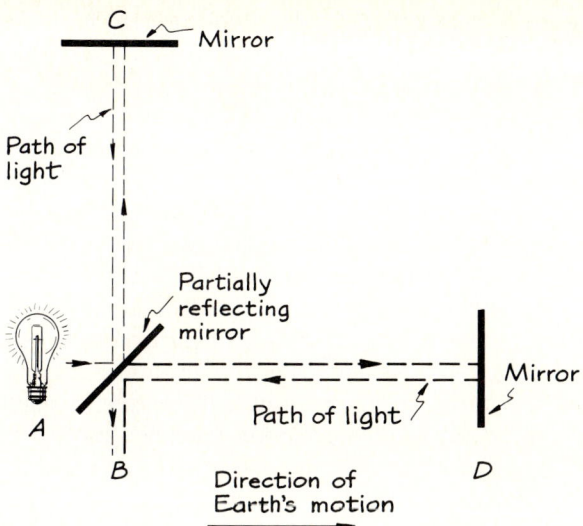

Figure 4-2
Michelson and Morley expected light to travel at a different speed to D and back than to C and back. Instead they found that light travels at the same speed along both paths.

pleted such an experiment in 1887 using the earth's motion as it revolves about the sun (30,000 m/s) to provide the moving reference frame.

A sketch of the **Michelson-Morley experiment** is shown in Figure 4-2. An observer at B is watching two light beams that travel identical lengths. The equipment is arranged so that the path ABD is in the direction of the earth's motion as it orbits the sun, while BC is perpendicular to the earth's motion. Since in one case the light is moving perpendicular to the motion of the observer, the analysis of relative speeds is not as simple as in our earlier examples. But we would still expect the speed of light to be different along the two paths. When actually measured, however, the speed of light was exactly the *same* along the two paths.

Because they contradicted our conventional view of relative speed, these results were staggering. We expect speeds to vary according to the motion of the reference frame from which they are observed, but the Michelson-Morley experiment tells us that the speed of light is always 300,000,000 m/s, regardless of the observer's motion. In terms of our example, light travels toward us at the same speed whether we are standing still or moving toward or away from the hamburger sign. The speed of light is independent of the observer's motion.

Special Theory of Relativity

By assuming that the speed of light is constant in all reference frames, the special theory of relativity offers us a solution to the apparent contradictions posed by the Michelson-Morley experiment. It begins with just two postulates:
1. The principles of physics are the same in all reference frames moving at a constant velocity relative to one another.
2. The speed of light in a vacuum is the same value regardless of the motion of the observer relative to the source of light.

The special theory of relativity provides new meaning to the concepts of relative speed, time, and length. At low speeds these concepts are essentially identical to the common-sense concepts developed in Chapters 1, 2, and 3. At speeds near the speed of light, however, the special theory alters these concepts radically. Since we have never traveled at such high speeds relative to nearby objects, we have no everyday experience against which to check the predictions. So don't be dismayed if these concepts seem strange.

When Einstein proposed the special theory of relativity in 1905, the only significant experience with high speeds had been the measurements of the speed of light. Lacking direct experience, Einstein and others developed their ideas using thought experiments. A **thought experiment** is an experiment conducted in the mind, so to speak: We set forth postulates and then consider what would happen under a variety of circumstances. In special relativity, a thought experiment deals with a situation involving motion at speeds very near the speed of light. While such an experiment is imaginary in the sense that we cannot actually perform it, it does provide us with a logical conclusion and suggests real experiments to verify its predictions. In the decades since the special theory of relativity was proposed, observations of high-speed particles and distant galaxies have provided real experimental verification for many of the thought experiments you will encounter throughout the remainder of the chapter.

RELATIVE VELOCITY AT HIGH SPEEDS

Measurements of the speed of light in a variety of experiments confirmed Einstein's postulate that the speed of light is constant in all reference frames moving at constant velocity relative to one another. Such measurements contradicted the concept of relative speed and velocity we encountered in low speed examples in Chapter 3. To explain these results, the special theory of relativity replaces our low-speed definition of relative velocity with one that can be applied to both low-speed and high-speed situations.

In Chapter 3 we developed a general rule for calculating the relative velocity between two objects given their velocities relative to the earth. When the robbers and stagecoach were moving toward one another, we added the magnitudes of their velocities relative to the earth. Suppose we replace the stagecoach and robbers with two spaceships, A and B, moving toward one another at speeds near the speed of light. Our low-speed definition of relative velocity predicts that the velocity of A relative to B is simply the sum of the magnitudes of their velocities relative to the earth. The special theory of relativity modifies this definition by adding a term that limits this sum at higher speeds. To get some feeling for how relative velocities change at high speeds, let's look at some specific examples.

Table 4-1 contrasts the low-speed and high-speed predictions of the relative velocity between our two spaceships, A and B. The low-speed column lists a relative velocity that is the sum of the magnitudes of the velocities of the two spaceships. The high-speed column shows that these relative velocities have to be modified at higher speeds. At ordinary walking speeds, 1 m/s, both the low-speed and high-speed definitions of relative velocity predict the same result—the velocity of one spaceship relative to the other is 2 m/s. The two definitions continue to predict roughly the same relative velocities up to 10,000 m/s (10^4 m/s). At higher speeds the relative velocity predicted by the high-speed definition begins to be smaller than that predicted by the low-speed definition. The difference between the relative velocities predicted by the two definitions seems small up to about one-third the speed of light. At still higher speeds, however, the difference becomes substantial. If both spaceships move toward one another at two-thirds the speed of light (2×10^8 m/s), their relative velocity is 2.77×10^8 m/s rather than 4×10^8 m/s.

If we replace spaceship B by a beam of light, then we see how the high-speed definition of relative velocity correctly predicts Einstein's second postulate. If observers in spaceship A could move at the speed of light, 3×10^8 m/s, they would measure the speed of the light beam moving toward them to be 3×10^8 m/s—the same speed measured by a stationary observer. The speed of light is 3×10^8 m/s regardless of the motion of the observers in spaceship A. The high-speed definition of relative velocity is consistent with our low-speed observations, consistent with the results of the Michelson-Morley experiment and consistent with Einstein's second postulate.

Table 4-1 Relative Velocity

v_A (m/s)	v_B (m/s)	Low-Speed Definition (m/s)	High-Speed Definition (m/s)
1	1	2	2
10	10	20	20
100	100	200	200
1×10^3	1×10^3	2×10^3	2×10^3
1×10^4	1×10^4	2×10^4	2×10^4
1×10^5	1×10^5	2×10^5	1.99999×10^5
1×10^6	1×10^6	2×10^6	1.99998×10^6
1×10^7	1×10^7	2×10^7	1.99778×10^7
1×10^8	1×10^8	2×10^8	1.8×10^8
2×10^8	2×10^8	4×10^8	2.77×10^8
3×10^8	3×10^8	6×10^8	3×10^8

SELF-CHECK 4A

An observer in spaceship A, moving at a speed of 1×10^8 m/s, measures the speed of a beam of light coming toward him. The speed of light relative to the earth is 3×10^8 m/s. What do the low-speed and high-speed definitions of relative velocity predict for the speed of light the observer measures? Is the high-speed result consistent with Einstein's second postulate?

SIMULTANEITY AT HIGH SPEEDS

Our concept of time (derived from experiences at low speeds) is absolute. We imagine that time proceeds at the same rate for all observers. Two events that occur simultaneously in one reference frame will be simultaneous in other reference frames as well. At speeds near the speed of light, however, this concept of simultaneity changes. We examine the reason for these changes with the help of a thought experiment involving two space travelers, Barbara and Allen.

As shown in Figure 4-3(a), Allen stands on the surface of the earth, while Barbara sits in a spaceship. A lamp has been mounted in the center of the spaceship. Each time the lamp flashes, light moves toward both ends of the spaceship. Since the distance to each end of the spaceship is equal, we expect light to reach the ends simultaneously. While the spaceship is stationary, Barbara and Allen will agree that light reaches the end of the spaceship simultaneously. When the spaceship moves past Allen at speeds near the speed of light, they no longer agree. Let us look at what each sees.

Figure 4-3(b) shows what Barbara sees in her reference frame. The lamp is motionless relative to Barbara and the spaceship—they are all moving together. Since the lamp always remains in the middle of the spaceship, light travels the same distance in reaching the front and the back of the spaceship. As stated in the second postulate, light travels at the same speed in all reference frames. Consequently, light reaches the front and back of the spaceship simultaneously. Barbara's report is the same as when the spaceship was stationary.

As shown in Figure 4-3(c), the spaceship moves past Allen as light travels from the lamp to each end of the spaceship. After the lamp flashes, the back of the spaceship moves toward the light and the front moves away from the light. The distance traveled by the light moving to the back is half the length of the spaceship minus the distance traveled by the spaceship while the light is in transit. Allen sees the light hit the back of the spaceship before he sees it hit the front. Why? The light traveling toward the front must travel half the length of the spaceship plus the extra distance traveled by the spaceship while the light is in transit. Light traveling to the back of the spaceship travels a shorter distance than light traveling toward the front. If the spaceship could

Simultaneity at High Speeds   65

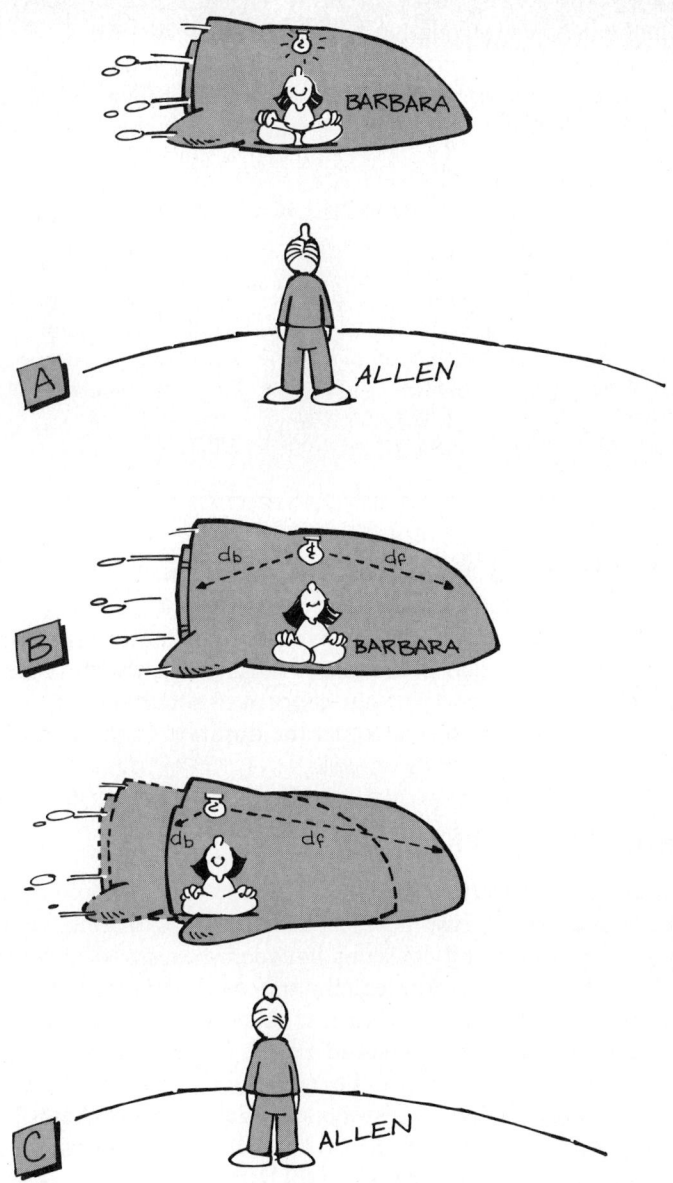

Figure 4-3
In Barbara's reference frame, light reaches the front and back of the spaceship simultaneously. In Allen's reference frame, light reaches the back before it reaches the front. Events that are simultaneous in one reference frame will not be in another that is moving at a constant velocity relative to the first.

travel at the speed of light (which it cannot, as we will see later), the light would never catch up to the front of the spaceship. As it is, Allen no longer reports that light strikes the front and back of the spaceship simultaneously.

Events that are simultaneous in Barbara's reference frame are no longer simultaneous in Allen's reference frame. Their disagreement cannot be resolved—both could produce measurements to substantiate their statements. They can, however, understand the reason for their disagreement. If the speed of light is the same in both reference frames, events that are simultaneous in one reference frame cannot be simultaneous in a second reference frame moving at a constant relative velocity. At low speeds, the difference is so slight that observers in both reference frames continue to report that the events are simultaneous. At speeds near the speed of light, however, the differences are considerable. Observers in the two reference frames no longer agree.

SELF-CHECK 4B

Barbara stands in the center of a supertrain. A lamp mounted in the center sends light toward both ends. As the supertrain moves past Allen at speeds near the speed of light, do Allen and Barbara agree as to whether light reaches the ends of the train simultaneously? Describe what each observes and why.

TIME INTERVALS AT HIGH SPEEDS

Most of our measurements of time involve duration rather than simultaneity. We measure the time needed to travel to Kansas City, the time for the earth to rotate on its axis and so forth. If our concept of simultaneous time depends on the reference frame, we would expect the duration of an event to be different in different reference frames as well.

An Experiment with Time

One measurement of duration, performed first in the 1940s and repeated in the 1960s, involved small, fast-moving particles called *muons,* created in the earth's upper atmosphere by collisions between atoms and high-speed particles from outer space. Once created, the muons do not exist very long; they spontaneously change into other particles, called *electrons* and *neutrinos.* The time of a muon's existence is measured by the *half-life*—the time it takes for one-half the muons in a group to change into electrons and neutrinos. On earth we measure the half-life of stationary muons to be about 1.53 microseconds (μs) (0.00000153 s, or 1.53×10^{-6} s). Thus, if we start with 1200 muons, 1.53 μs later we will have only 600 left; another 1.53 μs later we will have only 300 left; and so forth. The muons' behavior provides us with a convenient clock. Knowing the initial number of muons, physicists can measure duration in terms of the number of muons left at some later time.

As muons are created in the upper atmosphere, they move toward the surface of the earth at speeds of up to more than 99% the speed of light. By measuring the number of muons at various heights as they move toward earth, we can measure the half-life of the muons in this moving reference frame and compare it to the half-life of 1.53 μs measured in our stationary reference frame on earth.

In an experiment performed in the 1960s, physicists measured the number of muons found at the top of Mount Washington, New Hampshire, (altitude = 1907 m) and at sea level. The muons were traveling at 99.2% of the

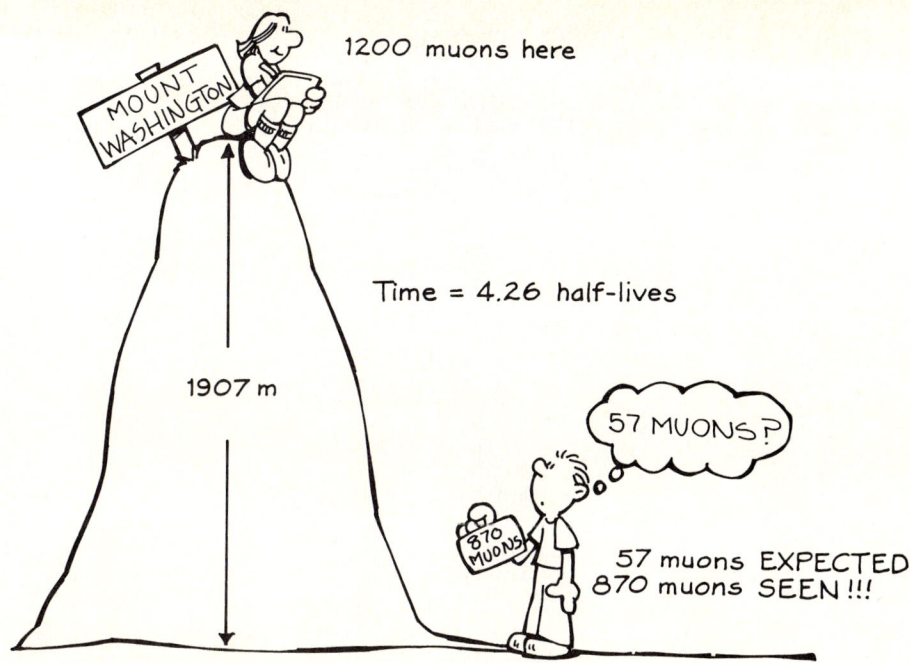

Figure 4-4

Muons traveled at over 90% the speed of light as they fell toward the earth's surface. If time is constant in the muons' and earth's reference frames, we should see 57 muons at the bottom of Mount Washington. Instead we see 870 muons. Time moves more slowly in the muons' reference frame.

speed of light. Since we know the speed of the muons ($0.992 \times 3 \times 10^8$ m/s = 2.976×10^8 m/s) and the distance traveled (1907 m), we can calculate the time required for the muons to make the trip by dividing the distance by the speed:

$$\text{Time} = \frac{\text{distance}}{\text{speed}} = \frac{1907 \text{ m}}{2.976 \times 10^8 \text{ m/s}} = 6.41 \times 10^{-6} \text{ s}$$

The time of travel, 6.41 μs, is more than four times the half-life of the muon particles, so we expect to see far fewer muons at sea level than at the top of the mountain. As shown in Figure 4-4, if 1200 muons are detected at the top of Mount Washington, only 57 should be detected at sea level if their half-life is the same as in the laboratory. Actual measurements, however, detected 870 muons at sea level. Clearly, the half-life of the muons in their moving reference frame is longer than the half-life measured in the laboratory. We investigate the reasons for this with another thought experiment involving Barbara and Allen.

Barbara proposes an experiment in which she and Allen measure the duration of an event—the time it takes light to travel from a lamp to a mirror and back to the lamp (Figure 4-5). Allen will measure the duration of this event in two reference frames: one in which Barbara's spaceship is stationary relative to him and a second in which she is moving past him at 80% the speed of light. This is analogous to measuring the muon's half-life when the muons are stationary and when they are moving toward the earth.

Figure 4-6 helps us compare what Allen will observe in the two reference frames. When the spaceship is stationary relative to Allen, the light travels to the mirror, is reflected, and travels back to the lamp along the vertical path shown in Figure 4-6(a). When the spaceship moves past Allen,

68 Chapter 4. Special Theory of Relativity

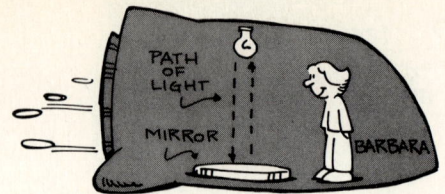

Figure 4-5
Allen will measure the time it takes for light to travel from the lamp to the mirror and back in two reference frames: one in which the spaceship is stationary relative to him and a second in which it moves past him.

Figure 4-6
When the spaceship moves past Allen, light travels from the lamp to the mirror and back along the diagonal in (b) rather than along the vertical path in (a). Since the distance is greater, the time interval will be greater.

light travels along the diagonal path shown in Figure 4-6(b). We have to add the horizontal motion of the spaceship to the vertical path of the light as it moves to the mirror and back. The distance along the diagonal path is greater than the distance along the vertical path. Since the speed of light is the same in both reference frames, Allen will report that it took the light longer to travel to the mirror and back in the moving reference frame than in the stationary reference frame.

Time Dilation

The same event occurs in both reference frames. Light is emitted by a lamp, reflected by a mirror, and returned to the lamp. However, the time between the beginning of the event (emission of light) and the end of the event (return of the light) is different. The duration of events is longer in the moving reference frame than in the stationary reference frame. This slowing of events is called **time dilation.** When observers in one reference frame measure time in another, they find that time varies with the relative speed of the two reference frames. Moving clocks run slower than stationary ones.

Table 4-2 allows you to compare the duration of events an observer measured in the observer's own stationary reference frame with the duration

Table 4-2 Time Dilation

Speed of Moving Frame Fraction of Light Speed	(m/s)	Time Duration in Stationary Observer's Frame (s)	Time Duration in Moving Frame as Measured by Stationary Observer (s)
0.0001	3×10^4	1.00	1.000000005
0.001	3×10^5	1.00	1.0000005
0.01	3×10^6	1.00	1.00005
0.1	3×10^7	1.00	1.005
0.2	6×10^7	1.00	1.02
0.4	1.2×10^8	1.00	1.09
0.6	1.8×10^8	1.00	1.25
0.8	2.4×10^8	1.00	1.67
0.9	2.7×10^8	1.00	2.29
0.99	2.97×10^8	1.00	7.09
0.999	2.997×10^8	1.00	22.37
0.9999	2.9997×10^8	1.00	70.71

observed in reference frames moving past at different speeds. The speed of the moving reference frame has been described in two ways: as a fraction of the speed of light and in meters per second. A reference frame that moves at 30,000 m/s is moving at 0.0001, or one ten-thousandth, the speed of light. At relatively low speeds, like 30,000 m/s, the difference between the two measurements of time is insignificant. We never notice time dilation in everyday life—not even when traveling by jet at hundreds of kilometers per hour. As objects begin to move at speeds near the speed of light, however, time dilation becomes impossible to ignore. A lamp that flashes once per second when motionless relative to you will flash once in 7.09 s when moving past you at 99% the speed of light. It is somewhat like watching a slow-motion movie. Compared to events in our stationary reference frame, events in the moving reference frame take longer. We conclude that time moves more slowly in the moving reference frame.

We can use these results to understand the surprising results of the muon experiment. When the muon is motionless relative to the experiment-

A STEP FURTHER—MATH

TIME DRAGS ON

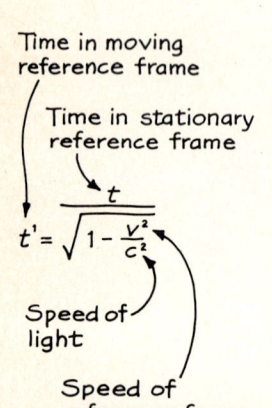

Using some geometry and algebra, we can derive an exact expression, called the time-dilation equation, that relates the time intervals measured in different reference frames.

$$\text{Time between two events in moving reference frame} = \frac{\text{time between events in stationary reference frame}}{\sqrt{1 - \dfrac{(\text{speed of frame})^2}{(\text{speed of light})^2}}}$$

Applied to our thought experiment, this equation tells us that the time Allen measures as Barbara moves past, called t', is equal to the time he would measure if Barbara were stationary relative to him divided by $\sqrt{1 - v^2/c^2}$.

We use this expression to compare the measurements Allen makes in the two reference frames. Suppose it takes light 3 s to travel to the mirror and back when the spaceship is stationary relative to Allen. When the spaceship moves past at 80% the speed of light ($v = 0.8c$), Allen measures a time of:

$$t' = \frac{t}{\sqrt{1 - \left(\dfrac{v}{c}\right)^2}} = \frac{3 \text{ s}}{\sqrt{1 - \left[\dfrac{(0.8)(3 \times 10^8 \text{ m/s})}{(3 \times 10^8 \text{ m/s})}\right]^2}} = 5 \text{ s}$$

An event that takes 3 s in a stationary reference frame takes 5 s if you watch it in a reference frame moving past you at 80% the speed of light. What if Barbara goes even faster? At 90% the speed of light, the 3 s stretch into 7 s. At 99.9% the speed of light, they become more than a minute. What about 99.99% the speed of light? Try it and see!

ers, its half-life is 1.53 μs. As the muons travel at 99% the speed of light relative to the observers, their half-life (as measured by the observers) is 7.09 times longer, 10.85 μs. Since the trip from the top of Mount Washington to sea level only lasted 6.41 μs, the muons had not been through even a single half-life. With the half-life of 10.85 μs predicted by time dilation, 870 of the 1200 muons would be left after 6.41 μs. This matched what had been measured at sea level!

SELF-CHECK 4C

An average cockroach has a lifetime of 3 months on earth. Suppose you observe the cockroach from the earth as it travels past you in a spaceship. What lifetime would you measure at each of the following relative speeds: 0.1, 0.4, 0.8, 0.9 times the speed of light?

Symmetry of Time Dilation

In thinking about time dilation, we must distinguish between what Barbara measures and what Allen measures. Our thought experiment described what Allen would see. If, for example, Allen sees an event take place in 3 s when Barbara is stationary relative to him, he will see the same event take longer—for instance, 5 s—when Barbara moves past him at 80% the speed of light. Consider the measurements that Barbara might make. In her own reference frame, the light will take 3 s to go down and back regardless of whether she is stationary or moving relative to Allen. Her measurements will agree with Allen's when she is at rest relative to him, but they disagree when she is moving relative to him.

If we ask Barbara to measure an event on earth, she will notice the same change in time duration as Allen. An event on earth that takes 3 s when she is stationary will take 5 s as she moves past at 80% the speed of light. When measuring events on earth, Barbara reports that earth time slows down as she moves relative to the earth. When measuring events on Barbara's spaceship, Allen reports that time slows down as she moves relative to the earth. Each views the other's time to be proceeding more slowly.

SELF-CHECK 4D

Allen sows a tomato seed. Three months later he picks the first ripe tomato. Barbara, traveling at 80% the speed of light relative to Allen, sows an identical tomato seed. In Allen's reference frame, how much time elapses before Barbara picks her first ripe tomato? In Barbara's reference frame, how much time elapses before Allen picks his first tomato?

Twin Paradox

While a bit bizarre, the fact that Allen sees the light take longer to bounce off mirrors does not seem impossible to believe. For most of us, the next step—from light clocks to biological clocks—is a much more difficult one to accept. If Allen sees time proceed more slowly in Barbara's reference frame as she speeds by, then he sees all events proceed more slowly—including the time between heartbeats. Allen sees Barbara age more slowly.

One of the more startling outcomes of time dilation is a thought experiment that came to be called the **twin paradox.** Jackie takes off in a rocket, travels to a nearby star at a speed near the speed of light, turns around, and comes back to earth. Her twin, Steve, stays home. During the trip out, each views the other's time to be proceeding more slowly than their own. Steve expects Jackie to be younger than he upon her return. Jackie expects Steve to be younger. When they meet again on earth, they cannot both be younger!

The solution to the paradox lies in the fact that the two reference frames did not move at a constant velocity relative to one another. Seen from Steve's earthbound reference frame, Jackie's spaceship accelerated to its cruising speed, continued at this speed until it reached the star, accelerated as it turned around and headed back to earth, cruised at a constant speed back to earth, and finally accelerated (negatively) as it landed. During the periods in which Jackie was moving at a constant velocity relative to Steve, each was seeing time proceed more slowly in the other's reference frame. During acceleration, however, Jackie's time was distorted, while Steve's was not. Jackie will return the younger twin. The symmetry of time dilation is broken once Jackie accelerates.

LENGTH CONTRACTION

Any measurement of length ultimately involves our concept of position. Typically, we place a meterstick along the side of the object and mark the position of each corner of the object relative to the meterstick. The length is then the difference between these two positions. While we think of position as a relative concept, most of us imagine length to be absolute. We expect the length of the object to be the same, regardless of the motion of the reference frame. But, as with our concept of time, the concept of length is altered in reference frames moving at high speeds.

Length in the Muons' Reference Frame

Consider a thought experiment with the muons that we introduced in the last section. At the top of Mount Washington, we counted 1200 muons; at the bottom, 870. Since the muons were moving at an enormous speed relative to us, we explained this surprising result in terms of time dilation. Relative to us in a stationary reference frame, time slows down in the moving reference frame.

Imagine that we can move into the muons' reference frame and travel at 99.2% the speed of light. Now we are motionless relative to the muons. We

Figure 4-7
Allen sees two parts of the pole simultaneously, but the pole is measured to be shorter than in Barbara's reference frame.

describe the muons and ourselves as stationary and Mount Washington as moving upward at 99.2% the speed of light. When we see the top of Mount Washington pass us, we count 1200 muons, and when we see the bottom (sea level) pass us, we count 870. A change in reference frame cannot cause muons to appear or disappear; the number we count must be the same in all reference frames. But here a contradiction arises. Since we are stationary relative to the muons, we once again measure their half-life to be 1.53 μs. To determine the duration of the descent, we divide the height of Mount Washington by our speed. Thus, we *expect* the descent time to be (1907 m)/(2.98 $\times$ 10^8 m/s), or 6.41 μs—more than four muon half-lives. If this time calculation is correct, only 57 muons should be left.

In both reference frames—the one in which the muons are stationary on earth and the one in which we are moving with the muons at 99.2% the speed of light—the half-life must be 1.53 μs. The number of muons counted at the top and at the bottom of Mount Washington must also be the same in all reference frames. Consequently the only incorrect reasoning is the calculation of the travel time. To arrive at this value we assumed that the height of Mount Washington (1907 m) is the same in our new (moving) reference frame as it was in our old (stationary) reference frame. Therein lies the problem. Our measurement of length depends on the motion of the reference frame.

Length Contraction

Normally, we expect objects to remain stationary as we measure their length. We lay the pole down next to the meterstick, mark the left end, look over, and mark the right end. Implicit in our actions is the belief that the left end does not move while we measure the right end. As long as we measure objects that are stationary relative to ourselves, our assumption is valid. But when the pole moves past us, we have to invent another strategy for measuring its length. We must locate the two ends of the pole simultaneously, that is, at the same time. Ultimately, our measurement of length depends on our concept of simultaneity.

To understand the limitations that simultaneity places on our measurements of length, contrast Barbara's and Allen's measurements of the length of a pole. Figure 4-7 shows Barbara holding a pole as her spaceship moves past Allen. As she holds the pole up next to a meterstick, she sees the two ends of the pole simultaneously and measures its length. On earth, Allen looks up and sees the pole simultaneously as well. However, since the spaceship moves past him as he looks, Allen does not see the pole exactly as Barbara does. Light from the left end reaches him before light from the right end, for example. As shown in the figure, this effect causes Allen's measurement of the pole's length to be smaller than Barbara's measurement. Rapidly moving objects are shorter to the stationary observer. The change in length that occurs when objects move at speeds near the speed of light is called **length contraction.**

Table 4-3 allows you to compare the length of the object an observer measures in his or her own stationary reference frame with the length he or she observes in reference frames moving past at different speeds. The speed of the moving reference frame has been described in meters per second and as a fraction of the speed of light. At relatively low speeds, like 30,000 m/s, the difference in length is not noticeable. Clearly, we never notice length contraction in everyday life. At speeds near the speed of light, however, length contraction becomes significant. An object that is 10 m long in a reference frame stationary relative to the observer will be 6 m long when moving past the observer at 80% the speed of light.

Table 4-3 Length Contraction

Speed of Moving Frame Fraction of Light Speed	(m/s)	Length in Stationary Observer's Frame (m)	Length in Moving Frame as Measured by Stationary Observer (m)
0.0001	3×10^4	1.00	0.999999995
0.001	3×10^5	1.00	0.9999995
0.01	3×10^6	1.00	0.99995
0.1	3×10^7	1.00	0.995
0.2	6×10^7	1.00	0.98
0.4	1.2×10^8	1.00	0.92
0.6	1.8×10^8	1.00	0.80
0.8	2.4×10^8	1.00	0.60
0.9	2.7×10^8	1.00	0.44
0.99	2.97×10^8	1.00	0.14
0.999	2.997×10^8	1.00	0.04
0.9999	2.9997×10^8	1.00	0.01

A STEP FURTHER—MATH

FISK'S DISK

As we did for time dilation, we can develop an equation that relates the lengths measured in the different reference frames.

$$\begin{matrix}\text{Length} \\ \text{measured for} \\ \text{moving object}\end{matrix} = \begin{matrix}\text{length of} \\ \text{stationary} \\ \text{object}\end{matrix} \sqrt{1 - \left(\frac{\text{speed of object}}{\text{speed of light}}\right)^2}$$

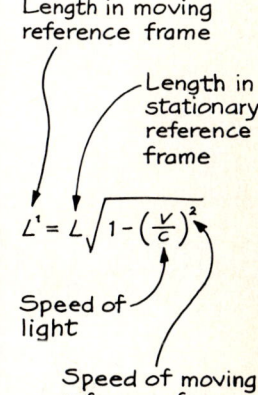

Called the Lorentz-FitzGerald contraction, after two physicists who first proposed it, this expression tells us that the length Allen measures as Barbara moves past, called L', is equal to the length he would measure if Barbara were stationary relative to him multiplied by $\sqrt{1 - v^2/c^2}$.

We use this expression to compare the measurements Allen makes for the two reference frames. Suppose Barbara has a 2-m rod in her spaceship. When Barbara is stationary relative to Allen, he reports that the rod is 2 m long. When the spaceship moves past him at 80% the speed of light ($v = 0.8c$), Allen measures a length of

$$L' = L\sqrt{1 - v^2/c^2} = (2\text{ m})\sqrt{1 - \left(\frac{(0.8)(3 \times 10^8 \text{ m/s})}{(3 \times 10^8 \text{ m/s})}\right)^2}$$
$$= 1.2\text{ m}$$

A rod that is 2 m long in a stationary reference frame is 1.2 m long if you measure it in a reference frame moving past you at 80% the speed of light. What if Barbara goes even faster? At 90% the speed of light, the 2-m rod contracts to 0.88 m. At 99.9% the speed of light, it measures 0.08 m. What do we have left at 99.99% the speed of light?

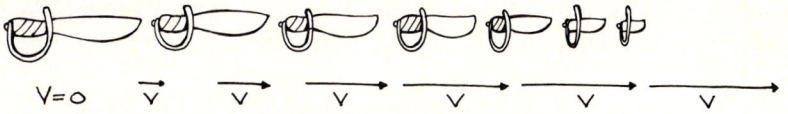

    There was a young fencer
        named Fisk
    Whose thrust was
        exceedingly brisk
    So fast was his action
    The Lorentz-FitzGerald contraction
    Reduced his rapier
        to a disk

Figure 4-9

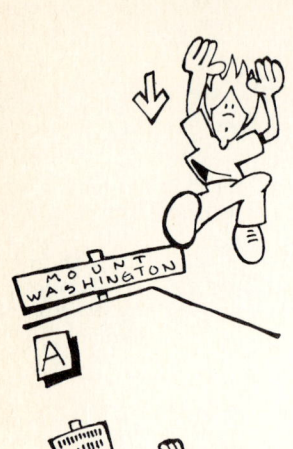

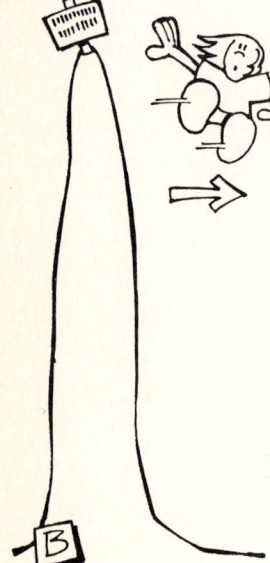

Figure 4-8
Only the dimensions along the direction of motion are affected.
(a) In a reference frame moving down, the height of Mount Washington changes but the width remains the same.
(b) In a reference frame moving along the earth's surface, the width changes but not the height.

Like time dilation, length contraction is symmetric. If Allen reports that a 10-m pole is only 6 m long as Barbara moves past him at 80% the speed of light, Barbara reports that a 10-m pole in Allen's reference frame is only 6 m long as she moves past him at 80% the speed of light. If we return to our experiment with the muons, this tells us that Mount Washington is shorter as we move with the muons at 99.2% the speed of light. If we actually performed the calculation, Mount Washington would be only 241 m high. Now, let us determine the time it takes the muons to travel from the top of Mount Washington to sea level. The duration of the descent is the distance the muons travel divided by their speed—(241 m)/(3×10^8 m/s) = 0.81 μs. It would take 0.81 μs to make the complete trip from top to bottom, less than a single half-life of muons. This is consistent with the 870 muons we measure at the bottom of Mount Washington.

Length contraction occurs only in the direction in which the object (or the reference frame) is moving. As shown in Figure 4-8(a), Mount Washington appears shorter but the same width as we move downward toward the surface of the earth. If we were to move along the surface of the earth at the speed of a muon (Figure 4-8(b)) Mount Washington would appear thinner, but its height would remain the same. Only the dimension in the direction of motion varies with the relative speed of the object.

SELF-CHECK 4E

A window in Barbara's spaceship is 2 m long and 3 m high (see Figure 4-9). What length and height does Allen measure for the window if Barbara passes by him at the following speeds: 0.1, 0.4, 0.8, and 0.9 times the speed of light?

Figure 4-10
If a baseball could travel faster than light, the fielder would have to catch the ball before he knew that it had been hit.

HOW FAST IS FASTEST?

The speed of light is the greatest speed we have encountered. The special theory of relativity suggests that the speed of light is the upper limit to all motion. Nothing can move faster. To understand the reasons for believing that such a limit exists, consider another thought experiment.

Suppose you are playing center field in a baseball game. The batter swings and sends a fly ball out to you, as illustrated in Figure 4-10. In order for you to see the ball, sunlight must be reflected from the baseball into your eyes. Since light travels very, very fast relative to the baseball, the reflected light reaches your eyes almost instantaneously and you can track the ball along its path. Imagine what you would see if the baseball traveled faster than the speed of light. Light travels at a constant speed, so the light that travels the shortest distance will reach you first. Because the ball moves faster than light, the ball reaches you before the light reflected from the ball at points 1, 2, 3, or 4. In fact, you must catch the ball before you see it. After catching it you would see the light reflected from points 4, 3, 2, and 1—in that order. You would catch the ball, then see it travel back to the batter. Finally, the batter would hit it.

Absurd? Yes! This result violates the basic logic of causality. A ball traveling faster than light puts the cart before the horse. The effect (catching the ball) occurs before the cause (hitting the ball). Either we must reject cause and effect or conclude that all objects travel more slowly than light. So far, experiments agree with these expectations. No one has discovered an object that exceeds the speed of light—3×10^8 m/s, or 186,000 miles per second.

Faster speeds? Shorter blocks? Let's return to the bicyclist whose delightful argument opened the chapter. In writing this story, George Gamow imagined what the world would look like if light moved at ordinary speeds—such as 5 m/s. Relative to this speed of light, the bicyclist certainly could

"That 187,000 miles per second makes me a bit skeptical about the whole thing."

pedal hard enough to see the blocks get shorter or even time slow down. But do the blocks *really* get shorter and does time *really* slow down? Our concepts of the length of a city block and the duration of a second—of space and time—depend on motion. Gamow's bicyclist catches one glimpse of space-time. Standing on the street corner, we catch another. Neither space nor time is absolute.

CHAPTER SUMMARY

Our experiences with objects at low speeds suggest that while position, time, and motion are relative concepts, relative speed, duration, and length *seem* to be constant in all reference frames moving at constant velocity relative to each other. In an experiment performed in 1887, Albert Michelson and Edward Morley showed the speed of light to be constant in all reference frames. This contradicted our low-speed concept of relative velocity. In 1905 Albert Einstein proposed the *special theory of relativity*. By assuming that the principle of relativity (Chapter 3) is valid and that the speed of light is constant in all reference frames, Einstein revised our concepts of relative velocity, simultaneity, time duration, and length.

The changes made by the special theory of relativity are summarized below:

Concept	Low-Speed Definition	High-Speed Definition
Relative velocity	Relative velocity is the same in all reference frames.	Relative velocity changes with the motion of the reference frame. The relative velocity can never exceed the speed of light.

Simultaneity	Events that are simultaneous in one reference frame are simultaneous in another.	Simultaneity depends on the motion of the reference frame. At high speeds, events seen as simultaneous in one reference frame will not be simultaneous in a reference frame moving at a high speed relative to it.
Time	Duration of an event is the same in all reference frames.	*Time dilation:* An observer sees time moving more slowly when the event occurs in a reference frame moving past him or her.
Length	Length of an object is the same in all reference frames.	*Length contraction:* An observer measures the length of an object to be shorter in its direction of motion.

At low speeds these new concepts are identical to those introduced in Chapters 1, 2, and 3. At speeds near the speed of light, the new concepts predict radically different results. Since we have little experience at such speeds, the concept that time slows down or that objects contract seems strange. Measurements at high speeds, however, demonstrate the validity of the theory.

The special theory of relativity proposes that the speed of light is the limit beyond which objects cannot move faster. If we imagine that objects move faster than light, then we see an event after its cause, and our logic of causality is destroyed. Consequently, we accept the speed of light as the upper limit for motion in the universe.

ANSWERS TO SELF-CHECKS

4A. Low-speed definition: speed of light is 4×10^8 m/s. The high-speed definition predicts that the observer measures the speed of light to be 3×10^8 m/s, the same as the speed of light relative to the earth. It does agree with Einstein's second postulate.

4B. Barbara and Allen do not agree. Barbara reports that light reaches the two ends of the train simultaneously. Allen reports that it does not. Because of the motion of the train past him, Allen reports that light reaches the back of the train before it reaches the front.

4C. 0.1 speed of light, 3.015 months; 0.4 speed of light, 3.27 months; 0.8 speed of light, 5.00 months; 0.9 speed of light, 6.88 months.

4D. In Allen's reference frame, 5 months elapse before Barbara picks her first ripe tomato. In Barbara's reference frame, 5 months elapse before Allen picks his first ripe tomato.

4E. The spaceship moves in the direction of the length of the window. Consequently, the window remains 3 m high. The length of the window depends on the speed of the spaceship: 0.1 speed of light, 1.99 m; 0.4 speed of light, 1.84 m; 0.8 speed of light, 1.20 m; 0.9 speed of light, 0.88 m.

PROBLEMS AND QUESTIONS

A. Review of Chapter Material

A1. What was the importance of the Michelson-Morley experiment to the special theory of relativity?

A2. State the two postulates of the special theory of relativity.

A3. What is a thought experiment?

A4. Describe what happens to the relative velocity between two objects as their velocities relative to the earth approach the velocity of light.

A5. Describe a thought experiment which shows that two events which are simultaneous in reference frame A are not simultaneous in a second reference frame moving at a constant velocity relative to A.

A6. State the equation for time dilation. Describe how measurements of the duration of an event vary with the velocity of the reference frame relative to the observer.

A7. Describe both a thought experiment and an actual experiment which show that the duration of an event changes in reference frames moving at high speeds.

A8. What do we mean when we say that time dilation is symmetric?

A9. State the equation for length contraction. Describe how measurements of length vary with the velocity of the object relative to the observer.

A10. Why do we think that light cannot travel faster than the speed of light?

B. Using the Chapter Material

B1. Edward is flying past you at a velocity of 200,000 km/s, east. He reports that the lights on the front and back of his spaceship are flashing simultaneously. Do you see them flashing simultaneously? When Edward is directly north of you, which one will you see flash first—front or back?

B2. Fran looks at the lights on Edward's spaceship (Problem B1). She reports that the ship's lights are flashing simultaneously. What is her velocity relative to Edward? What is her velocity relative to you?

B3. Barbara and Allen are moving toward each other at speeds (measured relative to the earth) that are two-thirds the speed of light. What is their relative velocity?

B4. During the flights to the moon the Apollo spacecraft averaged 55,000 m/s. If the astronauts had measured the speed of light reflected from the moon, what value would they have obtained?

B5. A muon moving at 80% the speed of light is located at the top of Mount Everest (altitude = 8848 m). In the muon's reference frame, how long is the distance to sea level?

B6. Suppose that the muon experiment had been conducted with muons moving at speeds of 0.6 the speed of light. What would be the half-life measured for these muons by an earthbound observer?

B7. Astronomers observe distant galaxies which are moving away from the earth at speeds of 270,000 km/h. At this speed an atom's vibration, which takes 1 μs on the earth, takes 2.29 μs when observed in the moving galaxy. Suppose an astronomer in the distant galaxy is looking back at the earth. What time measurement would she make for this same vibration on earth? What measurement would she obtain for an atom sitting motionless next to her?

B8. Three identical high-speed cars move past us. We see them as shown in Figure 4-B8. Which car is moving most rapidly? Which is moving most slowly?

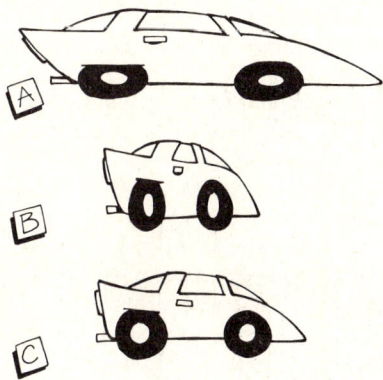

B9. A circle with a diameter of 1 m is moving to your right. Sketch the shape of the circle for each of the following speeds (given as a fraction of the speed of light): 0.4, 0.8, 0.9, and 0.999.

B10. An inventor claims that he can build an airplane which will travel from New York to Los Angeles at three times the speed of light. He asks you to invest money in his company to build his airplane. Explain why this might be a bad investment.

C. Extensions to New Situations

C1. Ned is holding a very long board as shown in Figure 4-C1. As Mary speeds by at 0.9 the speed of light, Ned drops the board so that in his reference frame both ends reach the ground at the same time.

a. What is the length of the board in Mary's reference frame?

b. In Mary's reference frame, do both ends reach the ground simultaneously? Why or why not?
c. If the answer to (b) is no, which end reaches the ground first?
d. Sketch a strobelike drawing of how the board must look in Mary's reference frame as it falls.

C2. Time dilation causes a difference in aging between high-speed space travelers and earthbound people. Consider a spaceship traveling at 0.99 the speed of light relative to the earth.

a. The space traveler's heart beats once per second as measured by the traveler. What time elapses on earth between the heart beats?
b. In 1 h, earth time, whose heart beats the most—the earthbound person or the space traveler?
c. The number of times the heart has beat is a good measure of aging. With each heartbeat we get a little older. In the earthbound reference frame, which person is aging more quickly—the earthbound observer or the space traveler?
d. Answer questions (a), (b), and (c) from the reference frame of the space traveler.

C3. The twin paradox is related to the result of Problem 4-C2. Jackie remains on earth while her twin brother Steve takes a high-speed space trip. He travels away from earth at 90% the speed of light, turns around, and returns to earth at 90% the speed of light. While traveling at a constant speed, Jackie and Steve each observe the other to be aging more slowly. Using the special theory of relativity, each would conclude that the other would be younger when they meet again. Thus, we have an impossible situation. Experiments with very sensitive clocks show that Steve will be younger. This result does not disagree with the special theory of relativity because one of its postulates is violated during the trip. Which one? How is it violated?

C4. Here is another paradox. A very high-speed airplane with a brave (or foolish) pilot is to fly through a tunnel in a mountain. When measured in a reference frame stationary relative to the tunnel, the airplane is longer than the tunnel. From the airplane's reference frame, the tunnel becomes even shorter at high speeds. In the earth's reference frame, an observer closes doors on both ends of the tunnel while the airplane is inside. This event could never happen in the airplane's reference frame. To see how this paradox is resolved, answer the following questions.
   a. The two doors are closed simultaneously in the earth's reference frame. Do the doors close simultaneously in the airplane's reference frame?
   b. Which door closes first in the airplane's reference frame?
   c. Use the results of (a) and (b) to describe the order of events as the airplane flies through the tunnel.
   d. How does the dependence of simultaneity on reference frame resolve the paradox?

C5. The volume of a box is its length times its width times its height. In a reference frame stationary relative to a box, its dimensions are: length = 2 m; width = 4 m; height = 6 m.
   a. What is the volume of the box in this reference frame?
   b. What volume do you measure when the box moves at 0.8 the speed of light relative to you in the direction of its length?
   c. What volume do you measure when the box moves at 0.8 the speed of light relative to you in the direction of its height?

C6. The nearest star, Alpha Centauri, is 4.3 light years (4.1×10^{16} m) from the earth.
   a. How long does light require to travel from Alpha Centauri to earth?
   b. Suppose space travelers move from earth to Alpha Centauri at the top speed of the space shuttle, 350 km/h. How long in earth time would the trip to Alpha Centauri take?
   c. How long, earth time, would the trip require if the travelers moved at 0.8 the speed of light?
   d. How far would Alpha Centauri be in their reference frame?
   e. How long, in the space traveler's time, would the trip take?
   f. Do trips to nearby stars seem feasible at speeds close to the speed of light?

C7. To see if trips to other galaxies might be possible, answer the questions in Problem 4-C6 for the Andromeda Galaxy, which is 2,200,000 light years (2.1×10^{22} m) away.

C8. Suppose two muons are coming toward each other. One is moving relative to the earth at 0.8 the speed of light, east, and the other is moving relative to the earth at 0.8 the speed of light, west.
   a. What is the relative velocity between the two muons?
   b. What will be their relative velocity if they move away from each other rather than toward each other?

D. Activities

D1. The quotation which opened this chapter is taken from "Mr. Tompkins in Wonderland" by George Gamow. In this story, Mr. Tompkins dreams that the speed of light is 10 mi/h. Describe some of the effects Mr. Tompkins will see as he pedals his bicycle around Wonderland. Then, read the story and compare your description with Dr. Gamow's. (The story has been published as part of *Mr. Tompkins in Paperback* (Cambridge University Press, 1969.))

D2. Gene Rodenberry, the creator of the science-fiction television and film series *Star Trek* was accused of not understanding physics because starships in the series moved at speeds greater than the speed of light. Rodenberry replied that he had no choice. If they did not move faster than light, the starships could never travel between galaxies. Discuss his answer using the results of Problem 4-C7.

INTERLUDE 1
Past, Present, Future, and Elsewhere

Reflecting on the conclusions of the special theory of relativity, our concepts of time become influenced by space. Most of us divide time into three regions—past, present, and future. The present is not really so much a region of time as it is a "dividing line" between the past and the future (Figure 1). Your future is any event you can still influence. Before Einstein's introduction to the special theory of relativity, we assumed that we shared a common past with our contemporaries. Along a common "present" line, all of us share the same past and the same opportunity to influence the future. No restriction is placed upon where in space events occur—only on when, relative to the present, the event occurs.

The special theory of relativity forces us to modify these ideas. When Einstein showed

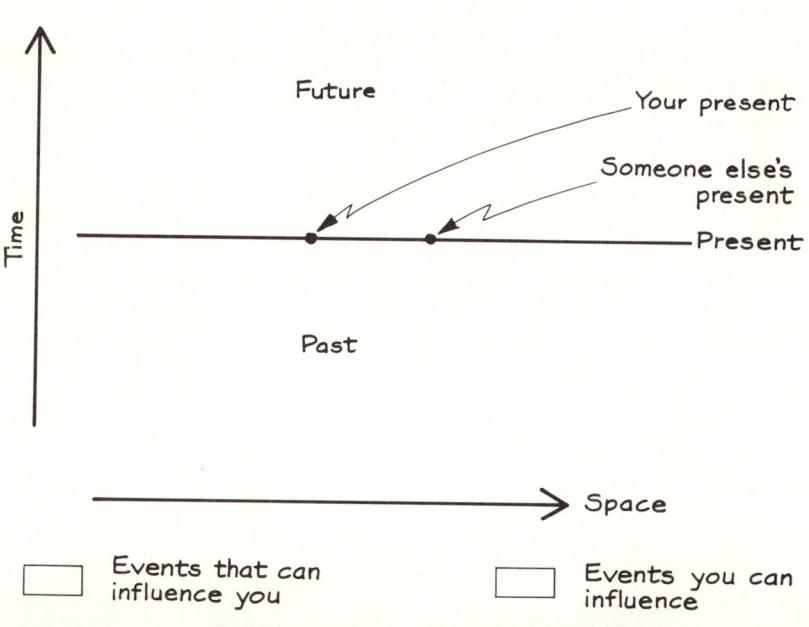

that the speed of light was the maximum speed for all objects, our personal concept of past, present, and future changed. To be influenced by an event which occurred in the past, we must have received information about that event in our present. An event cannot influence us if it occurs so far from us that information traveling at the speed of light cannot reach us. An event that occurs now in a galaxy 1 light-year away cannot affect us today, tomorrow, or even next month simply because we will not receive any information about it until next year. It may, however, affect us next year. By the same token, events simultaneous to our present cannot affect us at this moment. You cannot be influenced nor can you influence something that is happening at this very moment—be it 100 meters away, 100 kilometers away or 100 light-years away. Perhaps at some later time, when knowledge of the event has reached you, it can affect you and, thus, become part of your past. Your present, however, cannot be affected by an event simultaneous to it.

This realization restricts the events both in the past and in the future that are related to your present. As shown in Figure 2, the finite speed of light actually restricts your past and future to two cones that enclose all events in contact with your present. Only those events in the past about which you could have received information can affect your present. Similarly, only those events in the future that can receive information about your present can be affected by it. The other events exist in what we call your "elsewhere"—a region of space-time that includes events too distant in space or too near in time to influence you.

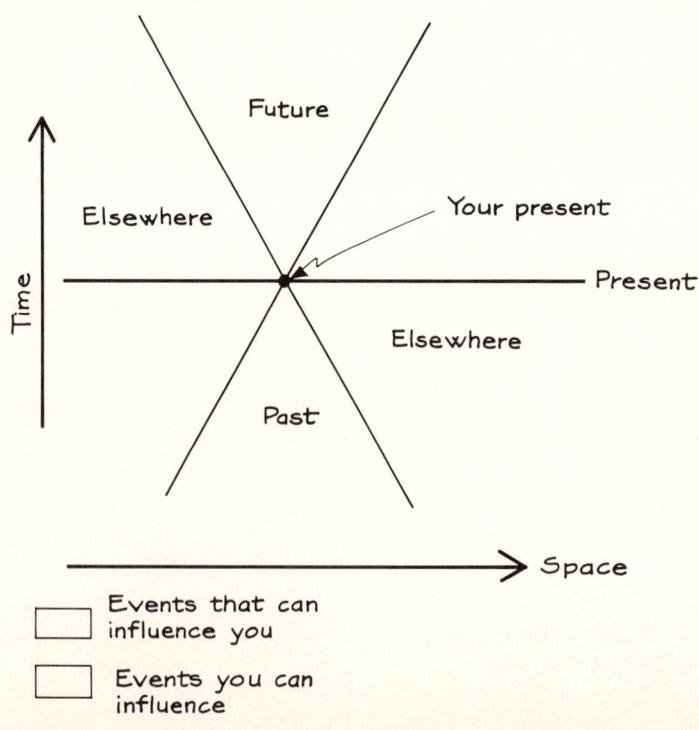

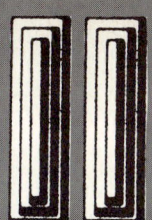

Interaction and Force

CHAPTER 5
Interaction and Momentum

CHAPTER 6
Interaction and Force

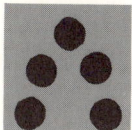

CHAPTER 7
Newton's Three Laws

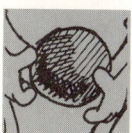

CHAPTER 8
The Fundamental Interactions

When two billiard balls collide on a billiard table, we easily see that each influences the other. We often use the concept of momentum to describe the interaction. When one billiard ball bounces off the side of the table, however, it is hard to see just how the billiard ball influences the table. The interaction seems one-sided, so we shift points of view and talk about the force the table exerts on the ball. Momentum and force both describe interactions.

But how, exactly, does one object influence another? Psychologists tell us that parents, friends, and acquaintances change our thoughts and lives in seemingly complex ways. Objects also seem to influence or act upon one another in complex ways. Physicists have been able to sort through much of this complexity and now identify just four fundamental types of interactions— gravitational, electromagnetic, strong nuclear, and weak nuclear. In the future lies the prospect that these four interactions can be unified into two, or perhaps a single theory of interactions.

Interaction and Momentum

I woke to myself and looked about me, and said to the folk of Troizen, "I have had the sign of Poseidon. He will shake the earth and soon. Warn them in all the houses to come out of doors. Send word to the Palace."

Theseus, King of Athens

Geologists can demonstrate that at least eight major earthquakes have occurred (on the southern San Andreas fault) in the past 1200 years with an average spacing in time of 140 years, plus or minus 30 years. The last such event occurred in 1857.... The aggregate probability for a catastrophic earthquake in the whole of California in the next three decades is well in excess of 50 percent.

Federal Emergency Management Agency

Change, whether abrupt like earthquakes or gradual like aging, is our constant companion. For Greek and Roman civilizations, change belonged to the gods. When displeased or angered, Poseidon shook the earth, Zeus hurled thunderbolts, and the Furies exacted punishment. A host of gods and goddesses governed all of life. Men like Theseus, who received signs directly from a god, or priests and priestesses skilled at interpreting oracles predicted the actions of the gods.

Our present model of earthquakes explains them in terms of the motion of crustal plates, which alternately stick and slip along their boundaries. Greece, the land of Theseus, lies along one such boundary; California, along another. Patterns of past earthquake activity and continual measurements of the motion of the plates are beginning to allow scientists to make rough predictions of future earthquakes. For science, change is a natural process. By observing carefully, we can build models or explanations that eventually allow us to predict change. Physicists describe change in terms of *interaction*. This chapter introduces the concept of *momentum* and describes its use in explaining interactions.

INTERACTIONS

When one object influences another, we say that the two objects *interact*. In order to analyze an interaction, we must be able to see some change. Fallen houses and displaced trees tell us that two crustal plates have interacted along their boundaries. The building of mountain ranges and eruption of volcanoes is evidence for the slow collision of continents brought about by the movements of plates over millions of years. **Change,** then, provides us the *evidence that an interaction occurred*. Measurements of the amount of change enable us to see patterns and build models that predict future change.

Because the details of an interaction can be extremely complex, we often look for change simply by comparing the "before" with the "after." To see how this works, compare before with after for each situation in Figure 5-1. The pictures in (a) show a standard advertising gimmick: "Interact with our product and you'll see a change!" A change in shape like (b) implies an interaction—presumably an unwanted one. Another change can be a change in velocity, such as that experienced by the ball in (c). An enormous variety of changes are possible, including changes in shape, size, volume, velocity, and temperature, to name a few. Simply comparing the "before" with the "after" enables us to identify and categorize these interactions.

In this chapter we will restrict ourselves to interactions in which a change in velocity has occurred, such as that shown in Figure 5-1(c). If an object slows down, speeds up, or changes direction, it must have interacted with something. You might ask why we have linked interactions with changes in velocity, since motion at a constant velocity also involves change—a change of position. But, objects that are stationary in one reference frame can be moving to observers in other reference frames, as you saw in Chapter 3. A change in position can occur when no interaction has occurred—the observer is simply

Figure 5-1
A change provides evidence that an interaction has occurred.

in a reference frame that is moving relative to the object. A change in velocity, however, does imply an interaction. We begin by looking at changes in velocity that occur when two objects interact.

SELF-CHECK 5A

Which pair(s) of before and after sketches of the ball in Figure 5-2 show(s) evidence that an interaction has occurred?

FACTORS AFFECTING INTERACTION

A tennis ball hitting a net, an egg striking the floor, a car colliding with a tree—in each case the motion of the object is abruptly stopped. An interaction has occurred. Yet tennis balls, eggs, and cars can experience other interactions that are not quite so abrupt. To understand interactions and the way in which they influence motion, we need to identify the characteristics of objects that affect their interactions.

Velocity Affects Interactions

A friend lobs a baseball, which you catch with your bare hand. It is an easy catch—you hardly feel it. Now your friend throws a fastball. If you catch it,

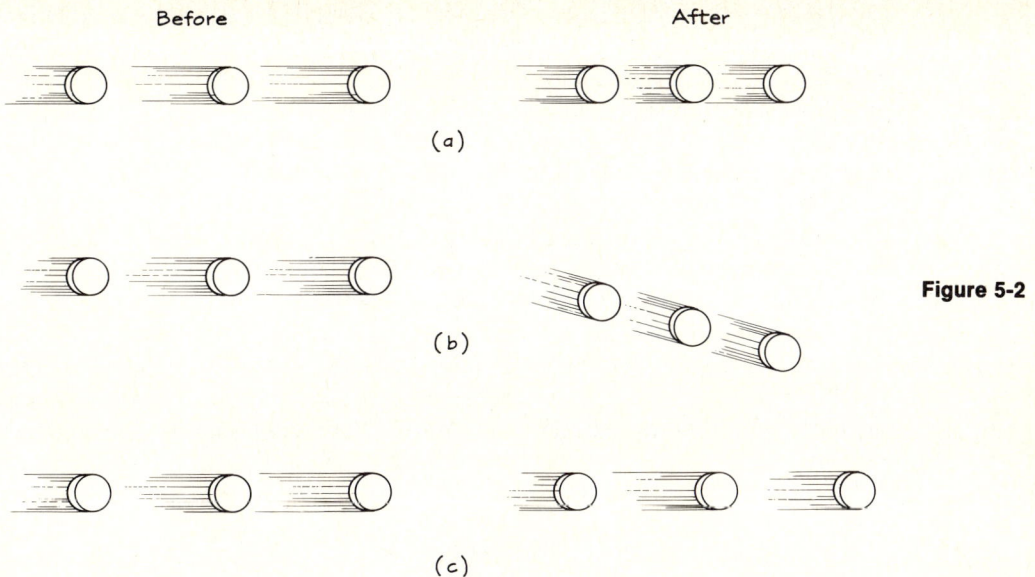

Figure 5-2

your hand stings; the speed of the ball affects the way your hand feels. Much the same thing happens when an automobile collides with a tree. An automobile moving 10 kilometers per hour (km/h) will not damage the tree nearly as much as an automobile moving 60 km/h. An egg dropped 1 centimeter (cm) is not moving as fast as an egg dropped 1 meter (m). The egg's speed before it interacts with the floor determines whether or not the egg breaks.

The *speed* of the object is not the only variable that affects the interaction. *Direction* is also involved. In automobile accidents, head-on collisions are much more damaging than rear-end collisions. Your hand will sting less if you move it away from the ball rather than toward it as you catch it. In the extreme case, the egg would not even interact with the floor if the egg were moving upward rather than downward. *Velocity,* which combines the concepts of speed and direction, affects the interaction.

Mass—A New Quantity

Velocity by itself does not explain all the differences that we see in interactions. A tennis ball and a hardball, both thrown at the same velocity, will leave markedly different impressions on your hand. A car is stopped by its collision with a tree, while a large truck may be only momentarily slowed as it knocks the tree down—even when the initial velocities of the car and truck are the same. The difference between a tennis ball and a hardball or between an automobile and a truck is the amount of matter each has.

The concept we use to describe the amount of matter in an object is its mass. Intuitively, we define **mass** as a measure of the amount of matter. To measure it, we establish standards and compare unknown masses to these standards. The fundamental unit of mass is the kilogram (kg). The mass of a tennis ball is about 0.064 kg, while the mass of a baseball is about 0.142 kg. Cars and trucks have an even larger range of masses—1000 kg for a small

Figure 5-3
The range of human masses is from a few kilograms to well over 100 kilograms.

car and 10,000 kg for a truck. Typical human masses are illustrated in Figure 5-3. Since direction has no meaning in describing the amount of matter in an object, mass is a scalar quantity.

In everyday conversation we use the terms *mass* and *weight* interchangeably. We say that the weight of a loaf of bread is half a kilogram, though we have really described the bread's mass. As you will see in Chapter 6, weight and mass have distinct meanings in physics. Mass refers to the amount of matter in an object, while weight describes the strength of the interaction between the object and the planet on which it is located. While weight does depend on the object's mass, it also depends on characteristics of the planet. Consequently, an object's weight is different on the moon than on the earth. An object's mass remains constant throughout space, while its weight varies with its location. Because we want to develop models that apply in space as well as on earth, we use mass to describe the effect that the amount of matter has on interactions.

SELF-CHECK 5B

Which variable, velocity or mass, affects these interactions?

a. You play catch with a basketball rather than a bowling ball.

b. In defense practice a football coach has the players tackling from in front of rather than behind the runners.

c. Your neighbor throws two identical eggs at a wall. One barely breaks; the other shatters completely.

MOMENTUM

Two variables, mass and velocity, help describe the different interactions we observe. Consider the role of each by using your hand to judge the strength of interaction with tennis balls and baseballs. You can mix the two variables, mass and velocity, in four ways: *low mass, low velocity; low mass, high velocity; high mass, low velocity;* and *high mass, high velocity.* From experience you can probably identify the extremes. A tennis ball lobbed toward you (low mass, low velocity) will sting very little compared to a baseball hurled by a fastball pitcher (high mass, high velocity). More difficult to judge is the difference between a tennis ball hurled by a pitcher (low mass, high velocity) and a baseball lobbed gently (high mass, low velocity). The reason it is more difficult to judge is that the two variables—mass and velocity—are actually combined in our perception of the interaction. In physics the concept of momentum combines the concepts of mass and velocity.

Momentum Defined

Momentum is defined as the product of mass and velocity:

momentum = (mass) × (**velocity**)

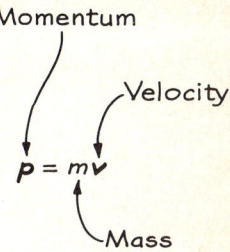

Mass is a scalar quantity. Velocity is a vector quantity. Consequently, momentum is defined as a vector quantity whose direction is the same as the direction of the velocity. The units in which momentum is measured are the units given by its definition: kilogram-meters per second (kg · m/s).

In one sense, momentum is a measure of the "influence" one object has on another in an interaction. We can explore this idea by calculating the momentum of each ball just before it interacts with your hand. The mass of a tennis ball is 0.064 kg and that of a baseball is 0.142 kg. We can estimate the speeds of a lob and a fastball to be 5 m/s and 50 m/s, respectively. Given these values, the momentum of the tennis ball lobbed toward you is (0.064 kg)(5 m/s, east) = 0.32 kg · m/s, east. With its larger mass, a baseball lobbed toward you has a momentum of (0.142 kg)(5 m/s, east) = 0.71 kg · m/s, east. A tennis ball hurled toward you has a still larger momentum, (0.064 kg)(50 m/s, east) = 3.2 kg · m/s, east. Finally, a baseball hurled toward you has the largest momentum, (0.142 kg)(50 m/s, east) = 7.1 kg · m/s, east. The ordering of momenta (plural of momentum) from least to most agrees with the feeling you have about how each one would hurt as it struck your hand. Moreover, the quantitative definition of momentum allows us to distinguish subtly different descriptions from one another—for example, low mass, high velocity from high mass, low velocity.

SELF-CHECK 5C

A ping-pong ball lies motionless on a table. Each of the balls listed below rolls toward and collides with it. Which has the most momentum? Which do you think will cause the greatest change in the motion of the ping-pong ball?

	Mass	Velocity
Glass ball	0.10 kg	1 m/s, E
Super ball	0.12 kg	0.5 m/s, E
Lead ball	0.20 kg	0.2 m/s, E

Momentum Before and After an Interaction

We can calculate a momentum for any object, even if that object does not interact with other objects. As you might expect, these kinds of calculations are

92 Chapter 5. Interaction and Momentum

Figure 5-4
After the interaction the second dancer steps back with a velocity less than that of the first dancer before the interaction. Her momentum before their interaction equals their total momentum afterwards.

not always particularly interesting. But when one object interacts with another, momentum becomes a very useful concept; it is the key that enables us to predict the outcome of the interaction.

To examine the role of momentum in predicting the outcome of interactions qualitatively, consider an example taken from classical ballet (Figure 5-4). One dancer runs toward the second, who is initially standing still. The second dancer catches the first and moves several steps backward in the process. The second dancer does not move backward with the same speed as the first; his speed is always less. Before the interaction, the first dancer has a momentum equal to the product of her mass and velocity. The second dancer has no momentum; his velocity is zero. After the interaction, the couple has a momentum equal to the sum of their masses times the velocity with which the second dancer steps backward.

If you compare the situation before with the situation after, momentum provides the link. The direction of motion is the same before and after the interaction—the dancers always move to the left. Greater mass compensates for the lower speed with which the second dancer steps backward, because he carries the first dancer. The momentum of the two dancers together immediately after is about the same as the momentum of the first dancer before the interaction occurred.

Momentum: A Constant of Motion

The example drawn from classical ballet actually involves interactions among more objects than just the two dancers. Both dancers interact continually with the floor, the air about them, and the earth. In addition, their motion is not simply horizontal, since the second dancer lifts up his partner. These additional interactions complicate our analysis and make it difficult to conclude anything more than that momentum seems to be involved in describing the outcome of the interaction.

Devices such as air hockey tables remove many of these complications. Consider an experiment with two air hockey pucks that can stick together upon impact. (Matching strips of velcro attached to each air hockey puck work well.) One lies essentially motionless, like the second dancer. The second puck moves toward the first, collides with it, and sticks to it. The two pucks then move backward, just as the second dancer stepped backward after lifting the first dancer. Both air hockey pucks float on a cushion of air, so their interactions with the table are minimized. Their motion is restricted to just one plane—the flat surface of the table. Thus, we can look at an experiment analogous to the two dancers, but one that minimizes the effects of all interactions except the interaction between the two pucks.

Figure 5-5 shows a strobelike drawing of the motion both before and after the interaction. The first puck, with a mass of 0.170 kg, is stationary. The second puck, also with a mass of 0.170 kg, has a velocity of 10 meters per second (m/s) to the right. After the interaction, the two pucks stick together and move on to the right. Their combined mass is 0.340 kg. Their velocity is 5 m/s to the right—half the initial velocity of the second puck. Using this information, we can compare the momentum before the interaction with the momentum after the interaction:

$$\textbf{Momentum before} = (0.170 \text{ kg})(10 \text{ m/s, right})$$
$$= 1.7 \text{ kg} \cdot \text{m/s, right}$$
$$\textbf{Momentum after} = (0.340 \text{ kg})(5 \text{ m/s, right})$$
$$= 1.7 \text{ kg} \cdot \text{m/s, right}$$

The two momenta are the same.

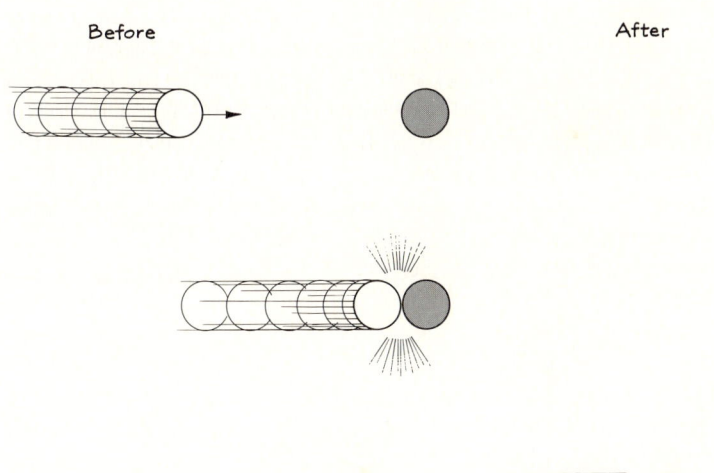

Figure 5-5

The second puck moves at a velocity of 10 m/s, right, toward a stationary puck. They stick together and move off at a common velocity of 5 m/s, right. With twice the mass, the combined pucks move at half the initial velocity of the single puck.

Momentum links the action before with the action after. Of all the results we can imagine for any given interaction, the result we actually see is that which keeps momentum constant.

> ### SELF-CHECK 5D
>
> You are standing motionless on a frozen pond. A friend whose mass is 70 kg slides toward you at a velocity of 2 m/s, east. The friend grabs you and the two of you slide on together. What is your friend's momentum before the interaction? What is the momentum of the two of you after the collision? Describe your motion after the collision.

CONSERVATION LOGIC

The notion that some quantities are constant throughout interactions is one that is deeply embedded in our view of everyday life. If you cannot find your house key, you search for it. You assume the key still exists—that it has not vanished into thin air. The continued existence of the key is something you take for granted. Yet, studies about how people develop intellectually show that you were not born with this sense of the constancy of things. The idea that certain things remain the same, which we call *conservation logic,* emerged from your experience with the world.

Ideas of Constancy Develop with Age

After years of careful study, researchers such as Jean Piaget have concluded that infants have no sense of constancy. Babies younger than about 6 months behave as though objects that are not visible do not exist. If you hold a favorite toy in front of a very young child, the baby will smile and reach for it. If you then hide the toy under a blanket as the child watches, the child immediately begins to cry. For the infant, the toy no longer exists. With experience, children form a concept of constancy of objects. Older babies will actively search for the toy, trusting that it continues to exist throughout the action of hiding. As we grow older, this sense of constancy broadens to include a number of other quantities.

Two simple tests often used to assess the reasoning skills of young children involve our concepts of substance and number. In one, a ball of clay is rolled into a long, thin snake as the child watches. The child is asked whether the snake has more clay, less clay or the same amount of clay as the ball. In a second test, three pennies are placed side by side on a table. As the child watches, the pennies are moved farther apart. The child is asked whether the second configuration has more pennies, fewer pennies or the same number of pennies as the first.

These questions may seem rather silly to you, but to a child they are not. Young children (4–6 years old) frequently respond that there is more clay in the snake and that there are more pennies in the second configuration. With experience, they gradually come to realize that the amount of clay and the number of pennies remain constant throughout the interactions.

Both of these exercises examine the development of conservation principles. A **conservation principle** states that a quantity does not change as a result of certain interactions. The amount of clay remains constant during the interaction with your hand—this is called **conservation of substance.** The number of pennies remains constant during changes in spacing—this is called **conservation of number.** In studying interactions, we examine yet another conservation principle, conservation of momentum.

Systems and Conservation Logic

Conservation principles are valid only when we are careful to define the objects involved in the interaction. Conservation of substance, as illustrated with the clay, is valid only when we do not add or take away any clay. Conservation of number remains valid only when pennies are not added or subtracted.

Another example is conservation of mass. Generally we expect the operation of sawing a board into two pieces to have no effect on the mass of the board; that is, the mass of the board is conserved. But if we compare the mass of the original board before with the combined mass of the two pieces after, we find mass is not conserved. Why? Because of the sawdust on the floor. Once the mass of the sawdust is included with the mass of the two pieces, mass is indeed conserved. Conservation principles are valid only when we are careful to keep track of all objects involved in the interaction.

The concept of a *system* helps identify and keep track of objects. A **system** is any set of objects we wish to study. Once the system has been identified, interactions can be divided into two groups:

1. interactions between objects in the system; and
2. interactions between an object in the system and an object outside the system.

In Figure 5-5 we defined our system to be the two air hockey pucks. When the two pucks collide with each other, an interaction occurs between objects within the system. If one puck hits the edge of the table, the interaction involves an object outside the system. When we choose our system so that all the interactions we are concerned with occur between objects in the system, we have chosen a **closed system.** The pucks compose a closed system as long as they interact only with each other.

Conservation principles are valid only in a closed system. Momentum is conserved in the closed system made up of the two air hockey pucks. Once the pucks bounce off the sides of the table, they have interacted with an object outside the system, and the momentum of the pucks will not be conserved. If we were to define a larger system that includes the two pucks and the table, momentum would be conserved for all objects in the system. But then we would have to include the mass and motion of the table (though its motion would be very minute) in addition to the mass and motion of the two pucks in our analysis. One way to be sure that you can apply conservation principles is to define your closed system to be the entire universe. But this presents us with a different problem—it is impossible to keep track of all the objects in the universe and most objects are not relevant to a given interaction anyway. (The snow in Moscow does not affect a basketball game in Poughkeepsie.) That is why we limit ourselves to the objects actually involved in the interaction.

As we gain experience with objects, our list of quantities that are conserved grows. Most of these conservation principles, like conservation of mass and number, become commonsense and we're surprised to learn that we didn't recognize them from birth. Others, like conservation of energy and momentum, we may realize at an intuitive level but don't verbalize until we study a specific field, like physics. Still others have yet to be discovered.

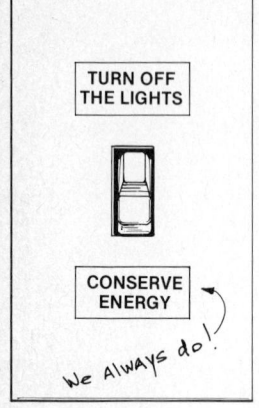

Figure 5-6

WARNING

In normal usage (especially today!), to conserve means to save. In physics, to conserve means to keep constant. Confusing *conserve* in everyday language with *conserve* in physics may be hazardous to your understanding.

CONSERVATION OF MOMENTUM

While most of us have some intuitive sense of momentum, applying conservation of momentum often seems rather formal. Choosing a closed system, identifying the objects that interact, separating the action before from the action after, determining the momentum of the objects involved in the interaction, finding the total momentum of the system—these steps are all part of the process of using conservation of momentum to predict the outcome of interactions. Let's look at how this process works.

Principle of Momentum Conservation

The **principle of momentum conservation** states that in any closed system, the total momentum of the system does not change even though objects within the system interact with one another. We can express this principle in equation form as

$$(\text{total } \mathbf{momentum})_{\text{before}} = (\text{total } \mathbf{momentum})_{\text{after}}$$

If the closed system consists of just two objects, this becomes

$$(\mathbf{momentum}\ 1 + \mathbf{momentum}\ 2)_{\text{before}} = (\mathbf{momentum}\ 1 + \mathbf{momentum}\ 2)_{\text{after}}$$

Sum of momenta before interaction
$P_{t\ (\text{before})} = P_{t\ (\text{after})}$
Sum of momenta after interaction

If the closed system involves more than two objects, we simply add more terms, one for each object.

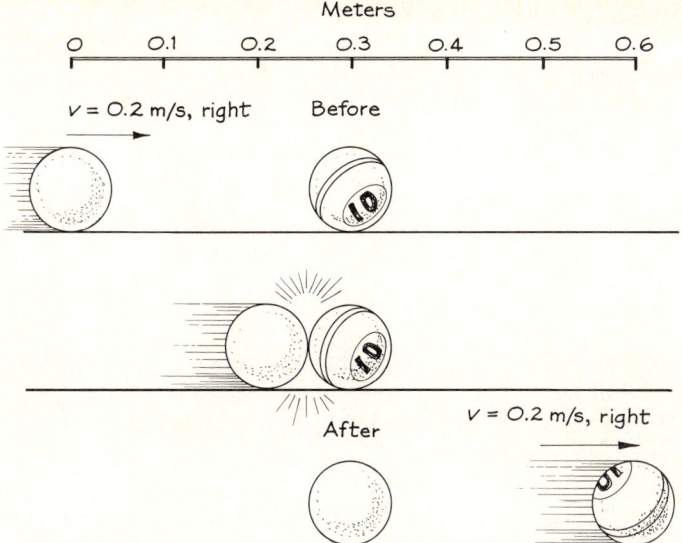

Figure 5-7
In this interaction, the cue ball transfers its momentum to the 10-ball. The cue ball moves in; the 10-ball moves off.

In applying this principle, it is important to distinguish between the momentum of individual objects in the system and the total momentum of the system. As objects interact, their individual momenta certainly do change. The momentum of each of the two dancers after their interaction was different from before. It was the *sum of their momenta* that remained constant before, during, and after the interaction. If, in applying the principle of momentum conservation, we find that the total momentum is not conserved, we conclude that other objects have been involved—that our system was not closed.

Collisions

When an air hockey puck collides so that it sticks to an identical puck, we see an example of momentum conservation. When the two pucks stick together, the mass of the moving object doubles. Consequently, the two pucks move off at a speed that is one-half the initial speed of the first puck alone. The momentum before the collision equals the momentum after the collision.

Momentum is also conserved when objects do not stick together. A common example of this type of collision is the interaction between the cue ball and another ball during a game of billiards. When the cue ball is shot without spin, it strikes another ball and stops. The second ball then moves off, as shown in Figure 5-7. In this type of collision, the object moving before the interaction is not the object moving after the interaction.

Let's apply the principle of momentum conservation to a closed system consisting of two billiard balls—the cue ball and the 10-ball. The mass of the two balls is identical, 0.170 kg. Before the interaction, the cue ball has a velocity of 0.2 m/s, right. The 10-ball is motionless. After the interaction, the cue ball is motionless and the 10-ball moves away at a velocity of 0.2 m/s,

Figure 5-8
The momentum of the system was zero before the interaction. As the person steps forward, the boat must move backward with a momentum equal to the person's forward momentum.

right. Expressed in terms of equations,

$$\text{momentum before} = (0.170 \text{ kg})(0.2 \text{ m/s, right}) + (0.170 \text{ kg})(0 \text{ m/s})$$

$$= 0.034 \text{ kg} \cdot \text{m/s, right}$$

$$\text{momentum after} = (0.170 \text{ kg})(0 \text{ m/s}) + (0.170 \text{ kg})(0.2 \text{ m/s, right})$$

$$= 0.034 \text{ kg} \cdot \text{m/s, right}$$

The momentum of the system remains constant during the interaction. The momentum of the cue ball is simply transferred to the 10-ball.

We have treated interactions of billiard balls ideally—that is, we have pretended that no interactions occur except the ones that interest us. In reality the billiard balls interact with the table, thus introducing an interaction outside the closed system. Additionally, billiard players do not simply roll the cue ball toward the 10-ball; they use a variety of spins to control the motion of the cue ball after the interaction. Essentially though, good billiard players develop an intuitive sense of momentum conservation in both one and two dimensions.

Stepping out of a Rowboat and Steering a Spaceship

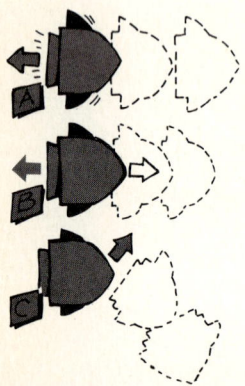

Figure 5-9
Spaceships use conservation of momentum in the design of their steering and control mechanisms. **(a)** When gas is ejected out the back, the spaceship speeds up. **(b)** When gas is ejected out the front, the spaceship slows down. **(c)** When gas is ejected upward, the spaceship moves downward.

A more subtle example of momentum conservation occurs when you try to step out of a rowboat (Figure 5-8). Again, our system involves just two objects—you and the rowboat. (We assume that the boat's interaction with the water can be ignored.) Initially, you are standing in a stationary rowboat. Assume that your mass is 60 kg and the mass of the rowboat is 40 kg. Since neither you nor the rowboat is moving initially, the total momentum of the system is zero. Conservation of momentum tells us that the total momentum of the system as you step out must still be zero. If you move forward at a speed of 1 m/s, your momentum is 60 kg · m/s, forward. In order for the total momentum of the system to remain zero, the boat must move with a momentum of 60 kg · m/s, backward. Your forward momentum must be canceled by the boat's backward momentum.

Spaceship designers take advantage of this when developing steering mechanisms. On earth we speed up, slow down, or change direction by interacting with the ground. In space, pilots use small gas outlets placed along the outside of the spaceship (Figure 5-9). To change velocity, the pilot ejects a large number of high-speed gas molecules from a selected outlet. Each molecule has a small momentum. Like the boat that moves backward as you step forward, the spaceship gains a momentum equal to but in the direction opposite to the ejected gas molecules. If the pilot wants to speed up, he or she fires the gas in a direction opposite to the present motion. To slow down, the pilot fires the gas in the same direction as the spaceship is moving. Other orientations will turn the vehicle.

SELF-CHECK 5E

A loaded rifle is initially motionless. The trigger is pulled and a 0.02 kg bullet leaves the rifle with a velocity of 300 m/s, N. What is the velocity of the 10 kg rifle after the bullet is fired?

A STEP FURTHER—MATH

A VERY STUBBORN EARTH

Momentum conservation allows us to do more than say that the boat moves backward with a momentum of 60 kg · m/s. If we write the equation that describes momentum conservation,

$$\mathbf{p}_{T(before)} = \mathbf{p}_{T(after)}$$

$$0 = m_{you}\mathbf{v}_{you} + m_{boat}\mathbf{v}_{boat}$$

we can rearrange it to solve for the velocity of the boat.

$$m_{boat}\mathbf{v}_{boat} = -m_{you}\mathbf{v}_{you}$$

$$\mathbf{v}_{boat} = -\frac{m_{you}}{m_{boat}}\mathbf{v}_{you}$$

Your mass is 60 kg and the mass of the boat is 40 kg. If you step forward with a velocity of 1 m/s, forward, then the velocity of the boat is

$$\mathbf{v}_{boat} = -\frac{60 \text{ kg}}{40 \text{ kg}}(1 \text{ m/s, forward})$$

$$= 1.5 \text{ m/s, backward}$$

You step forward at a speed of 1 m/s; the boat moves backward at a speed of 1.5 m/s. Why faster? Because the boat is less massive than you.

Take a step on land. Does the earth move backward as you step forward? You bet it does—but it is a bit harder to notice. Substitute the earth for the boat in the equation we just derived. The earth's mass is about 6×10^{24} kg. If you step forward with a speed of 1 m/s, how fast does the earth move backward? How about when you and a friend step forward in the same direction? How about when you and a thousand friends step forward in the same direction? It's a mighty stubborn earth!

INTERACTIONS WITH LARGE MASSES

Conservation of momentum is easy to notice when the two interacting objects have about the same mass. We have no difficulty recognizing it in interactions between billiard balls or collisions between air hockey pucks. But when the two objects differ greatly in mass, conservation of momentum is no longer obvious. A person catching a tennis ball, an egg striking the floor, an automobile colliding with a tree—in each case, the system had momentum before the interaction. After the interaction, however, nothing *appears* to move. The momentum of the system *seems* to have disappeared.

One response to this dilemma would be to suggest that momentum simply is not conserved in these situations. However, conservation logic has been such a powerful tool in understanding interactions that physicists are unwilling to abandon it. Instead, they look for an explanation in terms of momentum conservation. We now examine their explanation by first considering a series of ideal interactions between train cars.

Increasing the Mass Decreases the Velocity

Figure 5-10 shows what happens before and after collisions between two identical train cars when one of them is loaded with successively heavier masses. In each case, car X is moving initially, while car Y is stationary. When they collide, the two cars lock and move off to the right as a single unit. The drawings in (a) show the motions before and after a collision between two equal (empty) cars, while (b)-(e) show the motions as successively heavier loads are added to car Y. Consequently, in each drawing, car X interacts with a larger mass than in the previous drawing.

In each collision, the closed system consists of two train cars and any cargo we place in car Y. Before the collision, the momentum of the system is the momentum of car X. This quantity is the mass of car X times the velocity of car X. After the collision the momentum of the system is the total mass (car X + car Y + cargo in car Y) times their common velocity. Conservation of momentum states that the momentum of car X before the collision should equal the total momentum of car X, car Y, and the cargo after the collision. Table 5-1 shows the velocity of the cars after the collision for different loads of car Y, as predicted by momentum conservation.

Just by looking at the sequence of drawings, we can get some idea of what occurs as car X interacts with an increasingly more massive car Y. As the total mass of car Y (mass of car Y + mass of the cargo) increases, the velocity of the two cars after the collision gets progressively smaller. By (e), the velocity of the two cars is so small that it cannot be noticed on the scale used for the drawings. If you compare the column labeled "Mass Y" with the column labeled "Velocity After" in Table 5-1, you see the same result. As the mass of car Y increases, the velocity of the system after the collision decreases. By the time the mass of car Y is 20 times that of car X, their velocity after the collision becomes too slow to notice.

Interactions With Large Masses 101

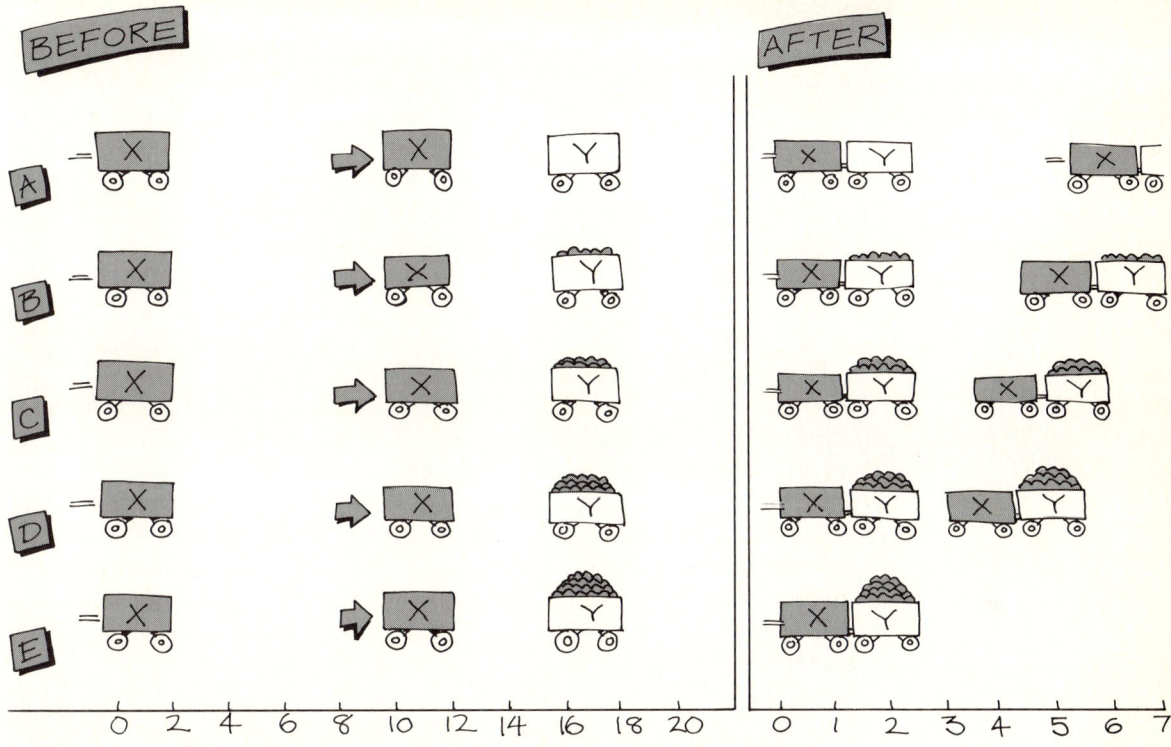

Figure 5-10 As the mass of car Y increases, the magnitude of the velocity after the interaction decreases. When the mass is extremely large, the motion may be so small that we do not perceive it.

Table 5-1 Interacting with Large Masses

Mass X	Velocity X	Momentum Before/After	Mass Y	Velocity After
1,000 kg	10 m/s, right	10,000 kg · m/s, right	1,000 kg	5 m/s, right
1,000 kg	10 m/s, right	10,000 kg · m/s, right	1,500 kg	4 m/s, right
1,000 kg	10 m/s, right	10,000 kg · m/s, right	2,333 kg	3 m/s, right
1,000 kg	10 m/s, right	10,000 kg · m/s, right	3,000 kg	2.5 m/s, right
1,000 kg	10 m/s, right	10,000 kg · m/s, right	19,000 kg	0.5 m/s, right

Tennis Balls, Eggs, and Automobiles

Physicists think that momentum is conserved in all interactions, providing the system is closed. When objects of vastly different masses interact, motion after the interaction may be too small to detect, but it is there. When you catch a tennis ball, the ball interacts with you and you, in turn, interact with the earth. The closed system includes you, the ball, and the earth. The mass of the tennis ball is so small compared with the total mass of the earth and you that we simply do not see any motion once the ball is caught. If you remove one of these masses, such as the earth, then the objects that interact become more comparable and you begin to see motion. For example, your interaction with the earth is minimized when you stand on ice. Try catching a fastball while standing on ice and notice your motion afterwards!

When the egg interacts with the floor, the floor is attached to the building, the building is attached to the ground, and the ground is part of the earth. The mass of the egg is negligible compared to the mass of the earth. The same is true when a car collides with a bridge. Interactions with massive objects result in motion too slow to be detected. Momentum is still conserved.

Steering a spaceship is like stepping out of a boat. Dropping an egg is like running into a brick wall. Continents collide just as cars do. Air molecules bounce around, exchanging momenta in much the same fashion that billiard balls interact on a billiard table. Conservation of momentum provides a powerful tool with which to understand a number of incredibly diverse interactions. In some cases, however, convenience dictates a change in perspective. When we look at motion from the perspective of just one of the objects, we must introduce the concept of force. Chapters 6, 7, and 8 consider motion from this point of view.

CHAPTER SUMMARY

Physicists describe change in terms of interaction. When one object influences another, we say that the two objects *interact* with one another. Measurable *change* provides the *evidence for interaction*. This chapter examines interactions in which a change in velocity occurs.

Two variables, mass and velocity, help describe many of the different interactions we observe. *Mass* is a measure of the amount of matter in an object. The unit of mass is the kilogram. Unlike weight, mass does not depend on the interaction of an object with a planet; mass is the same everywhere. Mass and velocity are combined in our perceptions of interactions. We use the concept of momentum to describe this combination. *Momentum*, defined as the product of mass and velocity, is a vector quantity whose direction is the same as the direction of the velocity.

When two (or more) objects interact only with each other, their combined momentum is the same before and after the interaction. We say that *momentum is conserved* in interactions within a closed system. A *closed system* is a group of objects that interact with each other but not with objects outside the system. Our experience has led us to expect certain quantities to be conserved—that is, to remain the same. We call this expectation *conservation logic*. Momentum is one of the quantities identified through our use of conservation logic.

Interactions between objects of about the same mass are easy to perceive in terms of momentum conservation. When one of the objects is much larger than the other, however, the momentum of the system *seems* to disappear. Momentum is still conserved, but the motion of the larger object is too small to measure after the collision.

ANSWERS TO SELF-CHECKS

5A. a. The velocity of the object decreases.
b. The velocity of the object changes because of a change in direction.

5B. a. mass
b. velocity
c. velocity

5C. Glass ball: 0.1 kg · m/s, E

5D. Friend's momentum before interaction = 140 kg · m/s, E. Total momentum after interaction = 140 kg · m/s, E. You and your friend slide eastward at a velocity less than 2 m/s. The magnitude of the velocity depends on your mass.

5E. $(0.02 \text{ kg})(300 \text{ m/s, N}) = -(10 \text{ kg})(v)$; $v = 0.6$ m/s, S

PROBLEMS AND QUESTIONS

A. Review of Chapter Material

A1. Define each of the following terms:
Interaction System
Mass Closed system
Momentum Conservation principle

A2. How is change related to interaction?

A3. In what units is momentum measured?

A4. If you know the mass and velocity of an object, how do you determine both the magnitude and direction of its momentum?

A5. Why is momentum a useful concept in describing interactions which involve a change in velocity?

A6. Under what conditions are conservation principles valid?

A7. How does our everyday use of the term *to conserve* differ from the way it is used by physicists?

A8. A system consists of two objects. Suppose that you know the mass and velocity of each object. Describe how you would determine the total momentum of the system.

A9. In a closed system, how does the momentum before the interaction compare with the momentum after the interaction?

A10. A closed system consists of two objects. During an interaction, the objects collide and stick together. Suppose you know the momentum of each object before the interaction. Describe how you would find the momentum of each object after the interaction.

A11. Why is momentum conservation difficult to observe when one of the objects is much more massive than the other?

B. Using the Chapter Material

B1. Two identical bowling balls are sitting motionless on the floor. Each ball is struck by a sledge hammer. After the interaction, ball A is moving more rapidly than ball B. If the hammers have equal masses, which one was moving more rapidly before the interaction? If they had equal speeds before the interactions, which one was more massive? Explain your answers.

B2. What is the momentum of a 70 kg sprinter moving at a velocity of 8 m/s, south?

B3. A 10,000 kg railcar is coasting on level ground at 5 m/s, west, when 1000 kg of snow falls vertically into it. What is the horizontal velocity of the car-snow system after the interaction?

B4. One quantity studied in nuclear interactions is parity. Before an interaction, parity is -1; after, it is $+1$. Physicists studying the interaction can draw one of two conclusions. What are they?

B5. An old circus trick is firing a person from a cannon. A 75 kg circus performer is fired north at 10 m/s from a 750 kg cannon. What is the velocity of the cannon after it is fired?

B6. A system consists of two ice skaters. The mass of each skater is 60 kg. Skater A is traveling west at 5 m/s; skater B, east at 4 m/s. What is the total momentum of the system?

B7. A 1 kg steel ball moving at 2 m/s, left, hits an identical ball that is not moving head-on. What is the speed of each ball after the interaction if:
  a. They do not stick together.
  b. They do stick together.

B8. Two people, one with a mass of 150 kg and the other with a mass of 75 kg, are walking toward you at identical speeds. They are so deeply involved in a conversation that they do not see you. A collision is inevitable. With which one would you choose to collide?

B9. The Army asks you to test a new cannon which has a mass of 50 kg and shoots 50-kg shells. Will you stand behind the cannon and pull the trigger?

B10. You throw a 0.25 kg snowball at a tree with a velocity of 2 m/s, SE. The tree is rigidly attached to the earth. What is the closed system? Why do you not notice the tree move as a result of the interaction?

B11. Admirals of 200 years ago fought battles by shooting broadsides—the simultaneous shooting of all cannons on one side of a ship—at their enemies. Using interaction and momentum conservation, explain why these cannons were usually mounted on wheels rather than rigidly attached to the ship.

C. Extensions to New Situations

C1. Collisions involving insects and cars occur frequently. Usually the driver notices the collision only when washing the bugs off. The insects, by that time, are no longer. As an example of this collision, consider a 1000 kg car moving east at 25 m/s and a 0.001 kg bug moving west at 1 m/s.
  a. What is the total momentum of the bug-car system before the collision?
  b. What must be the total momentum of the bug-car system after the collision?
  c. Why does the driver of the car seldom notice the collision?

C2. Large cylinders of highly compressed gas are used to carbonate soft drinks. If a valve breaks on one of these cylinders, the gas escapes rapidly; the cylinder acts like a rocket. Such an accident happened at an Indianapolis sports arena in the early 1960s. A cylinder valve broke, the cylinder was pushed over, and it exploded, killing several people. The preliminary report on this tragedy stated that the cylinder moved because the escaping air pushed on a nearby wall. Use conservation of momentum to argue that the cylinder would move even if the wall were not there. (Safety laws now require that these cylinders be chained to the wall.)

C3. Inside a box is a marble that is free to move. You cannot see into the box, but you have been asked to describe the location and velocity of the marble in it. You can, however, roll other marbles into the box and watch them when they come back out.
  a. How could rolling marbles into the box help you locate the marble in the box?
  b. What would happen to each marble when the two collided?
  c. How would the answer to (b) increase

the difficulty of knowing the location of the marble in the box?

C4. A rocket is propelled by high-speed gases, which shoot out the back of the rocket (Figure 5-C4). In most rockets the amount of exhausted fuel each second is about the same throughout the flight of the rocket. About 75% of the total mass of the rocket is its fuel.
  a. How does the total mass of the rocket near burnout of the fuel compare with the total mass at the beginning of the flight?
  b. How will the momentum of the rocket change each second if the momentum of the exhausted fuel each second is 2500 kg · m/s, N?
  c. When will the change in speed of the rocket be greater: near the beginning of the flight or near the burnout of the fuel?

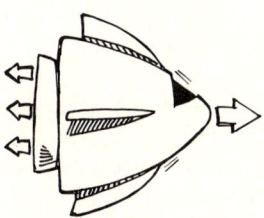

C5. On television a police officer will frequently shoot at a fleeing villain by holding the pistol in one hand. Actually, a real police officer seldom fires a weapon unless the officer has firmly gripped it with two hands. A major difference between stage guns and real guns is that the stage gun shoots a very small piece of paper rather than a lead slug. Does this explain the difference in the way real and stage pistols are held?

C6. When you walk, you interact with the earth in order to move forward. You change your speed. Determine how momentum is conserved in this situation.
  a. What objects are in the system for which momentum is conserved?
  b. When you are standing still, what is the total momentum of the system? (State the momentum relative to the earth.)

C7. In this chapter, we did not discuss what happens to momentum as we change reference frames. This problem allows you to fill this gap. Suppose a billiard ball with a mass of 0.5 kg is moving left relative to the earth at 2 m/s. It strikes an identical second billiard ball, B, which is stationary.
  a. What is the velocity relative to the earth of each ball after the interaction?
  b. A second observer is moving with a velocity of 3 m/s, right, before and after the interaction. In this reference frame, what are the velocity and momentum of each ball before the interaction?
  c. What is the total momentum of the system containing the two balls before the interaction?
  d. What are the velocity and momentum of each ball after the interaction?
  e. What is the total momentum of the system after the interaction?
  f. Use your answers to (b)–(e) to argue that momentum is conserved in the moving system.

C8. A popular toy, sometimes called clackers, is shown in Figure 5-C8. Five identical balls are suspended by strings. If you hold out one ball and release it, it falls and strikes the remaining balls. Eventually the ball at the far right moves outward.
  a. Use momentum conservation to predict the velocity of ball E compared to the velocity of ball A.
  b. If you pull balls A and B back and release them, balls D and E move outward with the same velocity that A and B had. Would momentum still be conserved if ball E had moved outward with twice the velocity of balls A and B?
  c. List other possibilities like (b) in which the momentum of the system is still conserved. (We investigate in Chapter 9 why these other possibilities do not occur.)
  d. Does momentum conservation by itself uniquely determine the outcome of interactions with the steel balls?

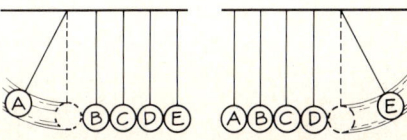

C9. When we drop a rock it is pulled downward by an interaction with the earth called gravity. The rock's speed increases continually as it moves toward the earth. Thus the rock's momentum is constantly changing. To determine how momentum is conserved in this situation, answer the questions below.
  a. What objects are included in the system for which momentum is conserved?
  b. What is the total momentum of the system as the rock is released?
  c. What must be the total momentum of the system as the rock starts falling?
  d. In order for momentum to be conserved, how must the earth move once the rock starts falling?
  e. Why do we not notice the motion of the earth?
  f. What would happen if everyone in the world dropped rocks at the same time?

D. Activities

D1. Watch a movie or play. Describe how fight scenes include or ignore momentum conservation. Then develop a series of stage directions for a realistic-looking fight. Describe how momentum conservation is important in your directions.

D2. Momentum conservation is evident in contact sports such as hockey, rugby, soccer, or football. Watch a game and describe some plays where you saw evidence of momentum conservation.

D3. Get on roller skates, ice skates, or a skateboard. Try to move about by throwing things of different masses. Describe the results and explain them in terms of momentum conservation.

Interaction and Force

Your father graduated from West Point. Your grandfather graduated from Annapolis. The Army-Navy game is today. Where do you sit? Which side do you support? Aaaaaahhhhhh!

Your political science instructor has just assigned two chapters to be read by tomorrow. You have a quiz in history and a theme due in English composition. You cannot possibly get all this work done in one evening. Help!

Forces—pushes and pulls—are so much a part of the physical world around us that we use these words to describe social situations too. You are caught in the middle, pulled by your father's allegiance to West Point and your grandfather's allegiance to Annapolis. Too many instructors are pushing you to get things done. With all these forces pulling and pushing you in different directions, you are going crazy. In social as well as physical interactions, a force is something exerted by one interacting object on another—on you, in the situations just described.

In Chapter 5 we described physical interactions in terms of the concept of momentum. While the momentum of each object changes during an interaction, the total momentum of a closed system remains constant. In this chapter we shift our attention from the system to individual objects within the system. We introduce the concept of *force* to explain the change in momentum of an individual object. When more than one force acts on the object, these forces are combined to produce a *net force*. We discuss *Newton's second law,* which relates the net force acting on the object to the mass and acceleration of that object.

FORCES—A FIRST LOOK

We often build an intuitive understanding of concepts long before we encounter their more formal definition. A concept like momentum, for example, emerges while playing billiards. When the cue ball strikes the 10-ball, "something" is exchanged. People may describe this thing differently, but they share an intuitive understanding gained from a common experience. Eventually we agree on a concept, like momentum, that supplies the needed vocabulary. Physicists then develop a more formal definition consistent with other concepts or principles in their field. When faced with these more formal definitions, we have to modify our intuitive ideas.

Force, like momentum, is a concept for which we have an intuitive definition. Our bodies provide sensations that report both when we exert a force and when a force is exerted on us. Before tackling the formal definition of force, we examine these intuitive ideas.

Pushes and Pulls

Force is commonly defined as a push or a pull. Physiologically, we detect forces through contractions or extensions of our muscles. When you open a door, you pull on it. The contractions of your biceps let you know that you are exerting a force. When you lift a glass of water, push on a piano, or pull a sled, similar sensations contribute to your intuitive sense of force. Other pushes and pulls do not involve us. A tow truck pulls a stalled car; gravity pulls a rocket back to earth; socks just removed from the dryer cling (are pulled) to one another. All these objects are involved in pushing and pulling—in exerting forces.

Pushes and pulls enable us to measure force. The instrument most commonly used to measure force is the **spring scale.** Forces either compress or extend a spring, which, in turn, moves a dial. When you stand on a bathroom scale, the pull of gravity on your body (also called your weight) causes a spring to be compressed. This compression results in a change in the dial on the scale. The force of gravity on your body, measured in pounds, appears on the scale. In the metric system, force is measured in units called *newtons*. One **newton** (N) is roughly equivalent to the force of gravity on a stick of butter, or about one quarter of a pound. Figure 6-1 shows several common forces and their approximate size in newtons.

Forces—A First Look

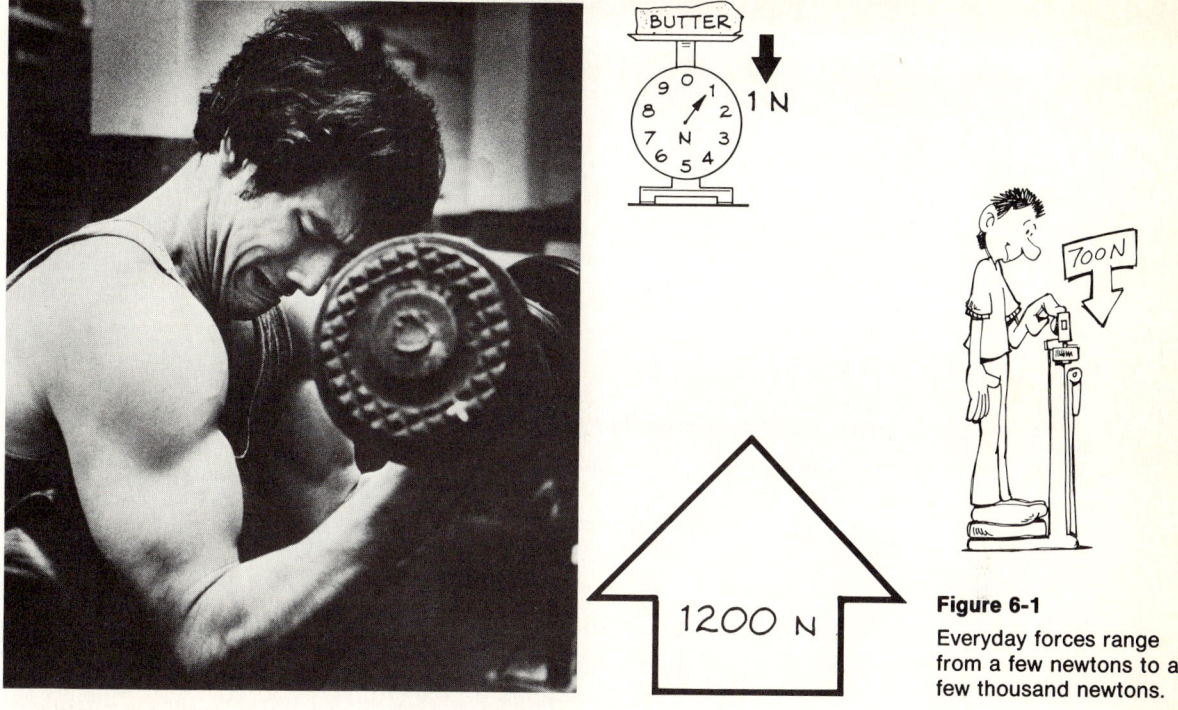

Figure 6-1
Everyday forces range from a few newtons to a few thousand newtons.

Pain, Pleasure, and Pressure

While muscular compressions and extensions provide the physiological sensations that we notice as we exert forces, nerve impulses from the skin tell us when forces are exerted on us. These impulses range from pleasure to pain—from the sense of loving communicated by a kiss to the sense of anger communicated by a slap.

A small force applied through a pin or needle is painful, while the same force applied by the touch of a hand is pleasurable. The difference between these two situations is the surface area of the skin over which the force is exerted. A pin applies a small force over a very small area, resulting in pain. A hand can apply the same force over an area 400,000 times greater, resulting in pleasure. Rather than sensing just the force, we feel a combination of force and area, called pressure.

Pressure is defined as the ratio of the applied force to the area over which it is applied:

$$\text{pressure} = \frac{\text{magnitude of force}}{\text{area of application}}$$

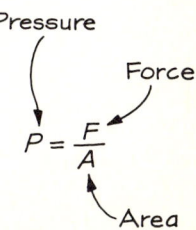

Pressure is measured in units of newtons per square meter, which is usually written N/m^2. The English unit of pressure is pounds per square inch ($lb/in.^2$).

Table 6-1 Some Common Pressures	
Event	Approximate Pressure (N/m^2)
"Peck" on the cheek	300
Kiss	500
Pat on the back	600
Slap on the face	5000
Auto shoulder harness during quick stop	2×10^6
Hypodermic syringe as it punctures skin	5×10^6
Blow that fractures a skull	5×10^7

We can use the definition of pressure to compare sensations of pain and pleasure. The pressure exerted by a 10 N force applied over the area of a pinhead (8×10^{-7}) square meters (m^2) is (10 N)/(8×10^{-7} m^2) = 1.25×10^7 N/m^2. The pressure is enormous! The same force applied over the area of a hand (0.02 m^2) is (10 N)/(0.02 m^2) = 500 N/m^2. Spreading the force out over a larger area decreases the pressure substantially. Table 6-1 lists common pressures we experience.

SELF-CHECK 6A

Which will result in greater pressure: 5 N applied by the pointed end of a pencil (area, 7×10^{-7} m^2) or by the end containing the eraser (area, 2×10^{-4} m^2)? If you try it, **be careful.** High pressures will puncture your skin.

FORCE AND CHANGE IN MOMENTUM

The strobelike drawings in Figure 6-2 represent a man wearing shoes while sliding across an ice-covered lake. He slides easily, traveling about five meters before coming to a stop. Collisions between his shoes and thousands of small bumps in the ice surface gradually slow him down. When we apply momentum conservation to each collision, we conclude that the ice accelerates to the right as the man slows down. The mass of the ice-earth system is so large, however, that we simply do not notice this motion.

Figure 6-2

The ice exerts a force that slows the man to a stop.

Now, cover the bottom part of the figure by placing a piece of paper along the dotted lines on each side of the drawing. You can still see successive images of the man, but you can no longer see the contact between the man and the ice. Yet you still know that the man is interacting with something. As successive images of the man get closer together, you imagine something slowing him down. While we do not know what other objects are involved, we do know that an interaction has occurred.

By covering the ice with the paper, we focus our attention on just one part of the interaction—the man. We explain the change in his motion by saying that something is exerting a force on him. In so doing, we replace the many collisions between the man's shoes and the surface of the ice with a single abstract quantity—a force. Let's examine how force is defined in terms of the man's change in momentum.

Force and Time Interval Affect Change in Momentum

Figure 6-3 represents the same man sliding across an ice-covered lake and across a waxed tile floor. In each example the man wears the same shoes, begins with the same momentum, and eventually slides to a stop. The change in his momentum is identical on the two surfaces. Only the surface on which he slides is different. This, in turn, affects how long and how far he slides. On ice he slides further and longer than on tile.

We can use the concept of force and the stopping time to describe the man's change in momentum. Both the ice and the tile floor exert forces that slow the man down. From experience and our intuitive feeling for forces, we expect the force exerted by the tile floor to be greater than the force exerted

Figure 6-3

The tile floor exerts a greater force than the ice. The man slides for a shorter distance and a shorter time on the tile.

by the ice. As illustrated by Figure 6-3, the greater force exerted by the tile causes the man to slide a shorter time, and hence a shorter distance, than the lesser force exerted by the ice. Given identical starting momenta, a large force leads to a short stopping time, while a small force results in a long stopping time.

This relationship is generalized in an equation first suggested in the seventeenth century by Isaac Newton:

force × time = change in **momentum**

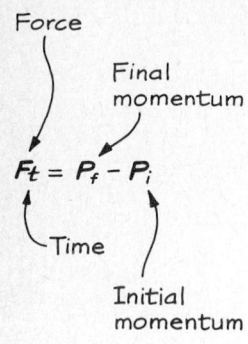

The change in momentum of an object is equal to the product of the force applied to it and the time interval during which the force is applied. When the force is measured in newtons and the time in seconds, the change in momentum is measured in kilogram-meters per second (kg · m/s).

Design Affects Force

To see how well this relationship among force, time, and momentum fits our experiences, we apply it to catching a baseball, first barehanded, then with a catcher's mitt, and finally with a net like that used to protect the fans behind home plate. The ball moves in one direction; you or the net exert a force in the opposite direction to stop it. Suppose a 0.15 kg ball moves at a velocity of 33.3 m/s, east, before it is stopped. Its initial momentum is (0.15 kg)(33.3 m/s, east) = 5 kg · m/s, east. Its final momentum is 0 kg · m/s. Consequently the change in momentum is −5 kg · m/s, east. We could just as easily say that the change in momentum is 5 kg · m/s, west.

Table 6-2 lists the force exerted and the time interval required to stop the ball. Comparing these quantities confirms our experiences. The pain is much smaller with a catcher's mitt because the force we must exert is much less. The padding sinks when struck by the ball, increasing the time interval over which the ball's momentum must be decreased to zero. This, in turn, decreases the force we need to exert. The net, because it sinks back even

Table 6-2 Force and Stopping Time in Catching a Baseball

Method	Change in Momentum of Baseball	Time Interval	Force
Bare hand	5 kg · m/s, W	0.0006 s	8333 N, W
Catcher's mitt	5 kg · m/s, W	0.003 s	1667 N, W
Net screen	5 kg · m/s, W	0.03 s	167 N, W

Force and Change in Momentum   113

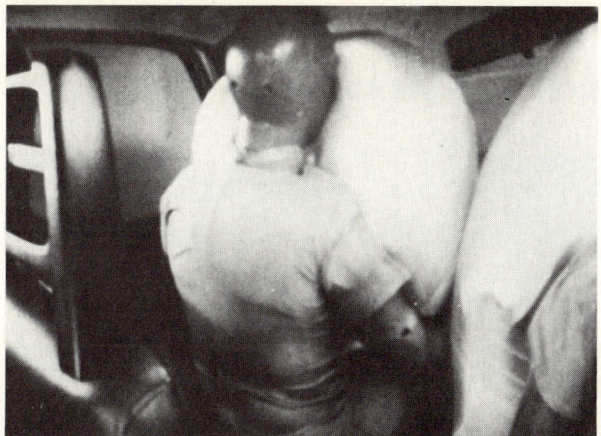
Courtesy Insurance Institute for Highway Safety.

Figure 6-4
Air bags increase the stopping time, thus decreasing the force exerted in stopping the passenger's forward motion.

further when struck by the ball, must exert an even smaller force. As the time interval required to stop the ball increases, the force that must be exerted decreases.

As another example, try dropping an egg on the floor. The hard floor "gives" very little, so the momentum of the egg must change in a very short time interval. Short time interval, large force: scrambled egg. If, on the other hand, we wrap foam rubber around the egg and then drop it, the egg does not break. The foam gives quite a bit, increasing the time interval in which the momentum of the egg must go to zero. Long time interval, small force: whole egg.

Air bags used in some automobiles are designed with the same concept in mind. When an automobile collides with some obstacle, its momentum is abruptly brought to zero. Passengers who are not wearing seat belts or shoulder harnesses collide with the dashboard, windshield, or steering wheel. The forces these objects exert depend on the time during which the collision takes place. Padding added to dashboards increases the stopping time (and safety) slightly, but passengers often collide with the windshield anyway. The air bag (Figure 6-4) is designed to provide additional protection. As soon as a switch on the automobile senses a sudden change in momentum, the air bag inflates and the partially filled bag provides a soft place to stop the passenger's motion gradually. Because a long stopping time decreases the force that must be exerted on the passenger, fewer broken bones result.

SELF-CHECK 6B

Use the relationship between force and stopping time to explain why the Post Office recommends using lots of padding for packages containing fragile objects.

LENGTHENING THE COLLISION TIME WITH PEDESTRIANS

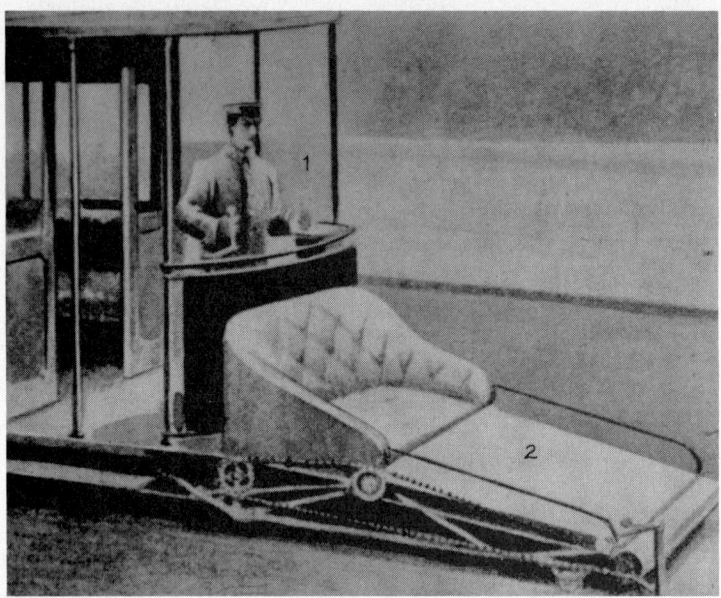

At the end of the nineteenth century, the accident rate between pedestrians and street cars was quite high. Since the pedestrians seemed uninterested in getting out of the way of the trams, an alternative solution was invented. The tram (1) would push a padded couch (2) in front of it. When an unobservant walker wandered in front of the moving tram, the person would collide with the front of the couch and fall into it. Because of the padding, the interaction time with the couch would be large. Thus, the momentum of the person would change gradually, keeping the force small. This device was tested in Los Angeles. Unfortunately for the inventor, the first test case fell and missed the couch. Instead, he struck the pavement, where the interaction time was very small and the force quite large. After that, Los Angeles pedestrians were on their own and eventually learned to watch out for trams and other horseless vehicles.

A Formal Definition of Force

The relationship between force, mass, and momentum first suggested by Newton provides us with a definition of force that is a bit more specific than the pushes or pulls with which we opened the chapter. This definition can be expressed in several different ways (see the box on page 115), but the most common form relates force to acceleration. The **force** exerted on an object is equal to its mass times its acceleration:

$$\text{force} = \text{mass} \times \text{acceleration}$$

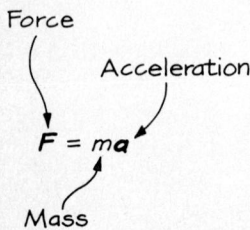

When the mass is measured in kilograms and the acceleration in (meters per second) per second ((m/s)/s), the force is given in newtons. (We define 1 N to be 1 kg · (m/s)/s. Force is a vector whose direction is the same as the direction in which the object accelerates. A force that acts in the same direction as the object's motion produces a positive acceleration—the object speeds up. A force that acts in the opposite direction produces a negative acceleration—the object slows down. Forces in other directions can cause changes in both the magnitude and direction of the velocity.

To see how to apply this definition, look back at the man who slides to a stop (Figure 6-3). On ice, the man slows down at a rate of 0.625 (m/s)/s; his acceleration is 0.625 (m/s)/s to the left. Given that the man's mass is 70 kg, the force exerted by the ice is (70 kg)(0.625 (m/s)/s, left) = 43.75 N to the left. On the tile floor, the man's acceleration is greater—about 1.25 (m/s)/s to the left. Again applying the definition of force, we find that the tile floor exerts a force of (70 kg)(1.25 (m/s)/s, left) = 87.5 N to the left. The force exerted by the tile floor is twice that exerted by the ice. A concrete sidewalk would exert an even greater force—the man would slide to a stop even sooner.

In applying the concept of force, it is important to remember that mass and acceleration describe characteristics of the object itself, while force de-

A STEP FURTHER—MATH

A FORCE IS A FORCE IS A FORCE

When a law is broad enough to explain diverse phenomena, it is often written in several different ways to suit a variety of purposes. Newton's relationship can be used to describe force in at least three different ways. We begin with:

$$\textbf{force} \times \text{time} = \text{change in } \textbf{momentum}$$

We can rearrange this to isolate force on one side of the equation:

$$\textbf{force} = \frac{\text{change in } \textbf{momentum}}{\text{time}}$$

This was the expression with which Isaac Newton actually introduced the concept of force. Since the momentum of an object depends on its mass and velocity, either or both can change as a result of the force exerted on the object. Most of the time, however, the mass of an object remains constant and only its velocity changes. This allows us to write Newton's relationship in yet a third way:

$$\textbf{force} = \text{mass} \frac{\text{change in } \textbf{velocity}}{\text{time}}$$

$$\textbf{force} = \text{mass} \times \textbf{acceleration}$$

All three describe the relationship between force and the motion of the object upon which the force acts. A force is a force is a force!

Figure 6-5

Copyright 1983, Universal Press Syndicate. Reprinted with permission. All rights reserved.

scribes an interaction that occurred with some other object. Objects do not accelerate by exerting forces on themselves (Figure 6-5). Other objects exert forces on them, causing them to accelerate in predictable ways. Sometimes these other objects are obvious, and we are able to associate forces directly with the objects involved. In other situations we see an object accelerating and infer that a force must be present—even when it is not at all obvious what object exerts that force.

Force due to Gravity

The most common application of our definition of force is the calculation of the force due to gravity acting on an object near the earth's surface. As described in Chapter 2, all objects experience an acceleration due to gravity near the surface of the earth. In the absence of other forces, this acceleration has been measured to be 9.8 (m/s)/s, down. We can use the definition of force to calculate the force due to gravity acting on a mass of 1 kg:

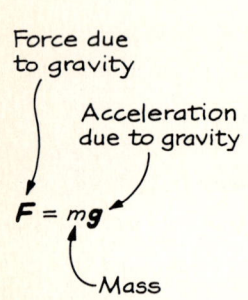

Force due to gravity = mass × **acceleration** due to gravity

= (1 kg)(9.8 (m/s)/s, down)

= 9.8 N, down

An object with mass of 1 kilogram experiences a gravitational force of 9.8 N, down, near the surface of the earth. An object with mass 2 kg experiences twice the gravitational force, or 19.6 N, down.

We can use this same procedure to determine the gravitational forces exerted by other planets or moons. For example, we can contrast the force due to gravity on the earth with the force due to gravity on the moon. On the earth, the acceleration due to gravity near the surface is 9.8 (m/s)/s, down. On the moon the acceleration due to gravity near the surface is 1.6 (m/s)/s, down. On the earth, an astronaut with a mass of 70 kg experiences a force due to gravity of

force = mg = (70 kg)(9.8 (m/s)/s, down) = 686 N, down

Table 6-3 Acceleration Due to Gravity on Various Planets and the Moon

Solar Body	Acceleration Due to Gravity (m/s)/s, down
Moon	1.6
Mercury	3.3
Venus	8.1
Earth	9.8
Mars	3.6
Jupiter	24.3
Saturn	10.3

When the astronaut reaches the moon, the force due to gravity is:

force = $m\mathbf{g}$ = (70 kg)(1.6 (m/s)/s, down) = 112 N, down

Table 6-3 lists the acceleration due to gravity for several different planets in the solar system, as well as the moon.

The force with which gravity acts on an object is called the object's **weight**. As illustrated in the example, an object's weight depends on the acceleration due to gravity on the planet on which the object is located. The astronaut has the same mass (70 kg) on both the earth and the moon. By contrast, the astronaut weighs 686 N on the earth but only 112 N on the moon. Even on the same planet an object's weight can vary slightly due to variations in the acceleration due to gravity. On the earth, for example, the acceleration due to gravity varies from 9.78243 (m/s)/s, down, near the equator to 9.82534 (m/s)/s, down, in Greenland. The astronaut's weight on the earth can vary from 684.8 N to 687.8 N. Admittedly small, such variations remind us that weight depends on location, while mass does not. Consequently, most physical laws are described in terms of mass.

Figure 6-6
On the moon the force due to gravity is one-sixth that on the earth.

_____ **SELF-CHECK 6C** _____

The acceleration due to gravity on the surface of Jupiter is 24.3 (m/s)/s, down. What is the weight of the 70 kg astronaut on Jupiter?

KINDS OF FORCES

At any given moment, literally billions and billions of forces are acting. Some, like those you exert when you push or pull on something, are relatively simple to identify. Others, like the force the floor exerts on you, become obvious only when they are no longer present. Stepping through a rotten floorboard reminds you that there is usually something pushing back as you walk along the floor. Still other forces, like gravity, seem almost magical, since they need not involve any contact between objects at all. We now examine the general kinds of forces present in our environment.

Contact Interactions

The **forces from contact interactions** include pushes, pulls, and friction. Normally we are able to identify pushes and pulls simply by looking at the situation. When you open a door, the contact between your hand and the door allows you to identify a pull acting on the door. When a bulldozer pushes the dirt, the contact between the shovel and the dirt allows us to identify the push. When a person stands on the ice, the contact between the ice and the person allows us to identify the upward push exerted by the ice.

Friction also involves contact between objects, but it is more easily overlooked. Whenever two objects are in direct contact, either while moving or stationary, a frictional force acts between them. These forces always oppose any motion.

Frictional forces can be divided into two categories: static friction and kinetic friction. **Static friction** arises when two objects are stationary relative to each other. When you stand on a steep hill, for example, the ground exerts a frictional force along the surface that keeps you from sliding down the hill. When you push a car, the force you need to exert to get the car started rolling is more than the force you exert to keep it moving. Static friction opposes any start of motion. **Kinetic friction** arises when one object slides relative to another. When the man slides across the ice, the ice exerts a frictional force that eventually brings him to a stop, though this force is much smaller than the frictional force exerted by the tile. Kinetic friction is also present as objects move through gases or liquids. The most common example is the air resistance which acts to slow all moving objects in the earth's atmosphere. As the man slides across the ice, he is slowed by the kinetic friction of the air as well as that of the ice.

Interaction at a Distance

A second category is **forces that act at a distance.** So far, we have discussed only one such interaction, gravity. The force due to gravitational interactions is present between all masses, but the magnitude of the force depends on the size of the masses and their distance from each other. On the earth, we notice only gravitational interactions between a small mass (like ourselves) and a large mass (like the earth). Gravitational interactions between two small masses (such as yourself and a desk) are much too small to notice. On earth,

Kinds of Forces   119

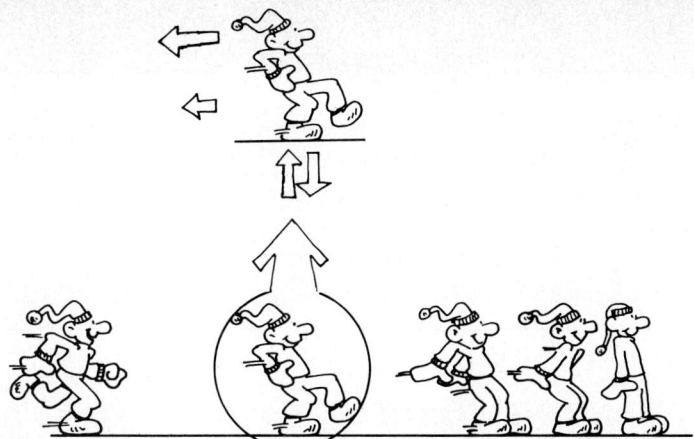

Figure 6-7

Four forces act on the man. The force due to gravity pulls the man downward. The ice exerts an upward force that balances the force due to gravity. The ice surface and surrounding air both exert frictional forces to the left.

the gravitational force always acts downward, pulling the small mass toward the earth's surface.

Three other kinds of interactions at a distance exist: electrical forces, strong nuclear forces, and weak nuclear forces. We discuss each of these in some detail in Chapter 8 and look at situations in which they act in later chapters dealing with electromagnetic, atomic, and nuclear interactions.

Identifying Forces

In analyzing interactions in terms of force, we are faced with the task of identifying the forces acting on an object. The two kinds of forces, contact forces and forces due to interaction at a distance, provide us with a two step process for accomplishing this task. Let's apply the process to the man sliding across the ice.

Although we described the man's motion in terms of a single force exerted by the ice, several kinds of forces are actually present, some of them exerted by the ice and some not. We first look for contact forces. The man is in direct contact with the ice and the surrounding air. His contact with the ice leads to the exertion of two forces—one to the left and a second upward. The ice exerts a frictional force that opposes the man's motion—as he slides to the right, the ice pushes back to the left. The ice also exerts a force upward to support the man. (As in the case of the broken floorboard, we often become aware of this upward force only when it is no longer present. We discuss the nature of this force in the next chapter.) The man is also in direct contact with the surrounding air, so air resistance acts to the left as he slides to the right. The only force due to interaction at a distance is the force due to gravity. The earth pulls the man downward toward its surface with a force equal to the man's weight.

As illustrated in Figure 6-7, four forces act on the man—the force due to gravity pulling downward, a force exerted by the ice pushing upward, and two frictional forces acting to the left. As you will see in Chapter 7, the force due to gravity is balanced by the upward force exerted by the ice, so the man does not move up or down. The two frictional forces, that due to air resistance and that exerted by the ice, slow the man down.

SELF-CHECK 6D

Identify the forces acting on an apple when it:

a. hangs from a limb on the tree

b. falls downward

MORE THAN ONE FORCE

Suppose you are standing by a fence. You see the tops of two refrigerators move by, as shown in Figure 6-8. Using only the information contained in the drawings, can you state which refrigerator is interacting with other objects?

If you use only these drawings and the ideas we have discussed so far, you must conclude that refrigerator B is interacting with another object, while refrigerator A is not. The images of refrigerator B become more closely spaced over time. Its momentum is decreasing, so something must be exerting a force to slow it down. The images of refrigerator A are equally spaced. Its momentum remains constant, suggesting that no force acts on it. Of course, your common sense tells you otherwise. Refrigerators don't just float down the street!

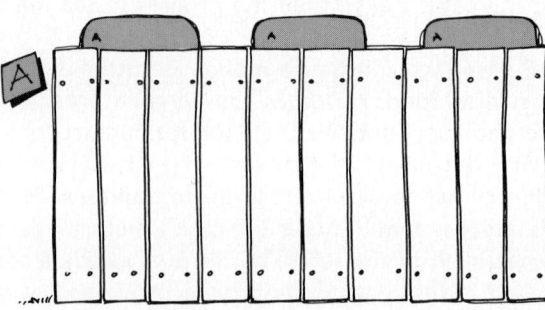

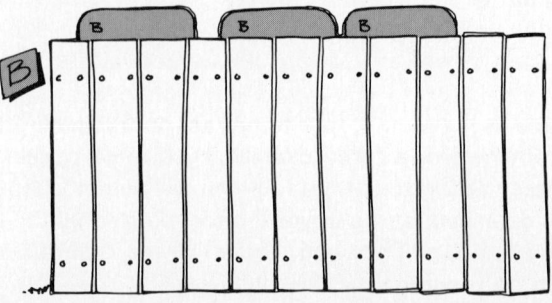

Figure 6-8
Based only on the drawings, which refrigerator is interacting with other objects?

Figure 6-9
Both refrigerators are interacting with a number of different objects—the woman, the floor, and the surrounding air. In (a) the forces sum to zero. In (b) the net force acts to the left, so the refrigerator slows down.

From experience, we know that any moving refrigerator has someone pushing on it. As shown in Figure 6-9, people are pushing both refrigerators. To understand how forces can lead to no change in momentum as well as to some change in momentum, we must examine what happens when more than one force acts on the same object.

What Forces Act?

We can identify the forces acting on both refrigerators using the process described in the last section. First look for contact forces—pushes, pulls, and friction; then look for forces due to interaction at a distance.

As shown in Figure 6-9, five forces act on each refrigerator. Four contact forces exist: one force (a horizontal push) due to the contact provided by the person pushing the refrigerator, two forces (an upward push and friction) due to the contact between the refrigerator and the ground, and one force (friction) due to the contact between the refrigerator and the surrounding air. The fifth force, which acts at a distance, is the force due to gravity exerted by the earth. We can look at the way these forces combine by considering the forces that act vertically separately from those that act horizontally.

Looking at the forces that act vertically, we have the force due to gravity pulling the refrigerator downward and the contact force between the refrigerator and the ground pushing back upward. As in the case of the man on the ice, these vertical forces cancel. The refrigerator does not move vertically. Looking at the forces that act horizontally, we have the woman pushing to the right and two frictional forces, that of the ground and that of the air, pushing to the left. For refrigerator B, the woman pushes to the right with a force that is less than the two frictional forces exerted to the left. The refrigerator gradually slows down, much as the man sliding along the ice did. For refrigerator A, however, the woman pushes to the right with a force that is equal to the two frictional forces exerted to the left. Refrigerator A continues to move at a constant velocity.

When more than one force acts on an object, the combination of forces can produce motion at a constant velocity, motion with acceleration, or no motion at all. For both refrigerators, the two vertical forces combined to produce no vertical motion. For refrigerator A, the horizontal forces combined to produce motion at a constant velocity. For refrigerator B, the horizontal forces combined to produce a deceleration.

Net Force

To describe situations in which more than one force acts, we introduce the concept of *net force*. When we talked about the force that the ice exerts on the man, we really meant the net force—the sum of all the forces exerted by the ice. We can modify our definition of force accordingly, substituting net force for force:

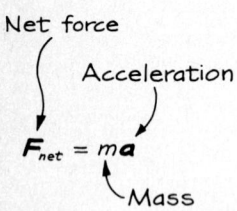

net force = mass $\times$ **acceleration**

Known as **Newton's second law,** this relationship can be used to predict the acceleration that results from one or several forces acting on an object.

The net force is the sum of all forces acting on an object. Since force is a vector quantity, we must keep track of directions as well as magnitudes. The concepts of vector addition introduced in Chapter 1 can be applied to forces as well as displacements. The net force that results from two or more forces acting in the *same* direction, Figure 6-10(a), is the sum of the magnitudes of the vectors. Three forces of 100 N, east, combine to exert a net force of 300 N, east. The magnitudes of forces that act in *opposite* directions must be subtracted, as shown in Figure 6-10(b). Forces of 150 N, east, and 150 N, west, exert a net force of 0 N. Forces that act at angles relative to one another, as in Figure 6-10(c), can be added using the tail-to-tip method. While the tail-to-tip method provides an excellent strategy for adding vectors, it is always a good idea to put your past experience to work in estimating the net force exerted. In Figure 6-10(c), for example, most people predict that a force of 10 N, north, and a force of 10 N, east, combine to exert a net force to the northeast.

To see how to apply Newton's second law, we return to the motion of the two refrigerators in Figure 6-9. No net force is acting on the refrigerators vertically. The upward force exerted by the ground just equals the downward force due to gravity. The net vertical force is zero; the refrigerators experience no vertical acceleration.

Each of the refrigerators has three forces acting on it horizontally. However, the net force acting on refrigerator A is zero, while that on refrigerator B is not zero. The force of friction exerted by the ground and surrounding air

More Than One Force   123

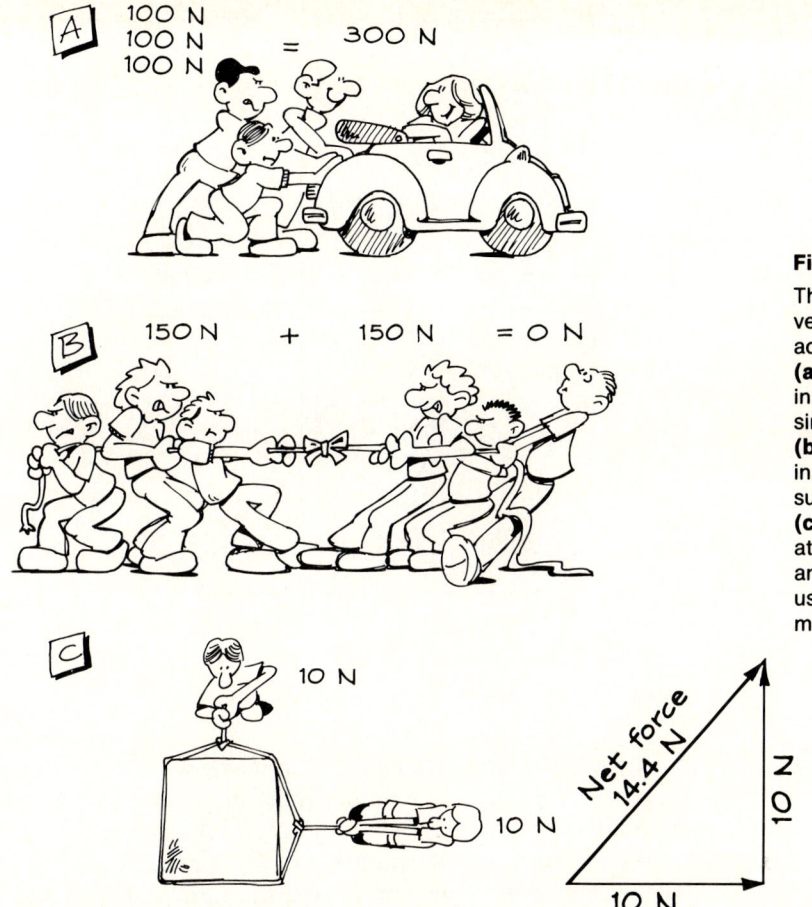

Figure 6-10
The net force is the vector sum of all forces acting.
(a) Forces that act in the same direction simply add.
(b) Forces that act in opposite directions subtract.
(c) Forces that act at right angles to one another can be added using the tail-to-tip method.

exceeds the force with which the woman pushes refrigerator B. The net force acts to the left, and refrigerator B slows down. By contrast, the woman pushes refrigerator A with a force that equals the force due to friction exerted by the ground and the surrounding air. The net force acting on the refrigerator is zero, and refrigerator A continues to move at a constant velocity.

Zero Net Force

As is evident from the two refrigerators, zero net force can result in either no motion or motion at a constant velocity. Intuitively, we know that no net force results in no motion. The cartoon in Figure 6-11 is funny because it seems absurd to us that the characters do not grasp this fact. The idea that zero net force can also result in motion at a constant velocity is not so obvious. From experience, most of us associate motion at a constant velocity with a force. We have to push a refrigerator in order to keep it moving at a constant velocity, and we generally assume that no other force is acting on it. What we have ignored, of course, is the frictional force against which we are actually pushing. The force we exert is balanced by the frictional force exerted by the

Figure 6-11
Why doesn't the rock move?

ground and the surrounding air, so that the net force on the refrigerator is zero.

Of course, to start a refrigerator moving we must exert a net force greater than zero. Each woman in our example must have initially exerted a force to the right greater than the net frictional force exerted to the left. When the refrigerators were not moving horizontally, a zero net force in the horizontal direction resulted in no motion. A zero net force in the horizontal direction also results in motion at a constant velocity. To keep a refrigerator moving once it has started, we need only balance the opposing forces.

You learned earlier in the chapter that the net force on an object equals the object's mass times its acceleration. A zero net force results in zero acceleration. This results either in no motion or motion at a constant velocity.

SELF-CHECK 6E

A child pushes a 10 kg wagon with a force of 10 N. The force of friction exerted by the ground and the surrounding air is 8 N. What is the net force acting on the wagon? What is the wagon's acceleration?

FORCE, ACCELERATION, AND THE SPEED OF LIGHT

Our definition of force must be modified somewhat to describe motion at speeds near the speed of light. If we apply a net force in the same direction as

OFF TALL BUILDINGS IN A SINGLE, SAFE BOUND

Net force and its relationship to momentum change and time were foremost in the mind of Thomas M. Prentiss when he designed the personal fire escape in 1873. When trapped by a fire in the upper levels of a multistory building, the owner of a personal fire escape could simply jump out the window. A parachute (*A*) attached to the head would act to slow the fall. The upward force of the air as it tried to move through the parachute would cancel some of the downward pull of gravity. Thus the net force downward on the person escaping the fire would be less than the force due to gravity. The acceleration would result in a sufficiently small speed to cause little harm on landing. To assure a safe landing, Mr. Prentiss included a pair of elastic padded overshoes (*B* and *C*). These shoes would increase the time of the collision with the ground and, thus, decrease the force applied on the person by the ground.

an object moves, the object's velocity will increase. In most everyday situations, forces act for just a short time. But what if the force continues to act for a long time—say days or months? For example, if we push continuously on a 1 kg mass with a net force of 100 N (approximately the force you exert to hold a bowling ball), the low-speed definitions of force and acceleration tell us that the mass would be moving at the speed of light after 34 days. However, according to the special theory of relativity, objects cannot exceed the speed of light. Consequently, our definition of force must be altered somewhat to place a limit on the extent to which an object can be accelerated by a constant force.

The modification we make is in our concept of mass. At ordinary speeds, we imagine the mass of an object to be independent of the object's speed. A 10 kg wagon has a mass of 10 kg at rest, at 10 m/s, or at 100 m/s. Einstein suggested that such a concept simply does not work, although its limitations are not apparent until the object's speed nears the speed of light. He proposed that the mass of an object does depend on the speed at which it moves. The mass of a moving object, called the **relativistic mass,** increases as the object's speed increases.

This new definition of mass produces results that are consistent with our low-speed experiences but that alter our concepts at high speeds. Table 6-4 compares the low-speed and high-speed definitions of mass as an object moves at higher and higher speeds. At everyday speeds, the difference in mass is not noticeable. At around half the speed of light, however, we could start detecting the difference. The mass of a 1 kg object increases substantially as the object approaches the speed of light and becomes almost 71 kg at 0.9999 the speed of light. As objects travel at higher speeds, their masses increase.

This definition of mass places a natural limit on the extent to which an object can be accelerated. If a constant force, say 10 N, acts on an object, the object will accelerate. As the object's speed increases, however, its mass increases. Since the force is constant, an increase in mass must cause the acceleration of the object to decrease. As the object begins to move at speeds approaching the speed of light, its mass increases at an ever greater rate, and its acceleration becomes increasingly small. This means that though the velocity gets closer and closer to that of light, it never quite equals it. The relativistic definition of mass establishes a natural limit to the extent to which we can accelerate objects. No object can be accelerated to speeds that exceed the speed of light.

Table 6-4 Relativistic Mass at Various Speeds

Speed of Moving Frame Fraction of Light Speed	(m/s)	Rest Mass (kg)	Relativistic Mass (kg)
0.0001	3×10^4	1.00	1.000000005
0.001	3×10^5	1.00	1.0000005
0.01	3×10^6	1.00	1.00005
0.1	3×10^7	1.00	1.005
0.2	6×10^7	1.00	1.02
0.4	1.2×10^8	1.00	1.09
0.6	1.8×10^8	1.00	1.25
0.8	2.4×10^8	1.00	1.67
0.9	2.7×10^8	1.00	2.29
0.99	2.97×10^8	1.00	7.09
0.999	2.997×10^8	1.00	22.37
0.9999	2.9997×10^8	1.00	70.71

$$\text{mass} = \frac{\text{rest mass}}{\sqrt{1 - \frac{(\text{speed of mass})^2}{(\text{speed of light})^2}}}$$

CHAPTER SUMMARY

The concept of force is used to describe situations in which we wish to focus our attention on just one of the objects involved in an interaction. A force is a description of the effect of an interaction on an object. Intuitively we define *force* as a push or a pull. When we exert forces, we feel the extensions and contractions in our muscles. When forces are exerted on us, we sense the pressure exerted on our skin. Force is measured in units called newtons; pressure is measured in units of newtons per square meter. *Pressure* is defined as the force exerted divided by the area over which the force acts.

When we focus our attention on just one of the objects involved in an interaction, the force times the time interval during which the force acts equals the object's change in momentum. More formally, *force* is the product of the mass of an object and its acceleration. Force is a vector quantity whose direction is the same as the direction of the object's acceleration.

Forces can be divided into two categories: contact forces and forces that act at a distance. *Contact forces* include pushes and pulls that arise from contact between objects, as well as frictional forces that oppose motion. *Forces that act at a distance* include gravitational and electrical forces and strong and weak nuclear forces. All objects exert gravitational forces on each other, but the gravitational pull exerted by small masses is too small to be noticed. A large mass, such as the earth, exerts a noticeable gravitational force downward on small objects near its surface. This force is called an object's *weight*.

We use the concept of net force to describe situations in which more than one force act. The *net force* acting on an object is the vector sum of all forces acting on it. When the mass of the object remains constant, the net force acting on it is the product of its mass and its acceleration: $\mathbf{F}_{net} = m\mathbf{a}$. This relationship is known as *Newton's second law*. When the net force acting on an object is zero, the object experiences no acceleration. It either remains stationary or continues to move at a constant velocity. When the net force is not zero, the object accelerates.

At speeds near the speed of light, the mass of an object does not remain constant. The mass of a moving object, called its *relativistic mass,* equals its rest mass divided by a quantity that depends on the object's speed compared to the speed of light. As objects move faster, their relativistic masses increase to the extent that no object can be accelerated beyond the speed of light.

ANSWERS TO SELF-CHECKS

6A. The smaller the area, the greater the pressure. The pointed end exerts a greater pressure.

Pointed end: $(5 \text{ N})/(7.0 \times 10^{-7} \text{ m}^2) = 7.1 \times 10^6 \text{ N/m}^2$

Eraser end: $(5 \text{ N})/(2 \times 10^{-4} \text{ m}^2) = 2.5 \times 10^4 \text{ N/m}^2$

6B. When a box is thrown into a truck, the box experiences a change in momentum. According to the definition of force, the greater the time interval in which the momentum changes, the less the force exerted on the object. The padding will increase this time interval.

6C. Weight = (70 kg)(24.3 (m/s)/s, down) = 1701 N, down

6D. When the apple is hanging from a limb on the tree, two forces act on it. The contact between the apple and the limb gives rise to an upward force that keeps the apple from falling. The other force is the force due to gravity pulling the apple downward toward the ground. When the apple is falling toward the ground, two forces act on it. The force due to gravity is pulling the apple downward toward the ground. The surrounding air exerts a frictional force upward, slowing the rate at which the apple falls.

6E. net force = 10 N, right − 8 N, left = 2 N, right;
acceleration = (2 N, right)/(10 kg) = 0.2 (m/s)/s, right

PROBLEMS AND QUESTIONS

A. Review of Chapter Material

A1. Define the following terms:
Force
Contact interaction
Interaction at a distance
Net force
Pressure
Static friction
Kinetic friction
Rest mass
Relativistic mass

A2. We defined force formally in terms of momentum and in terms of acceleration. State these two definitions. Which is more general? Why do we introduce two definitions?

A3. How do we intuitively define force?

A4. How does the concept of force enable us to concentrate on one object in an interaction?

A5. Describe how the change in momentum of an object depends on the force applied to it and on the time during which the force is applied.

A6. Use the definition of force to describe how air bags protect passengers in automobile collisions.

A7. List the two categories of force we encounter in most everyday situations.

A8. If more than one force acts on an object, how do you combine them to determine the net force?

A9. An object moves along at a constant velocity. What two possibilities exist for explaining this motion?

A10. Describe how you would calculate the force due to gravity on an object when you know its mass.

A11. Why must our concept of mass be changed at speeds near the speed of light?

B. Using the Chapter Material

B1. In the early 1960s women wore high-heeled shoes, called spike heels, which had a very small area at the heel. Typically, they had an area of 0.0001 m^2. What pressure could a force of 600 N apply on one heel? On two heels? (While spike heels were difficult to walk in, they were great for self-defense.)

B2. A bicycle brake applies a 20 N force for 10 s.
  a. What is the change in momentum of the bicycle?
  b. How much force would be needed to make the same change of momentum in 2 s?

B3. Helmets are used in a variety of situations to protect people from head injuries. As illustrated in Figure 6-B3, the wearer can endure forces that could normally kill a person. The protection offered by a helmet arises from two sources:
  a. Without a helmet, the force exerted by the hammer is simply exerted on the skull over the area of the hammerhead.

The helmet spreads this force over the entire area of the helmet. Estimate the area of the hammerhead and the area of the helmet. How does this reduce the pressure exerted on the skull?
b. The interior of most helmets is padded. How does this padding add further protection?

B4. Calculate the force acting:
a. when a 1000 kg drag racer accelerates from zero to 170 m/s in 5 s
b. when a 50 kg runner accelerates from zero to 12 m/s in 6 s
B5. Identify all forces acting on the rock in the first and last frames of the cartoon in Figure 6-11.
a. Why does the rock not accelerate if only one person pushes on it?
b. Why does the rock not accelerate when all four people push on it?
B6. While you hold a book motionless in your hand, what are the forces acting on it? Which are due to contact interactions and which are due to interaction at a distance?
B7. Three children are pulling on a toy. Kim pulls with a force of 30 N, north; Kevin, with a force of 30 N, south; and Julie with a force of 15 N, east. What is the acceleration of the toy?
B8. A 500 N force acts on a 2000 kg car in the direction in which the car is moving. Air resistance and friction supply a 300 N force in the opposite direction. What is the net force acting on the car? What is the car's acceleration?
B9. Shooting paper wads with rubber bands is a common pastime among junior high students. The force on the paper wad is applied through the rubber band. What are the net force and acceleration of a 0.01 kg paper wad in the situation illustrated in Figure 6-B9?

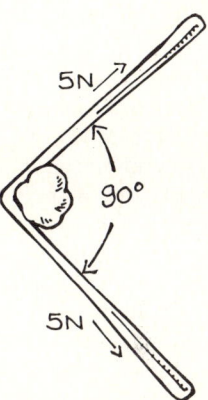

B10. The brakes on a small truck are designed to apply a force that uses the tires to stop the truck. The force results in an acceleration that stops the truck in a safe distance. Adding a camper body to the truck can increase its mass by 50%. The camper is frequently added without changing the brake system. In this situation, why may the truck no longer be able to stop within a safe distance?
B11. A small spaceship has a rest mass of 2000 kg. What mass would you measure for the ship when it moves by you at 50% the speed of light? 60% the speed of light? 90% the speed of light?

C. Extensions to New Situations

C1. Unpleasant forces on the human body occur in accidents in which the body changes its momentum from some value to zero. Determine the forces acting in each situation.
a. People have survived falls in which their momentum changed from 210 kg · m/s to 0 kg · m/s in 0.02 s.
b. A typical fall into a fireman's net gives a momentum change from 150 kg · m/s to 0 kg · m/s in 0.1 s.

130 Chapter 6. Interaction and Force

    c. If a standing person just fell over and hit his or her head on a hard surface, he or she could be killed. The momentum change of the person's head in this situation is from 120 kg · m/s to 0 kg · m/s in 0.004 s.

    d. During an accident, momentum changes of a head protected by a motorcycle helmet are typically from 80 kg · m/s to 0 kg · m/s in 0.02 s.

C2. During World War II the Soviet Union decided to attempt a surprise raid on the German Army by dropping soldiers onto battlefields without parachutes. Each soldier was placed in a large sack filled with straw and then dropped into snowdrifts from heights ranging from 5 to 15 m.

    a. Why would anyone think that such an operation would work?

    b. In terms of the change in momentum, what is the purpose of the straw and snow? (Only about one-half of the soldiers were able to fight after landing, so this type of air-to-ground delivery was stopped.)

C3. Frictional interactions between tires and the road cause vehicles to stop when the brakes are applied. The friction can occur with rolling tires or sliding ones (a skid). To see which stopping method works better, we consider measurements on identical cars of mass 1000 kg. One stops by rolling with its brakes applied; the other skids. Both cars are initially moving at a city speed limit of 15 m/s.

    a. Guess which one will stop more quickly.

    b. The car with rolling tires stops in 1 s. What is its acceleration? What force is applied to it?

    c. The skidding car stops in 1.5 s. What are its acceleration and the applied force?

    d. Which way is a better way to stop in an emergency?

C4. Apollo astronauts needed to learn to move under the lower gravitational force of the moon while still on earth. One method of training was to attach a cable to the astronauts, as shown in Figure 6-C4.

    a. Why could such a cable pulling up be used to simulate situations in lunar gravity?

    b. What force due to gravity acts on an

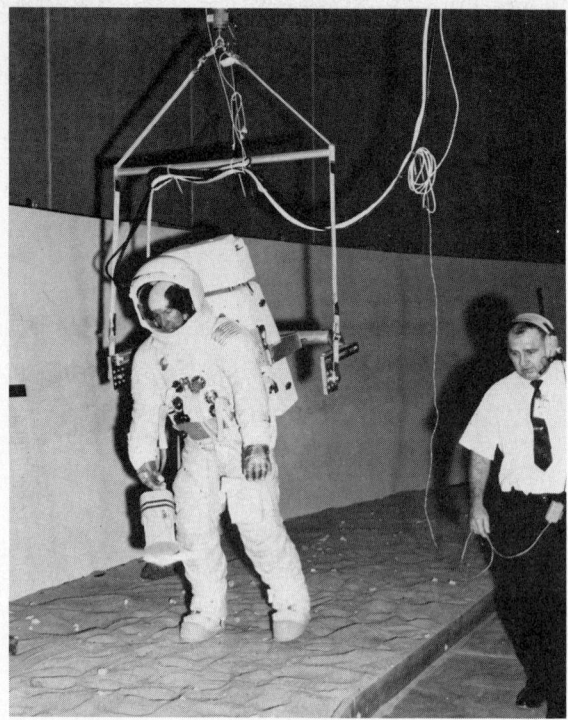

80 kg astronaut on the earth? On the moon?

    c. With what force must the cable pull upward on the 80 kg astronaut to simulate lunar gravity?

C5. One way of teaching people to ski is the Graduated Length Method (GLM). Novices begin on very short skis (usually 1 m in length). As they learn the various techniques, they graduate to longer lengths. Proponents of this method state that on short skis, beginners travel more slowly. Use the ideas of pressure, sinking into snow, and contact interactions to argue that this statement is correct. (While the skis change in length, they do not change in width.)

C6. In homes, paintings are frequently hung from the walls with wires, as shown in Figure 6-C6(a). However, art museums usually attach wires as shown in (b).

    a. For a 2 kg painting, what is the magnitude of the gravitational force acting downward on it?

    b. If the painting is to remain hanging, with what force must the wires pull upward?

c. What is the force applied by each wire in (b)?
d. To have this force acting up in (a), an equal force must act horizontally. With what force must the wire pull in (a) to achieve the same vertical force as in (b)?
e. Why can art museums use wire that is less strong than homes?

C7. When you take a shot in basketball, the ball follows the path illustrated in Figure 6-C7.

a. What contact forces act on the ball after it leaves the player's hand and before it strikes the basket?
b. What forces from interaction at a distance act on the ball?
c. In what direction does the ball accelerate?
d. Use the answers in (a)–(c) to describe why the ball follows the motion shown in the figure.

C8. See if you can figure this out for yourself before we discuss it in the next chapter. The force on a car moving around a curve is toward the center of the curve. That means that the car is being pulled inward. Yet guardrails to prevent cars from leaving the road are placed on the outside. Can you determine why?
a. Using the concept of force, guess why the guardrails are on the outside.
b. What would happen to a car's momentum if no force acted on it as it entered the curve?
c. Sketch a road with a curve on it. Now, remembering that momentum is a vector, show the path followed by a car if no force acts on it when it reaches the curve.
d. If no force acts as the car enters the curve, why will it interact with the guardrail?
e. If your guess is different from your answers to (c) and (d), try to explain the differences. If you are still uncertain, wait until we finish Chapter 7.

C9. A bicycle and rider with a total mass of 80 kg apply a force of 240 N to the road. The bicycle accelerates at 2 (m/s)/s, W.
a. What is the size of the force applied in the forward direction on the bicycle by the road?
b. What is the net force acting on the bicycle?
c. Why is the net force different from the 240 N?
d. What is the force of friction acting on the bicycle?

C10. Use the concept of relativistic mass to discuss:
a. how momentum varies as a particle approaches the speed of light
b. why the equation force = (change in momentum)/time is valid at all speeds

D. Activities

D1. Design a package for a fresh egg which can be dropped three stories without breaking the egg (no parachutes allowed.) Explain your design in terms of momentum change and force. Try it.

D2. Find locations on highways where accidents, and thus injuries, could result from rapid changes in momentum. Use the concepts of force and change in momentum to suggest improvements.

Gjon Mili, Life Magazine, © Time Inc.

Figure 7-1

Newton's Three Laws

While standing in front of a camera with its shutter held open, the late Pablo Picasso experimented with drawing in air. As he drew, Picasso applied a variety of forces to the penlight—pushing it upward, downward, to the left, and to the right. At the same time, gravitational forces alternately opposed or augmented his efforts. Frictional forces continuously opposed any motion. Though it would be difficult, we could use the concept of force to analyze the creation of each and every line. Such an analysis would tell us little about the creation of art, but it would demonstrate the role of forces in the world of motion around us.

Though motion can be complex and forces are numerous, the relationship between motion and force can actually be described quite simply. In Chapter 6 we began to consider this relationship in terms of our everyday experiences. In this chapter we examine it in terms of three laws of motion—simple but profound descriptions developed by Isaac Newton in the late seventeenth century. Newton's *first law* describes what occurs when there is no net force action on an object. His *second law,* which we introduced in Chapter 6, relates accelerated motion to a net applied force. The *third law* of motion describes the symmetrical relationship between two interacting bodies. These three laws, brief as they seem, provide a single theory with which to explain an enormous range of observations.

NEWTON'S FIRST LAW

Intuitively, we relate force to motion. Whether we push on the bike pedals to get up the hill, push on the ground to walk to class, or pull on a stuck drawer to make it open, the force we exert makes things move. As you saw in Chapter 6, however, motion can occur without force. Unless restrained by a seat belt, you keep moving when your car stops suddenly. Forces are acting on the car but not on you—until you hit the windshield. Forces can also act without motion, as you saw. Any architect will tell you that a variety of forces act on a high-rise building, but fortunately it does not move. Forces without motion and motion without force are the subjects of Newton's first law.

When Forces Disappear

In Chapter 6 we compared a man's motions as he slid on ice and on a waxed tile floor. He wore the same shoes, began with the same initial velocity, and slid to a stop on both surfaces. But there was one difference—his momentum changed over a longer time interval on ice than on tile. We explained this in terms of the size of the frictional force exerted by the two surfaces. The tile surface exerts a larger force than the ice. Large force, short time interval; small force, long time interval.

This trade-off between the size of the force and the length of the time interval allows us to imagine what would happen if the frictional force were to disappear. Suppose we could make the frictional force exerted by the ice even smaller—for example, by putting ice skates on the man. We would expect him to slide even longer. If we could continue to reduce the frictional force all the way to zero, our ice skater would be doomed to slide at a constant speed along a straight line forever. His momentum would never change.

Newton's First Law and Inertia

This argument is such a fundamental part of our concept of force that it is considered a major principle, called **Newton's first law:**

> When the net force acting on an object is zero, the object's momentum does not change.

In most interactions an object's mass remains constant, so that any change in momentum arises from a change in velocity. Consequently, Newton's first law tells us that when zero net force acts, the object's velocity must remain constant. If the object is standing still, it continues to stand still. If it is moving initially, it continues to move in a straight line at a constant speed. Neither the magnitude nor the direction of the velocity can change.

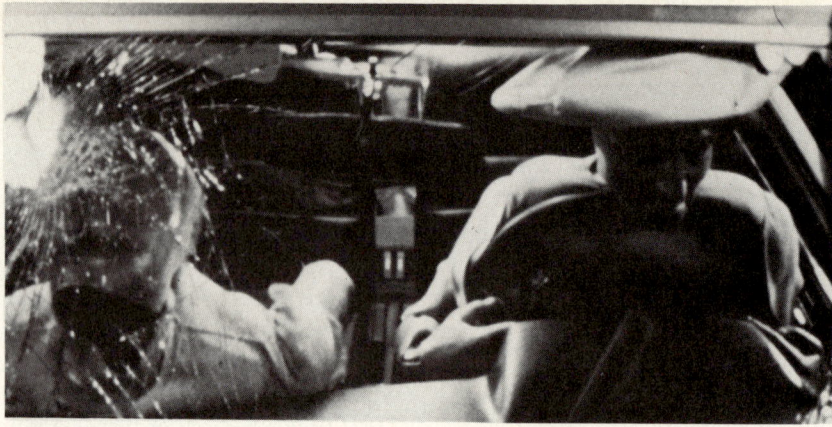

Figure 7-2
In an automobile collision, unrestrained passengers continue to move forward as the car stops. Seat belts safely transfer the force exerted on the car to the passengers.

Newton's first law was originally stated in somewhat different terms by Galileo in the early seventeenth century. Galileo introduced the term **inertia** to describe the tendency of an object to keep moving in a straight line at a constant speed or to remain stationary when zero net force acts on it. The greater the inertia of an object, the greater the force needed to cause a noticeable change in its motion. Consequently, Newton's first law is often referred to as the *law of inertia*.

Both inertia (a characteristic of an object) and Newton's first law (a statement about motion) express what occurs when a net force is absent. For systems in which no forces act, Newton's first law seems almost trivial. If no forces act, no interactions can occur. Of course, an object's momentum does not change. However, for systems in which forces act but their effects sum to zero, Newton's first law changes people's intuitive ideas about the relationship between force and motion.

On earth, all moving objects experience friction. To keep a bicycle moving at a constant velocity, you must apply a force by pushing on the pedals. For centuries, scientists made this rather natural link between force and motion at a constant velocity, believing that a force was always necessary to maintain a state of motion. Most of us develop this same sort of intuitive link between force and motion; it seems to us that we exert a force on the bicycle to keep it moving. In fact, we push on the pedals to overcome yet another force—friction. The two forces—our pushing and friction—sum to zero, enabling the bicycle to continue to move at a constant velocity. On earth, most situations in which the net force is zero arise because two or more forces act on the object simultaneously. It is rare that objects experience no force.

Newton's First Law and Seat Belts

Automobile accidents involve two collisions. The first occurs when the automobile strikes an object, such as a telephone pole. The pole provides the force needed to change the car's momentum, eventually bringing it to rest. A second collision, which occurs shortly after the first, involves the passengers. If they are not in some way attached to the car, the passengers do not experience the force exerted by the telephone pole. The car may stop, but the passengers continue moving forward at a constant velocity. According to Newton's first law, their forward motion will continue until they experience a force. Unfortunately, this force is usually exerted by the dashboard or windshield, and serious injuries result.

Car manufacturers conduct a great deal of research and testing aimed at reducing the severity of the injuries that result from this second collision. Padded steering wheels, padded dashboards, steering columns recessed from the wheel, and head rests are a few of the modifications made as a result of this research. Ultimately, however, seat belts offer one of the best solutions. In simulations conducted at automobile testing grounds, the role of seat belts in reducing injury was explored extensively. Figure 7-2 shows a car that had dummies used as passengers. The dummies did not wear seat belts. The photograph shows the continuing forward motion of the dummies until the second collision occurred. You can contrast this to the motion that occurred when seat belts were worn. When car and passenger are connected, their motion is stopped by the same force at the same time.

SELF-CHECK 7A

Lap belts restrain passengers' lower bodies but not their shoulders. Use Newton's first law to describe the motion of the lower body, head, and shoulders of a passenger who is wearing only a lap belt while involved in an automobile accident.

© 1981 United Feature Syndicate, Inc.

Figure 7-3

AP/Wide World Photos

Equilibrium

Towers that transmit television signals frequently have many cables attached to them. Each cable applies a force to the tower. In addition, wind, gravity, and the ground exert forces on the tower. Despite the action of all these forces, the tower usually does not move. This lack of acceleration is called **equilibrium,** a state in which the net force acting on an object is zero.

In designing structures that remain in equilibrium, architects are generally concerned with gravitational forces. Buildings must be designed so that the upward forces exerted by the skeletal structure balance the downward force due to gravity. When equilibrium is not maintained, disaster can occur. During a severe storm, the roof of Kemper Arena in Kansas City collapsed. A study of the design later revealed that water, which had accumulated on the flat roof during the storm, had caused the collapse. The building design had been adequate to support the weight of the roof but clearly not adequate to support the weight of the roof and the water. Figure 7-3 shows the collapsed roof of the Kemper Arena.

NEWTON'S SECOND LAW

Newton's second law, presented in the last chapter, tells us that a net force results in an acceleration.

> The net force acting on an object is equal to the product of the object's mass and its acceleration.

$$\textbf{net force} = \text{mass} \times \textbf{acceleration}$$

We see a variety of accelerations—cars as they move away from a stop sign, balls as they fall toward the earth, grocery carts as they gradually coast to a stop. These examples involve only a change in speed; the objects continue to move in a straight line. Motion along a curved path, such as traveling around on a merry-go-round, also involves acceleration. Here the speed of the object might remain constant, but its direction changes. In examining Newton's second law, we consider examples of acceleration along both straight-line and curved paths.

Hammer and Feather

In Chapter 2 we described the classic physics experiment in which Apollo Astronaut David Scott dropped a hammer and feather simultaneously on the moon's surface. The hammer and feather reached the lunar surface at the same time—their accelerations were identical. This seems strange to us because the same experiment performed on the earth's surface gives a different result. The hammer always reaches the ground before the feather. We can use Newton's second law to understand the two different results.

If we rearrange Newton's second law, we find that the acceleration that any object experiences is equal to the net force acting on it divided by its mass:

$$\textbf{acceleration} = \frac{\textbf{net force}}{\text{mass}}$$

The net force acting on a falling object is the force due to gravity, or the weight of the object. We have defined weight as the product of the object's mass and the acceleration due to gravity at its location. On the moon, the acceleration due to gravity is 1.6 (meters per second) per second [(m/s)/s], down. A 1 kilogram (kg) hammer will weigh (1 kg)[1.6 (m/s)/s, down], or 1.6 Newtons (N), down. A 0.001 kg feather will weigh (0.001 kg)[1.6 (m/s)/s, down], or 0.0016 N, down. The force due to lunar gravity is the only force on each object, so both accelerate at 1.6 (m/s)/s, down. Both hit the ground at the same time.

On the surface of the earth, two forces act on the hammer and feather. As on the moon, a force due to gravity acts downward, pulling objects toward the surface of the earth. But the earth's atmosphere adds a frictional force, called *air resistance,* that acts upward. Consequently the net force acting on each object is the vector addition of the force due to gravity and the contact force due to air resistance. This second force, the air resistance, leads to the difference between the results on the earth and those on the moon.

Unlike the force due to gravity, the force exerted by the surrounding air varies with the speed at which the object falls. As an object begins to fall, the force exerted by the surrounding air begins to increase. It continues to increase until the object no longer accelerates—that is, until the force exerted by the surrounding air equals the force due to gravity. The net force is then zero and the object stops accelerating. From then on the object continues to fall at a constant velocity—the velocity it reached the instant before the net force became zero.

Table 7-1 lists the forces acting on the hammer and feather, as well as the net force and acceleration at several times during the fraction of a second after each is released. The force due to air resistance is quite small compared to the force due to gravity acting on the hammer. This means that the hammer's acceleration decreases very little and the hammer accelerates rapidly to the ground. By contrast, the force due to air resistance is initially one-third the force due to gravity acting on the feather, and within a few hundredths of a second, the two forces on the feather are equal. The net force acting on the feather is zero, and it floats gently to the ground. The velocity at which the feather falls is the velocity it reached within that first few hundredths of a second after it was released.

Motion in a Circle: or How David Slew Goliath

We applied Newton's second law to motion along a straight line. Now we look at motion along a curved path. The simplest curved motion is that of a small rock as it is being twirled at a constant speed on the end of a string. As shown

Table 7-1 Hammer and Feather

	Hammer			
Time (s)	Force due to Gravity (down) (N)	Force due to Air Resistance (up) (N)	Net Force (down) (N)	Acceleration (down) (m/s)/s
0.00	9.8	0.000	9.8000	9.8000
0.01	9.8	0.0003	9.7997	9.7997
0.03	9.8	0.003	9.7970	9.7970
0.05	9.8	0.009	9.7910	9.7910

	Feather			
Time (s)	Force due to Gravity (down) (N)	Force due to Air Resistance (up) (N)	Net Force (down) (N)	Acceleration (down) (m/s)/s
0.00	0.0098	0.000	0.0098	9.8
0.01	0.0098	0.0003	0.0095	9.5
0.03	0.0098	0.003	0.0068	6.8
0.05	0.0098	0.009	0.0008	0.8

Figure 7-4
A strobe drawing of the motion of a rock shows it moving at a constant speed. Its direction, however, changes constantly. The rock accelerates.

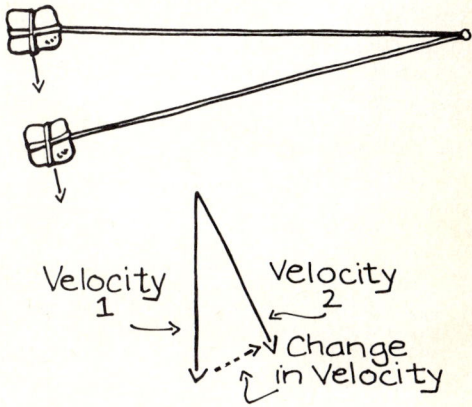

Figure 7-5
The change in velocity lies toward the center of the circular motion. Both the acceleration and the net force are directed toward the center.

in the strobelike drawing in Figure 7-4, the images of the rock are equally spaced. The rock's speed is constant throughout its motion around the circle. The direction in which the rock moves, however, changes constantly. At A, the rock moves to the left; at C, it moves downward; and so forth. This change in direction means that the rock is accelerating—that a net force must be acting on it.

Newton's second law states that the net force acts in the direction in which the object accelerates. When the object moves in a straight line, it is relatively simple to see the direction in which the object accelerates—it either slows down or speeds up. When the object moves in a circle, however, the situation is more difficult. We can determine the direction in which the rock accelerates by looking at the change in its velocity over a very short time interval. Figure 7-5 shows the rock initially moving downward. A short time later it is moving down and slightly to the right. Subtracting these two vectors shows that the change in velocity is a vector that points toward the center of the circle. This acceleration is called a **centripetal acceleration.** (The word *centripetal* is derived from a Greek word meaning center-seeking.)

The net force that leads to a centripetal acceleration is called, naturally enough, a **centripetal force.** A centripetal force is a force that acts toward the center of the circle, causing the object to move about in a circle at a constant speed. When you throw a rock, it moves in a straight line in the direction it was traveling at the moment you released it, in accordance with Newton's first law. But if you then apply a force by pulling on an attached string, the rock will continuously be deflected from its straight-line path, in accordance with Newton's second law. The rock's inertia—its tendency to keep moving in

Figure 7-6
When David released his stone, it moved off along a straight line in accordance with Newton's first law. By choosing the right release point, David slew Goliath.

a straight line—keeps it from being pulled back to you, but the centripetal force prevents it from moving farther away from you. The result is that the rock moves in a circle. If you have ever twirled something on a string, your first impression might be that the force you exert is not centripetal. In order to keep the object moving, you usually have to flip your hand. This flip is actually doing more than providing a centripetal force. It also provides the force necessary to balance gravity and air resistance as the object moves about in a circle. To feel just the centripetal force, let the string attached to the object wrap around your finger. As it does so, your finger will constantly provide a centripetal force to maintain the circular motion.

If a centripetal force suddenly disappears, as when the string on your rock breaks, Newton's first law continues to operate and the object continues to move in a straight line. When David slew Goliath, he used a slingshot that consisted of a small pouch on the end of a string. After placing a rock in the pouch, David applied a centripetal force to the string. In accordance with Newton's second law, the rock began moving in a circle at a constant speed. When David released the string that held the pouch shut, the rock no longer had any centripetal force acting on it. Newton's first law took over, and the rock moved off in a straight line in the direction it had at the time of release (Figure 7-6). By selecting the release point carefully, David was able to direct the rock so that it hit Goliath's head.

A modern example occurs with automobiles. In order to travel around a curve, a car must have a centripetal force acting on it. This force is usually supplied by the frictional interactions between the tires and the road. But if the road is icy or the tires are bald, the frictional force is decreased and cars sometimes cannot make the curve. In the absence of a centripetal force, the car continues to move in a straight line—right off the road!

ACCELERATED REFERENCE FRAMES

Sometimes we are in reference frames that are accelerating while we are not. As a car goes quickly around a tight curve, passengers not wearing seat belts often slide across the seat. Or, when a car stops suddenly, passengers not

Accelerated Reference Frames

Robert J. Bennett, Bridgeville, Del.

Figure 7-7

SELF-CHECK 7B

The gymnast, Ferris wheel, and car in Figure 7-7 each move at a constant speed. What is the direction of the net force and acceleration in each case? What object provides the centripetal force?

---

wearing seat belts continue to move forward. The car accelerates, but the passengers do not. Because the passengers use the car as a reference frame, they describe their motion relative to this accelerating reference frame.

Straight-Line Accelerated Reference Frames

If the car in which you are riding suddenly stops, you continue moving forward until your seat belt stops you. You might describe yourself as being "thrown forward." If the car suddenly accelerates forward, you say you were "thrown backward." The word *thrown* implies a force acting on you, and in these situations you often use words implying a force to describe what happened to you. Because no force really acts on you, such a description refers to what are called *fictitious forces*.

Fictitious forces are not real forces, but they arise from the acceleration of the reference frame. The real force in the case just described acts on the car; it is due to friction between the tires and road or to collisions with another object. The car accelerates; the passengers do not. However, the passengers treat the car as if it were a stationary reference frame and describe their motions in terms of a fictitious force. The fictitious force always acts in a

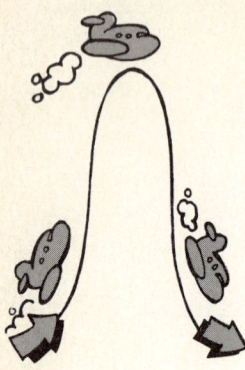

Figure 7-8
Using the path shown, NASA's reduced-gravity airplane can provide future astronauts with about 15 seconds of apparent weightlessness.

direction opposite to the direction in which the real force acts on the reference frame. When the car accelerates forward, the passengers describe a "force" acting to throw them backward. When the car stops suddenly, the passengers report that a "force" has thrown them forward.

Fictitious forces also arise when everything in a reference frame accelerates together, producing events that seem abnormal. Suppose, for example, that someone puts you in a closed box, takes the box up in an airplane, and drops it. If you drop an object while in the box, the object will accelerate downward at the same rate as you and the box. It will seem to float in front of you. Normally objects do not float in front of us, so we say that a force pushes upward on the object. Such a force arises from the motion of the reference frame and not from any real interaction with other objects.

Being placed inside a box and dropped toward the earth may seem absurd, yet astronauts are prepared for space travel in a similar way. A large airplane is flown along the path shown in Figure 7-8. The airplane moves up and down rapidly, creating segments of the trip in which the airplane and astronauts are "falling" downward together. This creates an apparent weightlessness and approximates the experience which the astronauts will have in space. During this time, astronauts practice the tasks they are assigned to complete in space. (They also feel side effects. NASA calls the airplane the "reduced gravity environment"; the astronauts refer to it as the "vomit comet.")

Rotating Reference Frames

As a car goes around a curve, passengers often slide across the seat. Barrel rides at amusement parks provide much the same sensation. As the barrel rotates faster and faster, people say they are pressed against the side of the barrel. Such forces are often called *centrifugal* (center-fleeing) forces.

Centrifugal forces are fictitious forces that arise from the rotation of the reference frame. Like the fictitious forces that act in straight-line accelerated reference frames, centrifugal forces appear to act in a direction opposite to that of the real force that is causing the reference frame to accelerate. As a car goes around a curve, friction between the tires and the road provides the centripetal force needed to cause the change in direction. Passengers, however, report a centrifugal force pushing them outward, away from the center of the circle. They reach this conclusion because the reference frame accelerates while they do not. Using the accelerating car as the reference frame, the passengers conclude that a force has been applied to them (Figure 7-9).

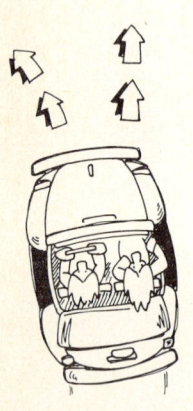

Figure 7-9
As the car turns the corner, an unrestrained passenger continues to move along a straight line in accordance with Newton's first law. Using the car as a reference frame, the passenger reports that a centrifugal force pushes her outward.

SELF-CHECK 7C

Suppose you are holding a drink while riding in a car with a seat belt on. The car suddenly turns to your left. What happens to the liquid in the glass? Explain this result in terms of fictitious forces.

HELPING "MOM"

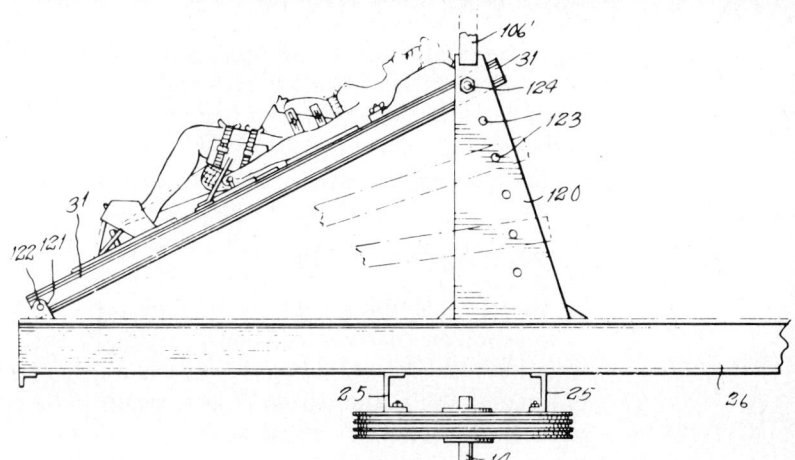

Giving help in the delivery of a baby was of concern to George B. and Charlotte E. Blonsky when they invented the Spinnet, a device that takes advantage of fictitious forces in accelerating reference frames. The woman who is about to deliver is strapped to a platform (31), which is rotated by a shaft (14) connected to a motor. During rotation, the mother and baby are in an accelerated reference frame in which a centrifugal force acts outward—in the same direction the soon-to-be mother needs to push. The two forces add, so the woman can exert less force than she would if the platform were not rotating. Seen from the nonrotating (hospital) reference frame, Mom's help comes from Newton's first law. While rotating, the baby will tend to move in a straight line rather than turn with the acceleration of its mother. A straight line brings the baby right out into the world.

Invented in 1965, the Spinnet has not been widely accepted by modern hospitals. However, the Blonskys did anticipate another practice that is being used in the Western world today—and has been used for centuries throughout the world. When they tilted the expectant mother downward, they let gravity lend a hand!

NEWTON'S THIRD LAW

Newton's first and second laws completely define force and describe the motion that results when a net force acts on a single object. Forces arise from interactions, however, and our discussion would not be complete without a look at the entire interaction. Newton's third law describes the forces on all objects involved in an interaction.

Action and Reaction

A skydiver is plummeting toward earth. At a certain altitude she opens a parachute to increase air resistance and reduce her acceleration. When she hits the ground, the ground exerts an upward force on her, abruptly reducing her downward momentum to zero. This force can be quite large—large enough to break bones if the skydiver is falling too quickly. But what about the ground? Just as the earth exerts a force on the skydiver, the skydiver exerts a force on the earth. The earth is so massive, we do not see its momentum change; but we see a small dent or a bit of mashed grass. Both objects involved in the interaction experience a force.

As a more subtle example, suppose you push on a wall while standing on a skateboard. You exert a reasonably large force on the wall, but the wall does not budge perceptibly. Instead you accelerate backward. We can only explain your motion by saying that the wall exerts a force on you. Both objects—you and the wall—exert forces during the interaction.

Forces Appear in Pairs

These two examples illustrate an important, yet subtle, characteristic of force. The skydiver exerts a force on the ground and the ground exerts a force on the skydiver. The skateboarder exerts a force on the wall and the wall exerts a force on the skateboarder. Forces occur in pairs. Newton stated this characteristic of force in his **third law:**

> Every applied force results in a reaction force on the applier.
> This reaction force is equal in magnitude but opposite in direction to the applied force.

The skydiver pushes downward on the ground; the ground pushes upward on her. The skateboarder pushes to the right on the wall; the wall pushes to the left on him. Both objects involved in the interaction exert forces.

Newton's third law seems at once both simple and perplexing. The cliché "for every action there is an opposite and equal reaction" is a common expression of this principle. Yet many people will, after a little thought, ask the natural question: If for every applied force a reaction force of equal magnitude and opposite direction occurs, how can motion ever happen? Don't the two forces just cancel?

The answer lies in the relationship between the concept of force and the concept of interaction. When the skateboarder pushes against the wall, he interacts with it. The very nature of the interaction requires that the wall do something to the skateboarder. Newton's third law reminds us that this single interaction results in two forces. The two forces, however, act on different objects. The wall pushes the skateboarder, and the skateboarder accelerates. The skateboarder pushes on the wall, and the wall accelerates. Generally, the wall is so much more massive than the skateboarder that we do not see its ac-

Figure 7-10
The two forces are more obvious when both objects are of about the same mass. The boy pushes on the box; the box pushes on the boy. Both move apart.

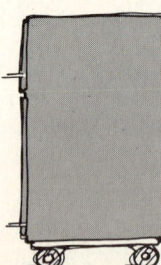

celeration. If we replaced it with a box on another skateboard (Figure 7-10), the box would accelerate visibly.

Essentially, Newton's third law reminds us that momentum is conserved in a closed system. If we ignore the friction between the skates and the ground, the skateboarder and the box, along with their skateboards, form a closed system. Before the interaction, the skateboarder and the box are stationary. The total momentum of the closed system is zero. After the skateboarder pushes on the box, both are moving. Conservation of momentum requires that the total momentum of the system remain zero. As the skateboarder moves to the left, the box must move to the right. Newton's third law predicts the same result. As the skateboarder pushes on the box, the box pushes back.

Either Action Can Be the Reaction

The wall applied a force to the skateboarder when the skateboarder applied a force to it. This sounds strange, because we think of pushing as a voluntary act. The skateboarder decides to push the wall in order to accelerate backward. You might well ask how the wall "knows" when to push back. The answer lies in what happens to an object during an interaction. If you step on a bed, you see and feel a change in it. As you sink down, the springs compress until they create a force that balances your weight. It is easy to see that a very flexible object like a spring pushes back on you as you push on it. On a soft, plush carpet you feel a similar, but smaller, compression as you walk. On a wooden floor you barely feel the compression, but compression forces are still present, pushing back on you. Whenever an object is compressed or stretched, it pushes or pulls on whatever is compressing or stretching it. The wall pushes back on the skateboarder because it is being compressed slightly.

When two objects interact, the forces they exert on each other occur at the same time. Newton's third law does not specify which of the two is the action force and which is the reaction force—as far as Newton's third law is concerned, there is no difference. We ordinarily describe forces exerted by people as action forces because this fits our image that forces arise from voluntary acts. However, when calculating the effects forces have, we could just as well say that the wall pushing on the skateboarder is the action and the skateboarder pushing on the wall is the reaction. Action-reaction pairs are symmetrical.

Walking Away

Newton's third law explains the process by which we walk. Every time you take a step, your foot exerts a force on the floor. When you push on the floor to the left, the floor distorts slightly and pushes you to the right. If this is the only force acting on you, you accelerate to the right. The floor is quite massive, so its acceleration to the left is too small to be noticed. However, if you have ever walked on a treadmill or stepped on a loose roller skate, you know how difficult motion is when the floor does accelerate!

The force that the floor exerts on your foot arises from frictional interactions. We can push on the floor and the floor can push back because friction causes our foot to "stick" to the floor and, thus, push back on it. When you walk on waxed surfaces or ice, the importance of friction becomes more apparent. If friction is small, our foot just keeps moving backward as we try to step forward rather than compressing the surface.

SELF-CHECK 7D

In Chapter 6 we used conservation of momentum to describe what occurs when a person tries to step from a boat. Use Newton's third law to describe the same event.

ALL THREE LAWS

While we have discussed each of Newton's laws separately, realistically we must combine them when we analyze most interactions. As an example, consider a common event—throwing a ball up and catching it.

The first step is to identify all the forces acting on the ball. There are three: (1) contact interaction between the ball and your hand, (2) contact interaction between the ball and the air, and (3) interaction at a distance between the ball and the earth. While all three forces act at some point during the ball's motion, at no point do all three forces act simultaneously. The force due to gravity acts on the ball continuously. The contact force you exert with your hand acts only while the ball is in direct contact with your hand—while you are holding it, throwing it up, or catching it. The third force, air resistance, acts only while the ball is actually moving through the air. Figure 7-11 and Table 7-2 summarize the various stages of the ball's motion and which forces act during each stage.

Newton's first law tells us that when the net force acting on an object is zero, the object does not accelerate. This occurs at two points—right before you throw the ball up and right after you catch it. As you hold the ball in your hand, your hand pushes up with just enough force to balance the downward force due to gravity. The net force is zero and the ball remains at rest.

Newton's second law tells us that when a net force acts on an object, the object accelerates. The ball accelerates as it is thrown upward, as it moves through the air, and as it is caught. Each of these stages involves a different set of forces, so we examine each stage separately.

While holding the ball (I), your hand exerts an upward reaction force which balances the downward force on the ball due to gravity. To throw the ball up (II), you use your muscles to increase the force you exert with your hand, so that the upward force is greater than the downward force. The ball accelerates upward because a net force acts upward. When you catch the ball, the same forces are involved. In order to stop its downward motion, a net force again must be applied upward. When the downward-moving ball touches your hand (VI), it pushes your hand down, a small distortion occurs, and your hand exerts a reaction force upward on the ball. This upward force exceeds the downward force due to gravity. Since a net force that opposes the motion of an object will cause it to decelerate, the ball slows to a stop. Then, the force supplied by your hand once again balances the force due to gravity (VII).

Once the ball leaves your hand, the only forces acting on it are the force due to gravity and the force of air resistance. As the ball moves upward (III), both forces act downward. The net force opposes the motion of the ball and it gradually slows down. In fact, it stops momentarily at the top of its path (IV). Once the ball starts moving downward, the force of air resistance acts upward, while the force due to gravity acts downward. The downward force due to gravity is greater than the upward force of air resistance. The net force is still downward and the ball accelerates downward (V).

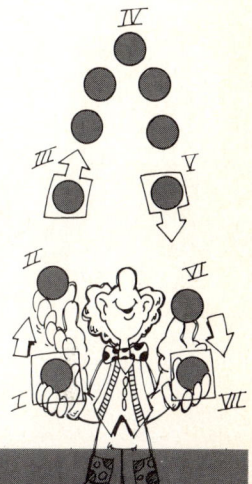

Figure 7-11
We can look at the forces acting during several stages in the ball's motion.

SELF-CHECK 7E

We can add some numbers to our discussion of the ball's motion. Let the mass of the ball be 2 kg; the force due to gravity be 19.6 N; and the force of air resistance be 0.5 N. What is the ball's acceleration as it moves upward? Downward? Why are these two values different?

Table 7-2 Summary of Ball's Motion

Stage	Interaction Force(s) up	Force(s) down	Net Force	Acceleration	Description of Motion
I Resting in hand	Contact (hand)	Gravity	None	None	Ball has no motion.
II During throw	Contact (hand)	Gravity	Up	Up	Ball accelerates upward as it leaves your hand.
III Moving up	None	Gravity Air resistance	Down	Down	Ball's upward velocity decreases.
IV At top of flight	None	Gravity	Down	Down	Ball's velocity is momentarily zero as it stops moving upward and begins to move downward.
V Moving down	Air resistance	Gravity	Down	Down	Ball's downward velocity increases.
VI During catch	Contact (hand)	Gravity	Up	Up	Ball's velocity decreases to zero.
VII Resting in hand	Contact (hand)	Gravity	None	None	Ball has no motion.

So far we have looked at a variety of interactions that could be described in terms of force. All the interacting objects exerted forces. But we have avoided an important question—just what causes these forces? Cement floors, rugs, tile floors, grass—all exert frictional forces on moving objects, but just what is friction? Are there as many forces as there are objects to exert them or can we categorize these forces into just a few major kinds? The attempt to simplify the classification of observed forces is centuries old. It is also the subject of Chapter 8.

CHAPTER SUMMARY

Newton's three laws describe the motion of any object upon which a force acts. *Newton's first law* states that when the net force acting on an object is zero, the object's momentum does not change. If the object is stationary, it remains stationary. If the object is moving, it continues to move in the same direction at a constant speed. This behavior is sometimes called *inertia,* which is a property of an object. Objects which have more than one force acting on them but a net force of zero are said to be in *equilibrium.*

Newton's second law states that the net force acting on an object is equal to the product of the object's mass and its acceleration. This means that when the net force is not zero, the object must accelerate. A net force can result in a change in speed as the object moves along a straight-line path. A net force can also produce motion along a curve—that is, a change in the direction of motion. A *centripetal force* is a net force that acts toward the center of a circle, causing an object to move at a constant speed along a circular path.

Because we experience motion in terms of a reference frame, the acceleration of a reference frame can make us feel that forces are exerted when in reality no force exists. These forces, called *fictitious forces,* arise from the acceleration of the reference frame without acceleration of the object in the reference frame. When a car stops suddenly, we say that we are "thrown" forward. The real force acts on the reference frame, the car. The "force" on us is fictitious—we continue moving forward because of inertia, or Newton's first law. A fictitious force is always perceived as acting in the opposite direction to the real force. *Centrifugal force* describes the fictitious forces we feel when a centripetal force actually acts on the reference frame.

Newton's third law states that forces always occur in pairs. When you push on the wall, the wall pushes back on you. An applied force is always accompanied by a reaction force equal in magnitude but in the opposite direction. The applied force and the reaction force always act on different objects.

ANSWERS TO SELF-CHECKS

7A. The seat belt enables the force that stopped the car to be transmitted to your lap. Your head and shoulders, however, have no net force acting on them. According to Newton's first law, they continue to move forward until a net force acts on them, typically when your head strikes the dashboard.

7B. In all three cases, the net force and the acceleration are toward the center of their circular motion. The force exerted on the gymnast is applied by the bar. The force exerted on the Ferris wheel is supplied by the axle. The force exerted on the car is supplied by the friction between the tires and the road.

7C. When the car suddenly turns left, the drink moves to the right relative to the car. In a reference frame attached to the road, the drink continues to move forward while the car turns to the left. Observers in the car will report that a centrifugal force pushes the drink away from the center of the turn.

7D. In stepping from the boat, the person applies a force that pushes the boat backward. According to Newton's third law, the boat exerts a force of equal magnitude, but in the opposite direction, on the person. The person moves forward while the boat moves backward.

7E. Moving upward: **Net force** = 19.6 N, down, + 0.5 N, down, **Net force** = 20.1 N, down = (m)(**a**); **a** = (20.1 N, down)/(2 kg) = 10.05 (m/s)/s, down. Moving downward: **Net force** = 19.6 N, down, + 0.5 N, up, **Net force** = 19.1 N, down = (m)(**a**); **a** = (19.1 N, down)/(2 kg) = 9.55 (m/s)/s, down. The acceleration is different because the force due to air resistance acts with the force due to gravity as the ball moves upward and against the force due to gravity as the ball moves downward.

PROBLEMS AND QUESTIONS

A. Review of Chapter Material

A1. Define the following terms:
Inertia
Equilibrium
Centripetal acceleration
Centripetal force
Centrifugal force
Fictitious force

A2. State each of Newton's three laws.

A3. How are inertia and Newton's first law related?

A4. What is the acceleration of an object that is in equilibrium?

A5. Must an object's velocity be zero if no forces are applied to it? Why or why not?

A6. Why must an object moving in a circle be accelerating even if its speed is not changing?

A7. Why do people say that they are thrown to the outside of a car traveling around a curve?

A8. What must be happening to a reference frame if a fictitious force is present?

A9. If Newton's third law is correct, why is every force not canceled by its reaction force?

A10. How is Newton's third law related to momentum conservation?

B. Using the Chapter Material

B1. A loose item on the shelf behind the back window of a car can become a deadly missile in a collision. Explain why in terms of one or more of Newton's laws.

B2. As you are turning a corner on a bicycle, a strap holding a package on the back breaks. Draw and explain the path of the package after it leaves the bicycle.

B3. A spaceship is motionless relative to a nearby star. The engines on the rear of the spaceship are fired for 3 minutes and then turned off. Describe in words the forces on the spaceship, the acceleration of the spaceship, and the changes in velocity that the spaceship experiences while the rockets are being fired and immediately after they are turned off.

B4. If the spaceship in Problem B3 has a mass of 2000 kg and its engines apply a force of 8000 N to the rocket, what is the acceleration of the rocket during the first 3 minutes? What is its acceleration after the first 3 minutes?

B5. To study the effects of weightlessness on equipment to be used in space, NASA drops it down a six-story shaft near Cleveland. While the equipment is falling, experiments are performed. Why does this procedure allow NASA to evaluate how the equipment will perform in space?

B6. One of the authors of this book was eating a meal on an airplane when the airplane suddenly dropped about 500 m. When describing the event later, the author said, "The plate flew up in the air." Describe the reference frame he used and its motion.

B7. When fuel is expelled from a rocket, interactions inside the rocket push the exhaust out the back of the rocket. Use Newton's laws to explain why the rocket moves forward.

B8. When you are standing on the floor, you are being pulled downward by gravity. What is the size of the force pulling down? However, you are not accelerating. What is the size of the force pushing up? What object applies this force?

B9. Momentum conservation tells us that a direct collision between a stationary and moving billiard ball will result in momentum exchange. The stationary ball begins to move and the moving billiard ball stops. Apply Newton's laws (particularly the third law) to this interaction to explain the observation.

C. Extensions to New Situations

C1. Whiplash is a serious neck and back injury which occurs in rear-end collisions. This injury usually happens to people who are in a stationary car that is hit from behind by another car.
   a. When a car is hit from behind, what are the directions of the net force and acceleration on it?
   b. Suppose a person is sitting in seat I (Figure 7-C1) during the collision. How does the force on the head differ from the force on the rest of the body?

   c. How will the motion of the head differ from the motion of the rest of the body? (This difference causes the injury.)
   d. Use Newton's laws to describe how whiplash injuries are decreased for passengers in seat II.

C2. During the spin-dry cycle of a washing machine, water leaves the clothes as they spin at a very high speed.
   a. What is the direction of the force acting on the spinning clothes? What object applies this force?
   b. If the clothes were not spinning, in which direction would they move?
   c. The washing machine tub contains holes large enough to let water out but too small to allow the clothes out. Use Newton's laws to explain why the water separates from the clothes.

C3. The best way to use Newton's laws in analyzing two-dimensional motion is to treat each dimension independently. As an example, consider a bale of hay dropped from an airplane to cattle stranded by bad weather. The airplane travels horizontally at a constant velocity of 20 m/s, east. The bale has a mass of 20 kg.
   a. While the bale is attached to the plane, what are the magnitudes of the net horizontal and vertical forces acting on it?
   b. What are the horizontal and vertical

speeds of the bale just before it is released?
c. After the bale has been released, does a horizontal force act on it? A vertical force?
d. If we neglect air resistance, what will be the bale's horizontal speed throughout its flight? Explain why.
e. In what direction will the bale accelerate?
f. Use your answers to sketch the path of the bale as it moves from the airplane to the ground.

C4. On Saturday morning cartoons, characters frequently run off cliffs. Usually the animator shows a character running horizontally until it notices that it has no support. Then, the character falls straight down, as shown in Figure 7-C4. Use Newton's laws to explain why this motion is impossible.

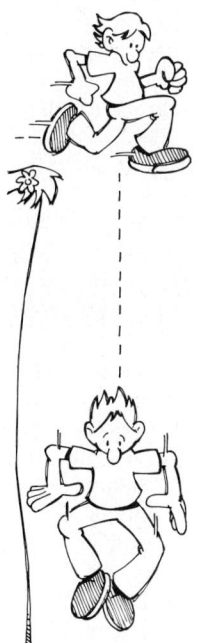

C5. The questions below ask you to apply Newton's laws to see if you could "pull yourself up by your own bootstraps." (Ignore the downward pull of gravity because it is balanced by the reaction force of the floor.)
a. Suppose that you grab the shoes you are wearing and pull upward with a force of 50 N. What is the force up?
b. What is the reaction force down?
c. What is the net force on your shoes?
d. Why do you not move?

C6. Figure 7-C6 shows the motion of two balls. One is dropped straight down; the other is given a push to the right at the start of its motion. Once the balls leave the table, gravity is the only force acting on them.
a. After the balls leave the table, what is the net force on each ball in the horizontal direction?
b. Why does the ball on the right keep moving to the right?
c. Why do the two balls accelerate downward at the same rate even though the ball on the right was given a little push to the right?

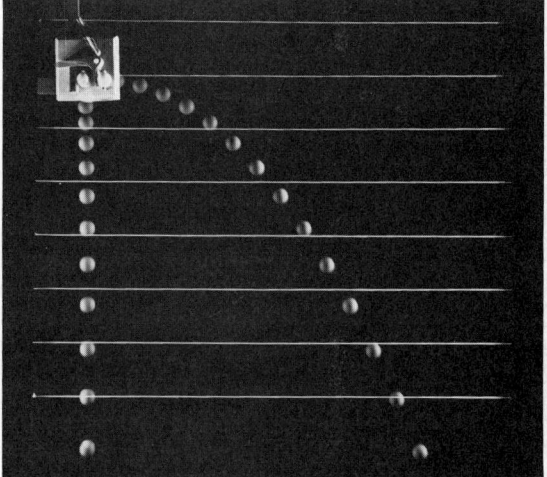

C7. Instead of being a perfect sphere, the earth bulges near the equator. That is, the distance from the earth's center to the poles is slightly less than the distance from the earth's center to the equator. Use Newton's laws to explain why.

C8. Rotating reference frames are considered the most likely way to simulate gravity in future space stations. In one design, the space station would be a big, spinning donut.
a. Use Newton's laws to describe the motion of objects inside the space station.
b. Which direction relative to the center of the space station would be up? Explain your answer.

152 Chapter 7. Newton's Three Laws

C9. One of life's small frustrations is that an empty beverage glass is likely to remain intact when dropped. However, a full one almost always breaks. To explain why, consider two identical glasses, each with a mass of 1 kg. One glass is empty; the other one contains 1 kg of water. Suppose both glasses are dropped so that they are in the air for 0.5 s.
   a. Use the equation speed = magnitude of acceleration × time to determine the speed of each glass just as it reaches the floor. What is the momentum of each glass just as it reaches the floor?
   b. What object applies the force to stop the glasses?
   c. If the glasses are stopped in 0.01 s, what force is applied to each? (Give both the magnitude and direction of the net force.)
   d. Why is the full glass more likely to break?

C10. To see the connection between momentum conservation and Newton's third law, consider once again the collision between moving and stationary billiard balls. For the purposes of this question, assume that each ball has a mass of 0.5 kg and that they are in contact for 0.01 s. Before the collision, ball A has a velocity of 1 m/s, east; ball B has a velocity of 0 m/s.
   a. Describe the horizontal forces on each ball before, during, and after the collision.
   b. What is the momentum of ball A before and after the collision?
   c. What is the momentum of ball B before and after the collision?
   d. What is the magnitude and direction of the force applied to ball B during the interaction? What object applies this force?
   e. What is the magnitude and direction of the force applied to ball A during the interaction? What object applies this force?
   f. In terms of Newton's third law, why must the decrease in momentum of ball A always equal the increase in momentum of ball B?

C11. When you jump off a chair and land on the earth, you could apply a force of 1000 N to the earth. In return the earth pushes back with a force of 1000 N. To see why you do not notice the earth moving backward, answer the questions below.
   a. If your mass is 60 kg, what acceleration do you experience when the earth pushes on you with a force of 1000 N?
   b. The mass of the earth is 6×10^{24} kg. What is its acceleration when you push on it with a force of 1000 N?
   c. Suppose all 5×10^9 people on earth jumped right now. What would be the earth's acceleration?
   d. If, somehow, we moved all these people to one side of the earth and they all jumped together, what would the earth's acceleration be?

D. Activities

D1. How must realistic stage or movie fight scenes include Newton's laws? Develop a set of stage directions for a fight scene that includes proper accelerations for each force.

D2. Stand on a bathroom scale in an elevator. Describe and explain the changes in your weight as the elevator starts and stops.

D3. The force due to gravity has an effect on the evolution of life. For example, a hopping animal such as a kangaroo would require smaller leg muscles on the moon than on the earth. Design animals or plants that could live in the gravitational environment of Jupiter and those that could live on the moon. Explain your designs in terms of Newton's laws.

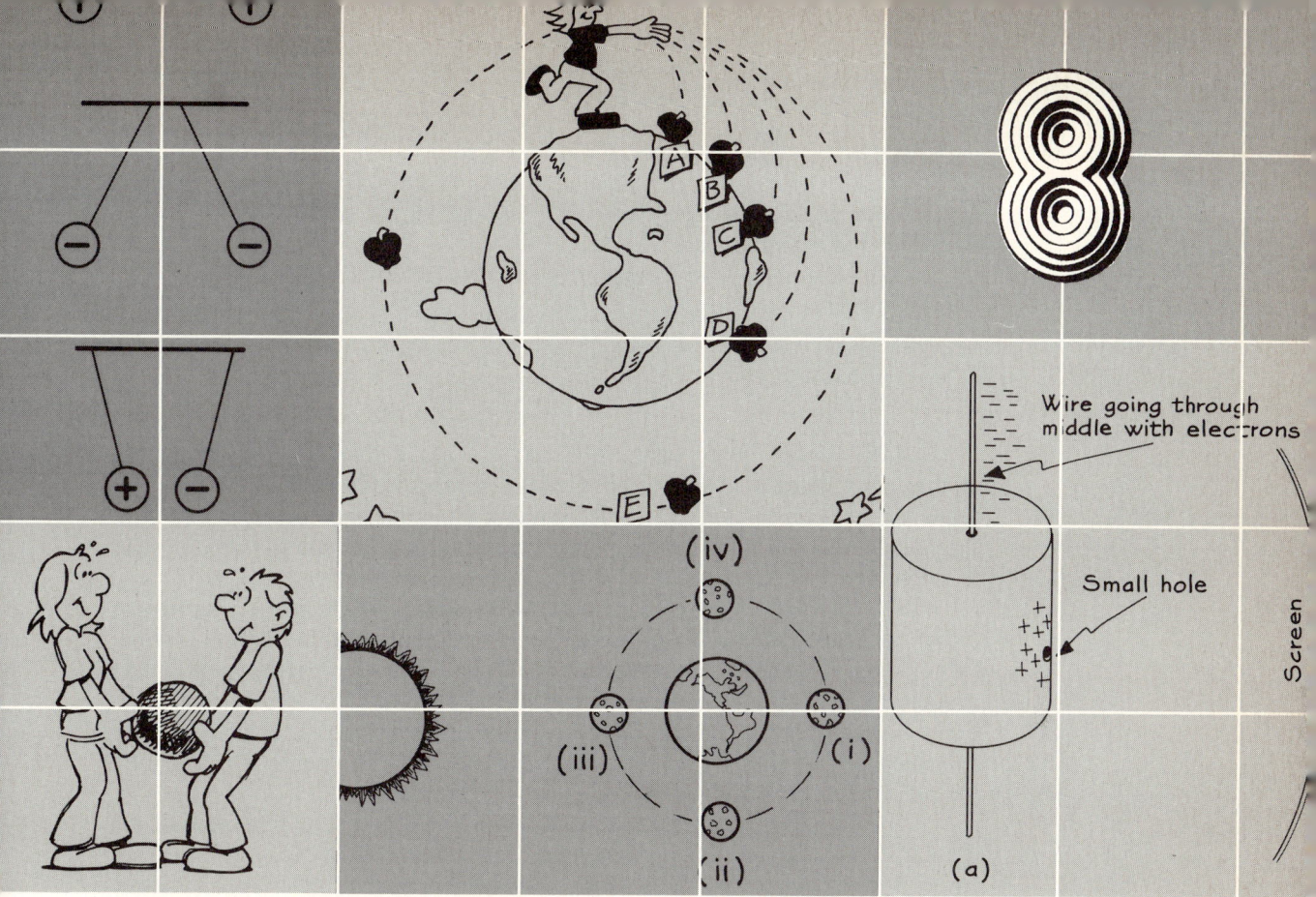

The Fundamental Interactions

To the North American Indians, the number four had magical properties, and they often classified things into groups of four. To Europeans, the magic number was three. The number seven, too, has been very significant in human legend and ritual. Psychologists suggest that seven is about the largest number of single bits of information we can process simultaneously. When we need to process more information, we lump seven bits into a larger bit and then process seven of those larger bits, and so forth.

The need to simplify seems almost instinctive. Faced with millions of life forms, biologists organize everything into five kingdoms: animal, plant, and three other kingdoms encompassing microscopic organisms. Faced with countless rocks, geologists organize the diversity they see into three groups: sedimentary, igneous, and metamorphic. Faced with the diversity of chemical elements, chemists organize the more than 100 elements into families and periods.

The drive to simplify has motivated physicists to look for just a few fundamental interactions that could embrace the enormous range of forces we experience daily. Physicists now identify just four mechanisms, called the *four fundamental interactions,* to explain all forces. These mechanisms are the *gravitational interaction,* the *electromagnetic interaction,* the *strong nuclear interaction,* and the *weak nuclear interaction.* In this chapter we examine each of these interactions, as well as current attempts to explain them in terms of a *unified theory of interaction.*

GRAVITATIONAL INTERACTIONS

In 1665, the Great Plague killed 10% of the population of London. By the fall of that year, fears that the disease would spread further caused officials to close the University of Cambridge, sending students and faculty home until the plague had run its course. One young student of mathematics, Isaac Newton, returned to his parents' home in rural Lincolnshire.

Legend relates that Newton's great inspiration came as he was resting under an apple tree on his parents' farm, contemplating the moon's motion. Why, he asked himself, does the moon revolve around the earth? How can its motion be explained? As Newton pondered, an apple fell beside him. Suddenly he realized the answer. Gravity! The moon revolves about the earth for the same reason that the apple falls to the ground. This realization, it is said, led Newton to propose the law of universal gravitation.

While an apple tree does grow near Newton's childhood home today, we do not know whether the story is true. Years later, when Newton described his work on universal gravitation to a friend, he used the falling apple to draw an analogy between the moon's motion and the motion of objects near the surface of the earth. So perhaps the apple really had been his inspiration. Regardless, the legend helps us understand how seemingly different motions can be linked by a single interaction—gravitation.

Satellites and Gravity

We can understand Newton's analogy between falling objects and the moon's orbit by considering Figure 8-1. Suppose you drop an apple. It accelerates downward along path A, landing directly beneath where you released it. Now suppose that instead of dropping the apple, you toss it away. The apple accelerates downward as before; but it also travels outward away from you, as shown by path B. Now throw the apple harder (path C). It lands even farther from you. The harder you throw the apple, the farther away from you it lands.

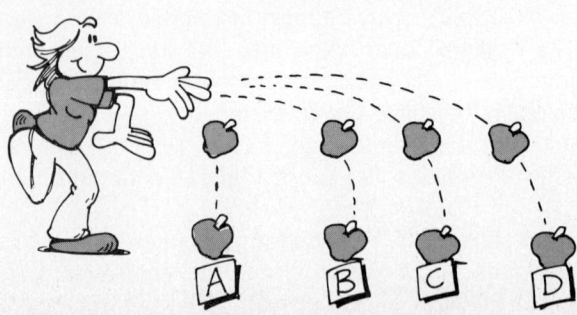

Figure 8-1

The harder you throw the apple, the farther away from you it lands.

Gravitational Interactions   155

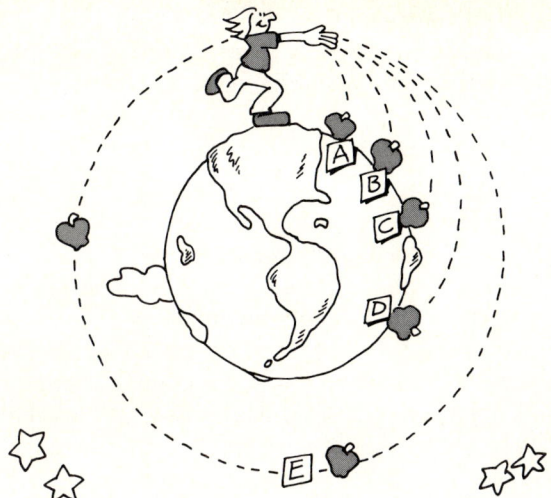

Figure 8-2
If we could actually throw the apple hard enough, it would miss the earth entirely. The apple would continue falling but would never quite catch up with the curvature of the earth.

If the surface of the earth were flat, there would be little more to say. The distance the apple travels outward before hitting the ground would simply depend on how hard you threw it. Since the earth is curved, however, the apple falls just a bit further each time it is thrown, as shown in Figure 8-2. The vertical distance it must travel increases because of the earth's curvature. If we could throw the apple hard enough so that it traveled from North America to, for instance, Africa (path D), it would have an enormous drop before hitting the ground. If we could throw it even harder still, it would eventually miss the earth entirely (path E). The apple now keeps falling and falling, but it never quite catches up with the earth's surface. Your apple behaves like the moon.

With this analogy, Newton conceived the artificial satellite 300 years before technology was advanced enough to launch one. He concluded that the same gravitational interaction that causes an apple to fall holds the moon in its orbit. With no horizontal velocity, the apple falls directly beneath its tree. With a horizontal velocity of over 1000 meters per second (m/s), the moon continues to fall toward the earth in an orbit 3.8×10^8 meters (m) above the earth's surface. The downward force of gravity becomes the centripetal force needed to hold the moon in its orbit about the earth. We can use the same concept to explain the motion of the earth around the sun. Just as the moon falls toward the earth, the earth falls toward the sun. Gravitational attraction holds our planet in orbit, just as it holds all the other planets, moons, and comets of the solar system.

Law of Universal Gravitation

Having realized that a single type of force, gravitation, is responsible for the acceleration of heavenly bodies as well as the downward acceleration of objects on earth, Newton developed a complete definition of the gravitational force. He found that the force increased as the mass of either object increased and as the objects came closer together. More precisely, the gravitational force is directly proportional to the masses of the two objects and inversely proportional to the square of the distance separating them. To calculate gravitational

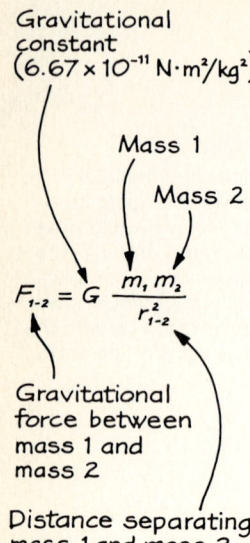

force, Newton used a constant, $G = 6.67 \times 10^{-11}$ N · m^2/kg^2, to convert units of mass and distance into units of force. The **law of universal gravitation** states that the magnitude of the force is:

$$\frac{\text{Force of gravity}}{\text{exerted on 1 by 2}} = G \frac{(\text{mass}_1)(\text{mass}_2)}{(\text{distance between 1 and 2})^2}$$

The force acts along a line joining the centers of the two objects. When object 2 pulls object 1 toward it, object 1 pulls object 2 toward it.

We can gain some insight into this law by using it to determine the size of the gravitational force exerted on the apple by the earth. The earth's mass is 5.98×10^{24} kilograms (kg). A very large apple would have a mass of about 1 kg. The distance separating the apple and the earth is taken to be the distance between their centers—approximately the radius of the earth, 6.38×10^6 m. (The distance from the earth's surface to the apple is too small to make any detectable difference.) Substituting these values into the law of universal gravitation, we find that the force on the apple is $(6.67 \times 10^{-11}$ N · m^2/kg$^2)(1$ kg$)(5.98 \times 10^{24}$ kg$)/(6.38 \times 10^6$ m$)^2 = 9.8$ N, toward the earth. The earth pulls the apple toward its center with a force of 9.8 N. If either the earth or the apple were twice as massive, the gravitational force exerted on the apple would be twice as great. If we moved the apple out into space, so that it was twice as far away from the center of the earth, the gravitational force exerted on it would be one-fourth as great. Three times as far away, the gravitational force exerted on the apple would be one-ninth as much (Figure 8-3).

Newton's third law assures us that if the earth attracts the apple, then the apple attracts the earth with a force equal in magnitude but opposite in direction. The law of universal gravitation provides us with the magnitude of both forces. The earth exerts a force of 9.8 N on the 1 kg apple. The apple exerts a force of 9.8 N on the earth. The acceleration experienced by each object as a result of these forces, however, depends on the object's mass. With a mass of 1 kg, the acceleration experienced by the apple is (9.8 N, down)/(1 kg) = 9.8 (m/s)/s, down. With a mass of 5.98×10^{24} kg, the earth experiences an acceleration equal to (9.8 N, up)/(5.98×10^{24} kg) = 1.6×10^{-24} (m/s)/s, up. The apple experiences an acceleration of 9.8 (m/s)/s, down, while the earth experiences an acceleration of 1.6×10^{-24} (m/s)/s, up. The earth's acceleration is much too small to notice.

SELF-CHECK 8A

Use the law of universal gravitation to calculate the gravitational force you exert on a table. Assume that your mass is 70 kg, the table's mass is 5 kg, and that you and the table are separated by 2 m. The earth exerts a gravitational force of 49 N, down, on the table. How does the force you exert on the table compare to that of the earth?

A STEP FURTHER—MATH

WHAT IS IT LIKE ON MARS?

Newton's law of universal gravitation allows us to predict the acceleration due to gravity on planets other than Earth. Newton's second law tells us that the magnitude of the force due to gravity (F_g) acting on an object of mass (m) is

$$F_g = mg$$

where g is the acceleration due to gravity. Newton's law of universal gravitation describes the magnitude of the force due to gravity (F_g) in terms of the mass of the planet (M), the mass of the object (m), and the radius of the planet (r):

$$F_g = G \frac{Mm}{r^2}$$

If we set these two expressions for the force due to gravity equal to each other,

$$F_g = F_g$$

$$mg = G \frac{Mm}{r^2}$$

$$g = G \frac{M}{r^2}$$

we find that the acceleration due to gravity, g, depends on the mass of the planet (M) and the radius of the planet (r).

To check to see if this relationship really works, let's calculate the acceleration due to gravity on earth. The earth's mass is 5.98×10^{24} kg and its radius is 6.38×10^6 m. Substituting these values into the relationship derived above:

$$g = G \frac{M}{r^2} = 6.67 \times 10^{-11} \frac{N \cdot m^2}{kg^2} \frac{(5.98 \times 10^{24} \text{ kg})}{(6.38 \times 10^6 \text{ m})^2}$$

$$= 9.8 \text{ (m/s)/s}$$

Newton's law of universal gravitation predicts the same value for the acceleration due to gravity as the measured value!

Now let's head out to Mars. Mars has a mass of 6.34×10^{23} kg and a radius of 3.43×10^6 m. What is the acceleration due to gravity at the surface of Mars?

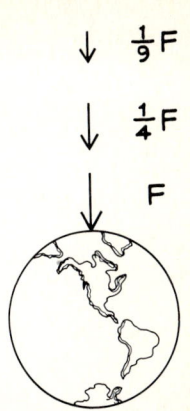

Figure 8-3
The gravitational force decreases quickly as we move away from the earth. Twice as far from the earth, the apple feels one-fourth the gravitational force. Three times as far away, the apple feels one-ninth the gravitational force.

The Future of the Universe

Gravity plays a major role in building models of the universe. The most widely accepted model of the universe is the **big bang model**. According to this model, all the matter in the universe was concentrated in an unimaginably small amount of space. Some 20 billion years ago an explosion, called the big bang, sent matter flying off in all directions. In some regions of space, the concentration of matter became high enough to allow gravitational forces to form planets, stars, and galaxies. These are the objects we see in the sky today. Other regions of space were very sparsely populated with matter, so bits of matter simply continued to move away from the point at which the big bang occurred. This matter makes up the interstellar dust and debris astronomers often mention.

The future of the universe depends on a struggle being waged daily between the outward motion that originated with the big bang and the inward forces due to gravitational attractions between masses. At present, all matter—the interstellar dust and the planets, stars, and galaxies—continues to move apart. If the gravitational forces that exist are large enough, this expansion will gradually come to a stop and the masses will once again be drawn back together. If the gravitational force is too weak, the universe will continue to expand forever.

Ultimately, the size of the gravitational force that exists depends on the amount of mass in the universe. Using Newton's law of universal gravitation, astronomers estimate that a minimum of 10^{-30} kg of mass per cubic meter of space is required to stop the present expansion. At present, the average known mass per cubic meter in the universe is less than this—about 10^{-31} kg/m^3. What is not known, however, is the amount of matter present in the interstellar dust and debris that we simply cannot see. This debris holds the key to our knowledge of the future of the universe.

ELECTROMAGNETIC INTERACTIONS

As humankind went about looking for problems to solve, it happened upon static cling. Clothes, particularly those made from synthetic fibers, stick together after being tumbled in a clothes dryer. Socks seem forever bound to shirts; pants wrap themselves tightly around your legs. An equally serious crisis is fly-away hair. Vigorous brushing, particularly during winter months, leads to hair that stands out on end. Rather than clinging to one another, the individual strands fly apart.

These experiences are modern examples of what the Greeks had noticed with amber, the hardened sap from some softwood trees. When the Greeks polished amber by rubbing it vigorously with a cloth, it attracted small bits of straw, paper, or seed. The amber exerts a force large enough to overcome the force due to gravity, pulling the bits of paper upward. To the Greeks, this effect seemed magical. Today we explain it in terms of electrical interactions which we have named from the Greek word *elektron*, meaning amber.

Experiments performed during the nineteenth century revealed that electrical interactions and magnetic interactions are related. Magnetism, also known to the Greeks, displays some of the same seemingly magical properties of acting across a distance. Our present model of forces unites electrical and magnetic phenomena under a single term—electromagnetic interactions. Here we will discuss only examples of electrical interactions; we consider electromagnetic phenomena further in later chapters.

Electromagnetic Interactions 159

Figure 8-4
Each plastic strip has been rubbed with its cloth. The two plastic strips repel each other. The two cloths repel each other. Each plastic strip attracts a cloth.

Rubbing Leads to Electrical Interactions

The Greeks rubbed amber as they polished it. Clothes rub against one another in the dryer. Bristles rub against strands of your hair as you brush them. It seems that rubbing things can cause them either to stick together or fly apart. Let's examine this phenomenon in more detail.

We begin with a system consisting of a plastic strip and a piece of cloth. Before we do anything to them, no interaction occurs between them. If we now rub the plastic strip with the cloth, they attract one another. This attraction must have been produced by the rubbing. We conclude that the rubbing transferred something between the plastic strip and the cloth. This "something" leads to the attraction we observe.

Now consider two identical systems, each consisting of a plastic strip and a piece of cloth. Before the strips are rubbed with the cloth, no interaction occurs within either system or between the two systems. After each plastic strip is rubbed with its cloth, the two plastic strips repel each other, the two pieces of cloth repel each other, and either piece of cloth attracts either plastic strip (Figure 8-4). Whatever is exchanged between each plastic strip and its cloth during the rubbing process can cause an interaction not only between objects within a system but also between objects in systems that are initially isolated from one another.

We conclude that a force, which we call an electrical force, results from rubbing two objects together. In an isolated system, the initial interaction of the two objects (rubbing) results in a force of attraction between them. In a nonisolated system, the initial interactions can result both in forces of attraction and forces of repulsion.

Electrical Charge and Static Electricity

In some ways, electrical interactions are remarkably similar to gravitational interactions. Both involve interactions at a distance. Both involve attractive forces. Mass is associated with gravitational force; something else must be present in order for electrical interactions to occur. This thing is a property of matter called **electrical charge,** measured in units called **coulombs** (C). Because we see both attractive and repulsive forces, two types of electric charge must exist. By convention, these two types of charge are called **positive** (+) and **negative** (−).

Electrical interactions occur only when electrical charges are present. The rubbing process somehow results in electrical charges on the two rubbed objects. Two possibilities exist: (1) rubbing creates electrical charge from nothing, or (2) rubbing causes the objects to exchange electrical charge. Creating something from nothing violates our basic belief in conservation. So, we imagine that the rubbing process moves electrical charge from one object to another. When one of the objects becomes positively charged, the other becomes negatively charged to the same degree, so that the total charge of the system remains constant. In a closed system, electrical *charge is conserved.*

In its normal state, matter has an equal number of positive and negative charges. The sum of these charges, or the net charge of the object, is zero. Before rubbing, each object (the plastic strip and the cloth) has a zero net charge, and no electrical interaction occurs between them. During the rubbing process, electrical charge is moved from one object to another. In this case, negative charges are rubbed off the cloth and transferred to the plastic strip. The plastic strip then has more negative charge than positive charge, giving it a net negative charge. The cloth, having lost negative charges to the plastic strip, has an excess of positive charge—a net positive charge. The attraction between the cloth and the plastic strip arises from their net electrical charges. As shown in Figure 8-5, objects with opposite net charges (one positive and one negative) attract one another. Objects with the same net charge (both positive or both negative) repel one another. Like charges repel; unlike charges attract. All these attractions and repulsions are examples of static electricity—the interaction of localized regions of net positive and net negative charge.

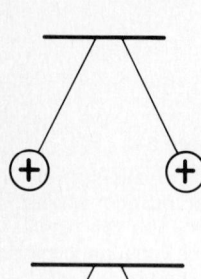
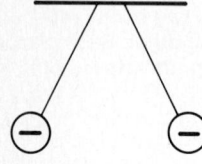
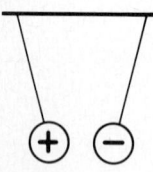

Figure 8-5
Like charges repel; unlike charges attract.

SELF-CHECK 8B

When rubbed with silk, a glass rod becomes positively charged.

a. Does the silk cloth have a net charge? If so, is it positive or negative?

b. Will the glass rod attract or repel the plastic strip in Figure 8-4?

Coulomb's Law

The similarities between electrical interactions and gravitational interactions led Joseph Priestley (1733-1804) to hypothesize that the electrical force law was similar to Newton's law of universal gravitation. He expected the size of the electrical force to change with the size of the electrical charge on the two objects and with the distance between the centers of the two objects. We can examine the qualitative effect of both these variables, amount of charge and distance between centers, in a simple experiment with a comb and paper.

To observe the effect of charge, rub a comb through your hair a few times; then place your comb near some bits of paper and watch how strongly they are attracted. Now rub the comb through your hair a few more times. Increasing the rubbing should increase the net charge on the comb. Now when you place the comb near some bits of paper, watch the paper jump. The force of attraction is much greater. The electrical force increases as the net charge on the comb increases. (If you are wondering why a comb with a net electrical charge attracts paper with no net charge, work Problem C6.)

We can examine the way in which electrical force varies with distance by charging the comb again and holding it at various distances from the bits of paper. When the comb is relatively far from the paper, no interaction seems to occur. The electrical force is not large enough to overcome the gravitational force holding the paper to the table. As the comb is brought nearer to the paper, we see an interaction. The electrical force increases until it is large enough to overcome the gravitational force. The paper jumps to the comb. As the separation between the comb and the bits of paper decreases, the electrical force between them increases.

A careful study of these two variables, net charge and separation between the objects, was first completed by Benjamin Franklin. However, Franklin's interests turned to politics, and the equation defining the electrical force was discovered by Charles Coulomb in 1789. Called **Coulomb's law,** this equation states that the electrical force that one charged object exerts on a second charged object is directly proportional to the product of the two net charges and inversely proportional to the square of the distance between the two objects. To calculate the electrical force, Coulomb used a constant, $k = 9 \times 10^9$ N $\cdot$ m^2/C^2, to convert units of charge and distance into units of force. The relationship between the magnitude of the force and charge is

$$\frac{\text{Force exerted on}}{\text{object 2 by object 1}} = k \frac{(\text{charge 1})(\text{charge 2})}{(\text{distance between 1 and 2})^2}$$

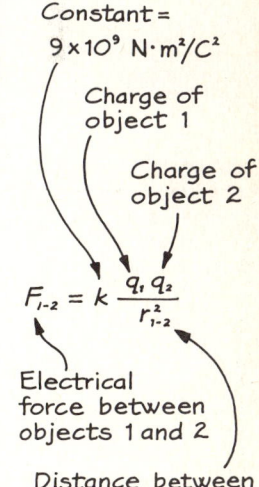

The direction of the force depends on the two charges. If charge 1 and charge 2 are the same (both positive or both negative), the force is positive and object 1 repels object 2. If charge 1 and charge 2 are different (one positive and the other negative), the electrical force is negative and object 1 attracts object 2. Newton's third law tells us that if object 1 attracts object 2, then object 2 will attract object 1 with a force of equal magnitude; likewise, forces of repulsion will be equal. The two charged objects exert mutual forces on one another.

Coulomb's law has exactly the same *form* as Newton's law of universal gravitation. Electrical charge has replaced mass. Two types of electrical charge result in two types of electrical force—attraction and repulsion. One type of mass results in one type of gravitational force—attraction. You can imagine the elation among physicists as they discovered a second interaction that behaved in almost exactly the same fashion as gravitation.

SELF-CHECK 8C

Each of the plastic strips in Figure 8-4 has a net charge of 10^{-7} C. What is the force that one strip exerts on the other if the two strips are separated by 10 m? By 1 m? By 0.01 m?

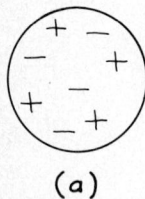

(a)

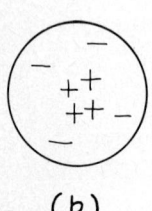

(b)

Figure 8-6
Charges could be distributed in the atom in two ways: **(a)** evenly, or **(b)** in charge concentrations.

Probing the Atom

The properties of the electrical interaction provide us with a powerful tool for investigating the structure of matter. By the turn of the twentieth century, physicists knew that matter could be broken down into atoms and that atoms contained both positive and negative charges. What remained a mystery, however, was how these charges were distributed. The charges could be randomly distributed throughout the atom or they could be concentrated in densely positive and negative regions (Figure 8-6). Ernest Rutherford and his students resolved this question by shooting positively charged particles toward thin sections of gold.

The positively charged particles, called alpha particles, are given off spontaneously by certain atoms. They are small enough to penetrate thin sections of matter, and since they are electrically charged, their motion could reveal the distribution of electric charge in matter. Let's examine how the alpha particles would behave in each of the models proposed for the atom (Figure 8-7).

First, suppose that the electrical charges are randomly distributed throughout the atom. As the alpha particles move through the gold foil, they would constantly experience both attraction and repulsion. The two forces would generally balance out and there would be no net force on the alpha particles. They would move almost straight through the gold foil (Figure 8-7(a)).

Now suppose that the electrical charge in matter is concentrated in densely positive or densely negative regions. In this case, the motion of the alpha particles would depend on the relative mass of the charge concentration. If the charge concentrations have about the same mass as the alpha particles, then Newton's second and third laws require that they accelerate equally as a result of the electrical interaction between them. But if the concentrations of electrical charges are much more massive than the alpha particles, then only the alpha particles would change their motion measurably. Figure 8-7(b) shows the various paths predicted when the alpha particles approach fairly massive, positively charged regions. If the alpha particles approach head-on, the repulsive force would slow them to a stop and send them back along the same path. If slightly off center, the alpha particles would move on by but would be deflected from their original courses. At a large

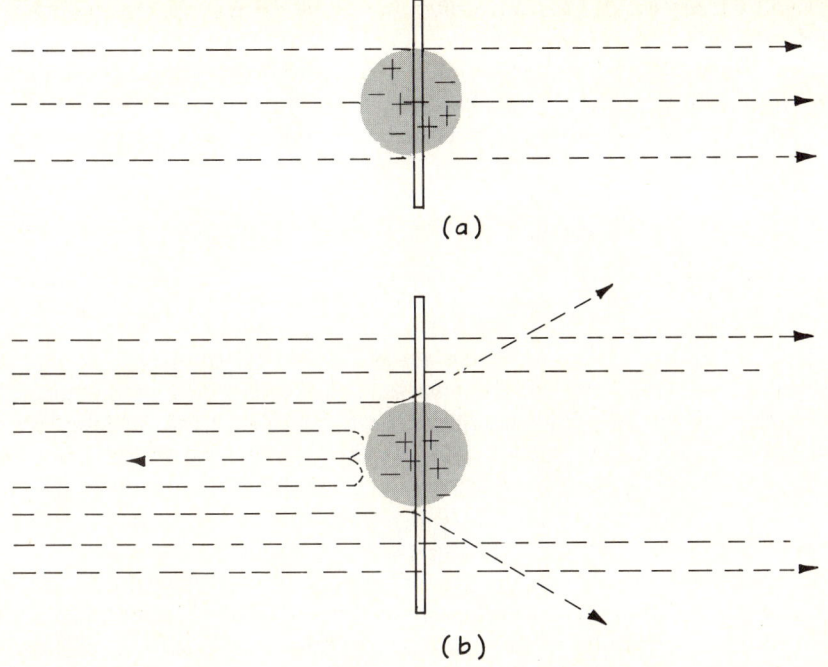

Figure 8-7
The paths followed by the alpha particles depend on the way in which electrical charge is distributed in the atom. **(a)** When the electrical charges are evenly distributed, the alpha particles continue along their original paths. **(b)** When the alpha particles encounter massive charge concentrations, some alpha particles will be deflected at large angles.

enough distance, the repulsive force would be too small to affect the alpha particles, which would continue moving along their original paths.

When he actually performed the experiment, Rutherford saw the patterns shown in Figure 8-7(b). Most of the alpha particles went through the gold foil undeflected, but a few were repelled at large angles. Still fewer were actually reflected—bounced back along their initial path. Only fairly massive concentrations of positive charge could explain these reflected alpha particles. Rutherford concluded that the gold foil contained massive regions of positive charge that were relatively far apart. The negative charges were probably scattered randomly through the foil. By applying his knowledge of Newton's laws and electrical interactions, Rutherford was able to construct a picture of the atom, an object far too small to see directly.

Rutherford's experiment and Coulomb's law provided the basis upon which the first workable model of the atom was built. As a result of Rutherford's discovery of massive concentrations of positive charge, Niels Bohr suggested that atoms are like miniature solar systems. A massive particle at the center, called the **nucleus,** is positively charged. Many small particles, called **electrons,** circle the nucleus much as planets circle the sun. The electrons are negatively charged; consequently they are held in orbit by the electrical force of attraction exerted by the nucleus. Atoms contain enough electrons to balance the positive charge of the nucleus, so each atom is left with zero net charge.

Bohr's model fit Rutherford's experimental results and prior experience with static electricity. The small, massive, positively charged nucleus causes alpha particles to rebound like tennis balls off a brick wall. The negatively charged electrons in orbit about the nucleus produce a fairly random distribution of negative charge in matter. Since the electrons have very small masses, they can be transferred from one object to another as a result of rubbing. This leaves one object positively charged and makes the other object negatively charged. Eventually Bohr's model gave way to a more sophisticated one, but the major features of negatively charged electrons with small masses and positively charged nuclei with large masses have remained unchanged.

NUCLEAR INTERACTIONS

Rutherford's experiment revealed the existence of the nucleus—a small, relatively massive, positively charged part of matter. Naturally enough, the next question asked was: What is inside the nucleus? Experiments with natural radioactivity provided part of the answer, and experiments similar in design to Rutherford's actually allowed physicists to penetrate the nucleus. By the 1930s, two additional particles—the **proton** and the **neutron**—had been found in matter. These two particles, collectively called **nucleons,** were found to exist in the nucleus. Table 8-1 summarizes the mass and electric charge of these two particles and the electron.

Subsequent research on the atomic nucleus has led physicists to believe that two more fundamental interactions, strong and weak nuclear interactions, exist in addition to gravitational and electromagnetic interactions. The strong nuclear interaction describes the mechanism that holds the nucleus together. The weak nuclear interaction describes a way in which the nucleus sometimes falls apart. Since research that probes the nucleus is a scant 50 or 60 years old, we still have much to learn.

Strong Nuclear Interactions Bind the Nucleus Together

Atoms have a diameter of about 10^{-8} to 10^{-10} m. The nucleus occupies a tiny fraction of this space—its diameter is estimated to be only about 10^{-14} m. Once physicists realized that the nucleus contains protons and neutrons, an interesting contradiction arose. Protons are positively charged; neutrons carry no electrical charge. Atoms consist of anywhere from 1 to over 100 protons and 0 to about 150 neutrons. We know that gravitational interactions exist among all these particles, but this interaction is very small (see Problem 8-B5).

Table 8-1 The Constituents of the Atom

Particle	Electrical Charge (C)	Mass (kg)
Electron	-1.6×10^{-19}	9.11×10^{-31}
Proton	$+1.6 \times 10^{-19}$	1.672×10^{-27}
Neutron	0	1.675×10^{-27}

Thanks to Electrical Charge...

The crackle and snap of static electricity is the most obvious example of electrical interactions in our everyday life. However, because all matter is composed of electrically charged particles, many other common interactions are electrical in nature. For example, electrical interactions within and between atoms are responsible for nearly all contact interactions—the forces you exert with your body, the force of friction offered by surfaces, the reaction forces provided by solid surfaces.

The electrical interactions that hold atoms and molecules together provide the resistance we have called a reaction force. When you push on a table, the molecules in your hand interact with the molecules of the table. Your hand is trying to push through the table. However, the only way for the table to give would be for these electrical interactions between atoms to break. They resist being pulled apart, so you feel a hard surface. Electrical forces form almost a miniature net—giving slightly but never enough to let you slip through.

For gases and liquids, the electrical interactions among molecules are not as strong as they are among the molecules of solids. An object can break the weak electrical forces holding a molecule to its neighbors. Thus, you can move through liquids and gases but not solids.

The frictional force that arises when one object moves along another also results from electrical interactions. When one object touches another, its molecules become distorted because of electrical forces from molecules of the neighboring object. These forces cause changes in the motions of the molecules in each object. Because the molecules move randomly in all directions, these forces do not cause motion of the whole object. Instead, these electrical forces are the force we call friction.

These electrical interactions can differ in strength. Anyone who has tried to move a heavy object knows that it takes more force to get the object moving than to keep it moving. We say that static friction is greater than moving friction. We can explain these observations in terms of electrical attraction between the object and the floor, for example. When an object sits in one place, even for a short time, some of its atoms "sink" into the floor. Because the atoms of the object and floor are now closer together, it takes more force to break their mutual electrical forces than when the atoms are moving past one another. Because of a decrease in distance, static frictional forces are greater than moving frictional forces.

Electrical forces sometimes cause atoms to be transferred from one object to another. Each time an automobile tire rotates, one layer of rubber molecules is actually "ripped" from the tire and left on the pavement. Take a step and you leave a bit of shoe leather behind. The electrical force required to transfer these atoms is the force we call friction.

Thanks to electrical charge, we can take that step forward, push that chair out of the way, or not fall through the floor. Electrical forces are, in fact, the glue that holds us all together.

Because of their electrical charge, however, protons should repel one another. Coulomb's law predicts a force of 2.3 N between two protons separated by the diameter of the nucleus. While this force does not seem very large, it will produce enormous accelerations (1.4×10^{27} (m/s)/s) on masses as small as protons. If gravitational forces are too small to hold the nucleus together and enormous electrical forces are driving the protons apart, just what is it that keeps the nucleus together?

Physicists believe that a third type of interaction, called the **strong nuclear interaction,** holds the nucleus together. Like electrical forces, the strong nuclear force can be either attractive or repulsive. Its range of effectiveness, however, is very limited. Beyond separations of 10^{-12} m, the strong nuclear force is, for all practical purposes, zero and electrical forces dominate all interactions. Within a distance of 10^{-12} m, however, the strong nuclear force produces an interaction between protons that is 100 times as great as the electrical force of repulsion at that same separation. At separations between 10^{-12} and 10^{-16} m, protons attract protons, neutrons attract neutrons, and protons attract neutrons. Within still shorter distances, 10^{-16} m or less, the strong nuclear force becomes repulsive and the nucleons all repel one another. Consequently, the nucleus does not collapse on itself.

Weak Nuclear Interactions

Even before Rutherford discovered the nucleus, Henri Becquerel and Marie and Pierre Curie had seen the results of nuclei falling apart. Some materials spontaneously emit electrically charged particles, collectively called radioactivity. Some radioactive emissions eventually were explained in terms of electrical interactions between protons. Too many protons packed into too small a space can result in an electrical repulsion large enough to cause the nucleus to fly apart, emitting some of its nucleons. Other emissions could not be explained so easily.

One emission that seemed particularly difficult to explain involves the neutron. A neutron left by itself will spontaneously turn into a proton, an electron, and a third particle called an antineutrino. These particles have electric charges of plus, minus, and zero, respectively. When this happens within the nucleus, the electron and antineutrino are ejected, while the proton remains inside. Neither electrical nor strong nuclear forces shed any light on why the neutron disintegrates.

The disintegration of the neutron and a variety of other nuclear interactions can be explained only if we introduce another interaction—the **weak nuclear interaction.** For many years this interaction has been the least understood of the four fundamental interactions. We will do little more here than simply list it among the fundamental interactions.

INTERACTION AT A DISTANCE

The introduction of the two types of nuclear interactions brings the number of fundamental interactions to four. All four are described as interactions at a distance. Objects exert forces on one another but do not actually touch. While

Figure 8-8
Seen from afar, a game of catch might look like interaction at a distance. The more massive the "ball," the stronger the force.

we are able to explain many observations in terms of interaction at a distance, we have not approached the question: How does an object *here* "communicate a force" to an object *there*? To address that question, we begin with an analogy.

Interaction at a Distance and a Game of Catch

Suppose you and a friend play a game of catch. The two of you stand several meters apart and throw a baseball back and forth. The distance between you and your friend will depend on your ability to throw, but it will never exceed some practical maximum. You and your friend will remain in some well-defined region of space.

Imagine how this game would look to observers who are close enough to see the two players but not the ball. The players would move about, sometimes moving closer together but never going farther apart than some maximum throwing distance (Figure 8-8). Being unable to see the ball, the observers might conclude that an attractive force holds the two players near one another. While the exact nature of the force would not be clear, it could explain both your friend's and your behavior.

Now imagine how the game would look if you played with balls of vastly different masses—for instance, a baseball and a bowling ball. In order to play catch with a bowling ball, you and your friend would have to stand rather close to one another. The observers would report a force with a very small range. By contrast, you and your friend could stand much farther apart if you played with a baseball. The observers would report a force with a much larger range. The range of the force seen by the observers is related to the mass of the ball you use to play catch. As the mass of the ball increases, the range of the force decreases.

Forces and the Exchange of Particles

The game of catch is analogous to an explanation, first suggested by H. Yukawa in the 1930s, of how the strong nuclear interaction works. Yukawa suggested that the strong nuclear force arises from the exchange of a particle between nucleons. From the known range of the strong nuclear force, Yukawa calculated the mass of this supposed particle, which he called the **pi-meson** or **pion,** to be about one-seventh the mass of the nucleon. Such a particle had never been seen when he offered this explanation, but shortly thereafter it was detected. Yukawa had been right.

The discovery of the pion led to a generalization of Yukawa's model to the other three fundamental interactions. In each case, an **exchange particle** can be used to explain the process by which objects exert forces on other objects across a distance. The mass of the exchange particle determines the range of the interaction.

Both gravitational and electrical interactions extend over all space. Newton's law of universal gravitation and Coulomb's law both predict that the forces become increasingly smaller as the separation between objects increases, but neither force actually reaches zero. For a force to extend over an infinite distance, the exchange particle must have zero mass. Such a particle, called the **photon,** has been discovered in electrical interactions. The photon has no mass, but it can move between electrically charged objects. The exchange particle for gravitational interactions, called the **graviton,** has yet to be detected. Like the photon, the graviton must have zero mass. In order to explain gravitational interactions, the graviton must be exchanged by all objects having mass.

The weak nuclear interaction presents a different situation. The range of the weak nuclear interaction is known to be quite short, on the order of 10^{-17} m. If our model of exchange particles is correct, then the exchange particle for these interactions, called the **W,** must have a mass considerably larger than the pion. In January 1983 experimenters announced that such a particle had been detected. A summary of the four interactions, their rela-

Table 8-2 The Fundamental Interactions

Interaction	Relative Strength	Approximate Range (m)	Exchange Particle	Mass of Exchange Particle (kg)
Strong nuclear	100	10^{-15}	Pion	2.5×10^{-28}
Electromagnetic	1	No limit	Photon	0
Weak nuclear	10^{-11}	$<10^{-17}$	W	$>1.5 \times 10^{-25}$
Gravitational	10^{-40}	No limit	Graviton*	0

*The graviton has not yet been observed.

tive strengths, their ranges, their exchange particles, and the mass of each particle is given in Table 8-2. The electromagnetic force is arbitrarily assigned a strength of 1. The strong nuclear force is 100 times greater, the weak nuclear force only 10^{-11} as large, and so forth.

A Unified Theory of Interactions

The four fundamental interactions and their properties have been known since the discovery of the pion in the 1930s. In recent years, new particles ranging in mass from zero to several times the mass of the nucleon have been discovered. In all cases the forces among these particles could be explained in terms of the four fundamental interactions.

The reduction of a multitude of forces to four interactions provides a remarkably simple model of events. Yet, we might hope for an even simpler picture. The uncanny similarity between Newton's law of universal gravitation and Coulomb's law led to speculation that these two interactions are somehow variations of a single interaction. Albert Einstein spent a major portion of his life unsuccessfully trying to unify these two interactions. The discovery of the pion and a host of other particles smaller than either protons or neutrons has led to a new attack on the question of a unified theory of interactions.

Once considered a single unit, matter was eventually broken down into atoms. Atoms, in turn, were broken down into protons, neutrons, and electrons. Naturally, physicists wondered whether these particles could be broken down into still smaller particles. In the 1960s, a major theory proposed that nucleons (protons and neutrons) are composed of three smaller, charged particles called **quarks,** held together within each nucleon by the strong nuclear force. To date, no one has been able to "pull" a quark out from inside the nucleon. However, experiments similar in design to Rutherford's show three distinct locations of electric charge within each nucleon. Consequently, the quark model is regarded quite seriously.

In contrast to protons and neutrons, the electron seems to be a fundamental entity. Nothing suggests that it can be broken down into still smaller particles. However, a number of particles similar to electrons have been discovered. For example, the positron has the same mass as the electron, but it has a positive electric charge. Taus and muons behave like electrons but have larger masses. Together with still other similar particles, electrons, positrons, taus, and muons form a family of particles called **leptons.**

With quarks and leptons as the fundamental particles, physicists are now looking for a single interaction that unifies strong nuclear, weak nuclear, and electrical interactions. This single interaction is believed to act at distances less than 10^{-31} m. According to this unified theory of interaction, exchange particles, called X particles, move among quarks and leptons. At distances greater than 10^{-31} m, the X particle is replaced by pions, W particles, and photons; consequently, we see three separate interactions: strong nuclear, weak nuclear, and electrical interactions. (The gravitational interaction remains unreconcilable in this model.)

The success of a **unified theory of interaction** awaits the observation of events at distances less than 10^{-31} m. Today we are not yet able to ob-

© 1979 by Sidney Harris.

serve such events. Some indirect evidence in support of this theory comes from interactions at larger distances, and the model seems promising. Ultimately, however, our commitment to such a model is as much a reflection of our belief in simplicity as it is a reflection of reality. For physicists, the promise that three interactions can be replaced by one is indeed compelling.

Today we think that change occurs through four, and possibly only two, fundamental interactions. With them we can explain the fall of a pencil, the motion of the universe, the reason for static cling, and the mechanism by which atoms and nuclei are held together. Yet amid all this knowledge, fundamental questions still persist. What causes gravity? Why does an electric force exist? The concept of exchange particles provides a partial answer. Gravity and electrical forces are our way of describing the behavior of objects that exchange gravitons and photons. But then another question arises: Why do exchange particles go back and forth? Were we to answer that question, there would be another. Like the child who endlessly asks Why, we have an insatiable appetite for understanding.

CHAPTER SUMMARY

All forces in nature can be classified into four fundamental interactions: gravitational, electromagnetic, strong nuclear, and weak nuclear interactions.

The *gravitational interaction* explains the motion of falling objects on the surface of the earth, the motion of the moon about the earth, the motion of the planets about the sun, and the motion of galaxies. Every object attracts every other object with a force that is directly proportional to the product of the objects' masses and inversely proportional to the square of the distance between their centers. This is known as Newton's *law of universal gravitation.*

The description of the *electrical interaction* (one part of the electromagnetic interaction) is identical in form to the description of the gravitational interaction. The electrical force between two objects is directly proportional to the product of the objects' electrical charge and inversely proportional to the square of the distance between their centers. There are two kinds of *electrical charge,* called *positive* and *negative,* and the direction in which the force acts depends on the type of charge on each object. Like charges repel; unlike charges attract.

The two nuclear interactions describe forces that occur over the very short distances within the nucleus of the atom. The *strong nuclear interaction* is responsible for holding the nucleus together. The *weak nuclear interaction* is needed to explain why certain nuclei fall apart.

Each of the four fundamental interactions involves action at a distance. Action at a distance can be explained in terms of *exchange particles,* which move back and forth between two objects that are interacting. The range of the force depends on the mass of the exchange particle. *Pions,* the *W particle,* and *photons,* the exchange particles for strong nuclear, weak nuclear, and electrical interactions, respectively, have been detected. The *graviton,* yet to be observed, is postulated to be the exchange particle for the gravitational interaction.

The weak nuclear, electromagnetic, and strong nuclear interactions are now believed to be different manifestations of one interaction. This *unified interaction theory* will be tested when experiments can be performed to study interactions within distances of 10^{-31} m.

ANSWERS TO SELF-CHECKS

8A. $F_{\text{you on table}} = G \dfrac{M_{\text{you}} M_{\text{table}}}{(r_{\text{you-table}})^2}$

$= 6.67 \times 10^{-11} \dfrac{\text{N} \cdot \text{m}^2}{\text{kg}^2} \dfrac{(70 \text{ kg})(5 \text{ kg})}{(2 \text{ m})^2}$

$= 5.8 \times 10^{-9}$ N, toward you

The gravitational force we exert on the table is tiny compared to the gravitational force exerted by the earth. In a tug of war, the earth wins—hands down!

8B. a. The silk must have a net negative charge.
b. It attracts the plastic strip.

8C.
$$F = k \frac{(\text{charge 1})(\text{charge 2})}{(\text{distance})^2}$$

At 10 m: $F = 9 \times 10^9 \dfrac{\text{N} \cdot \text{m}^2}{\text{C}^2} \dfrac{(10^{-7}\,\text{C})(10^{-7}\,\text{C})}{(10\,\text{m})^2} = 9 \times 10^{-7}\,\text{N}$

At 1 m: $F = 9 \times 10^9 \dfrac{\text{N} \cdot \text{m}^2}{\text{C}^2} \dfrac{(10^{-7}\,\text{C})(10^{-7}\,\text{C})}{(1\,\text{m})^2} = 9 \times 10^{-5}\,\text{N}$

At 0.01 m: $F = 9 \times 10^9 \dfrac{\text{N} \cdot \text{m}^2}{\text{C}^2} \dfrac{(10^{-7}\,\text{C})(10^{-7}\,\text{C})}{(0.01\,\text{m})^2} = 9 \times 10^{-1}\,\text{N}$

PROBLEMS AND QUESTIONS

A. Review of Chapter Material

A1. Briefly describe each of the four fundamental interactions.

A2. How is the fall of an apple related to the motion of the moon about the earth?

A3. State how the gravitational force varies with mass and distance.

A4. What is the unknown quantity in determining if the universe will expand forever? How is this variable related to Newton's law of universal gravitation?

A5. How does the electric force vary with the type of electric charge, the size of the charge, and the distance between the charged objects?

A6. How are the electrical and gravitational interactions similar? How do they differ?

A7. Explain how Rutherford used the electrical interaction to learn about the structure of the atom.

A8. Explain why a strong nuclear interaction is necessary to explain the existence of the nucleus.

A9. What types of events are explained by the weak nuclear interaction?

A10. How do particle exchanges describe interaction at a distance?

B. Using the Chapter Material

B1. Two identical planets are orbiting a star. Planet A is 4000 km from the sun; planet B is 8000 km. On which planet does the sun exert a greater gravitational force of attraction? What is the ratio of the force felt by planet A to the force felt by planet B?

B2. Satellite C has twice the mass of satellite D. Both are orbiting the earth at the same altitude. On which satellite does the earth exert the greater gravitational force? What is the ratio of the gravitational force exerted on satellite C to the gravitational force exerted on satellite D?

B3. Skylab had a mass of 90,606 kg; the earth, 5.98×10^{24} kg. When it began orbiting, Skylab was 432 km above the earth's surface. The radius of the earth is 6380 km. What gravitational force did Skylab exert on the earth?

B4. The Apollo spacecraft felt gravitational attractions from both the earth and the moon. Was there any time during the flight when one of these forces was zero? Could a place exist where the net force from the two planets was zero?

B5. The mass of a proton is 1.672×10^{-27} kg. What is the force due to gravitational attraction for two protons separated by a distance of 10^{-15} m? Compare this to the electrical force between the same two protons.

B6. Which sets of charged objects have a larger electrical force acting on them?
a. two objects with charges of +12 C each and 3 m apart or two objects with charges of +24 C each and 3 m apart
b. two objects with charges of −14 C each and 6 m apart or two objects with charges of −14 C each and 20 m apart

B7. The electrical charge on the electron is -1.6×10^{-19} C. On the hydrogen nucleus the charge is $+1.6 \times 10^{-19}$ C. The distance between the nucleus and electron is 5.3×10^{-11} m. What is the electrical force on the electron? On the nucleus?

B8. A plastic comb begins with a zero net charge. After being used, it has a net charge of -3×10^{-6} C. What is the charge on the hair? Explain how you reached your answer.

B9. The net attractive force between two protons inside the nucleus is less than that between two neutrons. Why?

B10. We observe results of both electrical and gravitational interactions every day, but we do not notice direct evidence of nuclear interactions. Why?

B11. How would the mass of the pion need to change to decrease the range of the strong nuclear force?

C. Extensions to New Situations

C1. In Chapter 5 we noted that momentum was conserved for all interactions in a closed system. One such system involves an object falling near the surface of the earth. Suppose a 0.5 kg book is released from a height of 2 m and falls to the earth.
   a. What objects form the closed system for this interaction?
   b. Which of the fundamental interactions is involved in this interaction?
   c. Why does the momentum of the earth change as the book falls?
   d. Why is this momentum change not noticeable?

C2. When Newton introduced his law of universal gravitation, he successfully explained the ocean tides. The questions here will help you produce part of his explanation. Consider the earth-moon system shown in Figure 8-C2.
   a. Why does the moon exert a force on the earth?
   b. On which of the surface points—A, B, C, or D—is the force exerted by the moon greatest? Why?
   c. Suppose that an ocean is located at the point you chose in part (b). Draw an exaggerated picture of how the ocean would look at this point. Explain your drawing.
   d. Now imagine that the earth is rotating while the moon stays fixed. Why will the ocean rise as each area reaches the point of maximum force?
   e. Use the answers to (a)–(d) to explain why tides occur.

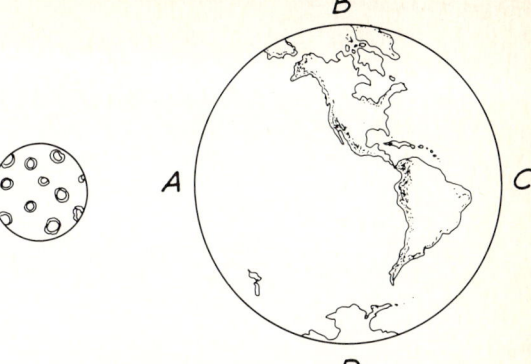

   f. A second tide occurs each day when the location is opposite the first tide point. Can you explain this tide by comparing the force on the earth and the water farthest from the moon?

C3. Tides arise from gravitational interactions between the earth and moon. Because the moon rotates about the earth once every 28 days, the high tides do not occur at the same time each day. Also, the magnitude of the high tide will vary from day to day. To understand why, consider various positions of the moon relative to the sun and earth as illustrated in Figure 8-C3.
   a. For which of these configurations will the high tide be the greatest? Explain your answer.
   b. For which of these configurations will the high tide be the least? Explain your answer.
   c. Describe the sun's role in determining the magnitude of the high tides.

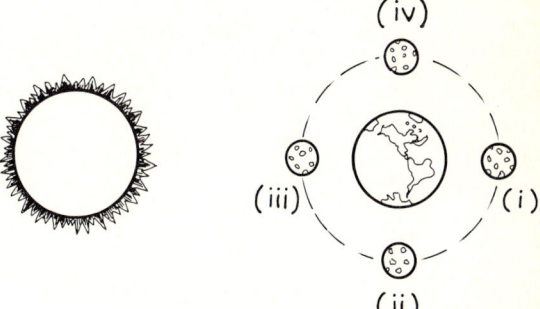

C4. The presence of the outermost planets in the solar system was detected by noticing small irregularities in the orbits of the known planets. Their orbits deviated slightly from those predicted by the known gravitational forces. Explain why such irregularities would be considered evidence for another planet.

C5. Oscilloscopes and old-fashioned televisions use a series of electrical forces to create a picture. Electrons hit a screen and create light. The color of the light depends on the type of atom struck by the electron. Electrical forces are used to get the electron to the place it is needed.

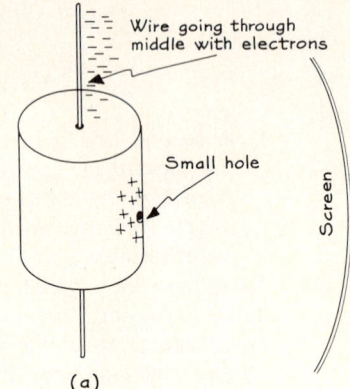

(a)

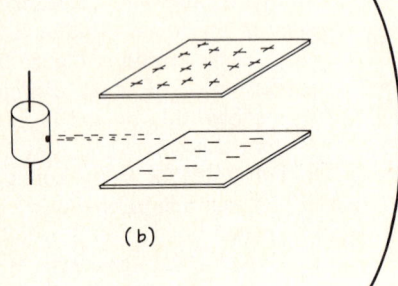

(b)

a. The first step is to get electrons moving. The basic setup is shown in Figure 8-C5(a). Why does this arrangement result in a bright spot on the screen as shown? (This arrangement is called an electron gun.)
b. Looking at bright spots on screens is not very exciting, so we must move the electrons. How does the arrangement in Figure 8-C5(b) change the motion of the electrons? Draw the path of the electron from the electron gun to the screen.
c. In television the charge on the deflecting plates changes from positive on the top and negative on the bottom to positive on the bottom and negative on the top in $\frac{1}{60}$ s. The change is gradual, not sudden. What will be the motion of the electrons?

C6. When a plastic comb with a net charge on it is placed near paper with zero net charge, the paper will still accelerate toward the comb. A net charge on one object is sufficient for electrical attraction to occur. To see why, consider a piece of paper in which no net charge exists but in which the electrons can move. A comb with a net positive charge is brought near the piece of paper.
a. How will the electrons in the paper move when the comb is brought near to but does not touch the paper? Explain your answer.
b. Draw the comb and the paper showing the positions of positive and negative charges in each. Explain why your drawing is correct.
c. What are the directions of the electrical forces on the paper's positive and negative charges?
d. On which set of charges, positive or negative, will the magnitude of the force be greater? Why?
e. Suppose that the comb had a net negative charge. Work through steps (a)-(d) and show that the paper will still be attracted to the comb.
f. Describe why an object with a positive or negative net charge can attract an object with no net charge.

C7. Nuclei come with a variety of different numbers of protons, ranging from 1 to about 110. The number of neutrons extends over a wider range, 0 to more than 150. However, nuclei which consist only of protons do not exist, except for hydrogen. Very massive nuclei exist only in forms for which the neutrons outnumber the protons.
a. What are the forces acting between two protons? Between a proton and a neutron? Between two neutrons?
b. How does the net force between a proton and a neutron differ from that between two protons?
c. Could an "all-proton nucleus" ever result in a net force that is repulsive for the protons? How might that occur?
d. Use the answers for (a)-(c) to explain why neutrons must be present in the nucleus.

INTERLUDE 2
Each Step Beyond

Newton, of course, did not create his laws of motion spontaneously while sitting under an apple tree. While in many ways his work was a revolutionary departure from previous ideas about motion, it owed an immense debt to Galileo's exhaustive study of motion. Newton and Galileo saw, as others had not, what motion would be like in a vacuum. To illustrate both the context from which Newton's ideas grew and the stumbling block posed by motion in a vacuum, we pause to consider the development of the laws of motion.

The first laws of motion were proposed by Aristotle (384–322 B.C.). In observing objects around him, Aristotle noticed that an object required a force in order to start moving. If that force was removed, the object eventually came to a stop. A large force resulted in a large velocity; a small force resulted in a small velocity. He concluded that force was related to velocity.

In some situations, however, force resulted in no velocity. We could push and push on a heavy boulder, but it would never budge. In order to explain these observations, Aristotle added the concept of resistance. Resistance to motion can arise from two sources. If the resistance is offered by the object we are trying to move, then resistance is a property of that object—like the later ideas of mass and inertia. If the resistance is offered by the material through which the object moves, then resistance is like friction—a force exerted by other materials. Aristotle chose the latter, calling resistance a property of the medium through which an object moves.

Combining the concept of resistance with the idea that force was related to velocity, Aristotle formulated his "second law:"

$$\text{Velocity} = \frac{\text{force}}{\text{resistance}}$$

It looks remarkably like Newton's second law, except that it relates force to velocity instead of acceleration, and it defines resistance as a property of the medium, not the object being moved. When Aristotle tried to imagine what would happen in a vacuum, he saw an absurdity. In a vacuum, the resistance would be zero and the object's velocity would increase to infinity. Aristotle concluded that vacuums did not exist.

In the years from Aristotle to Galileo, Philoponus, Avicenna, and a number of other scientists grappled unsuccessfully with this problem of motion in a vacuum. Philoponus (ca. A.D. 500) attacked this problem of motion in a vacuum by suggesting that Aristotle's relationship be modified to

$$\text{Velocity} = \text{force} - \text{resistance}$$

If the resistance were zero, the object would move with a constant velocity directly proportional to the force. Consequently, motion in a vacuum would be possible. By suggesting that an object's motion depends on the difference between force and resistance, Philoponus' modification introduced what we have come to call the *net force*. However, Philoponus still related force to velocity, not acceleration.

Avicenna (A.D. 980-1037) proposed a different modification of Aristotle's relationship. Avicenna suggested that objects themselves have a property, which he called *mail,* that resists a change in motion. Objects moving in a vacuum would continue moving forever, not speed up forever as Aristotle had supposed. Avicenna's concept is in many ways a forerunner of Galileo's concept of inertia or Newton's concept of mass. However, since Avicenna had never seen objects that moved at a constant velocity forever, he, too, concluded that vacuums do not exist. Like Aristotle, Avicenna failed to pursue the question of what would happen if. For these scientists, what had not been observed simply was not possible.

Galileo (1564-1642) eventually made the mental jump from the observed to the hypothetical—from motion in everyday experience to motion without resistance. Going down an incline, he reasoned, a ball accelerates. If he placed a second incline facing the first, the ball would move up the second incline almost to the same height from which it had been released on the first incline. Mentally, Galileo removed friction and concluded that the ball would continue up the second incline until it reached exactly the same height from which it had been released. Next, Galileo decreased the angle of the second incline. Each time, the ball traveled until it almost reached the same height from which it had been released. At lower inclines, however,

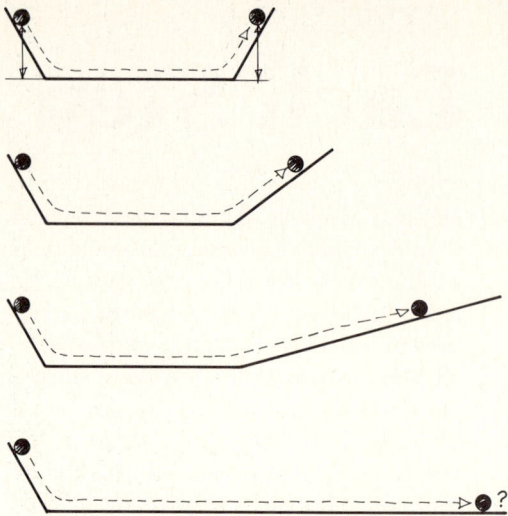

the ball had to travel farther (Figure 1). If the angle of incline were reduced to zero, Galileo reasoned, the ball would continue moving forever. It is only the resistance offered by the surface and the air that keeps this from being so. Called the law of inertia, Galileo's conclusion paved the way for Newton's first law of motion.

Because of friction, Galileo was never able to actually observe an object moving along a level board with an unchanging velocity. But he could mentally remove friction and imagine what would happen. His guide in performing these imaginary experiments was the simplicity of the mathematical relationships he had discovered from actual measurements. A commitment to experimentation and simplicity allowed Galileo to see what others had not seen— motion in a frictionless world.

Newton built upon Galileo's work, adding the concepts of force and mass to Galileo's descriptions of motion. His second law,

$$\text{Acceleration} = \frac{\text{net force}}{\text{mass}}$$

relates net force to acceleration and identifies the mass of the object as the source of inertia. The second law incorporates Aristotle's concept of the resistance provided by the medium into the concept of net force. The net force acting on an object that you push or pull is the force you exert minus any frictional force provided by the medium. Resistance to motion is indeed a property of the medium. But resistance to a change in motion is a property of the object (called mass). Because they take both these properties into account, Newton's laws could predict motion in a vacuum as well as motion on earth.

If you had been asked to write your own laws of motion before reading about Newton's laws, you might well have written something like Aristotle's or Philoponus' laws. Most of us would. We are easily influenced by our everyday experiences. Sometimes these laws turn out to be wrong—like Aristotle's. Sometimes they turn out to be right, but only for a limited number of situations.

Sometimes, they are broad enough to encompass a wide range of experiences, like Newton's laws, and they become part of the scientific heritage we hand down to future generations.

For over 200 years, Newton's laws were thought to describe all motion, observed and hypothetical. Once again, however, physicists had not considered a type of motion they had never seen—motion at speeds near that of light. Albert Einstein did. Wondering what he would see while riding along a beam of light, Einstein completely redefined our concepts of space and time. His work exposed the limitations of Newton's laws and presented an even broader model—the special theory of relativity. Like Newton, Einstein imagined something that he could not observe. Yet his ideas, like Newton's, have been borne out by actual experiments. By looking beyond our own experience—by taking that one step beyond—we find simpler and more powerful descriptions of nature.

Energy

CHAPTER 9
Energy

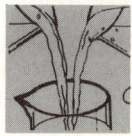
CHAPTER 10
Thermal Energy in Matter

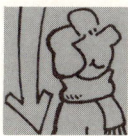
CHAPTER 11
How Thermal Energy is Transferred

CHAPTER 12
Thermodynamics

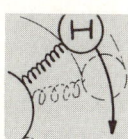
CHAPTER 13
Atoms, Molecules, and Thermal Energy

Energy has proven to be extraordinarily useful for explaining differences we often sense intuitively. A moving car has something more than a stationary car—kinetic energy. An airplane in flight has something more than an airplane on the ground—gravitational potential energy.

Our model of matter is, in most respects, a transposition of these experiences with macroscopic objects to the microscopic world of molecules. How is ice different from water? Ice has a different amount of potential energy stored in the bonds that hold its molecules together. How is hot water different from cold water? The molecules in hot water have more kinetic energy, on the average, than the molecules in cold water. The success of this model enables us to examine thermal processes in matter from a broader perspective and offers our first glimpse at the probabilistic view of matter.

Figure 9-1
© Gordon Alexander, Freelance Photographer's Guild.

Energy

Swinging back and forth (Figure 9-1). The motion seems endless in its repetition. But, as young children and their parents quickly learn, the swing's motion is neither spontaneous nor enduring. Left to itself, a swing hangs motionless, the downward force of gravity balanced by the upward force of the rope. We have to pull it back or push it forward to get it going. Once moving, the swing constantly exchanges height for velocity. At the top, the swing stops—momentarily suspended in midair. Remember the sensation? Increasing speed as it descends, the swing reaches its maximum speed as it hits bottom, only to slow back down to zero again at the top. Each time the swing moves upward, it reaches a lesser height; each time it swings downward, its velocity at the bottom is less. Unless we pull or push it again, the swing eventually stops.

We can describe the swing's motion in terms of the forces acting on it: the parent's push, the force due to gravity, and so on. But we often reach for other words to describe what we see. When we pull the swing back, we *give* it something that enables it to move. This something seems to be transformed—from position to motion to position—as the swing moves back and forth. Finally, this something is gradually lost as the swing slows down and stops. Whatever it is, seems irretrievably lost, for the swing never spontaneously starts to move again. The name we give to this something is energy.

Like momentum, *energy* is a commodity transferred during an interaction. In this chapter, we will look in detail at two forms of energy: *gravitational potential energy* and *kinetic energy*. We will see that when no friction exists, the sum of the gravitational potential energy and kinetic energy of a system remains *constant*. As we identify other forms of energy, we will generalize this constancy as the *law of conservation of energy*. When all parts of a system are identified and all forms of energy taken into account, the energy of a system is conserved.

INTERACTION, WORK, AND ENERGY

Change accompanies interaction. Whenever we observe a change, we know that an interaction has occurred. In Chapter 5 we introduced the concept of momentum to describe interactions in which objects change their velocities. Now we introduce energy, a concept that describes a much broader range of changes. Unlike momentum, there are many forms of energy.

The Energy Model

Intuitively, we define energy as the ability to make a change during an interaction. All objects possess energy—all are able to initiate change. We are not aware of this ability, however, until a change occurs. Consequently, we measure energy as it is transferred from one object to another. The giver is called the **energy source,** and the recipient is called the **energy receiver.** Energy transferred from a battery to a light bulb causes the bulb to emit light. Energy transferred from gasoline to an automobile engine causes the car to move. Energy transferred from you to the swing causes the swing to move. Energy transferred from the sun to a plant causes the plant to grow. In each case energy has been transferred from a source to a receiver, causing one of many possible kinds of change—production of light, motion, and growth, to name only a few.

In order to trace the transfer of energy from source to receiver, we need to be able to measure it. Because so many kinds of change can result from its transfer, measuring energy is not a simple matter. We begin by measuring energy in terms of work. Later, as we examine the many forms that energy can have, we will develop more direct ways of measuring it.

Energy and Work

Since real situations are often quite complex, let's begin with an idealized example. Imagine a box lying on a very well-waxed floor—so well waxed that we can ignore friction. Suppose you push the box with a force of 10 newtons (N) to the right while the box moves 2 meters (m) to the right (Figure 9-2). When you release the box, it will continue moving with the velocity it had the

182   Chapter 9.   Energy

Figure 9-2
A force of 10 N to the right moves the box 2 meters to the right. The work done is 10 N × 2 m or 20 joules.

moment you stopped pushing. There is no friction to slow it down. In this situation, you have acted as an energy source and the box as an energy receiver. The evidence that energy has been transferred is the change in the box's motion. Initially the box was stationary; now it is moving at a constant velocity.

We can measure the amount of energy that was transferred in terms of the work you did in pushing the box. **Work** is defined as the product of the magnitude of the force exerted on the object and the distance the object moves in the same direction as the force. This definition is summarized by the equation

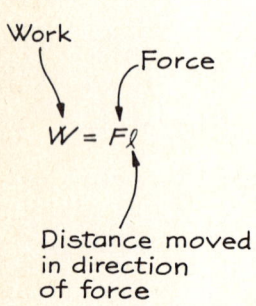

$$\text{Work} = \text{force} \times \text{distance}$$

Work is a scalar quantity. It depends only on the components of the force and distance that are in the same direction. The units in which work is measured are the units of force times the units of distance, newton-meters (N · m). This unit could be called a newton-meter, but it is given the name **joule** (J).

In moving the box, you exerted a force of 10 N to the right. The box moved a distance of 2 m to the right. Since the box moves in the same direction as you push, the work you do is equal to 10 N times 2 m, or 20 J. This work is a measure of the energy transferred from the energy source—you—to the energy receiver—the box. The box now has 20 J of energy that it did not have before, because you did 20 J worth of work on it. We are aware of this energy because the box is now moving.

To give you some feeling for the size of the joule, Figure 9-3 includes the energy content measured in joules for some common energy sources. You are probably used to measuring the energy content of some of these sources in other units, called calories. Various units are used for different kinds of energy, but all of them can be converted into joules. Today we use the joule as the standard unit with which to measure energy and work.

We commonly use the term *work* to describe any situation in which we exert forces, regardless of the result of our exertions. Most of us, for example, would claim that a weight lifter is doing work as she struggles but fails to lift a barbell (Figure 9-4). But physicists restrict their use of *work* to situations in which energy has actually been transferred. Until the barbell moves, the weight lifter has not transferred any energy to it. Consequently, she has done

Interaction, Work, and Energy   183

16,700 J/battery

160,000 J/oz
5280 J/12 oz jar

130,000,000 J/gal.

Figure 9-3
Energy content, in joules, of several energy sources.

Figure 9-4
The weight lifter has not done any work on the barbell until she actually lifts it, causing it to change position.

Figure 9-5

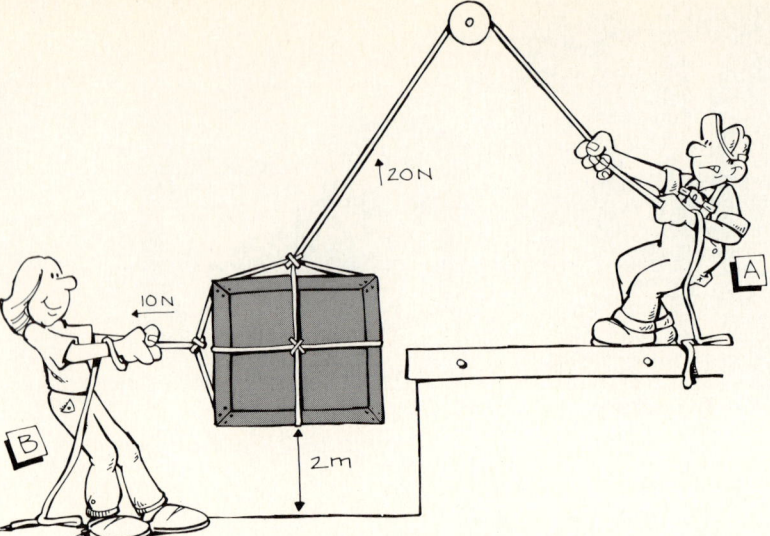

no work on it. This restriction is consistent with our definition of work. The struggling weight lifter exerts a force but does not move the weight any distance in the direction of the force.

SELF-CHECK 9A

Figure 9-5 shows two workers lifting a crate up to a loading dock. One lifts the crate upward with a force of 20 N, while the second pulls outward with a force of 10 N to keep the crate from swinging into the wall. The box moves upward a distance of 2 m.

a. Calculate the work done by each worker.

b. How much energy has been transferred to the box?

ENERGY OF POSITION

A house painter, startled by a wasp, jumps off a ladder. As he strikes the ground, energy is transferred from him (the energy source) to the ground (the energy receiver). While the energy is not actually transferred to the ground until he reaches it, the potential—or possibility—for this energy transfer existed as soon as he stepped on the ladder because of his height. Consider another example. We drop the pile driver shown in Figure 9-6. As it strikes the nail, energy is transferred from the hammerhead (source) to the nail (receiver). The energy is not transferred until the hammerhead hits the nail, but the potential for energy transfer existed as soon as the hammerhead moved upward. The painter and hammer both gain energy from their positions relative to the ground.

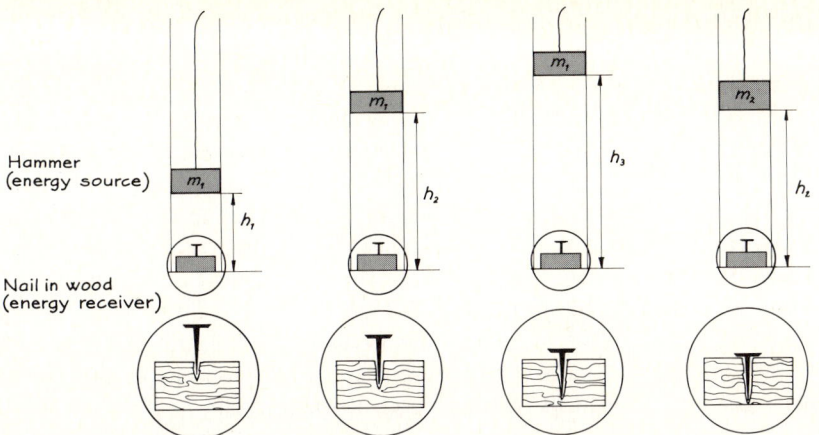

Figure 9-6
The hammer acts as an energy source and the nail as an energy receiver. The amount of energy transferred can be described in terms of the distance the nail is driven into the wood. The greater the height from which the hammer is dropped, the more energy it transfers to the nail. The greater the mass of the hammerhead, the more energy it transfers to the nail.

Gravitational Potential Energy

Gravitational potential energy is the energy stored in an object because of its position relative to a massive object, such as the earth. As the hammerhead is pulled upward, it gains gravitational potential energy because the earth is pulling it back downward. We must do work to raise the hammerhead against the force due to gravity. In the process of doing this work, we transfer energy to the hammerhead. The evidence that energy has been transferred is the new position of the hammerhead relative to the earth.

We can use the concept of work to develop a more complete definition of gravitational potential energy. The work done in moving an object away from the earth equals the gravitational potential energy gained by the object. To lift the hammer, for example, we must exert a force vertically upward. The magnitude of the force we exert is equal to the weight of the hammer—the product of its mass and the acceleration due to gravity. This is the force needed to keep the hammerhead moving at a constant velocity against the pull of gravity. The distance the hammer moves is its final height above the bottom of the pile driver. Since the force we exert is in the same direction as the hammer moves, the work we do is:

$$\text{Work} = \text{force} \times \text{distance}$$
$$= (\text{weight of hammer}) \times (\text{height of hammer})$$
$$= \binom{\text{mass of}}{\text{hammer}} \times \binom{\text{acceleration due}}{\text{to gravity}} \times (\text{height})$$

In lifting the hammerhead, we act as the energy source and the hammer acts as the energy receiver. The work we do becomes the gravitational potential energy stored in the hammerhead. Consequently, we define **gravitational potential energy** as the product of an object's mass, the acceleration due to gravity, and the object's height above some reference point:

$$\binom{\text{Gravitational}}{\text{potential energy}} = \binom{\text{mass of}}{\text{hammer}} \times \binom{\text{acceleration due}}{\text{to gravity}} \times (\text{height})$$

Gravitational potential energy — Mass, Height
$$\text{GPE} = mgh$$
Acceleration due to gravity

When mass is given in kilograms (kg), acceleration due to gravity in (meters/second)/second, and height in meters, gravitational potential energy is given in joules.

This definition allows us to calculate the gravitational potential energy stored in any object, regardless of how complex the energy-transfer process is. The house painter, for example, supplies his own energy as he climbs the ladder. Energy supplied by the food he eats is released in a series of complex chemical reactions; this enables him to climb the ladder. Without knowing any details of the process, we can calculate the increase in his gravitational potential energy from his mass (70 kg) and the height he climbs (4 m). He gains (70 kg)(9.8 (m/s)/s)(4 m) = 2744 J. As he stands at the top of the ladder, he has 2744 J of gravitational potential energy. When you eat $\frac{1}{20}$ tablespoon of yogurt, you gain about this much energy. It isn't much!

The Pile Driver

So far we have considered the hammerhead as an energy receiver. It is raised from the bottom of the pile driver and, in the process, acquires gravitational potential energy. Once raised, the hammerhead can act as an energy source. Its gravitational potential energy is eventually used to drive a nail into a piece of wood. The distance the nail is driven into the wood provides a rough estimate of the amount of gravitational potential energy the hammerhead had before its descent. We can use this relationship to gain additional insight into our definition of gravitational potential energy.

The definition of gravitational potential energy tells us that the hammer's energy depends on its mass, its height above the nail, and the acceleration due to gravity. Figure 9-6 describes the results of experiments in which the hammerhead's height and mass were varied. Dropped from different heights, the same hammer drives the nail different distances into the wood. Roughly speaking, when we double the height from which the hammer is dropped, we double the distance the nail is driven into the wood. Dropped from the same height, different hammerheads also drive the nail different distances into the wood. The larger the mass of the hammerhead, the greater the distance the nail is driven. The hammer's energy depends on its mass and on the height from which it is dropped.

A third variable is the effect of gravity. If our hammer were moved to an orbiting space station, it would never fall. Both the hammerhead and the space station fall toward the earth together; consequently, they do not move relative to one another. A less extreme example would be to take the hammer to the moon. There the gravitational attraction is about one-sixth that on earth. In an experiment identical to one performed on earth, the nail would be driven in only one-sixth as far on the moon. The hammer's energy depends on the strength of the gravitational force at its location.

The experiments with the pile driver increase our confidence in the definition of gravitational potential energy. The distance the nail is driven into the wood depends on the mass of the hammerhead, the height from which the hammerhead was dropped, and the acceleration due to gravity at the location

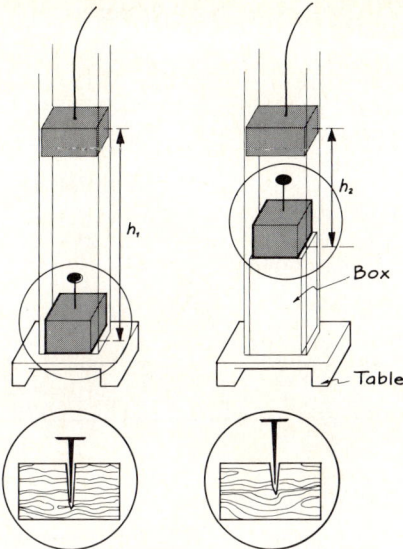

Figure 9-7
Relative to the table, both hammers have the same gravitational potential energy. Relative to the nail, the hammer on the left has more gravitational potential energy than the hammer on the right.

at which the experiment is performed—the same variables that define the gravitational potential energy of the hammerhead. The energy that the hammerhead transfers to the nail is equal to the energy the hammerhead gained when it was raised.

Gravitational Potential Energy is a Relative Concept

Gravitational potential energy is a relative, not absolute, concept. The acceleration due to gravity and mass are defined by the location and object, respectively; so these quantities are the same regardless of the energy receiver chosen. The height, however, depends on the position of the energy receiver chosen.

We can demonstrate the importance of the location of the energy receiver chosen with the experiment shown in Figure 9-7. The hammerhead has been lifted to a height of 1 m above the table. In (a) the nail is placed directly on the table. In (b) the nail is placed on a box halfway between the hammerhead and the table. While the height of the hammer relative to the table is the same in both cases, its height relative to the nail is not.

Once the hammer is dropped, the importance of the reference point chosen becomes apparent. Relative to the table, the hammer had the same gravitational potential energy in both situations. Relative to the nail, the hammer in (a) had the greater gravitational potential energy. The nail has been driven in further in (a) than in (b). The amount of gravitational potential energy possessed by the hammer is only of interest to us when it has been transferred to an energy receiver. Consequently, the vertical distance from the energy source to energy receiver (hammer to nail) is used to calculate the gravitational potential energy of the source.

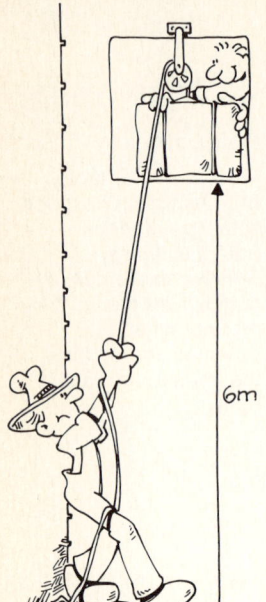

Figure 9-8

SELF-CHECK 9B

In the good old days, hay bales were lifted to barn lofts by pulleys and ropes, as shown in Figure 9-8.

a. What is the gravitational potential energy of the 10-kg hay bale relative to the farmer?

b. What is the gravitational potential energy relative to his son?

c. If the bale drops, who would receive more energy—the father or the son?

Cable Cars and Swings

Our civilization has many devices for changing the gravitational energy of people and objects. Ski lifts and elevators change the gravitational potential energy of people. Cranes change the gravitational potential energy of building materials as the cranes lift the materials high above the steel skeletons of giant skyscrapers. Usually other forms of energy—electricity, oil, or diesel fuel—are used to accomplish these changes. The San Francisco cable car system, however, uses the decrease in the gravitational potential energy of one object to increase the gravitational potential energy of another.

San Francisco, with its steep hills, presents a challenge to nearly all modes of transportation. In the 1870s its now-famous cable car system was designed and constructed. A cable car, as its name implies, is connected to long cables that move beneath the street. To move the car, the operator pulls a lever that connects the car to the cable. To stop the car, he or she releases the car from the cable and sets the brake. The cable for each car line forms a complete loop; so cars going uphill and cars going downhill are attached to the same cable (Figure 9-9).

The gravitational potential energy of the cars is constantly changing. Cars going up increase their gravitational potential energy, while cars going down decrease theirs. A car that is moving downhill does work on a car that needs to go uphill—it pulls it up. In this way the gravitational potential energy lost by a downhill car is gained by an uphill car—an extremely efficient design!

The chapter opened with the photograph of a swing. We were looking for concepts to describe its motion. Gravitational potential energy provides us with one such concept. Pulled back from rest (Figure 9-10), the swing gains gravitational potential energy. Though the swing is being pulled both outward and upward, only the change in vertical height contributes to this gain. Suppose an empty swing of mass 2 kg is pulled to a vertical height of 1.2 m above its position at rest. It will have gained (2 kg)(9.8 (m/s)/s)(1.2 m) = 23.5 J. Once the swing is released, this gravitational potential energy decreases to zero as the swing moves back to its original position.

© Gene Ahrens, Freelance Photographer's Guild.

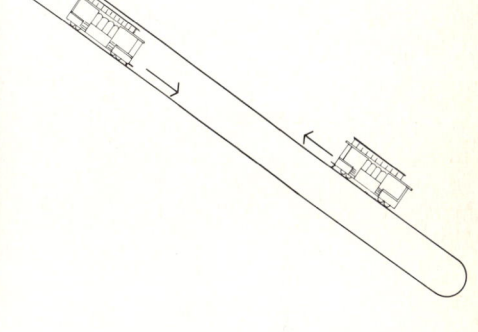

Figure 9-9
The cable forms a complete loop so that cars going downhill can transfer energy to the cars going uphill.

ENERGY AND MOTION

The swing's speed increases as its gravitational potential energy decreases. The same is true about the house painter, the pile driver, and a San Francisco cable car. The instant before he hits the ground, the house painter has no gravitational potential energy, but he is moving. The instant before it transfers its energy to the nail, the hammerhead has no gravitational potential energy, but clearly it has energy to transfer. The motion of the cable cars traveling downhill somehow supplies the gravitational potential energy needed to move other cars uphill. Energy must be present in moving objects.

Variables that Affect Energy of Motion

A convenient example with which to investigate the energy associated with motion is a car. A moving car has energy. This energy changes when the frictional force of the brakes does work to bring the car to a stop. When we apply the brakes, the wheels lock; increased friction between the tires and the road stops the car. The product of the constant force exerted by friction and the distance the car travels in coming to a stop, called the *stopping distance*, is the work done by friction. Like the energy gained by the box when we did work in pushing it, the energy lost as the car comes to a stop should equal the work done by friction. This relationship between work and energy enables us to identify the variables that describe the car's energy while moving.

From experience we know that the faster we drive, the greater the stopping distance we require. Figure 9-11 shows a diagram found in most driver-training manuals. The average stopping distance is shown for cars moving initially at different speeds. Since the frictional force applied by the surface is

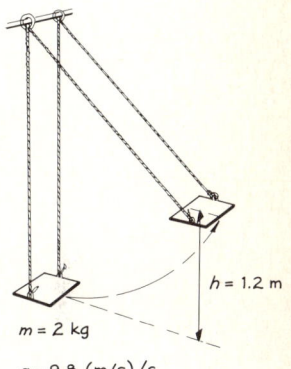

$m = 2$ kg
$h = 1.2$ m
$g = 9.8$ (m/s)/s

Figure 9-10
Pulled back from its rest position, a swing gains gravitational potential energy equal to its mass times the acceleration due to gravity times its vertical height above the rest position.

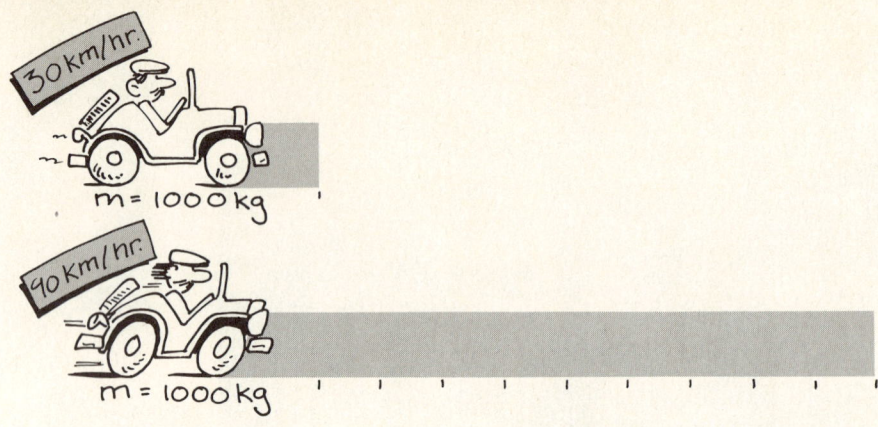

Figure 9-11
The stopping distance depends on speed. At 90 kilometers per hour, the stopping distance required is nine times that at 30 kilometers per hour.

roughly the same in each situation, the stopping distance provides a direct measure of the energy associated with the speed each car is moving. At 90 kilometers per hour (km/h), the stopping distance is four times that at 45 km/h and nine times that at 30 km/h. Doubling the speed quadruples the stopping distance required; tripling the speed increases the stopping distance by a factor of nine. Energy of motion is related to the square of the speed.

A second important variable is mass. Most owner's manuals include information about the stopping distance required when you apply the brakes to a car moving initially at 60 miles per hour (mi/h). This distance is listed for different loads—typically, light loads and maximum loads. If we were to examine the stopping distance for these different loads, we would find that doubling the mass of the car doubles the stopping distance required (Figure 9-12). The car's energy while moving is directly related to its total mass.

Kinetic Energy

Kinetic energy is the name given to the energy an object possesses by virtue of its motion. The swing, the house painter, the hammer, and the cable car all have kinetic energy as they move downward. Our experience with stopping distances for cars suggests that this energy is related to the mass of the object and the square of its speed. By combining Newton's laws with our definitions of work, velocity, and acceleration, we can develop a complete definition for kinetic energy:

$$\text{Kinetic energy} = \tfrac{1}{2} \times (\text{mass}) \times (\text{speed})^2$$

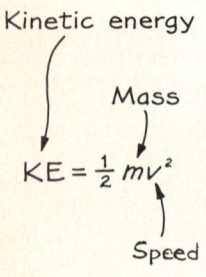

The **kinetic energy** (KE) of an object is one-half the product of the object's mass and the square of its speed. When mass is expressed in kilograms and speed in meters/second (m/s), the kinetic energy is given in joules.

This definition allows us to calculate the kinetic energy of any moving object, given its mass and speed. For example, we can compare the kinetic energy of a car moving along an interstate highway with that of the same car traveling in a school zone. Assume that the car has a mass of 1000 kg (subcompact size). The speed limit on an interstate highway is about 25 m/s (55 mi/h), while that in a school zone is about 9 m/s (20 mi/h).

Figure 9-12
The stopping distance depends on mass. Doubling the mass doubles the stopping distance required.

Interstate Highway
$$KE = \tfrac{1}{2}mv^2$$
$$= \tfrac{1}{2}(1000 \text{ kg})(25 \text{ m/s})^2$$
$$= 312{,}500 \text{ J}$$

School Zone
$$KE = \tfrac{1}{2}mv^2$$
$$= \tfrac{1}{2}(1000 \text{ kg})(9 \text{ m/s})^2$$
$$= 40{,}500 \text{ J}$$

Because kinetic energy depends on the square of the speed, the increase in kinetic energy is substantial. It is not hard to see why accidents are so much more devastating on the highway than in city traffic.

Kinetic energy is very similar to the concept of momentum introduced in Chapter 5. Both depend on the object's mass and motion. Momentum depends on the object's mass and its velocity. Kinetic energy depends on the object's mass and the square of its speed. When an object's velocity doubles, its momentum doubles, while its kinetic energy increases by a factor of four. Another important difference between the two concepts is that momentum is a vector quantity, while kinetic energy is a scalar. In a system consisting of two objects, the total momentum of the system can be zero even though both objects are moving. The total kinetic energy of this same system can never be zero. Kinetic energies always add.

SELF-CHECK 9C

The 2-kg swing in Figure 9-10 has a speed of 4.86 m/s at the bottom of its swing. What is its kinetic energy? How much work must be done to stop it?

ENERGY IS CONSERVED

Let's return once more to the motion of the swing. In pulling the swing back from rest, we do work on it—giving it the energy that enables it to move.

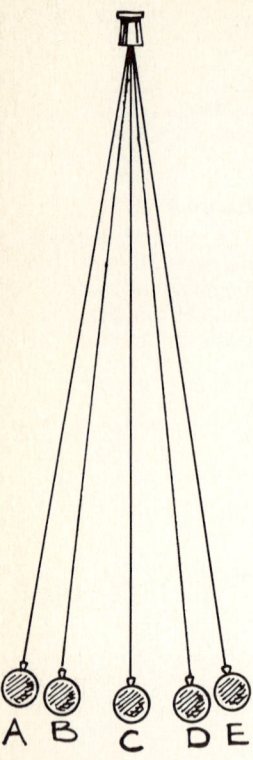

Figure 9-13
The motion of the pendulum bob captured in a strobe-like drawing shows the change in height and velocity of the pendulum ball.

Armed with the concepts of gravitational potential energy and kinetic energy, we can now describe this motion in more detail. To do so, we first examine the motion of a simplified version of the swing, the pendulum.

Potential Becomes Kinetic Becomes Potential

The strobe-like drawing in Figure 9-13 captures the motion of a pendulum bob through one complete swing. Successive images are closely spaced at A, become more widely spaced at C and then once again become more closely spaced at E. Using the methods developed in Chapter 2, we would describe the speed of the pendulum bob as being at a minimum at A and E and a maximum at C. By contrast, the height of the pendulum bob above its rest point is at a maximum at A and E and a minimum at C. Height seems to be exchanged for motion and then motion for height.

We can use the concepts of gravitational potential energy and kinetic energy to describe this interchange of height and motion. Figure 9-14 shows the motion of an idealized pendulum, one in which no friction acts to slow its motion. In order to calculate the gravitational potential energy of the pendulum bob at any point in its swing, we need to know the bob's mass, its height above the rest position, and the acceleration due to gravity. The kinetic energy of the pendulum bob depends on its mass and its speed at each specific location. Table 9-1 includes the height and speed of a 1 kg pendulum bob at several points along its path. The gravitational potential energy and kinetic energy have been calculated for each location.

As shown in Table 9-1, the pendulum bob has maximum gravitational potential energy at A. As it moves downward, its gravitational potential energy decreases steadily, while its kinetic energy increases. At the bottom of the swing, the bob has no gravitational potential energy but a maximum kinetic energy. As the pendulum moves upward, the process reverses itself: Kinetic energy decreases while gravitational potential energy increases. At E the gravitational potential energy is again at a maximum and the kinetic

Table 9-1 Gravitational Potential Energy and Kinetic Energy of Pendulum

Position (Figure 9-14)	Height (m)	Gravitational Potential Energy (J)	Speed (m/s)	Kinetic Energy (J)	Total Energy (GPE + KE) (J)
A	2.04	20	0	0	20
B	1.02	10	4.47	10	20
C	0	0	6.32	20	20
D	1.02	10	4.47	10	20
E	2.04	20	0	0	20

Throbbing Motion

The arteries and veins in our bodies form a network that is about 96,000 km long, through which about 5 liters of blood circulate. The rate at which this blood circulates depends on the energy supplied by the pumping action of the heart. We can use the concept of kinetic energy to compare this energy at both low and high levels of activity.

The heart rate increases as we increase our level of activity. This increased heart rate accomplishes two things: (1) it increases the mass of blood ejected with each contraction, and (2) it increases the speed with which the ejected blood moves. At a low level of activity, a single contraction of the heart ejects .06 kg of blood at an average speed of 0.3 m/s. At a higher rate of activity, each contraction ejects .12 kg of blood at an average speed of 0.6 m/s. The kinetic energy of the ejected blood associated with each level of activity is:

Low Level
$$KE = \tfrac{1}{2}mv^2$$
$$= \tfrac{1}{2}(.06 \text{ kg})(0.3 \text{ m/s})^2$$
$$= 0.0027 \text{ J}$$

High Level
$$KE = \tfrac{1}{2}mv^2$$
$$= \tfrac{1}{2}(.12 \text{ kg})(0.6 \text{ m/s})^2$$
$$= 0.0216 \text{ J}$$

Because both mass and speed are increased, the heart must supply considerably more energy at higher levels of activity.

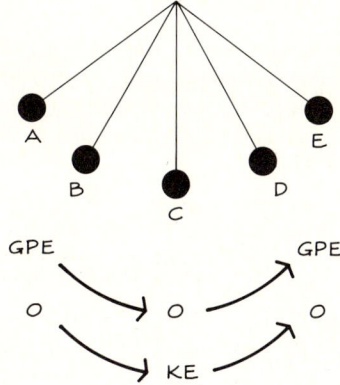

Figure 9-14
The gravitational potential energy is a maximum at each end and zero in the middle. The kinetic energy is a maximum in the middle and zero at each end.

energy is zero. Gravitational potential energy has been transformed into kinetic energy and back into gravitational potential energy.

Conservation of Energy

The transformation of gravitational potential energy into kinetic energy and back again can be understood in terms of a much broader concept—the law of conservation of energy. Notice that the gravitational potential energy of the pendulum bob at A is equal to the kinetic energy at C. In fact, if you add the gravitational potential energy and kinetic energy for each position of the swing, you'll see that it always totals 20 J. The sum of these two quantities is conserved.

REMINDER

In everyday use, to conserve means to save. In physics, to conserve means to keep constant. Confusing the everyday use with the physics definition may be hazardous to your understanding.

The **law of conservation of energy** states that the total energy of a closed system remains constant. In the case of our idealized pendulum, which does not interact with air or with any supporting mechanism, the closed system consists of the pendulum and the earth. In pulling the pendulum back from its rest position, we give the closed system 20 J of energy. Once we release the pendulum, the energy of that system remains constant. The interaction between the pendulum and the earth causes the energy to change form—from gravitational potential energy to kinetic energy and back again. As the gravitational potential energy decreases, the kinetic energy increases, keeping the total energy of the system constant at 20 J. We can summarize this with a single equation:

Total energy at one time = total energy at later time

$$(GPE + KE)_{time\ 1} = (GPE + KE)_{time\ 2}$$

Because the energy of the system is conserved, our ideal pendulum keeps swinging forever! Of course, you have never seen a pendulum do this because you have never seen a closed system consisting of just the earth and the pendulum. Energy exchanges on earth, as you will see, involve other energy receivers and other forms of energy.

SELF-CHECK 9D

In Figure 9-6 the model pile driver has a hammer mass of 5 kg and is raised to a height of 1 m above the nail. Ignore frictional interactions.

a. What is the total energy of the system?

b. At a later time, the moment before the hammer strikes the nail, what is the total energy of the system?

c. What is the hammer's kinetic energy right before it strikes the nail?

EASIER THAN WALKING?

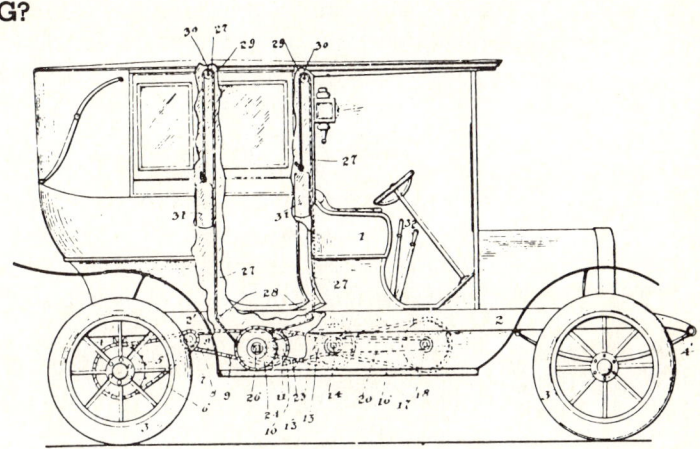

Why waste the energy provided by fossil fuels when we have all that gravitational potential energy around? That question was considered by Isaac Smyth in 1911 when he invented the gravity-powered automobile. Inside the walls of the car are weights (31), which can be raised to the roof level (30) by means of a crank (26). When released, the weights drop and transmit their gravitational potential energy through pulleys and cables (27 and 29) to the drive chains (9 and 20). Thus, the gravitational potential energy of the weights is converted into kinetic energy of the automobile. Unfortunately, the gravity-driven car had a couple of disadvantages. First, the back wheels had to be lifted off the ground while raising the weights. More importantly, the energy to raise the weights had to come from somewhere. The driver needed to supply a lot of energy in order to get to the grocery store!

ENERGY TAKES MANY FORMS

Like conservation of momentum, conservation of energy demonstrates our search for quantities that remain constant throughout interactions. Unlike momentum, energy can assume more than one form. Conservation logic has enabled us to identify these other forms of energy.

Energy Does Not Always Seem To Be Conserved

You do work—giving a child a push on a swing. The child seems to be enjoying thoroughly the change from gravitational potential energy to kinetic energy and back again. The next thing you know, the swing stops and the child is crying. The child has neither gravitational potential energy nor kinetic energy. Real swings and real pendulums eventually stop.

The ski lift takes you to the top of the mountain. Using energy supplied by the lift, you have increased your gravitational potential energy. As you ski down the hill, your kinetic energy increases—but never as much as your gravitational potential energy decreases. When you reach the bottom of the mountain, you gradually come to a stop. Both your gravitational potential energy and your kinetic energy are gone.

In real interactions, the sum of the gravitational potential energy and kinetic energy does not remain constant. If we watch the skier or the swing long enough, we see the sum of these two forms of energy gradually decrease. If energy is really conserved, this lost energy must be transferred to other receivers within our closed system. Let's look for these receivers and methods by which the "lost" energy could have been transferred to them.

Thermal Energy

The skier-earth and swing-earth systems are not closed. In both situations, interactions occur with other objects through frictional forces. The skier slides along the snow. The swing rubs against its point of suspension. Both the swing and the skier experience air resistance. These interactions identify the other energy receivers.

We can measure the amount of energy transferred to these objects in terms of the work done by frictional forces. The energy transferred from the skier to the snow, for example, is equal to the product of the frictional force exerted by the snow and the distance the skier travels.

$$E_s = F_f \ell$$

(Energy transferred to snow; Frictional force; Distance traveled)

Energy transferred to snow = work done by friction

$$= \begin{pmatrix} \text{force of friction} \\ \text{exerted by snow} \end{pmatrix} \times \begin{pmatrix} \text{distance} \\ \text{traveled} \end{pmatrix}$$

A frictional force of 0.1 N applied along a ski slope 100 m long results in 10 J of energy being transferred to the snow. This energy usually appears in the form of heat, or **thermal energy.** Careful measurements of the snow would show a slight increase in its temperature. We could repeat the analysis for the frictional force offered by air resistance. We expect to find that the surrounding air would be warmed slightly as energy is transferred from the skier to the air.

Energy is still conserved. Our law of conservation of energy is simply modified to take into account this new form of energy.

Total energy at one time = total energy at later time

$$(\text{GPE} + \text{KE} + \text{thermal})_{\text{time 1}} = (\text{GPE} + \text{KE} + \text{thermal})_{\text{time 2}}$$

The addition of a new form of energy means that we need to look for changes other than changes in position or motion. Chapter 10 deals with the changes that result from a transfer of thermal energy.

SELF-CHECK 9E

A skier whose total mass is 80 kg stood at the top of a ski slope whose vertical height was 100 m. At the bottom of the slope, the skier's kinetic energy was 50,000 J.

a. What was the skier's gravitational potential energy at the top of the slope?

b. Was all of this energy converted into kinetic energy? If not, where did it go?

Other Forms of Energy

The analysis we have just completed illustrates the process by which physicists identify other forms of energy. Conservation logic is such a compelling part of our experience that whenever conservation of energy seems to fail, we begin looking for a previously unknown form of energy. So far this procedure has always worked, and many forms of energy have been identified. All forms of energy can be categorized as potential or kinetic.

Potential energy describes the energy an object has by virtue of its position. It can be thought of as energy stored with the object. In the case of gravitational potential energy, the object's position is measured relative to the earth. The work done in opposing the gravitational force leads to an increase in an object's gravitational potential energy. Each of the other fundamental interactions—electrical, strong nuclear, and weak nuclear—also have forms of potential energy associated with them. Electrical interactions are the glue that holds matter together. Electrical forces bind electrons to atomic nuclei; they bind atoms to one another to form molecules; they bind molecules together to form cohesive materials. Because electrical interactions are such a fundamental part of matter, **electrical potential energy** can take many forms. A coiled spring, for example, has **elastic potential energy** that results from electrical interactions between molecules in the spring. Another example is the **chemical potential energy** in molecules of chemical compounds. This stored energy is due to the electrical interactions that hold atoms in specific positions within molecules of the compounds. Both elastic and chemical potential energy are the result of work—forces acting over distances to compress the spring or to arrange the atoms into molecules. **Nuclear potential energy** exists for both strong and weak nuclear interactions. The most awesome demonstration of this form of energy occurred when the nuclear bombs were dropped at Hiroshima and Nagasaki.

Kinetic energy is the energy of motion. While we have used it only to describe the energy of motion associated with large objects, we can use it to describe other forms of energy associated with the motion of atoms or molecules. For large objects we measure kinetic energy in terms of the mass and speed of each object. At the microscopic level, we measure energy of motion in terms of the temperature of the object. Atoms and molecules move constantly, transferring their energy to one another through collisions. We perceive this transfer as temperature change. This kind of energy of motion is called **thermal energy,** or heat. A third form of energy that involves motion is **wave energy.** Common examples include sound and light.

Mass is a Form of Energy Too

One of the more surprising forms energy takes is mass. Suppose we are exerting a net force on an object, causing it to accelerate. We act as the energy source and the object acts as the energy receiver. The work we do as we apply this net force appears as an increase in the object's kinetic energy. At usual speeds the object's kinetic energy quadruples each time its speed doubles. As discussed in Chapter 6, however, we cannot keep accelerating the object forever. As the object's speed approaches that of light, its mass increases. From our point of view, we have to exert even more force to accelerate it further. We are still delivering energy to the object, but the energy no longer produces the same change in speed as it did at lower speeds. At speeds near that of light, our energy appears in the increased mass of the object. Energy is being converted into mass.

Einstein described this **mass-energy equivalence** in what is probably the most famous equation of twentieth-century physics:

$$\text{Energy} = \text{mass} \times (\text{speed of light})^2$$

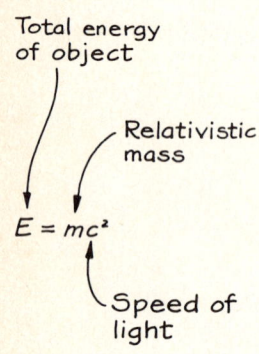

It states that the total energy of an object equals its relativistic mass times the speed of light squared. When mass is expressed in kilograms and the speed of light in meters per second, the energy is given in joules. As we saw in Chapter 6, the relativistic mass of an object varies with its speed. Now we see that this increase in mass reflects an increase in the amount of energy stored with that object.

The most surprising conclusion of mass-energy equivalence comes when an object is not moving. At zero speed, an object still has mass, its rest mass. Einstein's equation tells us that we can associate this mass with a certain amount of energy — the object's rest mass times the speed of light squared. If we could convert all the rest mass found in 2 kg of potatoes into energy, we would have $(2 \text{ kg}) \times (3 \times 10^8 \text{ m/s})^2 = 18 \times 10^{16}$ J of energy. Because the speed of light is such a large number, the energy equivalent of an object's rest mass is enormous!

Mass-energy equivalence requires that we modify our conservation principles. The total energy of a system must now include the energy equivalent of the rest masses of the objects found in the system. In everyday interactions, the rest masses of objects remain constant, so we can effectively ignore them. At high speeds or in interactions in which the total mass of the system changes, the energy associated with mass becomes important. Mass-energy equivalence has been dramatically demonstrated in nuclear interactions. Both nuclear reactors and nuclear weapons derive their energy from decreases in

CHANGING ENERGY FORMS: CANINE TO KINETIC

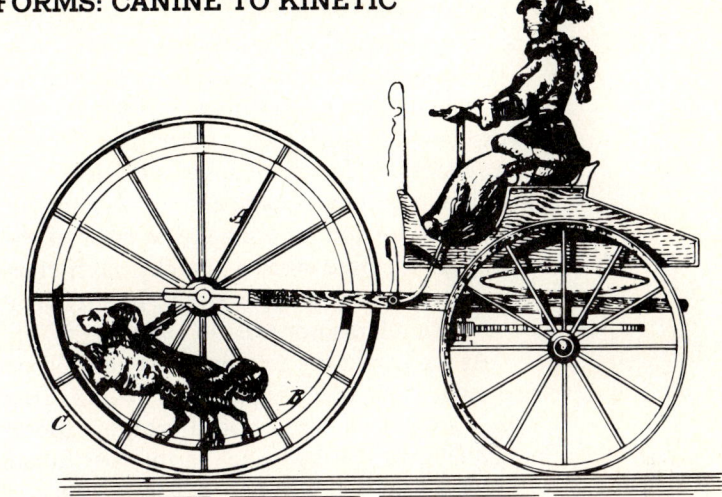

Modern vehicles use the chemical potential energy stored in gasoline to create the kinetic energy of the car. Just about the same time as the automobile was being invented, F. H. C. Mey came up with the idea of using the chemical potential energy stored in a different source—a source (*B*) available in many homes. The users feed their dogs lots of energy in the form of food. Then, the dogs are placed in the hollow wheel (*C*) of Mr. Mey's specially designed vehicle. With a little urging from the whip (unlabeled), the dogs start running and turn the wheel that pulls the vehicle and its passenger. The chemical potential energy of the dog food is converted into kinetic energy of the person and vehicle.

the mass of atomic nuclei. These interactions will be discussed in more detail in Chapter 22. When the total mass of the system changes, mass-energy equivalence must be included in order to conserve energy.

The process by which we developed definitions of gravitational potential energy and kinetic energy can be used to develop definitions of these other forms of energy. The remaining chapters in this book deal with many of these forms of energy in turn. Chapters 10-13 consider the electrical potential energy and thermal energy found in matter. Chapters 14-16 discuss wave energy. Chapters 17 and 18 look at the electrical energy involved in the atom. Chapters 19 and 20 discuss electrical energy, and Chapters 21 and 22 discuss nuclear energy. The breadth of phenomena encompassed by the various forms of energy make energy a powerful concept—one that unites the various fields of physics.

CHAPTER SUMMARY

Energy is the ability to make a change during an interaction. Like momentum and unlike force, it is a commodity that is transferred from one object to another during an interaction. Energy that is transferred from an *energy source* to an *energy receiver* results in an observed change in the energy receiver. Energy can be measured in terms of the work done on the receiver. *Work* is defined as the product of the force exerted on an object and the distance the object moves in the same direction as the force. Work and energy are measured in units called *joules*.

This chapter formally defines two of the many forms of energy. *Gravitational potential energy* is the energy due to the position of an object relative to a massive object such as the earth. It is formally defined as the product of the object's mass, its height relative to a selected reference point, and the acceleration due to gravity. *Kinetic energy* is energy associated with the motion of an object. It is defined as one-half the mass of the object times the square of its velocity. In a closed system in which no friction exists, the sum of the gravitational potential energy and kinetic energy remains constant. The *total energy* of the system is *conserved*.

As we apply the principle of energy conservation to more complex systems, new forms of energy are discovered. These forms of energy can be categorized as being energy due to position (potential energy) or energy due to motion (kinetic energy). Each of the four fundamental interactions has a form of potential energy associated with it: *gravitational potential, electrical potential, strong nuclear potential,* and *weak nuclear potential* energies. Energy of motion includes the kinetic energy we defined for large objects, *thermal energy* associated with the motion of atoms and molecules in matter, and *wave energy* associated with sound and light. At speeds near that of light, energy is stored in the increased mass of the object. Einstein demonstrated a *mass-energy equivalence,* in which the total energy of an object is the product of its relativistic mass and the speed of light squared. The energy of a closed system remains constant at all times in systems in which all forms of energy have been identified. This principle is called the *law of conservation of energy.* The concept of energy unites the various fields of physics.

ANSWERS TO SELF-CHECKS

9A. a. Worker A:

$$\text{Work} = \text{force} \times \text{distance}$$
$$= (20 \text{ N}) \times (2 \text{ m})$$
$$= 40 \text{ J}$$

Worker B does no work on the crate. The force B exerts does not result in any horizontal motion of the crate. Consequently, the distance the box moves in the direction of B's force is zero.

b. 40 J of energy have been transferred to the crate.

9B. a. The bale is 6 m above the farmer.

$$\text{GPE} = mgh = (10 \text{ kg})[9.8 \text{ (m/s)/s}](6 \text{ m})$$
$$= 588 \text{ J}$$

b. The bale is 0 m above his son.

$$\text{GPE} = mgh = (10 \text{ kg})[9.8 \text{ (m/s)/s}](0 \text{ m})$$
$$= 0 \text{ J}$$

c. The farmer would receive more energy.

9C.
$$\text{KE} = \tfrac{1}{2}mv^2 = \tfrac{1}{2}(2 \text{ kg})(4.86 \text{ m/s})^2$$
$$= 23.6 \text{ J}$$

Work done to stop the swing is equal to the kinetic energy lost by the swing: work = 23.6 J.

9D. a. Initially, the total energy of the system is the gravitational potential energy of the hammer.

$$\text{Total energy} = \text{GPE} = mgh$$
$$= (5 \text{ kg})[9.8 \text{ (m/s)/s}] (1 \text{ m})$$
$$= 49 \text{ J}$$

b. The total energy of the system is conserved.

$$\text{Total energy at one time} = \text{total energy at later time}$$

The total energy is 49 J.

c. The moment before the hammer strikes the nail, the gravitational potential energy is zero. The total energy of the system is now the kinetic energy of the hammer head.

$$\text{Total energy} = \text{KE} = 49 \text{ J}$$

9E. a.
$$\text{GPE} = mgh = (80 \text{ kg})[9.8 \text{ (m/s)/s}](100 \text{ m})$$
$$= 78{,}400 \text{ J}$$

b. The kinetic energy at the bottom was 50,000 J. Not all the gravitational potential energy has been converted into kinetic energy. (78,400 J − 50,000 J), or 28,400 J, went into thermal energy.

PROBLEMS AND QUESTIONS

A. Review of Chapter Material

A1. Define the terms listed below:

Energy	Gravitational potential energy	Thermal energy	Work
Energy source	Kinetic energy	Energy of position	Energy conservation
Energy receiver	Mass-energy equivalence	Energy of motion	

A2. In what units are work and energy measured?

A3. Why does the object have to move in the direction of the force in order for work to have been done?

A4. What variables affect the gravitational potential energy of an object?

A5. Why is gravitational potential energy a relative concept?

A6. What two variables affect the kinetic energy of an object?

A7. Which will cause the greater increase in an object's kinetic energy—doubling its mass or doubling its speed?

A8. How does common use of the term *to conserve* differ from the way it is used in physics?

A9. Describe the process by which new forms of energy are often discovered.

A10. What two categories describe the various forms of energy?

A11. How would you determine the energy equivalent when you know the mass of an object? How does this equivalence modify the law of conservation of energy?

B. Using the Chapter Material

B1. A crane exerts a constant force of 9800 N and lifts a slab of concrete onto a section of scaffolding 100 m above the ground. What is the work done by the crane? What form of energy is transferred to the concrete?

B2. If the slab of concrete (mass = 100 kg) in Problem B1 falls, what is its kinetic energy just before it strikes the ground? (Neglect friction.)

B3. Include friction and explain why the actual kinetic energy of the concrete slab will be less than the value calculated in Problem B2.

B4. In which of the following situations is no work done?
   a. A spaceship moves at constant velocity.
   b. A child slides down a playground slide.
   c. You push on a heavy box but cannot move it.
   d. You slam on the brakes and your car stops quickly.

B5. The acceleration due to gravity on Mars is 3.7 (m/s)/s. What is the gravitational potential energy of a 70 kg Martian who is 3 m above the surface of the planet?

B6. A 500 kg elevator is stopped on the second floor of a building. What is the gravitational potential energy relative to the first floor, 10 m below the second? Relative to the basement, 10 m below the first floor?

B7. What is the kinetic energy of a 50 kg skateboarder who is traveling at each of these speeds: 1 m/s, 2 m/s, 3 m/s, 4 m/s, 6 m/s? How does kinetic energy vary with speed?

B8. What is the kinetic energy of bicyclists who are traveling at 5 m/s and have the following masses: 40 kg, 50 kg, 80 kg, 100 kg? How does kinetic energy vary with mass?

B9. If you watch a gymnast revolving around a crossbar, you will notice that her speed is lowest when she is directly above the bar and greatest when she is directly below it. Explain this observation in terms of conservation of energy.

B10. Juliet is locked in a tower by her father. Romeo, standing directly below her open window, wishes to throw her a rock with a message attached. The window is 10 m straight up and the rock and message have a mass of 0.3 kg. How much kinetic energy must Romeo give the rock in order for it to reach Juliet?

B11. If you were converted entirely to energy, how much energy would you be?

C. Extensions to New Situations

C1. A child slides down a hill on a sled. At the top of the hill he has 800 J of gravitational potential energy. At the bottom he has 200 J of kinetic energy. How much energy was transformed into thermal energy due to the frictional interaction between the sled and the snow?

C2. Most roller coasters are pulled by a chain to the top of the first hill on the ride. At that point the speed of the roller coaster is approximately zero. After that no more energy is put into the roller coaster throughout the ride.
   a. What types of energy must designers of roller coasters consider as they plan a new roller coaster?
   b. Is it possible for any other hill to be higher than the first one?
   c. For most roller coasters, each hill is shorter than the previous one. Is this design necessary?

C3. When we first studied collisions, we saw that momentum was conserved for collisions in a closed system. Consider energy conservation for these same collisions. The collision involved is one in which a 5 kg ball moving at 2 m/s strikes a stationary 5 kg ball head on. For parts (a)–(c) we assume that the balls do not stick together.
   a. Use momentum conservation to determine the speed of each ball after the collision.
   b. What is the kinetic energy of each ball before the collision? After the collision?
   c. Is kinetic energy conserved in this collision?
   d. Now assume that the two balls stick together after the collision. Use momentum conservation to determine their speed after the collision.
   e. What is the total kinetic energy before the collision? After the collision?
   f. Is kinetic energy conserved in the sticky collision?
   g. Use the results of (a)–(f) to make a general statement about conservation of kinetic energy in collisions.

C4. A ball that strikes the floor changes its shape slightly as it interacts with the floor (Figure 9-C4). Usually some energy is changed from kinetic to other forms during this process.

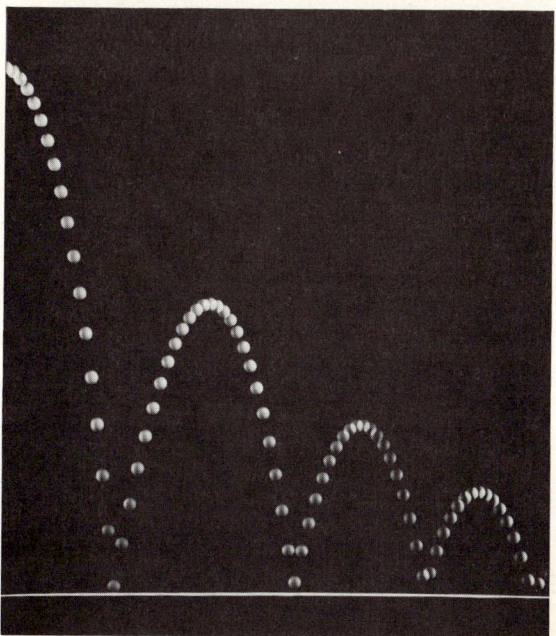

   a. In this situation the maximum height reached after the first bounce will be less than the height from which the ball was dropped. Explain why.
   b. How would the maximum height after each successive bounce compare to the one before it?

C5. A specific example of Problem C4 is a 0.5 kg ball that is dropped from a height of 4 m.
   a. What is its gravitational potential energy at the start of the fall?
   b. What is its kinetic energy just before it strikes the ground?
   c. On the first bounce the ball returns to a maximum height of 3.5 m. What is its gravitational potential energy when it reaches that height?
   d. What is its kinetic energy just as it leaves the ground after the first bounce?
   e. How much energy was transformed into other forms of energy during the bounce?
   f. What reasoning would you use to predict the approximate maximum height the ball would reach after the second, third, and fourth bounce?

C6. An interesting toy consists of five pendulums suspended as shown in Figure 9-C6 (a). The before-after pictures in Figure 9-C6(b) and (c) show what happens when we pull back and release different numbers of balls. Using momentum conservation alone, we could explain the after motion knowing only the before situation. But, momentum conservation allows events we never see. For example, in the situation in (c) we always see two balls go out after the collision. Yet one ball going out at twice the speed of the incoming balls would still conserve momentum. To understand what actually happens, consider an example in which two balls are moving before the collision. Each ball has a mass of 1 kg and, just before the collision, is moving at 1 m/s.
   a. Is momentum conserved if, just after the collision, each of two balls has a speed of 1 m/s? If one ball has a speed of 2 m/s? If each of four balls has a speed of 0.71 m/s?
   b. Is kinetic energy conserved if just after the collision each of two balls has a speed of 1 m/s? One ball has a speed

204   Chapter 9.   Energy

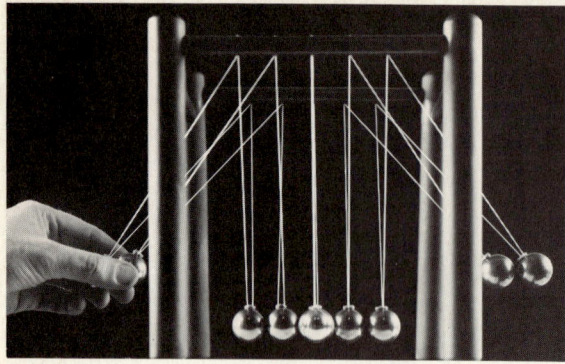
(a) © 1972, Fundamental Photographers.

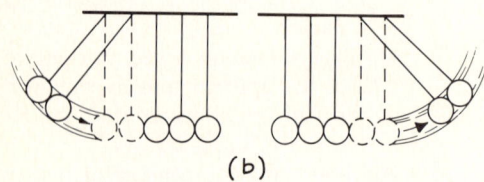
(b)

of 2 m/s? Each of four balls has a speed of 0.71 m/s?
  c. In which case are both momentum and energy conserved?
  d. Why are the results shown in the figure the only ones that occur in nature?
C7. You can get from the bottom to the top of the Empire State Building (a distance of approximately 400 m) in two ways, by walking or by riding the elevator. For this problem, assume that you have a mass of 75 kg.
  a. Does the amount of work done depend on the method you use to get to the top?
  b. What (or who) does the work in each case?
  c. How much is the change in gravitational potential energy?
C8. Airlines must continually worry about the fuel needed for their flights. Some regulations or practices are related to the energy they have to provide. Use both gravitational potential and kinetic energies when answering these questions.
  a. On many international flights each passenger is limited to 20 kg of luggage. For more mass, the passenger must pay an extra charge. Does this regulation make sense in terms of energy use?
  b. In a lighthearted moment, an airline executive suggested that overweight people should pay a higher price for airline tickets than people of average weight; underweight people should get a discount. In terms of energy, is this sensible?
  c. Suppose you were given the job of establishing fares based on the ideas in (b). How would you use energy concepts to do so?
C9. When energy costs increased rapidly, airlines began looking for ways to cut the use of fuel. For example, TWA stripped the paint off the exterior of some of its airplanes, removed pillows, and swept the planes more often. The paint removed from the exterior had a mass of 100 kg and each pillow had a mass of about 0.5 kg.
  a. How much work must be done to get the paint moving at a speed of 245 m/s (typical airline cruising speed)?
  b. How much work is done to get a pillow to that speed?
  c. What is the gravitational potential energy of the paint and pillow at a cruising altitude of 10,000 m?
  d. Why should the TWA management think that these actions would decrease fuel consumption?
C10. When kinetic or gravitational potential energy seems to appear or disappear, we look for another form of energy. Use the concept of conservation of energy to identify at least one new form of energy in each example below. Explain how you arrived at your answer.
  a. In a pinball machine you pull back and release a spring. Then the ball starts moving.
  b. An ancient way to start a fire is to rub two sticks together.
  c. A basketball player jumps up for a rebound.

d. A car moves by burning gasoline.
e. Elevators are lifted by electric motors.

C11. Some interactions result in a decrease in the total mass involved. The energy released in these interactions is the energy equivalent to the difference in total mass before and after the interaction. For each interaction described below, use Einstein's mass-energy equivalence relationship to calculate the energy released.
   a. Uranium (mass = 3.918×10^{-25} kg) splits into five particles with a total mass of 3.915×10^{-25} kg.
   b. Four hydrogen nuclei (mass of each nucleus is equal to 1.673×10^{-27} kg) combine to make one helium nucleus (mass = 6.6443×10^{-27} kg) and two other particles with masses of 0.0091×10^{-27} kg each.
   c. An electron and a positron, each with a mass of 0.0091×10^{-27} kg, combine, releasing energy only. No mass is left over.

D. Activities

D1. Keep a diary of the number of times you increase your gravitational potential energy during one day. Estimate how much your gravitational potential energy increases during the day.

D2. Look up the stopping distances, load, and maximum load in the owner's manual for your automobile. Are the stopping distances listed consistent with our definition of kinetic energy?

Thermal Energy in Matter

We rub our hands together; they become warmer. We rub two blocks of ice together; they melt. We rub two sticks together; they become hot—hot enough to ignite. For millenia people have known about the connection between motion and heat. Only within the last century, however, have we realized the reason for this connection. Heat and motion are both forms of energy. The energy of motion can be converted into the energy of heat.

Heat can cause a substance to change its temperature or to change its state from a solid to a liquid or a liquid to a gas—or even from a solid to a gas. In this chapter we look at interactions that involve changes in temperature and changes in state. Changes in temperature are described in terms of the different *specific heat capacities* of objects. Changes in state are described in terms of *latent heats of fusion* and *latent heats of vaporization*. The energy exchanged during these interactions comes from the energy stored internally in matter, called *thermal energy*.

Figure 10-1
Magnified 80,000 times, the necrosis virus protein shows an orderly arrangement of molecules.

THE ENERGY STORED IN MATTER

A glass of hot water looks the same as a glass of cold water. Yet if you put your finger in each glass, you can feel a difference. A piece of ice looks different from a puddle of water. Yet both can exist at the same temperature—your finger would not notice any difference. Hot water, cold water, ice, and puddles all differ in the amount of energy stored in them internally. To understand how this energy is related to temperature and state, we need to examine matter microscopically.

Just looking at a piece of ice, it is hard to imagine that it contains billions of tiny particles. Yet an overwhelming amount of indirect evidence convinced scientists a century ago that matter is made up of molecules. Today, electron microscopes allow us to see directly what those before us merely imagined. Consisting of one to hundreds of atoms, molecules range in size from 10^{-10} meters (m) to 10^{-5} m. Billions of neatly stacked molecules are found in the virus protein shown in Figure 10-1. Individual molecules in the virus protein are about 10^{-8} m in diameter. The neat stacks arise from electrical forces that hold molecules together. The irregularities arise from the fact that individual molecules move about constantly, occasionally venturing out of position. Electrical potential energy is associated with the electrical forces, and kinetic energy is associated with the molecular motion. Changes in the state of matter—from water to ice, for example—are related to changes in electrical potential energy. Changes in temperature are related to changes in kinetic energy.

Differences in the strengths of the electrical forces among molecules are responsible for the three **states of matter**—solid, liquid, and gas (Figure 10-2). Very weak electrical forces lead to the lack of shape and volume characteristic of gases. The molecules of a gas move about independently of one

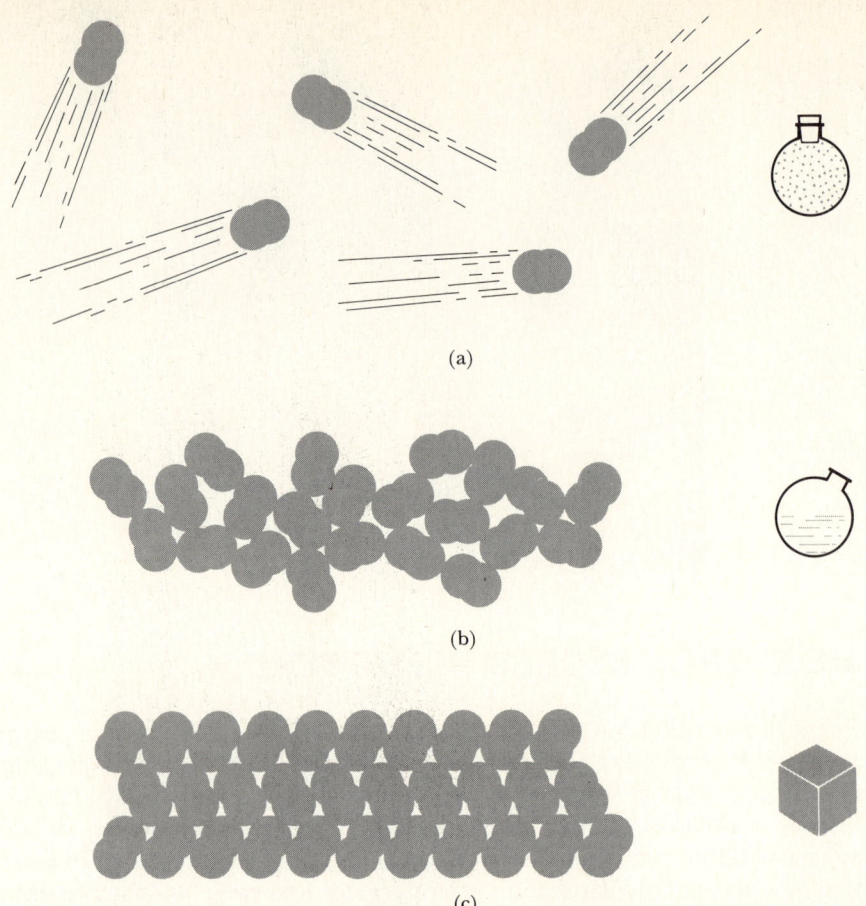

Figure 10-2
The three states of matter differ because the strength of the electrical forces among molecules varies.
(a) Very weak forces allow molecules in gases to move about almost at will.
(b) Slightly stronger forces allow the molecules in a liquid to move about some, but the liquid retains a definite volume.
(c) Stronger forces limit the molecular motions in solids to vibrations about a central location. A solid maintains a definite shape and volume.

another. Both the shape and the volume of a gas change easily. When a gas condenses into a liquid, the molecules move close enough to experience much larger electrical forces. The forces are large enough to keep molecules within a definite volume but not within a definite shape. A liquid, unlike a gas, stays confined in an open container; however, it does assume the shape of the container. As a liquid solidifies, its molecules move even closer together. The electrical force becomes large enough to keep the molecules in the orderly stacks that lead to a definite shape and volume. Molecular motions in solids are restricted to small vibrations about fixed centers.

The electrical potential energy found in matter depends on the strength of these intermolecular forces, much as the gravitational potential energy of an object depends upon the strength of the gravitational force acting on it. When we compared the gravitational potential energy of the pile driver on earth and on the moon, we found it to be greater on earth. Larger gravitational forces lead to larger amounts of energy stored in an object in a given position. By the same token, the amount of electrical potential energy stored in matter depends on the strength of the electrical force binding the molecules together. For any given substance, we will find more electrical potential energy in its solid state, less in its liquid state, and very little in its gaseous state. Changes in state involve changes in electrical potential energy.

Differences in the motion of molecules within matter are responsible for the temperature differences we feel with our hands or measure with a thermometer. Molecules move constantly. In gases like air, molecular motions are fairly obvious because they give rise to motions of visible particles. Dust particles, for example, do not simply fall to the ground. Air molecules collide with them constantly, keeping them suspended. Pollen grains suspended in a liquid also bounce around as the moving molecules of the liquid hit them. Molecules move within solids, too, although this is not so obvious. Kinetic energy is associated with these molecular motions. The faster the molecules move, the higher are their kinetic energies. A substance's **temperature** is proportional to the average kinetic energy of its molecules. Changes in temperature involve changes in kinetic energy.

The combination of the electrical potential energy and the kinetic energy associated with molecules is called the **thermal energy** of a substance. Interactions in which there is a change in temperature or a change in state are called thermal interactions. Many of the concepts used to describe thermal interactions were actually introduced before the molecular model of matter. In Chapters 10–12 we examine these concepts in connection with everyday situations. In Chapter 13 we discuss the molecular model of matter in more detail and see how these everyday events can be described at the molecular level.

SELF-CHECK 10A

Hot water and cold water differ in temperature. How do they differ in the thermal energy stored in their molecules? Ice at 32°F and water at 32°F differ in state. How do they differ in the thermal energy stored in their molecules?

TEMPERATURE

Suppose you are heating some chocolate milk on the stove. You wander off to watch a football game while thermal energy is being transmitted from the flame to the pot to the milk. You don't call it thermal energy, of course. You call it heat. Ordinarily we say that the flame heats up the milk. When you come back to the kitchen, you are not sure whether the milk is hot enough, so you test it with your finger. It is hot enough, all right! Thermal energy—heat—was transferred from the milk to your finger, raising your skin temperature noticeably.

Heat and Temperature

Heat and temperature—these two terms are so closely involved in describing interactions like making hot chocolate that we often confuse them. Heat and temperature, however, describe two very different concepts.

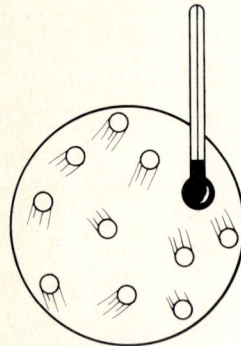

Figure 10-3
The temperature of a substance is a measure of the average kinetic energy of its molecules.

Heat is the name we give to thermal energy that is in the process of being transferred from one object to another. As happens in all energy transfer processes, one object acts as the energy source and the other as the energy receiver. The energy source loses thermal energy—molecular bonds tighten or molecules move around more slowly. The energy receiver gains thermal energy. Its bonds loosen or its molecules move about more rapidly. Heat simply describes the thermal energy as it is in transit from one to the other. Physicists now use the term *thermal energy* for heat as well as for the energy stored internally in the molecular bonds and motions found in matter.

Though everyone knows intuitively what temperature is, there is no simple definition. In everyday terms, **temperature** is the hotness or coldness of an object. Hot describes how we perceive high temperatures; cold describes how we perceive low temperatures. Scientifically, we define **temperature** as a measure of the average kinetic energy of an object's molecules. *Hot* describes objects in which the molecules are moving about rapidly. *Cold* describes objects in which the molecules are moving about much more slowly. The higher the temperature of a substance, the faster its molecules move about (Figure 10-3).

This rather general definition of temperature allows us to understand the distinction between heat and temperature. When two objects at different temperatures are brought into contact with one another, heat always moves from the hotter object to the colder one. Heat was transferred from the hot chocolate to your finger. The molecules in the hot chocolate begin to slow down, while the molecules in your skin begin to speed up. Heat is transferred until molecules in your finger and the hot chocolate are moving at about the same rate—until both are at the same temperature. Physicists say that thermal energy is transferred until molecules in your finger and the hot chocolate have about the same average kinetic energy.

Since thermal energy can be stored in molecular bonds as well as in the motion of an object's molecules, heat can be transferred even though an object does not change temperature. When you hold an ice cube, you hand gets colder and the ice cube melts. The molecules in your hand slow down, but the molecules in the ice move apart rather than speed up. The heat you supply goes into stretching the molecular bonds, thus decreasing the strength in the electrical forces between molecules. The ice cube stays at the same temperature, even though you constantly supply heat—thermal energy—to it.

Temperature Measurement

While your finger is a convenient temperature-measuring device, it is not very precise. The best it can give you is a qualitative assessment, such as warm. A variety of instruments have been invented to measure temperature more precisely, the most common of which is the thermometer. A thermometer consists of a small amount of fluid, usually mercury, inside a closed tube. A scale along the outside of the tube allows us to measure the expansion and contraction of the mercury as it gains or loses thermal energy.

When a substance is heated, its molecules start moving over a wider area (as well as faster), so the substance expands. When a substance is cooled, its

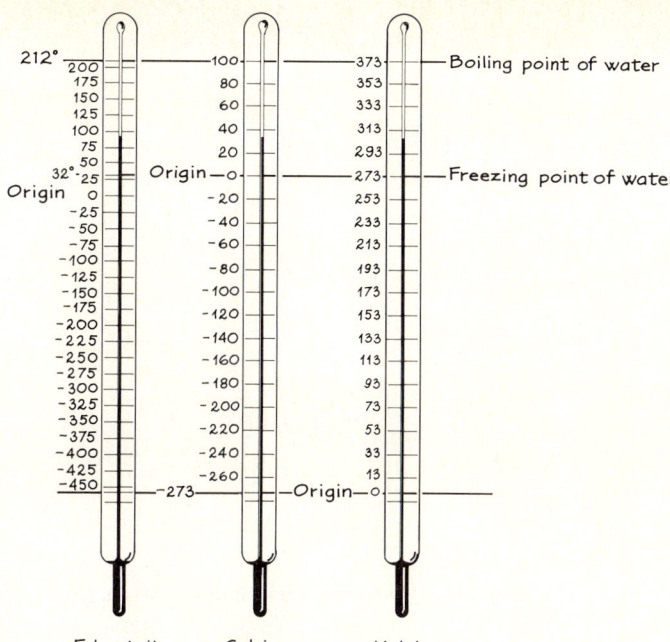

Figure 10-4
The Celsius, Kelvin, and Fahrenheit scales differ in the origin, degree size, and reference temperatures.

molecules move closer together again and the substance contracts slightly. This expansion and contraction allows us to measure temperature. When you place a thermometer in hot water, for example, thermal energy is transferred from the water to the mercury inside the tube. The mercury becomes hotter and expands, rising inside the glass tube. Thermal energy is transferred until the mercury reaches the same temperature as the hot water. We then associate the height of the mercury inside the column with the temperature of the water. If you then move the thermometer to a pan of cold water, thermal energy is transferred from the mercury to the water. The mercury cools, contracting, so its height in the glass tube drops. The mercury continues to lose thermal energy until it reaches the same temperature as the cold water. The scale along the outside of the glass tube allows you to associate a number with the height of the mercury column. You call this number the water's temperature.

Three different temperature scales are in use today—the Celsius, Fahrenheit, and Kelvin scales. These scales differ in their origin, or zero point, and in the sizes of their divisions, or degrees. The origin and degree size are both determined by the choice of reference temperatures used to establish the scale. Since water is an abundant and readily available substance, the freezing and boiling points of water are convenient reference temperatures. Figure 10-4 compares the origin, degree size, and reference temperatures for the Celsius, Fahrenheit, and Kelvin scales.

The **Celsius** scale is by far the most widely used temperature scale. Its origin (0°C) is the freezing point of water. The boiling point of water has been arbitrarily chosen to be 100°C. Since the two reference points are separated by 100 units, each unit is 0.01 the temperature difference between freezing

© 1980 by Sidney Harris.

and boiling water. Consequently, the Celsius scale is often referred to as the centigrade (centi = one-hundredth) scale. Since we often experience temperatures below the freezing point of water (0°C), both negative and positive temperatures are common.

The **Kelvin** scale assigns different values to the two reference temperatures. The freezing point of water is called 273 K and the boiling point of water is called 373 K. One hundred units still separate the freezing and boiling temperatures of water, so the size of the Kelvin degree is the same as the Celsius degree. But, as illustrated in Figure 10-4, the origin has been shifted. In the Kelvin scale, the origin (0 K) has been selected to be the temperature at which the molecules in a substance have the minimum possible kinetic energy. This means that the substance cannot get any colder. Consequently, negative Kelvin temperatures do not exist, and 0 K is an **absolute zero.** Additionally, the temperatures in the Kelvin scale represent actual proportions of kinetic energy. If the state of a substance has not changed, matter at 200 K has molecules with twice the average kinetic energy as matter at 100 K. Consequently, the Kelvin scale is frequently used in scientific work. The conversion between the Celsius scale and the Kelvin scale is given by:

$$K = °C + 273$$

In the United States, the **Fahrenheit** scale is still used extensively. Fahrenheit also chose to divide his scale into 100 degree intervals, but he chose as the origin (0°F) the freezing temperature of a particular mixture of salt and water. Apparently this temperature (equal to about $-18.9°C$) was the coldest he could produce. He then chose human body temperature as the second reference point, designated by 100°F. We have kept this reference point, though we know now that human body temperature is about 98.6°F. (Fahrenheit had either a hot body or a bad thermometer!) Translating the reference temperatures of the Celsius and Kelvin scales to Fahrenheit units, we find that the freezing point of water is 32°F and the boiling point of water is 212°F. On the Fahrenheit scale, there are 180 degree units between the two reference temperatures, while both the Celsius and Kelvin scales have only 100 degree units. Consequently, the Fahrenheit degree is smaller than the Celsius or Kelvin degree. While we have grown accustomed to the Fahrenheit system, its reference temperatures are difficult to reproduce accurately. Consequently, scientific work uses only the Celsius and Kelvin scales.

THERMAL ENERGY TRANSFER AND CHANGES IN TEMPERATURE

Cold milk is poured into hot coffee. The coffee cools down and the milk warms up. Cold air from an air conditioner mixes with the warm summer air in the room. The mixture is neither warm nor cold, but somewhere in between. Cold vegetables are plunged into a pot of boiling water. The water stops boiling and the vegetables get warm. Objects at different temperatures transfer thermal energy and, in the process, change temperature. In this section we investigate the relationship between thermal energy transfer and temperature change.

Specific Heat Capacity

Your pizza has arrived straight from the oven. Carefully, you pick up a slice. The crust seems warm, but not too hot to eat. Of course, the hands are not outstanding temperature sensors, so you carefully touch the crust to your tongue. The pizza is quite warm, but it does not burn. So . . . you take a big bite. Ouch!!! The sauce burns the roof of your mouth. What happened? The sauce and crust were at the same temperature, yet the sauce burned your mouth while the crust did not. Both sauce and crust contained thermal energy, which was then transferred to your mouth, raising your mouth's temperature. Evidently, though, the sauce contained a lot more thermal energy than the crust.

Figure 10-5
Think! . . . About specific heats?

Our experience with pizza illustrates the fact that substances heated to the same temperature may store and later transfer very different amounts of thermal energy. If we heat equal amounts of pizza crust and sauce to exactly the same temperature, the sauce will have absorbed more thermal energy than the crust. On the other hand, if we transfer exactly the same amount of thermal energy to the same mass of pizza crust and sauce, the temperature of the sauce will be lower than the temperature of the crust. More thermal energy is required to raise the sauce to the same temperature as the crust. We use the concept of specific heat capacity to describe the different amounts of thermal energy stored in different substances.

The **specific heat capacity** of a substance is defined as the amount of energy transferred to or from 1 kilogram (kg) of a substance for each 1° change in temperature. Specific heat capacities for a variety of substances are listed in Table 10-1. The specific heat capacity for tomatoes, for example, is 3.9 kilojoules per degree Celsius per kilogram (kJ/°C · kg). A kilojoule (kJ)

Table 10-1 Specific Heat Capacities of Common Substances

Substance	Specific Heat Capacity (kJ/°C · kg)
Water	4.2
Beer	4.0
Tomatoes	3.9
Cheese	3.4
Apples	3.3
Olive Oil	2.0
Wood (average)	1.7
Flour dough	1.7
Porcelain	1.1
Sugar	1.1
Air	1.0
Aluminum	0.9
Brick	0.8
Rock (average)	0.8
Glass	0.8
Iron	0.5
Copper	0.4
Silver	0.3
Tin	0.2

is 1000 joules. In other words, 1 kg of tomatoes must receive 3.9 kJ of thermal energy before its temperature will increase by 1°C. By contrast, copper has a specific heat capacity of 0.4 kJ/°C · kg. One kilogram of copper will increase its temperature by 1°C when only 0.4 kJ of thermal energy have been added. A material with a high specific heat capacity stores more thermal energy per kilogram than one with a lower specific heat capacity.

Specific heat capacity can help us understand the mouth-burning capabilities of pizza sauce. As shown in Table 10-1, tomatoes have a much higher specific heat capacity than dough. (This is primarily due to the fact that tomatoes have large quantities of water in them, and water has an extremely high heat capacity. By contrast, cooked pizza dough has a relatively small amount

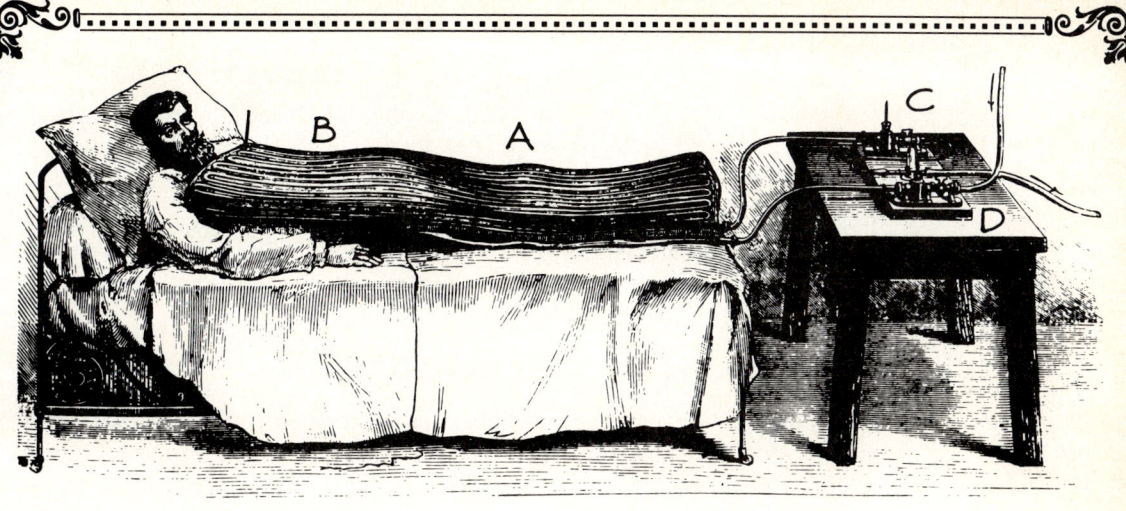

BLANKETS TO KEEP YOU COOL

We have electric blankets to keep us warm in the winter, so why shouldn't we have a cooling blanket for the summer? Such a blanket would have to take thermal energy from us and deposit it somewhere else. In 1879 these thoughts led to the invention of the refrigerating blanket (A). Something was circulated through tubes (B) in this blanket by means of pumps (C and D). To be efficient, the material that circulated had to take away the maximum amount of thermal energy for each kilogram and each degree decrease in temperature. Thus the circulating material needed to have the highest specific heat capacity possible. The inventor of the blanket chose water for this reason. The water also allowed the blanket to be reversible. Its high specific heat capacity allowed the user to heat the water before it was circulated and transfer a lot of energy to the sleeper. It could be warm in the winter and cool in the summer. Try that with an electric blanket!

of water in it.) When the pizza is placed in the oven, the sauce and dough both heat to the same temperature as the oven. Because of its high specific heat capacity, however, the sauce absorbs more thermal energy per degree change in temperature than the dough. When you put a piece of the hot pizza in your mouth, it cools to the same temperature as your mouth. To reach this temperature, the sauce must transfer larger amounts of thermal energy to your mouth than the crust. That burning sensation is a consequence of the high specific heat capacity of tomatoes.

Of course, just giving a phenomenon a name such as *specific heat capacity* does not explain it. You may be wondering how objects of the same mass and temperature can contain different amounts of thermal energy, or how objects of the same mass can require different amounts of thermal energy to reach the same temperature. In a substance with a high specific heat capacity, where does the "extra" thermal energy go? We examine this in more detail in Chapter 13.

SELF-CHECK 10B

Twenty kilojoules of thermal energy are transferred to each of two spoons of the same mass—one made out of wood and the other made out of aluminum. Use the specific heat capacities in Table 10-1 to determine which spoon will reach a higher temperature.

Specific Heats and Solar Heating Design

Specific heat capacity is a valuable concept to someone who is trying to store energy for future use. For example, a solar heating system stores energy, which is available on sunny days, for use at night or on cloudy days. This energy can be stored by converting the sun's energy into thermal energy stored in matter. Later, the thermal energy can be withdrawn and used to heat a home.

To be useful in storing solar energy, a substance must have several characteristics. Since the space devoted to home heating systems is rather small, the substance must have a small mass and volume. Additionally, since a material at a high temperature could boil away or become a fire hazard, the substance should remain at a fairly low temperature as its absorbs solar energy. A material with a high specific heat capacity stores more energy per kilogram and per degree increase in temperature than a material with a low specific heat capacity.

Water, with its extremely high specific heat capacity, is sometimes chosen for solar heating systems. Water will absorb a great deal of thermal energy before its temperature rises to the boiling point. However, since liquids can evaporate, cause rust in metal parts, or leak from their containers, water is not the most convenient material. Instead, many designers choose various types of rocks. Rocks are easily handled, inexpensive, and stable at everyday temperatures. These characteristics frequently outweigh the advantage of using water with its higher specific heat capacity (Figure 10-6).

How Much is Transferred?

Thermal energy is transferred from warmer objects to cooler ones. The flame transfers thermal energy to our chocolate milk. The oven transfers thermal energy to the pizza crust and sauce. Experience tells us something about the amount of thermal energy transferred in each of these processes. We can describe the amount of thermal energy transferred in terms of the energy receiver—the hot chocolate or the pizza. The hotter you want your hot chocolate, the more thermal energy you must supply. If you want to heat 2 cups of milk instead of just one to the same temperature, you have to supply more thermal energy. And, as illustrated by our pizza example, if you want different substances to reach the same temperature, you have to supply different amounts of thermal energy to them.

Figure 10-6 Solar heat storage systems use materials that have relatively high specific heats. **(a)** Tubes that are filled with water are placed in the window. Energy stored in the water can enter the room when the louvered doors are open. **(b)** The brick surface stores solar energy.

Consistent with these observations, the *thermal energy transferred* in interactions involving a *change in temperature* is the product of the substance's mass, specific heat capacity, and change in temperature:

$$\text{Thermal energy transferred} = (\text{mass}) \times (\text{specific heat}) \times \begin{pmatrix} \text{change in} \\ \text{temperature} \end{pmatrix}$$

To see how this definition can be applied, suppose you want to cool 0.35 kg of beer from room temperature (20°C) to 5°C. The amount of thermal energy you need to remove from the beer is the product of the beer's mass, its specific heat capacity, and the change in temperature—(0.35 kg)(4.0 kJ/°C · kg)(20°C − 5°C) = 21 kJ. We have to remove 21 kJ of energy from the beer in order to cool it from room temperature to 5°C. (Text continues on p. 220.)

$TE = mc\,\Delta T$

- Thermal energy transferred
- Specific heat capacity
- Mass
- Change in temperature

SELF-CHECK 10C

In cooling the beer, we ignored the fact that we also have to cool the can. How much additional thermal energy must be removed to cool a 0.02 kg aluminum can from 20°C to 5°C? (See Table 10-1 for the specific heat capacity of aluminum.)

A STEP FURTHER—MATH

HOT, COLD, AND LUKEWARM

Have you ever wondered how people think up equations like the thermal energy transfer equation? While you can probably convince yourself that it makes sense, a few relatively simple experiments can show you how scientists control variables one at a time in order to build a complete description of a relatively complex interaction. Let's examine a series of experiments in which a hot substance is mixed with a cold substance.

First, consider the effect of temperature. Figure 1 shows three experiments in which 0.050 kg of hot water at different initial temperatures has been mixed with 0.050 kg of cold water at 20°C. The magnitudes of the change in temperature of the hot water always equal those of the change in temperature of the cold water. In Figure 1(a), for example, the hot water cools from 40°C to 30°C ($\Delta T = -10°C$), while the cold water warms up from 20°C to 30°C ($\Delta T = 10°C$). In equation form, we say that

$$-\Delta T_{\text{hot}} = \Delta T_{\text{cold}}$$

Next, add the effect of mass. Figure 2 shows three experiments in which different masses of hot water at the same initial temperature (40°C) have been mixed with 0.050 kg of cold water at 20°C. When we compare the changes in temperatures, our simple relationship no longer works. In Figure 2(c), for example, the hot water cools from 40°C to 35°C ($\Delta T = -5°C$), while the cold water warms up from 20°C to 35°C ($\Delta T = 15°C$)—three times the temperature decrease in the hot water. But, we added three times as much hot water as cold water. If we compare the products of mass and temperature change for the hot and cold water, we find them to be equal. Try it!

$$-m_{\text{hot}}\Delta T_{\text{hot}} = m_{\text{cold}}\Delta T_{\text{cold}}$$

Finally, add the effect of specific heat capacity. Figure 3 (page 219) shows three experiments in which equal masses (0.050 kg) of water, copper, and aluminum at the same initial temperature (40°C) have been mixed with 0.050 kg of cold water at 20°C. Hot water caused a temperature increase of 10°C in the cold water; aluminum, an increase of 3.5°C; and copper, an increase of only 1.6°C. If

Figure 1
Equal masses of hot and cold water are mixed.

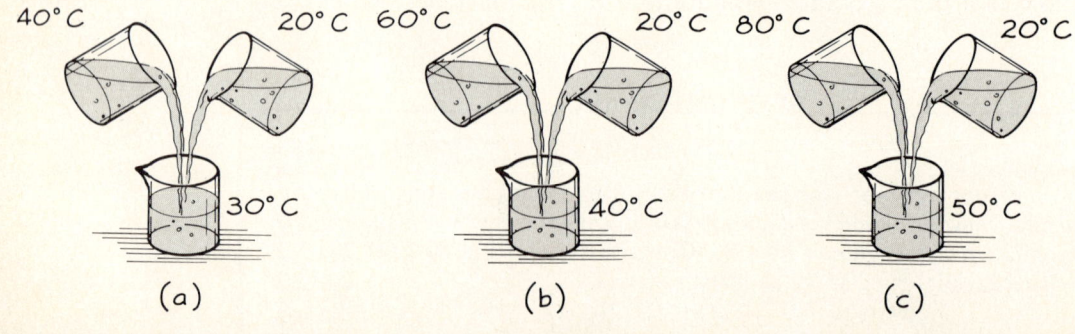

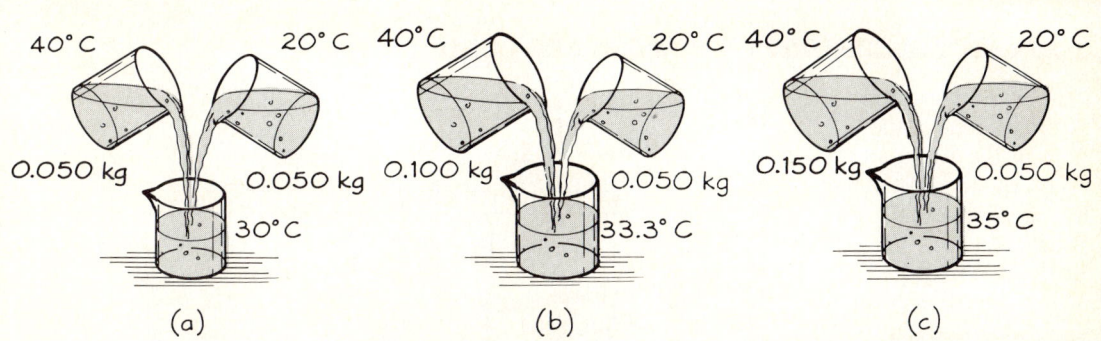

Figure 2
Different masses of hot water are added to 0.050 kg of cold water.

we compare the specific heat capacities of the three materials (Table 10-1), we find that water has the highest specific heat (4.2 kJ/°C · kg), aluminum has a significantly lower specific heat (0.9 kJ/°C · kg), and copper has the lowest specific heat (0.4 kJ/°C · kg). We modify our relationship one final time:

$$-m_{hot}c_{hot}\Delta T_{hot} = m_{cold}c_{cold}\Delta T_{cold}$$

Does it work? Try it and see!

Beneath this series of experiments lies a fundamental assumption—that energy is conserved. We assume that the hot object acts as the energy source and the cold object acts as the energy receiver. Whatever thermal energy is, we assume that the thermal energy lost by the hot substance equals the thermal energy gained by the cold substance. By manipulating the three variables one at a time, we could discover the role each plays in defining the amount of thermal energy transferred.

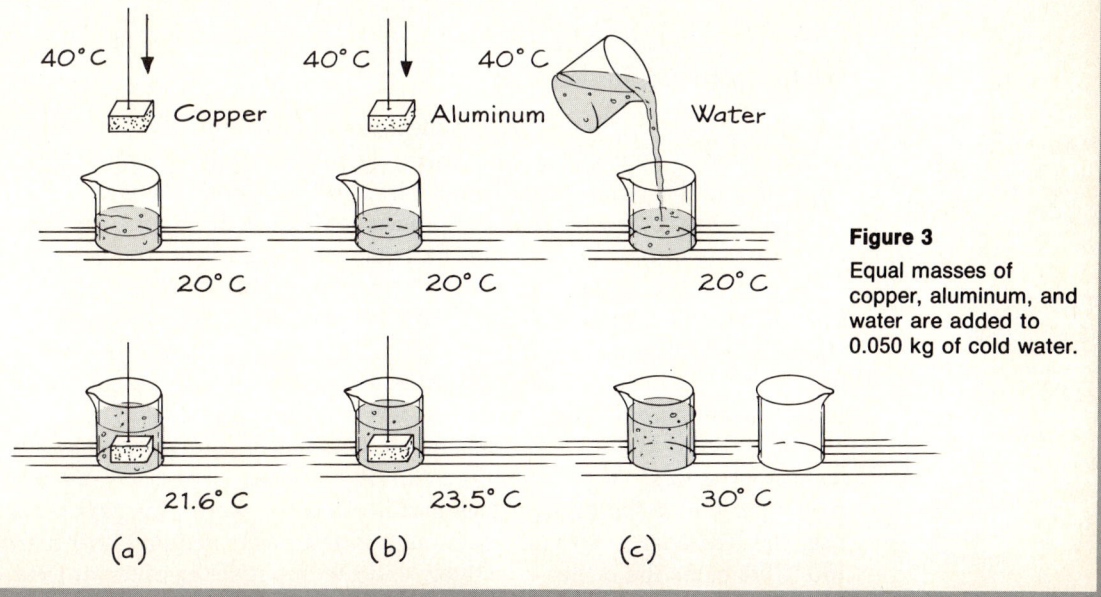

Figure 3
Equal masses of copper, aluminum, and water are added to 0.050 kg of cold water.

YOU CAN'T SLEEP IF THE FLOWERS ARE COLD

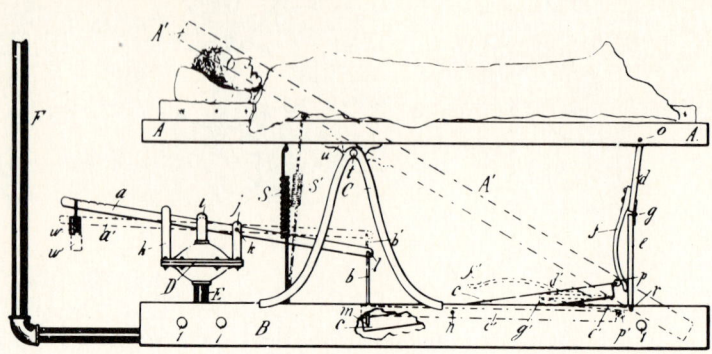

What would happen in the middle of the night if the steam heat in your greenhouse goes off? If you don't wake up, your roses might freeze. To avoid this possibility Ludwig Ederer invented, in 1900, the greenhouse-keeper's alarm bed. It relies on the volume changes that take place when a gas turns into a liquid. Under normal conditions, steam in the pipe (F) heats the greenhouse and applies a force to a series of levers, one of which (g) is holding up the bed (A). If the fire in the boiler goes out, the steam cools and turns into a liquid. Because the liquid takes up a smaller volume, it does not apply the same force to the levers. They collapse, and the bed moves to position A'. The force of gravity being what it is, the person is soon on the floor. The only way the bed can return to the horizontal position (A) is by changing the water back to steam. So, the greenhouse-keeper must get the fires burning. Then, both the keeper and the roses can sleep in comfort.

Dishpan or Dishwasher?

Figure 10-7

We can put this relationship to practical use in estimating the energy required for daily tasks—the energy for which we pay. The age-old argument about doing things the old-fashioned way versus with modern devices is bound to reoccur as energy becomes more expensive. One such argument could occur over washing dishes. Dishwasher manufacturers are quick to point out that a dishwasher actually uses less energy than washing dishes by hand. Of course, such an argument depends on just how far back you want to trace the energy required—do you include the energy required to construct the dishwasher, for example? But if we take the statement at face value, we can evaluate it using our definition of the amount of thermal energy transferred.

The basic argument is that a dishwasher uses less water than washing dishes by hand since you can do a full day's dishes all at once. Because the major portion of the energy goes into heating the water, the less water you use, the less energy you use. Automatic dishwashers require very hot water (60°C) to clean the dishes effectively and use about 40 kg of water for a complete cycle. If we assume that the water is initially at room temperature, then we must supply (40 kg)(4.2 kJ/°C · kg)(60°C − 20°C) = 6720 kJ. Washing dishes by hand usually requires water at about 38°C and uses 28 kg of water, two average-sized sinks half full (one for washing and one for rinsing). Again, if we assume that the water is initially at room temperature, then we must supply (28 kg)(4.2 kJ/°C · kg)(38°C − 20°C) = 2117 kJ. For a single washing, the difference is substantial. We did not even take into account the energy required if you use the dry cycle. Even if you wash dishes by hand three times a day, the dishwasher uses more energy—although not much more. But if you wash dishes by hand less often, the dishpan is clearly an energy saver!

How much energy do you save when you turn your thermostat down to 18°C? What kind of temperature rise will the waste heat from a power plant produce in a nearby lake? If you let the bath water cool to room temperature before letting it out, will it really heat your bathroom enough to matter? Like the dishwasher/dishpan issue, these questions often involve more factors than can be seen on the surface. But do not let the complexity keep you from making rough estimates like the ones we made for the dishwasher and dishpan. The definition of the amount of thermal energy transferred during changes in temperature allows us to check the validity of many statements made about energy.

THERMAL ENERGY TRANSFER AND CHANGE OF STATE

Place a 1 kg block of ice at −20°C on a stove. Turn on the burner so that a constant amount of thermal energy is transferred to the ice. Then monitor the temperature of the ice (Figure 10-8). As we expect, the thermal energy

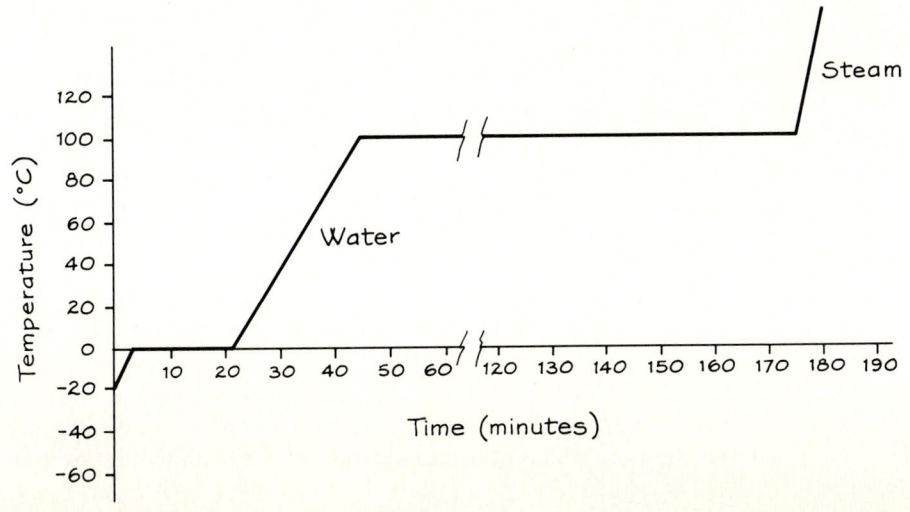

Figure 10-8

The temperature of a substance remains constant as it changes state, even though thermal energy is constantly being provided to it.

supplied by the burner causes an increase in the temperature of the ice. But, when the ice reaches 0°C, something surprising occurs. Its temperature stops increasing as it starts to melt, even though thermal energy is continually being supplied to it. Instead of seeing a change in temperature, you see a change of state. Once the ice has completely melted, the temperature of the water again begins to increase steadily. If you continue to heat the water, you see a similar result at 100°C, when it begins to boil. The temperature of the water remains constant until all the water has been converted to steam. At a substance's melting or boiling point, its temperature remains constant even though it is absorbing thermal energy. The thermal energy is not available for producing a temperature rise—it is needed to cause a change in state.

Latent Heat

Water changes from solid to liquid at 0°C and from liquid to gas at 100°C. Other substances undergo the same changes in state, though at different temperatures. If we measure the amount of thermal energy needed to melt 1 kg of ice with that needed to vaporize 1 kg of water, we find them to be considerably different. Only 320 kJ of energy will melt 1 kg of ice, while 2160 kJ are needed to vaporize 1 kg of water. If we compare the amount of energy needed to melt 1 kg of ice with that needed to melt 1 kg of iron, we see substantial differences again. It takes 320 kJ of energy to melt 1 kg of ice, while only 33 kJ melt 1 kg of iron. The amount of thermal energy needed to cause a change in state varies with whether the change is from a solid to a liquid or from a liquid to a gas and with the particular substance involved. The concept of latent heat is used to describe the thermal energy needed to produce a change of state in a specific substance while the temperature of the substance remains constant.

The **latent heat of fusion** of a substance is the amount of thermal energy per kilogram needed to change it from a solid to a liquid at its melting point. Ice has a latent heat of fusion of 320 kJ/kg. That means that we must supply 320 kJ of thermal energy to each kilogram of ice at 0°C in order to change it from a solid to a liquid. Latent heats of fusion for a variety of substances are listed in Table 10-2.

The **latent heat of vaporization** of a substance is the amount of thermal energy per kilogram of substance needed to change it from a liquid to a gas at its boiling point. Water has a heat of vaporization of 2160 kJ/kg. We must supply 2160 kJ of thermal energy to a kilogram of water at 100°C in order to change it from a liquid to a gas. Latent heats of vaporization for a variety of materials are given in Table 10-2. Generally, the latent heat of vaporization for a substance is considerably larger than its latent heat of fusion.

To change ice to water and water to steam, we must supply thermal energy to the substance. If we reverse the process, changing steam to water and water to ice, the substance must lose thermal energy. If 1 kg of steam condenses to form 1 kg of water, the steam loses 2160 kJ of thermal energy. Similarly, 1 kg of water must lose 320 kJ of energy in order to become ice at 0°C. The latent heat of fusion and vaporization of a substance define the

Thermal Energy Transfer and Change of State   223

Table 10-2 Latent Heat of Fusion/Latent Heat of Vaporization for Common Substances

Substance	Latent Heat of Fusion (kJ/kg)	Latent Heat of Vaporization (kJ/kg)
Water	320	2160
Freon	—	156
Alcohol	104	853
Oxygen	14	213
Hydrogen	59	452
Nitrogen	26	201
Mercury	12	213
Iron	33	—
Lead	25	871
Silver	88	2336
Gold	64	1578
Copper	134	5069

Figure 10-9

Some substances, like dry ice, change directly from a solid to a gas, absorbing energy along the way!

quantity of thermal energy involved in a change of state regardless of the direction in which the change is occurring. In one direction, the substance increases its thermal energy, acting as an energy receiver. In the other direction, the substance decreases its thermal energy, acting as an energy source.

Once again, giving a name like *latent heat* to the thermal energy transferred during a change of state does not really explain it. You may be wondering why the energy is needed and where it goes, since the temperature of the substance remains constant. In essence, the thermal energy is needed to change the average separation between molecules and, hence, the strength of their electrical bonds. As we see in Chapter 13, these bonds are stronger between some molecules and weaker between others, leading to the differences in latent heats found in Table 10-2.

SELF-CHECK 10D

Use the values given in Table 10-2 to determine the amount of thermal energy transferred when 0.5 kg of Freon (a substance used in refrigerators and air conditioners) changes from a liquid to a gas. Does Freon act as an energy source or energy receiver?

Change of State in Ice Boxes and Refrigerators

The thermal energy associated with a change of state has long been applied to the problem of preserving foods. Most foods spoil rapidly at room temperature. To retard spoilage, we place foods in cooler storage locations. Originally, families used underground cellars. Then came the icebox and, more recently, the refrigerator.

As its name implies, the icebox is simply a storage container that holds ice. The melting point of ice (0°C) is well below the temperature at which we like to store foods. As it melts, the ice acts as an energy receiver, absorbing 320 kJ of energy from the surrounding air for each 1 kg of ice melted. The air inside the icebox becomes cooler and the ice gradually melts. As long as fresh ice is added every few days to replace the ice that has melted, the inside of the box stays cool. The system is not very convenient, however, and the icebox has given way to the refrigerator.

Modern refrigerators also take advantage of the thermal energy associated with change of state. Instead of using ice as the energy receiver, refrigerators use a liquid called Freon, which changes to a gas as it absorbs energy from the surrounding air. The latent heat of vaporization of Freon is 156 kJ/kg, so each kilogram of Freon absorbs 156 kJ of energy.

The use of a liquid instead of a solid allows designers to circulate the Freon throughout the refrigerator. The complete cycling system for a typical

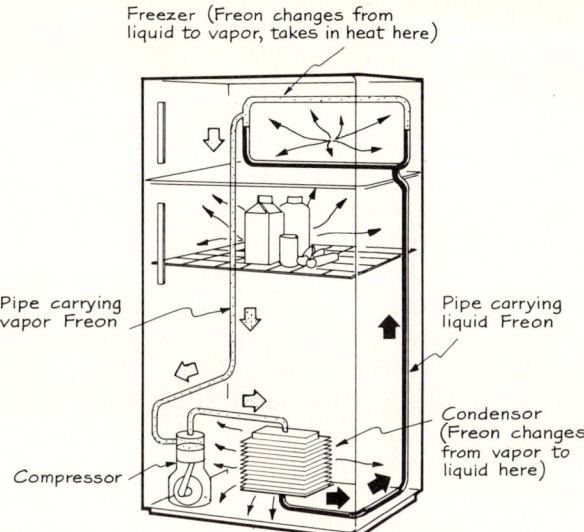

Figure 10-10 Liquid Freon is circulated through the refrigerator and freezer. It absorbs thermal energy, changing into a gas. A compressor at the bottom of the refrigerator converts the gaseous Freon back into a liquid, releasing energy. The thermal energy released at the condenser exceeds the thermal energy absorbed by the Freon. Freon is then recycled through the refrigerator.

refrigerator is shown in Figure 10-10. As the liquid circulates through the interior pipes, it absorbs thermal energy from the air inside the refrigerator. This cools the air and vaporizes the Freon. The gas is then pumped down to the condenser outside the refrigerator, where it is condensed back into a liquid. The thermal energy released by the gas as it changes back into a liquid is exhausted outside the refrigerator. (Most condensers are located at the bottom and toward the back of the refrigerator. The warm air you feel there is due to the energy released as the gas is converted back into a liquid.) The liquid Freon is then pumped back through the interior of the refrigerator to absorb more thermal energy.

How Much is Transferred?

It is a hot day and you want some cold tea. You have already made the tea and it is at room temperature. Now, how should you chill it? If you add a cooler substance, it will act as an energy receiver, absorbing thermal energy from the tea. From experience you know that ice is an effective coolant. But why not simply use cold water instead of ice? After all, the ice usually melts and dilutes the tea just as much as the cold water would. Why go to the trouble of making ice if cold water could serve the same function?

To answer this question, we need to determine the thermal energy lost or gained when a substance changes state. You already know that if you add cold water to the tea, you can determine the amount of thermal energy the cold water absorbs. All you need to do is to calculate the product of the mass of the cold water, the specific heat capacity of water, and the change in temperature of the cold water. We need to develop an equivalent expression for the thermal energy transferred when there is a change of state.

The latent heat of fusion and the latent heat of vaporization provide us with the expression we need. Both latent heats define the amount of thermal energy transferred per kilogram of a substance that changes state. Thus, the only variables needed to determine the **thermal energy transferred** are the mass of the substance changing state and its latent heat of fusion or vaporization. For a substance undergoing a **solid-liquid transition**:

Thermal energy transferred (solid-liquid)
{ Latent heat of fusion
$TE_{(S-L)} = mH_f$
Mass

$$\text{Thermal energy transferred (solid-liquid)} = \text{mass} \times (\text{heat of fusion})$$

For a substance undergoing a **liquid-gas transition**:

Thermal energy transferred (liquid-gas)
{ Latent heat of vaporization
$TE_{(L-G)} = mH_v$
Mass

$$\text{Thermal energy transferred (liquid-gas)} = \text{mass} \times (\text{heat of vaporization})$$

For our glass of tea, the amount of thermal energy that is absorbed by 0.050 kg of ice that completely melts is the product of the mass of the ice and the latent heat of fusion of water—(0.050 kg)(320 kJ/kg) = 16 kJ. The ice absorbs 16 kJ of energy as it melts.

SELF-CHECK 10E

How much thermal energy is released to the environment when 10 kg of water at 0°C freezes and becomes 10 kg of ice at 0°C?

Evaporation

Although a change in state occurs most rapidly when the substance is at its melting or boiling point, changes can occur more gradually. Puddles in the road gradually dry up. The water level in a flower vase gradually drops. **Evaporation** is the gradual change of a liquid into a gas that occurs at temperatures below the boiling point.

Evaporation is an important process in cooling the human body (Figure 10-11). When we become hot, either from exertion or from our surroundings, one of the body's mechanisms for reducing body temperature is perspiration. Water on the surface of the skin absorbs thermal energy from the body and evaporates. Removing this energy helps us keep cool.

A STEP FURTHER—MATH

REALLY COOL TEA!

Remember the problem of cooling the drink? Suppose we have two glasses of tea at room temperature (20°C), and we want to cool them. In one glass we place 0.050 kg of water at 0°C; in the second glass we place 0.050 kg of ice at 0°C. A short time later we measure the temperature of the two mixtures. The water-tea mixture is at 16°C, while the ice-tea mixture, in which the ice has melted completely, is at 2°C. As we know from experience, ice is a better coolant than cold water. To see why, we calculate the thermal energy absorbed from the tea in each glass. (For simplicity, we assume that no energy is transferred to the air in the room.)

When we added cold water, the temperatures of the water and the tea changed. Using the definition for the thermal energy transferred during a change in temperature, we find that the cold water absorbed 3.4 kJ of energy from the tea:

$$TE = (m) \times (c) \times (\Delta T)$$
$$= (0.050 \text{ kg})(4.2 \text{ kJ/°C} \cdot \text{kg})(16°C - 0°C)$$
$$= 3.4 \text{ kJ}$$

When we added ice to the second glass, two things happened. First, the ice melted. Secondly, the melted ice increased its temperature from 0°C to 2°C. Because both a change of state and a change in temperature occurred, we have to calculate the total thermal energy absorbed by the ice in two steps.

Change in state:
$$TE = (m) \times (H_f)$$
$$= (0.050 \text{ kg})(320 \text{ kJ/kg})$$
$$= 16 \text{ kJ}$$

Change in temperature:
$$TE = (m) \times (c) \times (\Delta T)$$
$$= (0.050 \text{ kg})(4.2 \text{ kJ/°C} \cdot \text{kg})(2°C - 0°C)$$
$$= 0.42 \text{ kJ}$$

The total thermal energy absorbed by the ice as it melts and warms up to 2°C is 16.42 kJ, more than four times the thermal energy absorbed when cold water is added instead. Most of that energy went into changing the state of the ice. We use ice to cool not just because it supplies cold water, but because the melting process absorbs a tremendous amount of energy.

Figure 10-11
Evaporation is an important process in cooling the human body.

The thermal energy associated with evaporation is approximately the same as the latent heat of vaporization. Our body must supply 2160 kJ (2.16×10^3 kJ) of thermal energy per kilogram of water evaporated from the skin. Measurements show that the human body dissipates about 10.5×10^3 kJ of thermal energy per day. Using the latent heat of vaporization of water, the amount of perspiration that would have to be evaporated in order to remove this much thermal energy is (10.5×10^3 kJ)/(2.16×10^3 kJ/kg) = 4.86 kg. This is equivalent to about 5 quarts of water! If evaporation were our only mechanism for cooling ourselves, we would need to drink 5 quarts of water per day just to replenish what is lost to perspiration. While evaporation is an important cooling mechanism, it is clearly not the only mechanism we use. We consider some of the other mechanisms in the next chapter.

In some respects we can view matter as an energy-storing device. The waters of the ocean store energy from the sun and maintain a mild temperature range within which life has evolved. Freon stores energy long enough to transfer it from inside to outside the refrigerator. Perspiration stores energy, removing it from our bodies and dispersing it throughout the atmosphere. The electrical potential energy and kinetic energy of billions of tiny molecules provide a convenient and efficient method for storing and transporting energy.

CHAPTER SUMMARY

Thermal energy is the energy stored internally in matter. It includes the electrical potential energy of the bonds holding the molecules together and the kinetic energy of molecular motion. The *state* of a substance—solid, liquid, or gas—is related to the amount of electrical potential energy stored in each molecular bond. The temperature of a substance is related to the average kinetic energy of its molecules.

Thermal energy that is being transferred from one place to another is called *heat*. The direction in which heat flows is always from the object at a higher temperature to the object at a lower temperature. *Temperature* is the perceived hotness or coldness of a substance and is measured with a thermometer. Temperature scales in use today include the *Celsius* scale, the *Kelvin* scale, and the *Fahrenheit* scale. These scales differ in the reference temperatures used to establish the scales' origins and in their degree sizes.

When a substance gains or loses thermal energy, it may change temperature. Two objects that undergo the same change in temperature may gain or lose different amounts of thermal energy. *Specific heat capacity* is defined as the amount of thermal energy required to change the temperature of 1 kg of a substance by 1°C. The greater the specific heat capacity of a substance, the more useful it is to us as a material with which to store energy. The *thermal energy* gained or lost when a substance *changes temperature* is given by:

$$\text{Thermal energy} = (\text{mass}) \times \begin{pmatrix} \text{specific} \\ \text{heat} \end{pmatrix} \times \begin{pmatrix} \text{change in} \\ \text{temperature} \end{pmatrix}$$

When a substance gains or loses thermal energy, it may change state. No temperature change occurs during a change in state. The *latent heat of fusion* of a substance is the amount of thermal energy required to change 1 kg of a substance from a solid to a liquid at its melting point. The *latent heat of vaporization* of a substance is the amount of thermal energy required to change 1 kg of the substance from a liquid to a gas at its boiling point. The *thermal energy* (TE) gained or lost when a substance *changes state* is given by:

(Solid-liquid) TE = mass × (heat of fusion)

(Liquid-gas) TE = mass × (heat of vaporization)

ANSWERS TO SELF-CHECKS

10A. The temperature of the hot water is higher than the temperature of the cold water. Since temperature is a measure of the average kinetic energy of the molecules, hot water has molecules with a greater average kinetic energy than cold water.

Ice is a solid and water is a liquid. Solids and liquids differ in the amounts of energy stored in the electrical bonds holding the molecules together. Water has more electrical potential energy stored in its molecules.

10B. Wood: specific heat = 1.7 kJ/°C · kg

$$\Delta T = 20 \text{ kJ}/1.7 \text{ kJ}/°C \cdot \text{kg}$$
$$\Delta T = 11.8°C \text{ per kilogram of wood}$$

Aluminum: specific heat = 0.9 kJ/°C · kg

$$\Delta T = 20 \text{ kJ}/0.9 \text{ kJ}/°C \cdot \text{kg}$$
$$\Delta T = 22°C \text{ per kilogram of aluminum}$$

The change in temperature of the aluminum is nearly twice that of wood. The aluminum spoon will reach a higher temperature.

10C. TE = $(m) \times (c) \times (\Delta T)$
= (0.02 kg)(0.9 kJ/°C · kg)(20°C − 5°C)
= 0.27 kJ

10D. The latent heat of vaporization of Freon is 156 kJ/kg. Since we have just $\frac{1}{2}$ kg of Freon, we need $\frac{1}{2}$ of 156 kJ, or 78 kJ.

The Freon is changing from a liquid to a gas, so thermal energy must be supplied to the Freon. It acts as an energy receiver.

10E. TE = (mass) × (heat of fusion)
= (10 kg)(320 kJ/kg)
= 3200 kJ

PROBLEMS AND QUESTIONS

A. Review of Chapter Material

A1. Define the terms listed below:
Thermal energy
Celsius scale
Kelvin scale
Fahrenheit scale
Absolute zero
Temperature
Specific heat capacity
Latent heat of fusion
Latent heat of vaporization
Evaporation

A2. With what form of energy is a change of state associated?

A3. With what form of energy is a change in temperature associated?

A4. How do temperature scales differ?

A5. Use the concept of specific heat capacity to explain why pizza sauce burns your mouth, while the crust, at the same temperature, does not.

A6. When substances at different temperatures are mixed, they exchange thermal energy until both substances reach a common temperature. In which direction is the thermal energy transferred: from the warmer substance to the cooler substance or from the cooler substance to the warmer substance?

A7. List the three variables that determine the amount of thermal energy transferred when a substance changes temperature but does not change state.

A8. Thermal energy is required to melt or vaporize a substance even though the substance remains at a constant temperature. Why is this energy required to change the state of a substance?

A9. What two variables affect the amount of thermal energy transferred when a substance changes state?

A10. What two changes provide evidence that thermal energy has been exchanged between substances?

B. Using the Chapter Material

B1. Use the molecular model to describe how the thermal energy stored in ice at $-50°C$ is different from the thermal energy stored in ice at $-20°C$.

B2. Use the molecular model to describe how the thermal energy stored in boiling water at $100°C$ is different from the thermal energy stored in steam at $100°C$.

B3. Suppose you establish your own temperature scale. You choose the freezing point of water to be $25°X$ and the boiling point of water to be $125°X$. Compare the size of the degree on your scale with the Celsius scale. How is your scale different from the Celsius scale?

B4. You have 0.1 kg of cooked apples sitting in a 0.1 kg aluminum pan. Both the apples and the pan are at the same temperature. In terms of the specific heat capacity of apples and aluminum, which is more likely to burn your hand?

B5. In the good old days, bedrooms were unheated. You could take hot objects to bed with you. Which would provide you with the most thermal energy: 1 kg of wood at $70°C$ or 1 kg of bricks at $70°C$?

B6. The specific heat capacity of the human body is about $3.3 \text{ kJ}/°C \cdot \text{kg}$. When you have a fever, how much thermal energy is needed to raise your temperature by $2°C$? The average person has a mass of about 70 kg.

B7. After sitting in the sun all day, the inside of a car reaches a temperature of $50°C$. The mass of air in the car is 1 kg. The car's air conditioner removes the air at $50°C$ and replaces it with air at $10°C$. How much air must be removed and replaced to obtain a final temperature of $30°C$?

B8. If you place your finger in water at $100°C$, it will be burned. However, the burn will be much more severe if you place your finger in steam at $100°C$. Why?

B9. How much energy is removed from a refrigerator when 0.3 kg of Freon change from a liquid to a gas?

B10. As water freezes, does the temperature of the air inside the freezer increase or decrease? What about when the ice melts?

B11. The human body can generate approximately 6000 kJ per hour during heavy physical activity. If the mass of the average person is 70 kg and the specific heat capacity of the human body is about $3.3 \text{ kJ}/°C \cdot \text{kg}$, how much would your body temperature rise if no thermal energy was released?

C. Extensions to New Situations

C1. A major use of chemical potential energy is the heating of water for baths, dishes, and cooking. With some care we can decrease the energy needed for this purpose.
   a. Water for baths or showers has a temperature of 40°C and is heated from 20°C. An average shower requires 20 kg less water than a bath. How much energy is saved by taking showers instead of baths?
   b. Usually the hot water used for a bath or shower is allowed to go down the drain, thus keeping the sewers warm. A bathtub will hold 165 kg of water. If this water were allowed to cool from 40°C to 20°C before leaving the tub, how much energy would be transferred to the air in the bathroom? How could this procedure decrease your utility bill?

C2. The amount of energy needed to heat an oven and its contents can be determined from the information in this chapter. Typically, an oven contains 0.4 kg of air, which is raised from 20°C to 170°C. The oven is enclosed in 50 kg of metal (specific heat capacity = 0.5 kJ/°C · kg). An average dinner includes 2 kg of food (average specific heat capacity = 3.4 kJ/°C · kg).
   a. How much energy is needed to heat the air in the oven?
   b. How much energy is needed to heat the metal in the oven from 20° to 170°C?
   c. How much energy is needed to heat the food?
   d. What is the total energy required to get the contents and oven from 20°C to 170°C? (As we will see in Chapter 11, more energy is actually needed for cooking.)
   e. A microwave oven heats only the food, not the oven or air. How much less energy is required with a microwave oven?

C3. Many items can be cooked in small, countertop ovens rather than the larger stove ovens. A countertop oven holds 0.01 kg of air, while a regular oven holds about 0.4 kg.
   a. How much energy is required to get the air in each oven from 20°C to a typical cooking temperature of 170°C?
   b. Which oven is more practical for small meals?

C4. For lack of something better to do, you wish to turn 2 kg of ice at 0°C into 2 kg of steam at 100°C.
   a. How much energy is required to change the ice at 0°C into water at 0°C?
   b. How much energy must you supply to increase the water's temperature from 0°C to 100°C?
   c. What is the energy required to change the water at 100°C into steam at 100°C?
   d. Which of the above processes took the most energy? The least energy?
   e. What is the total energy required?

C5. In recent years aluminum beverage containers have replaced steel and glass containers. One of the reasons for this change is related to the masses and specific heat capacities of the containers. Typical values are: aluminum, mass = 0.02 kg, specific heat capacity = 0.9 kJ/°C · kg; glass, mass = 0.20 kg, specific heat capacity = 0.8 kJ/°C · kg; steel, mass = 0.05 kg, specific heat capacity = 0.5 kJ/°C · kg. Beverages stored in these cans are usually cooled from 20°C to 5°C.
   a. How much energy must be removed from the beverage? Assume the mass of the beverage is 0.35 kg.
   b. How much energy must be removed from each of the containers?
   c. What is the total energy (container plus beverage) removed from each type of container?
   d. If you were the proprietor of a store and had to pay electric bills to cool drinks, which container would you prefer?

C6. If you drop a snowball from a high enough location, it will melt completely upon impact with the ground.
   a. Neglecting the frictional interaction with the air, explain why the snowball can melt.
   b. How much energy is needed to melt a 2 kg snowball completely?
   c. Which of the following heights will be great enough to melt the 2 kg snowball: 4 m, 400 m, or 40,000 m? (Remember the answer to (b) is in kilojoules!)

d. What would happen to the water after it melted if the snowball were dropped from a height greater than that needed to melt it?

C7. The coolant used in refrigerators, Freon, comes in many different varieties. Each type has a slightly different chemical form and has different thermal properties. For example, Freon-11 has a boiling point of 23.8°C, a specific heat capacity of 1.29 kJ/°C·kg, and a latent heat of vaporization of 181 kJ/kg; Freon-14 boils at 3.8°C and has a specific heat capacity and latent heat of vaporization of 1.02 kJ/°C·kg and 136 kJ/kg, respectively.
  a. Suppose the inside of your refrigerator is kept at a constant 4°C. In what state would each of these two Freons be when inside the refrigerator?
  b. How much energy would 1 kg of each absorb at this temperature if each entered the refrigerator as a liquid?
  c. Which would be a better coolant for a refrigerator?

C8. In a famous thermal energy experiment, Humphry Davy kept all items in a laboratory at 0°C. Then, handling two blocks of ice by long sticks so that he transferred no body heat to them, he rubbed the ice blocks together. The ice melted. Why should the ice melt?

C9. James Joule performed experiments to show that gravitational potential energy could be converted into thermal energy. In one such experiment, he connected a 26-kg mass to a mixer, which was submerged in water (Figure 10-C9). The mass was dropped 1.5 m. Each drop resulted in a 0.015°C increase in the temperature of the water.
  a. Why should we expect the water temperature to increase?
  b. If all the gravitational potential energy of the mass was transferred to the water, how much thermal energy was gained by the water?
  c. Joule stated that this experiment showed that heat was a form of energy. Why could he make that statement?

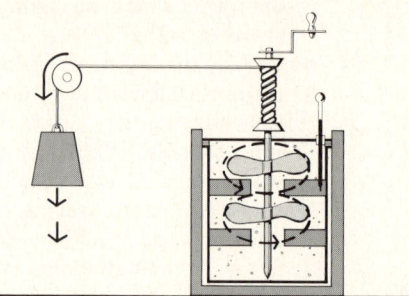

D. Activities

D1. Use specific heat capacities, latent heats of vaporization, and latent heats of fusion to estimate the total amount of energy needed to prepare your food for a day.

D2. Design a system that would store energy coming through a window on a sunny day. Explain how you would choose the material for the system.

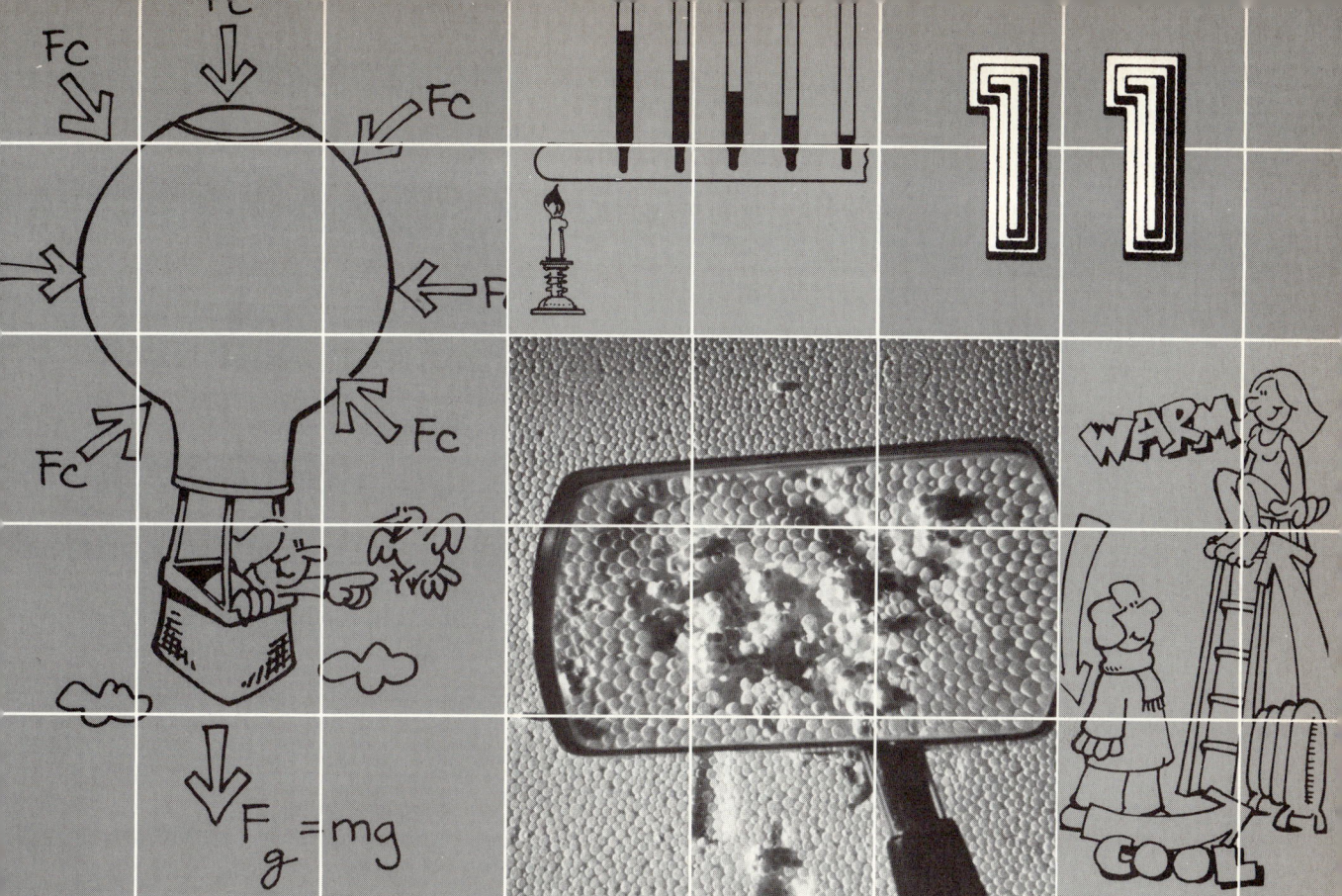

How Thermal Energy Is Transferred

Insulating your home can decrease your home heating costs. So can storm windows. But, using a fireplace can increase your heating costs by as much as 20% (not including the price of the wood!).

Comments like these occur frequently in advice we are given on decreasing the amount of fuel needed to heat or cool a home. In winter we try to keep thermal energy inside. During the summer we try to keep it outside. To evaluate the advice we are given, including surprising bits of advice like not using a fireplace in winter, we must understand the processes by which thermal energy is transferred from one place to another.

In the last chapter we described the changes that occur when thermal energy is transferred from one object to another without worrying about how the transfer actually occurs. In this chapter we look at the three transfer processes—convection, conduction, and radiation-absorption. *Convection* describes the transfer of thermal energy in liquids and gases. *Conduction* describes the transfer process that occurs predominantly in solids. *Radiation-absorption* is the process by which thermal energy is transferred through empty space. Home heating and cooling involve all three processes.

CONVECTION

One method of transferring thermal energy is to heat a substance and then allow the substance to move, carrying the thermal energy with it to a new location. This mechanism is called *convection*. Smoke from a fire, steam from a tea kettle, and hot exhaust fumes from automobiles all carry thermal energy from a source and disperse it throughout the surrounding air. This process, the primary mechanism by which thermal energy is transferred in liquids and gases, relies upon two characteristics of gases and liquids. The first is that they flow, carrying thermal energy with them. The second characteristic is that warmed gases and liquids rise in their cooler surroundings. Before looking at the transfer of thermal energy itself, we need to understand why warmed fluids rise.

Warmed Fluids Rise

One way to analyze the process by which warmed fluids rise is to apply Newton's laws. A hot-air balloon accelerates upward when the air inside the balloon is warmer than the surrounding air. According to Newton's second law, a net force must act upward on the balloon. To see how this net upward force arises, we must identify all forces acting on the balloon. Two forces act: a gravitational force due to the earth-balloon interaction and a contact force

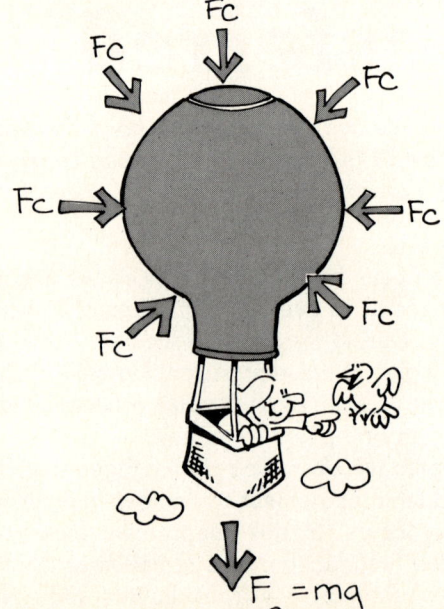

Figure 11-1
The interaction between the balloon and air leads to forces that act on the balloon in all directions.

arising from the interaction between the surrounding air and the balloon. The gravitational force acts downward. The contact force acts in all directions, as shown in Figure 11-1. On a still day, the horizontal forces cancel. Since the balloon rises, we know that the vertical forces do not cancel. The surrounding air must push the balloon up more than gravity pulls it down.

It may seem strange to you that the air below the balloon pushes upward. A model that helps us visualize this is illustrated in Figure 11-2. We can imagine that the atmosphere consists of a series of layers. Because of its weight, each layer presses down on the layers below it. To keep the upper layers from collapsing into a thin layer at the earth's surface, the lower layers must exert upward forces to balance gravity. These forces are exerted through collisions between molecules in two adjacent layers. Called **buoyant forces,** these forces are just large enough for each layer to support the layers above it. Force X is equal to the weight of layer A; force Y is equal to the weight of layers A and B; force Z is equal to the weight of layers A, B, and C; and so forth.

We can use this model to explain why objects rise, sink, or remain stationary in air. If we draw an imaginary line around any part of the atmosphere (Figure 11-3), the section enclosed by the line is in equilibrium. The weight of the enclosed air and the air above this section is balanced by the buoyant forces exerted by the layers of air below it. Now replace this section of the atmosphere with a filled balloon. If the balloon and its contents weigh the same as the air they replace, the balloon remains stationary. If they weigh more, the buoyant forces exerted by the lower layers are not large enough to support them, and the balloon sinks. If they weigh less, the buoyant forces are

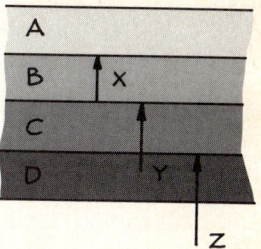

Figure 11-2

Each layer in the atmosphere must exert enough force upward to support the layers of air above it.

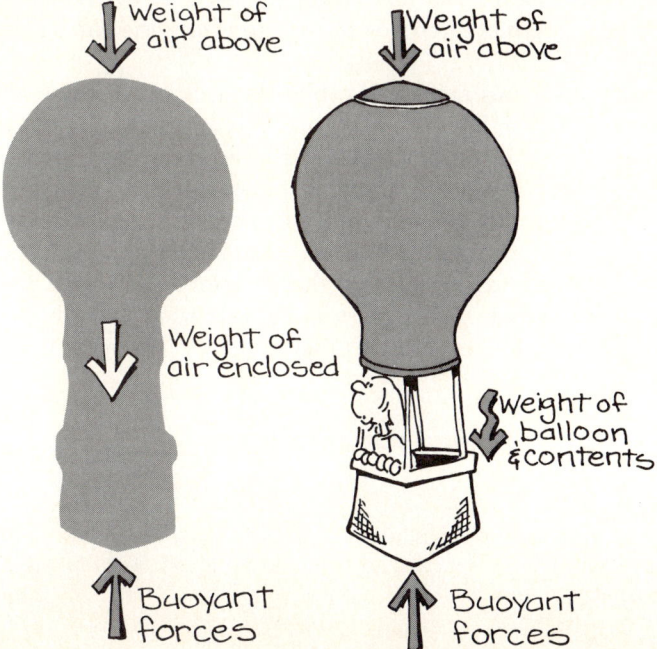

Figure 11-3

The balloon replaces one section of the atmosphere. If it weighs the same as the air that it replaces, the balloon remains stationary. If it weighs more, the balloon sinks. If it weighs less, the balloon rises.

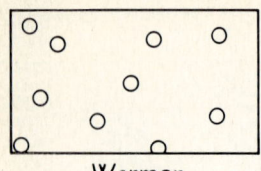

Warmer

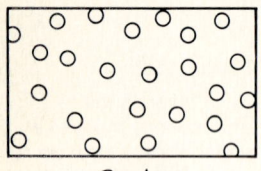

Cooler

Figure 11-4
A given volume of gas contains more molecules if it is cool than if it is warm.

larger than their combined weight, and the balloon rises. The weight of the balloon and its contents, compared to the weight of the air they replace, determines whether the balloon rises, sinks, or remains stationary. The same argument applies to any volume of fluid placed in another fluid. Its motion depends on its weight compared to the weight of the fluid it replaces.

This model can be used to explain the fact that warmer air rises when placed in cooler surroundings. If we compare equal volumes of cool and warm air at the same pressure, the cooler air always weighs more. As the molecules of a gas gain thermal energy, they start moving farther apart. As shown in Figure 11-4, a given volume of gas contains more molecules if it is cool than if it is warm and therefore weighs more. Hot-air balloons rise because the hot air inside the balloon weighs considerably less than the cooler air it replaces. Consequently, if we introduce a sample of warmed air into the atmosphere, it will rise. Similarly, a sample of air that is cooler than its surroundings will sink. The motion of warmer or cooler samples within a fluid provides a mechanism by which thermal energy can be transferred throughout the fluid.

Warm Air Rising Leads to Convection

Because warmed air rises, it can transfer thermal energy from an energy source to other locations. **Convection** is the process of transferring thermal energy by the movement of warmer gases or liquids in cooler surroundings. A candle, for example, transfers thermal energy to the air surrounding its flame. The heated air next to the flame weighs less than the cooler air surrounding it, so it rises. As the heated air moves upward, it is replaced by cooler air. This cooler air is then heated by the candle flame, rises, and is again replaced by cooler air. The process continues as long as the candle burns and cooler air is available to replace the air that has been warmed. Due to the upward motion of the warmed air, thermal energy is transferred from the candle flame to the surrounding air.

In a closed room, this rising and replacement process eventually leads to the closed cycle shown in Figure 11-5. Typically, a radiator acts as the energy source. The air near the radiator is warmed and rises. As it moves upward, this air interacts with the walls, with the people in the room, and with the furniture. Its molecules collide with molecules in the walls, people, and furniture. With each collision, thermal energy is transferred from the warmer air to the cooler surroundings. Gradually, the air cools, sinks, and is eventually drawn back to the radiator. The process repeats itself. Thermal energy has been transferred from the radiator to other objects in the room by the process of convection.

Figure 11-5
The air near the radiator is warmed and rises. After losing some of its energy to the rest of the room, this cooler air sinks and is drawn back over the radiator where it is heated once again.

SELF-CHECK 11A

Convection can occur in liquids as well as in gases. Draw the circulation of water that is established when an ice cube floats at the surface, as shown in Figure 11-6.

FLOATING FROM TOWN TO TOWN

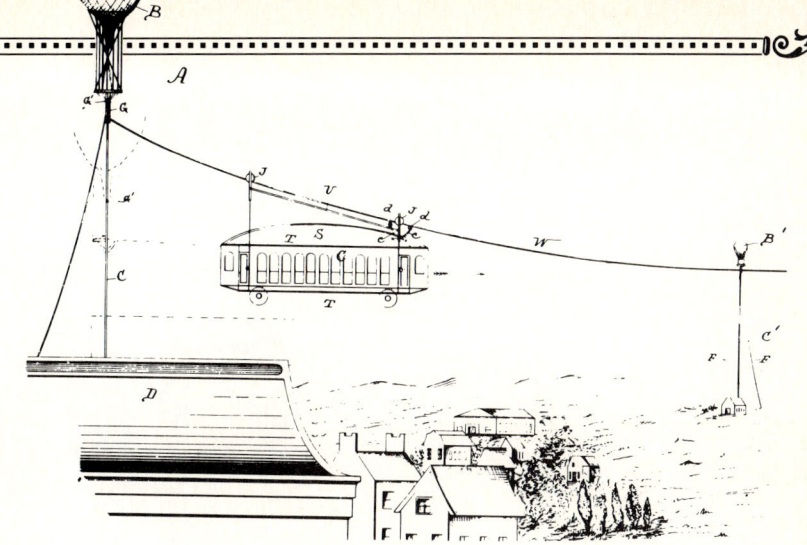

San Francisco's cable cars convert gravitational potential energy into kinetic energy. But, how can the people of the flatlands take advantage of this readily available form of energy? Andrew J. Morrison figured it out when he designed an aerial railroad that used the idea of hot air rising. The air in two balloons (B and B') are heated so that they weigh less than an equal volume of surrounding air. Thus, they feel a net upward force and move upward. To keep the balloons from floating away, they are held by cables (C and C'). Connected between the two balloons is a third cable, (W), from which a railroad passenger car (T) is suspended. Balloon B is kept on a longer string than balloon B' so the car starts at B with greater gravitational potential energy than it would have at B'. It is all downhill, so the car and its passengers roll to their destination. To come back they need only pull in balloon B and let balloon B' float higher in the air. Mr. Morrison was apparently concerned about the downward force that the railroad car would exert on the cable. To decrease this force he added an upward force on the car. The roof area (S) is a compartment that can be filled with warmed air. The upward force on this gas will decrease the net force applied on the cable. We can only wonder if Mr. Morrison thought about how the force on this gas would affect the travel time of the car.

Convection in Fireplaces

A fireplace behaves in the same way as a candle or a radiator. It is an energy source that can heat the surrounding air, establishing convection. Unfortunately, the process of convection works to our disadvantage at the hearth.

Like a radiator, a fire in the fireplace warms the air near it. This warmed air weighs less than an equal volume of cooler air, and it rises. Unlike the air warmed by a radiator, however, the air warmed by a fireplace contains poisonous gases that must be exhausted outside the house. A chimney is open at the roof, and the warmed air goes out the chimney rather than being recycled back into the room. Warm air going out the chimney does not do much to keep a cold body warm!

If that were the whole story, a fireplace would be neither advantageous nor disadvantageous—a nice, though useless, ornament. But it is worse than that. Air must be drawn from the room to replace the air exhausted up the

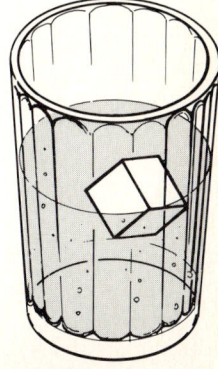

Figure 11-6

238   Chapter 11. How Thermal Energy is Transferred

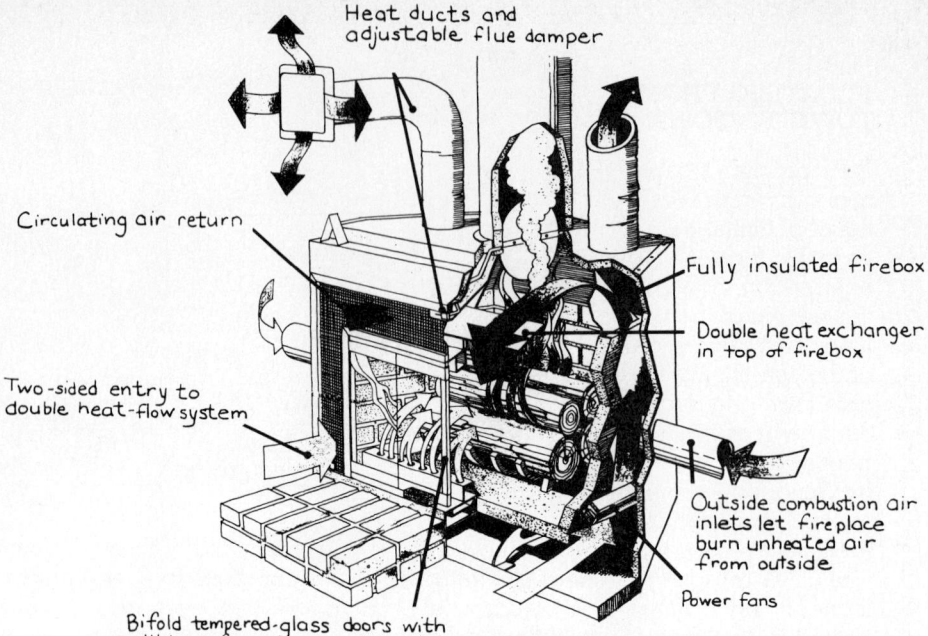

Figure 11-7
Devices that minimize the flow of warmed room air into the fireplace or that actually return warmed air to the room make the fireplace a much more useful energy source.

chimney, and in most homes this replacement air has been heated by a furnace. Oil, gas, or electricity has been used to heat air that ultimately must be exhausted out the chimney. The furnace must then heat more air that is again exhausted out the chimney, and so forth. Consequently, a fireplace will make the furnace run more, consuming more fuel instead of less.

You may object that fireplaces do make you feel warmer if you are close to them. True—some thermal energy does enter the room, but by the process of radiation-absorption rather than convection. The trick to turning a fireplace into an energy source rather than an energy drain is to minimize the flow of heated room air into the fireplace. A variety of designs are now available for this purpose (Figure 11-7).

CONDUCTION

You have probably used old wire coat hangers to cook hot dogs over campfires. You placed the hot dog on one end of the wire and held on to the other. Even though the end with the hot dog was the only part of the coat hanger actually in the fire, the end you were holding eventually got hot. Much the same process occurs when we place a pan on the stove. The bottom of the pan is the only surface in direct contact with the burner. Yet if the pan handle is metal, it quickly becomes too hot to touch. Thermal energy is transferred from one end of the wire to the other and from the bottom of the pan to the handle.

In each example, only one part of the metal is heated. Yet thermal energy is transferred throughout. Since solids do not flow like liquids or gases, we cannot use convection to explain the transfer of thermal energy. Instead we introduce a second process, called *conduction*.

Thermal Conduction and Insulation

Conduction is the process by which thermal energy is transferred without significant motion of the material's molecules away from their original locations. In convection, molecules carry thermal energy with them as they move about. In conduction, molecules transfer thermal energy without moving permanently from their original locations. When one end of a solid is placed near a heat source, for example, adjacent molecules gain kinetic energy and start to move faster and farther. They collide with neighboring molecules, transferring some of their kinetic energy. These molecules then interact with their neighbors and the thermal energy is gradually transferred along the solid. Thermometers placed along a solid (Figure 11-8) show evidence of the transfer of thermal energy throughout. Since molecules are free to move about in liquids and gases, thermal transfer by convection usually exceeds thermal transfer by conduction. Consequently, conduction describes thermal transfer in solids and is generally insignificant in describing thermal transfer in liquids and gases.

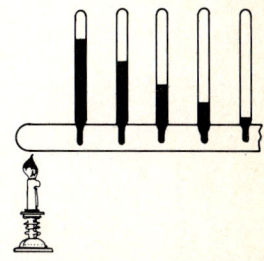

Figure 11-8
Thermometers placed along a solid show the transfer of thermal energy by conduction.

While all solids transfer some thermal energy by conduction, they do not do so equally well. If our hot-dog wire and pan become too hot to handle, we use hot pads to protect our hands. Likewise, when the hot dog is burned to our satisfaction, we wrap a bun around it before pulling it off the wire. Some solids transfer thermal energy more easily than others.

The terms *conductor* and *insulator* provide us with a way of categorizing materials according to their abilities to transfer thermal energy. **Thermal conductors** are materials that conduct thermal energy easily, like coat hangers or metal pans. Materials that do not conduct thermal energy easily, such as hot pads or hot-dog buns, are called **thermal insulators**. Differences in molecular structure and bonding lead to the differences between thermal conductors and insulators.

Thermal Resistivity

Insulating ourselves from the cold is a major goal in winter. When paying for thermal energy, we need a way of comparing the insulating ability of materials in order to determine which are the most cost-effective. The concept of thermal resistivity provides us a quantitative way of comparing the extent to which different materials transfer thermal energy.

The **thermal resistivity** of a material is a numerical value that describes the resistance of that material to the conduction of thermal energy. A high thermal resistivity means that thermal energy moves through the material very slowly. A low thermal resistivity indicates that thermal energy is easily conducted. Good thermal insulators have thermal resistivities in the range of 5 to 100 $m^2 \cdot s \cdot °C/J \cdot m$, while good thermal conductors have values of about 0.002 to 0.01 $m^2 \cdot s \cdot °C/J \cdot m$. (More will be said about these cumbersome units in the next section.) Thermal resistivities of typical materials are included in Table 11-1.

Table 11-1 Typical Values of Thermal Resistivities

Material	Thermal Resistivity (r_t)	
	Metric Units ($m^2 \cdot s \cdot °C/J \cdot m$)	Lumber Yard Units ($ft^2 \cdot h \cdot °F/BTU \cdot in.$)
Silver	0.0023	0.00033
Copper	0.0025	0.00036
Steel	0.0030	0.00043
Gold	0.0032	0.00046
Aluminum	0.0125	0.00180
Brick	1.4	0.02
Wallboard	4.8	0.69
Oak	4.9	0.71
Linoleum	5.1	0.74
Hard maple	5.5	0.79
White pine	9.1	1.31
Balsa wood	21.7	3.13
Asbestos	3.5	0.5
Fiberglass	22.3	3.2
Cork	24.8	3.6
Rock wool	26.7	3.8
Styrofoam	31.5	4.5
Polyurethane foam	43.3	6.2
Leather	6.3	0.91
Linen	11.4	1.64
Silk	17.4	2.51
Air	39.6	5.71
Water	1.6	0.24
Ice	0.58	0.08
Snow	1.4	0.20

Figure 11-9

Styrofoam is a good insulator because it traps air and uses the insulating ability of still air.

The thermal resistivities listed in Table 11-1 match our experiences pretty well. Styrofoam is a better insulator than aluminum. Wool is a better insulator than cotton. One surprise is the high thermal resistivity of still air. We seldom think of air as an insulator because air transfers thermal energy by convection as well as by conduction. Whatever insulating ability air has by virtue of its high thermal resistivity is offset by the large amount of thermal energy it transfers by convection. When air is trapped so that its molecules are held in relatively fixed locations, like solids, it becomes an excellent insulator. Some materials are good insulators because they trap air, thus taking advantage of its insulating abilities.

A close look at styrofoam (Figure 11-9) reveals lots of tiny pockets. These pockets trap small quantities of air, preventing it from circulating. Fur traps air near the surface of an animal's body; this still air insulates the animal from the cold. The colder it is, the more the animal fluffs out its fur. Many other insulators—goose down, fiberglass, even hot-dog buns—use the insulating ability of still air.

SELF-CHECK 11B

To keep warm in winter we are told to wear several layers of clothing. Why is layering advantageous, other than because you can take the layers off one at a time?

R-values

Home builders, lumber yards, and energy auditors usually describe the insulating ability of solids in terms of R-values. The R-value of a solid is the product of the thermal resistivity of the material and the distance through which the thermal energy moves, usually the thickness of the solid.

Thermal resistivity of material

R-value = $r_t d$

Distance energy travels

$$\begin{pmatrix} \text{R-value of} \\ \text{a solid} \end{pmatrix} = \begin{pmatrix} \text{thermal resistivity} \\ \text{of material} \end{pmatrix} \times \begin{pmatrix} \text{distance energy} \\ \text{travels} \end{pmatrix}$$

Using the thermal resistivities listed in Table 11-1, the R-value of a slab of styrofoam 0.1 meters (m) thick is (31.5 m² · s · °C/J · m)(0.10 m) = 3.15 m² · s · °C/J.

Using R-values, you can directly compare the insulating properties of different thicknesses of different materials, assuming that they will be used under identical conditions. Insulating materials to be used in attics and walls can be compared in this manner. When you buy insulation from a lumber yard, look at the R-values listed for each type of insulation—the higher the R-value, the better the insulator.

Many common applications involve solids that consist of several different materials. In dressing warmly for winter weather, we wear several different layers of clothing—shirts, sweaters, jackets, and coats. Walls typically consist of drywall, insulation, and wood or brick siding. To determine the R-value of a combination of materials, you add the separate R-values of each material:

$$\text{Total R-value} = \text{R-value}_1 + \text{R-value}_2 + \text{R-value}_3 + \ldots$$

For example, the wall of a typical house consists of 0.02 m of drywall, 0.09 m of fiberglass insulation, and 0.02 m of wood (pine) siding. Using the thermal resistivities in Table 11-1, the total R-value of the wall is:

$$\begin{aligned} \text{Total R-value of wall} &= \begin{pmatrix} \text{R-value of} \\ \text{drywall} \end{pmatrix} + \begin{pmatrix} \text{R-value of} \\ \text{insulation} \end{pmatrix} + \begin{pmatrix} \text{R-value of} \\ \text{wood siding} \end{pmatrix} \\ &= [(4.8)(0.02) + (22.3)(0.09) + (9.1)(0.02)]\ \text{m}^2 \cdot \text{s} \cdot °\text{C/J} \\ &= 2.28\ \text{m}^2 \cdot \text{s} \cdot °\text{C/J} \end{aligned}$$

SELF-CHECK 11C

When home heating fuels were not so expensive, contractors frequently left the walls hollow. Calculate the total R-value for a wall that consists of 0.02 m of drywall, 0.09 m of moving air, and 0.02 m of pine siding. (Moving air, regardless of its thickness, has an effective R-value of 0.18 m² · s · °C/J.)

Thermal Conduction Equation

The rising cost of energy has made nearly everyone concerned about utility bills. The primary way to decrease utility bills is to decrease the amount of energy conducted through the walls. A quick look at the variables that contribute to higher utility bills tells us what factors affect the conduction of thermal energy through a solid, like the walls of a house.

CLOSER IS WARMER

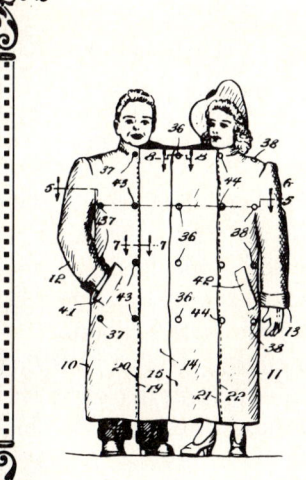

Because the conduction of thermal energy depends on surface area, we can decrease the loss of heat from our bodies by decreasing the surface area between us and the cold outside world. One common example of using this idea is the mitten, which has a smaller surface area than the glove. Howard C. Ross applied this same concept in 1953 when he patented the two-person overcoat. By standing close together, the two people can attach a panel (14) to the buttons (43 and 44) of each of their coats. Then, one side of each person's body is right next to the other's and does not present any surface area to the lower temperature air. Less surface area means less thermal energy conducted. Reaching out and touching someone can be warmer as well as friendlier.

If we always maintain the same temperature inside the house during the winter, utility bills will be higher for:

Colder weather than for warmer weather	(ΔT, the difference between the inside and outside temperatures)
Long winters than for short ones	(t, the total time)
Houses with greater outside wall area	(A, the outside wall area)
Houses with no insulation	(R-value)
Houses with thin walls	(d, the distance the heat has to travel)

When you are trying to maintain the inside of your home at a constant temperature, the amount of thermal energy transferred through the wall depends on the difference between the inside and outside temperatures and the total time during which this difference exists. Long, cold winters result in much higher heating costs than short, mild winters. The remaining factors include the area of the walls, the effective R-value of the materials from which the walls are made, and the thickness of the walls. Architects and contractors take these factors into account in designing and building energy-efficient homes.

The effects of these factors can be combined into a single equation, called the **thermal conduction equation,** that describes the amount of thermal energy conducted through a material.

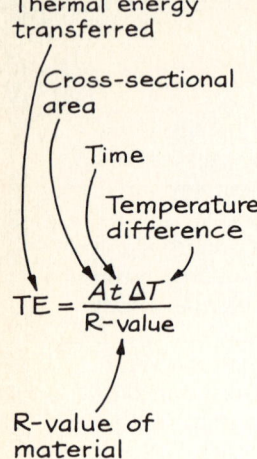

$$\text{Thermal energy conducted} = \frac{\begin{pmatrix}\text{cross-sectional}\\\text{area}\end{pmatrix} \times (\text{time}) \times \begin{pmatrix}\text{temperature}\\\text{difference}\end{pmatrix}}{\text{R-value}}$$

While we have justified this equation by examining an application from home heating, it is a general equation that applies to the conduction of thermal energy in any solid.

The variables included in the thermal conduction equation help explain the rather cumbersome units we introduced with the concepts of thermal resistivity and R-value. In order to make a fair comparison of the insulating abilities of two materials, we must measure the amounts of thermal energy they transfer when all the other variables involved in our equation are kept constant. We need to compare the amounts of thermal energy conducted during equal time intervals by solids equal in cross-sectional area and thickness and whose two sides are exposed to the same temperature difference. As illustrated in Figure 11-10, we generally use a cross-sectional area of 1 square meter (m^2), a time interval of 1 second (s), a temperature difference of 1°C, and a thickness of 1 m. Thermal resistivities are measured in square meters (area) times seconds (time) times degrees Celsius (temperature difference) divided by joules (energy) times meters (thickness). The equivalent units used in the United States are square feet (area) times hours (time) times degrees Fahrenheit (temperature difference) divided by British Thermal Units (BTU) (energy) times inches (thickness).

Thermal resistivities, R-values, and the thermal conduction equation give us a concrete way to evaluate advice given on decreasing home heating costs. There are lots of suggestions: Insulate your walls; keep your thermostat at 20°C (68°F) in the winter and 25.6°C (78°F) in the summer; buy an automatic timer to turn your thermostat down at night; heat rooms on only one floor; and so on. We can evaluate some of this advice by calculating the thermal energy transferred with and without the advised action and comparing the results.

Figure 11-10
Thermal resistivities are measured in terms of the thermal energy conducted by a slab of material with a cross-sectional area (A) of 1 m², a distance (d) of 1 m, a temperature difference (ΔT) of 1°C, and a time interval (t) of 1 s.

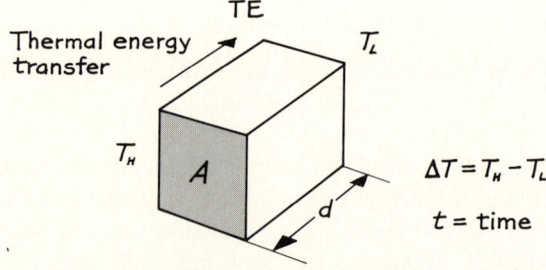

---- **SELF-CHECK 11D** ----

Use the thermal conduction equation to argue that large picture windows would be less energy efficient than wooden walls.

A STEP FURTHER—MATH

COST OF HEATING A HOME

Nearly everyone accepts the need to insulate walls. Let's use the thermal conduction equation to compare the amount of thermal energy transferred across insulated and uninsulated walls. We assume that the wall has a cross-sectional area of 120 m^2, that we are maintaining an indoor temperature of 20°C, and that the outdoor temperature is a moderate 0°C. As shown earlier, the total R-value for a wall insulated with fiberglass is 2.28 m^2 · s · °C/J, while the R-value for an uninsulated wall is 0.46 m^2 · s · °C/J. For comparison's sake, we calculate the thermal energy lost during 1 minute. Substituting these values into the thermal conduction equation, we have:

Insulated: $$TE = \frac{(A) \times (t) \times (\Delta T)}{R\text{-value}} = \frac{(120 \text{ m}^2)(60 \text{ s})(20°C - 0°C)}{2.28 \text{ m}^2 \cdot \text{s} \cdot °C/J}$$

$$= 63{,}160 \text{ J (each minute)}$$

Uninsulated: $$TE = \frac{(A) \times (t) \times (\Delta T)}{R\text{-value}} = \frac{(120 \text{ m}^2)(60 \text{ s})(20°C - 0°C)}{0.46 \text{ m}^2 \cdot \text{s} \cdot °C/J}$$

$$= 320{,}000 \text{ J (each minute)}$$

The thermal energy flow across uninsulated walls is five times that across insulated walls.

One recommendation acted upon by the federal government is that thermostats be set at 20°C (68°F) in winter and 25.6°C (78°F) in summer. Most people set their thermostats at 23°C in winter and 20°C in summer. We can use the thermal conduction equation to determine just how much energy is used at various thermostat settings.

Assume that your house has insulated walls and that you are trying to cool your house on a typical August day in Kansas. The outdoor temperature is 40°C (104°F). Using the thermal conduction equation, we can calculate the thermal energy conducted through the walls when the thermostat is set at 20°C:

$$TE = \frac{(A) \times (t) \times (\Delta T)}{R\text{-value}} = \frac{(120 \text{ m}^2)(60 \text{ s})(40°C - 20°C)}{2.28 \text{ m}^2 \cdot \text{s} \cdot °C/J}$$

$$= 63{,}160 \text{ J (each minute)}$$

We can make similar calculations for thermostat settings of 25°C, 30°C, 35°C, and 40°C. Table 11-2 includes the thermal energy transferred per minute for each of these inside temperatures when the outside temperature is 40°C. Significant savings occur for each 5°C increase in the thermostat setting. In winter, we could just reverse the advice and reduce our thermostat setting in order to save energy.

We can use the thermal conduction equation to investigate other recommendations. Of course, in real-life applications, more factors are often involved and have to be taken into consideration. Arguments continue about turning the thermostat up and down several times a day because of the additional energy required to bring a house up 10°C compared to maintaining it at a constant temperature all day. Regardless of this complexity, the thermal energy equation remains a useful tool for evaluating advice.

Table 11-2 Thermal Energy Conducted Per Minute for Various Thermostat Settings

Description	Inside Temperature	Outside Temperature	Temperature Difference	Thermal Energy Conducted/Minute (kJ/min)
Cool	20°C	40°C	20°C	6.3
Warm	25°C	40°C	15°C	4.7
Very warm	30°C	40°C	10°C	3.2
Hot	35°C	40°C	5°C	1.6
Very hot	40°C	40°C	0°C	0

Conduction at Boundaries

In the wall of a home, thermal energy is conducted from the air in the room through the inner wall to the insulation, to the outer wall, and then finally to the air outside. This conduction occurs within each material and between materials. Whenever two substances are in contact, thermal energy passes from the warmer substance to the cooler one by conduction—whether the substances are solids, liquids, or gases.

As another example, consider the interaction of air and a black asphalt road on a sunny day. The road is hot. Some of its thermal energy is transferred by conduction to the air immediately above the road. The warmed air then rises. Thermal energy must be conducted from the road to the air before convection can carry it away. But how did the road get hot to begin with? That is our next topic.

RADIATION-ABSORPTION

Your bicycle with its black seat has been sitting in the sun on a hot day. Wearing a pair of jogging shorts, you hop on the bike to go home. OUCH! Thermal energy has been transferred from the bike seat to you. The next day is equally hot. Wanting to avoid that uncomfortable sensation, you park the bike in the shade. That evening the ride home is much more comfortable. Since the air temperature was about the same on both days, the difference in the temperature of the seat must have been related to the seat's exposure to the sun. One day the seat was in the sun, and the next it was in the shade. Your bike gained thermal energy from the sun.

A bicycle seat gaining thermal energy from the sun is no surprise. Our sunburns, the growth of plants, and the evaporation of puddles all provide evidence that the sun is a significant energy source. The puzzle is how thermal energy is transferred from the sun, which is separated from us by some 155 million kilometers of essentially empty space. Both convection and conduction require matter with which to transfer thermal energy. Clearly, a third mechanism is needed to explain the transfer of thermal energy from the sun.

Transfer of Thermal Energy by Radiation-Absorption

Radiation-absorption is the process by which thermal energy is transferred by electromagnetic waves. As the name *electromagnetic* implies, these waves are best described in terms of electricity and magnetism, so we save a detailed discussion of them for Chapter 15. A thermal energy source, such as the sun, converts thermal energy into electromagnetic waves. These waves travel to an energy receiver, which converts the electromagnetic wave energy back into thermal energy. The energy source is said to **radiate** the energy and the energy receiver to **absorb** it. Consequently, the entire process is called radiation-absorption. Electromagnetic waves can be transmitted in materials as well as through empty space. While radiation-absorption describes the process by which thermal energy is transferred to us from the sun, it is also a common process of transferring thermal energy between objects on earth. Since the process does not require any matter, radiation-absorption acts separately from the processes of convection and conduction.

The thermal energy that warms us as we stand near a fireplace has been transferred by radiation-absorption. We feel this radiation in spite of convection (which is busy pulling thermal energy up the chimney) because radiation is transferred without using air as the vehicle. Lamps, toasters, electrical heaters—all radiate thermal energy, which our bodies absorb. We, in turn, also radiate thermal energy. Infrared film shows the thermal energy radiated by plants, animals, and even human beings.

Radiation-absorption is one of the processes by which objects exchange thermal energy with their surroundings. All objects radiate and absorb thermal energy continually. If they radiate more thermal energy than they absorb, the objects become cooler. If, like the bicycle seat in the sun, they absorb more energy than they radiate, the objects warm up. Most objects maintain an energy balance between absorption and radiation, always remaining at about the same temperature.

Dependence of Radiation-Absorption on Color

The ability to absorb and radiate thermal energy depends on one important factor—color. Consider an alternate solution to your problem of the bicycle seat. If you could not find any shade, another approach would be to cover the seat. A white seat cover keeps the seat cooler. Both black and white seats absorb radiated energy, but the white seat absorbs less than the black seat.

We can demonstrate the role of color in radiation-absorption with an experiment first performed by Benjamin Franklin. Franklin placed a black cloth and a white cloth on snow during a sunny day. After several hours he observed that the white cloth remained near the top of the snow, while the black cloth had sunk several centimeters. Dark-colored objects absorb radiation more readily than light-colored objects.

We take advantage of this difference in ability to absorb radiant energy when we choose clothing for different climates. In tropical climates, clothing is nearly always light in color. In more temperate climates, people wear light clothes in the summer and dark clothes in winter. Cold climates demand darker shades. Sometimes we even mix light and dark materials to obtain the best combination for maintaining a comfortable body temperature.

SELF-CHECK 11E

If you lived in Arizona, would you prefer to own a black car or a white one? Use the concept of radiation-absorption to explain your choice.

Good Absorbers are Good Radiators

Dark colors are good absorbers; light colors are not. How do both perform as radiators? A brief experiment provides us with an answer. We begin with two pans of fudge. One is vanilla flavored, so it is almost white. The other is chocolate flavored and dark in color. Both are removed from the oven at the same time and placed in a draft-free location to cool. Figure 11-11 shows the change in temperature in each as a function of time. The chocolate fudge cools much more rapidly than the vanilla. While some thermal energy can be carried away by convection or conduction, equal amounts should have been carried away from both the chocolate and vanilla fudge. The rate of temperature change differs because chocolate fudge is a better radiator of thermal energy than vanilla fudge. (Try it. You can eat the experiment when you finish!)

The difference between the two kinds of fudge is color. Dark colors are good radiators and good absorbers. Light colors are poor radiators and poor

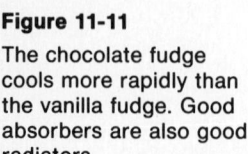

Figure 11-11

The chocolate fudge cools more rapidly than the vanilla fudge. Good absorbers are also good radiators.

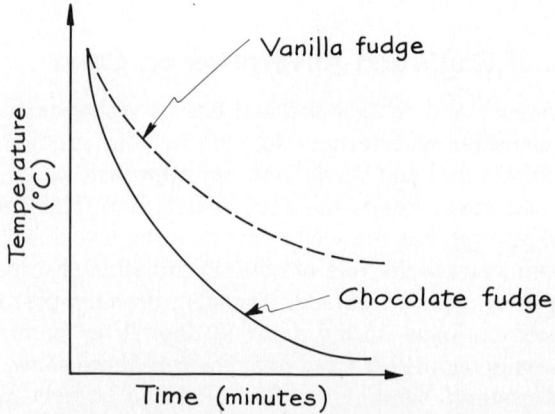

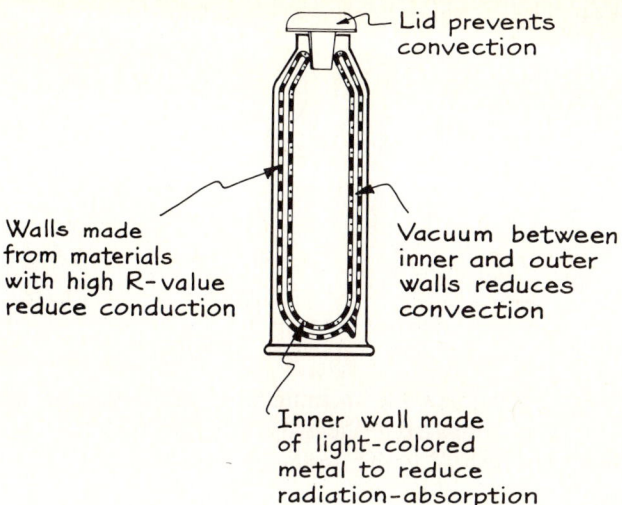

Figure 11-12
Thermos bottles are designed to minimize thermal energy transfer by all three processes.

absorbers. The two processes—radiation and absorption—go hand in hand. A good radiator is a good absorber; a poor radiator is a poor absorber.

The poor radiating ability of light colors is often combined with materials with high R-values to retain thermal energy. For example, styrofoam coffee cups are usually white. The relatively high R-value for styrofoam prevents thermal energy from being conducted through the walls of the cup. What little thermal energy is conducted to the outside of the cup will be radiated away more slowly because the cup is white.

ALL THREE PROCESSES

We have considered the three mechanisms by which thermal energy can be transferred from one place to another—convection, conduction, and radiation-absorption. Rarely will a transfer process involve only one of these mechanisms. A warm drink, for example, loses energy by convection as air along its surface is warmed and rises, by conduction through the walls of the container, and by radiation at its surface and along the outside walls of the container.

Thermos bottles are designed to minimize thermal energy transfer by all three processes, as shown in Figure 11-12. The lid prevents convection of the air above the liquid. The walls of the container are made of materials with high R-values, and many thermos bottles use a semivacuum between the outer and inner walls. Both of these characteristics minimize the energy loss due to conduction. Finally, the inner walls are made of a light-colored metal to reduce the thermal energy loss arising from radiation-absorption. This careful consideration of all three transfer processes has led to a very effective container that keeps hot drinks hot and cold ones cold.

Designers of solar energy systems must take all three transfer processes into consideration in their designs. The solar collector itself must operate using the radiation-absorption transfer process, since solar energy is radiated to us.

© 1979 Field Enterprises, Inc. Courtesy of News America Syndicate.

Once collected, the processes of conduction and convection are needed to transfer the thermal energy from the collector to the building. We can see how these processes are involved in the design of a simple solar collector used as a window heater (Figure 11-13).

The collector is essentially a wooden box with a metal strip down the center and transparent plastic across one side. The sun's radiation passes through the transparent plastic and is absorbed by the metal surface. The metal strip is painted black to increase the amount of radiant energy absorbed. Metal is chosen because its low specific heat capacity allows it to get much hotter than most other materials. An insulator placed below the metal strip limits the flow of thermal energy to the air above the plate. This establishes a path along which convection can occur. As shown in Figure 11-13, cool air is drawn in through the lower opening, circulated past the top of the metal plate (where it is warmed), and exhausted back into the room through the upper opening. The process continues as long as the metal plate absorbs solar radiation.

All three transfer processes—convection, conduction, and radiation-absorption—are involved. Solar energy is transferred to the metal by radiation-absorption. The air in contact with the metal strip is warmed by conduction. The flow of warmed air into the room and replacement air into the heater is accomplished by convection. Most solar collectors are more sophisticated, but even a design as simple as this can warm the air by some 50°F.

Figure 11-13

A solar collector, used here as a window heater, incorporates all three thermal-energy transfer processes in its design. Air circulates through the collector, where it is warmed by the black metal absorber in the center.

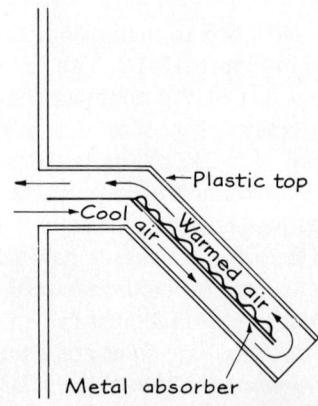

Fireplaces increase your home heating costs; insulating your walls decreases them. The three energy-transfer processes—convection, conduction, and radiation-absorption—enable us to audit our use of thermal energy. From the complexities of a solar power plant to the simplicity of hot bicycle seats, understanding the ways in which thermal energy moves from one object to another gives us control over what has become an increasingly expensive part of our lives.

CHAPTER SUMMARY

Thermal energy is transferred from one place to another by three processes: convection, conduction, and radiation-absorption.

Convection is the process of transferring thermal energy by the movement of warmed gases or liquids. It relies upon two characteristics of matter: (1) the ability of gases and liquids to flow, and (2) the ability of warmed gases and liquids to rise in their cooler surroundings. Thermal energy is transferred as the warmed gas or liquid circulates.

Conduction is the process by which thermal energy is transferred in solids, where the molecules remain in relatively restricted locations. Molecules in a solid transfer energy only to their immediate neighbors. *Thermal conductors* transfer thermal energy easily; *thermal insulators* do not. *Thermal resistivity* is a numerical value that describes the resistance of a material to the conduction of thermal energy. The higher the thermal resistivity of a substance, the less thermal energy it conducts. In most home heating applications, the thermal resistivity of a material and the thickness of the solid are combined to give an *R-value* for the solid:

$$\text{R-value} = \text{thermal resistivity} \times \text{distance energy travels}$$

If a solid consists of several different substances, its R-value is the sum of R-values for each substance:

$$\text{Total R-value} = \text{R-value}_1 + \text{R-value}_2 + \text{R-value}_3 + \cdots$$

The thermal energy transferred by a solid depends on the cross-sectional area of the solid, the time, the temperature difference between the two sides of the solid, and the R-value of the solid. The *thermal conduction equation* is

$$\text{Thermal energy conducted} = \frac{\begin{pmatrix}\text{cross-sectional}\\ \text{area}\end{pmatrix} \times (\text{time}) \times \begin{pmatrix}\text{temperature}\\ \text{difference}\end{pmatrix}}{\text{R-value}}$$

Radiation-absorption is the process by which thermal energy is transferred by electromagnetic waves. These waves can be transmitted through materials or empty space. An energy source radiates energy and an energy receiver absorbs this energy. The ability to absorb and radiate thermal energy depends on the color of the material. Black is a good absorber-radiator; white is a poor absorber-radiator.

ANSWERS TO SELF-CHECKS

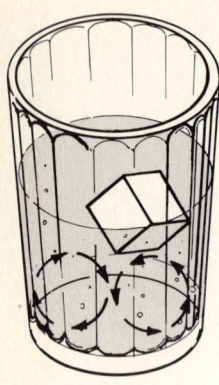

Figure 11-11A

11A. Water near the ice will be cooled. It will become heavier than the surrounding water and will sink. At the bottom of the glass this cooler water gains thermal energy conducted through the glass walls and from the surrounding molecules. As it warms up, this water rises back to the top to be cooled once again.

11B. Each layer of clothing traps air. Several layers of clothing provide layers of still air to insulate your body further.

11C. Total R-value = R-value$_1$ + R-value$_2$ + R-value$_3$

$$= [(4.8)(0.02) + 0.18 + (9.1)(0.02)] \text{ m}^2 \cdot \text{s} \cdot °\text{C/J}$$

$$= 0.46 \text{ m}^2 \cdot \text{s} \cdot °\text{C/J}$$

11D. The R-value of glass is lower than that of walls because of the thermal resistivity of glass and the fact that glass windows are usually thinner than walls. The thermal energy equation states that the thermal energy conducted by a solid is inversely proportional to the R-value. The lower the R-value, the more thermal energy conducted across the material.

11E. You would prefer a white car. White is a poor absorber of radiant energy. Since the environment in Arizona is so hot, you would want cars to absorb as little thermal energy as possible.

PROBLEMS AND QUESTIONS

A. Review of Chapter Material

A1. Define the following terms:
- Convection
- Conduction
- Radiation-absorption
- Thermal resistivity
- Insulator
- Conductor
- R-value
- Buoyant force

A2. How is the weight of a gas used to determine if it will rise or fall in its surroundings?

A3. Describe the process of convection.

A4. What variables affect the thermal energy transferred through a solid?

A5. How is the R-value related to thermal resistivity?

A6. How would you calculate the total R-value of a wall composed of several layers of material?

A7. How does radiation-absorption differ from the other two processes?

A8. What factor is important in determining if an object will be a good absorber?

A9. How is a good radiator related to a good absorber?

A10. Describe how the design of the thermos bottle takes all three thermal energy transfer processes into account.

B. Using the Chapter Material

B1. Our 5-year-old notices that Cheerios rise in milk and Granola sinks. If you had equal volumes of the three substances (Granola, Cheerios, and milk), which would weigh the most? The least? Explain how you arrived at your conclusion.

B2. The total weight of a balloon and the gas inside it is 3.5 newtons (N). The volume of air it replaces weighs 3.0 N. Will the balloon and its contents rise?

B3. Draw the circulation of water in a pan sitting on a hot stove burner.

B4. A desk lamp has small holes near the top of the metal lampshade. How do these holes help to keep the lamp cool?

B5. When installing insulation in a home, you should be sure that it is fluffy rather than packed tightly. Why?

B6. Why will a potato with a nail stuck in it cook more rapidly than one without the nail?

B7. Many electronic parts become warm when operating. Metal is used to conduct heat away from them as rapidly as possible. If cost were not a factor, which of the metals listed in Table 11-1 would you choose?

B8. A homemade ice chest is made of 0.02 m of pine and 0.03 m of styrofoam. What is its total R-value?

B9. Twenty years ago refrigerator walls were 0.10 m thick and were filled with fiberglass. Today many refrigerator walls are 0.03 m thick and filled with styrofoam. Which provides better insulation?

B10. The outside temperature is 30°C; the inside is 20°C. A wall consists of a material with a thermal resistivity of 6 m² · s · °C/J · m and is 0.10 m thick. In 100 s how much energy is transferred through a 1 m² wall? Through a 2 m² wall? Which direction does the energy move: inside to outside or outside to inside?

B11. The earth's atmosphere provides protection from much of the sun's radiation. Why do astronauts wear light-colored suits on the moon, where no atmosphere exists?

B12. When it snows in Montana and Pennsylvania, the state highway departments spread coal dust on top of the snow. When the sun comes out, the snow melts rapidly. Why?

C. Extensions to New Situations

C1. A thermos bottle is used to keep cold drinks cold or hot drinks hot. To do so, it must decrease all forms of thermal energy transfer. The bottle has a light-colored interior, a tight-fitting stopper, and styrofoam walls.
 a. Explain how each method of thermal energy transfer is decreased by the design.
 b. Before styrofoam was popular, the thermos bottle had a vacuum between its inner and outer walls. How did this help?

C2. Prior to the 1950s, most large homes were multistory dwellings. Then the ranch-style home was designed, so that every room was on the same floor. Consider two homes with the same living space, shown in the figure below.
 a. What is the effect of convection on heating and on energy losses in the two homes?
 b. If all the outside walls and the roofs in both houses have total R-values of 3 m² · s · °C/J, how much energy from each house is transferred in 100 seconds on a day when the outside temperature is 0°C and the inside temperature is 20°C?
 c. A larger area provides greater radiation-absorption. Which roof will radiate more energy at night? Which will absorb more energy during the day?
 d. Use the answers to parts (a)–(c) to discuss which house has lower utility bills.

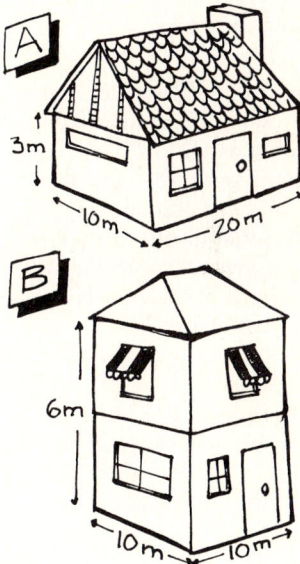

C3. On sunny days, pilots of hot-air balloons find that their balloons are pushed up when they are over an asphalt (black) road. This phenomenon can be understood in terms of thermal energy transfer processes.
 a. Why will the road be hot on a sunny day?

b. How will the energy from the road be transferred to the air immediately above the road?
c. What happens to the heated air? Why?
d. How is the balloon forced up?

C4. On a cold, sunny day you need to decide whether you have a better energy balance with the curtains open or closed. For a 1 m² window (no storm window) on the south side, use the following information:

R-value of glass $= 0.16 \text{ m}^2 \cdot \text{s} \cdot °C/J$

R-value of air trapped between glass and curtain $= 0.18 \text{ m}^2 \cdot \text{s} \cdot °C/J$

R-value of curtain $= 0.17 \text{ m}^2 \cdot \text{s} \cdot °C/J$

Inside temperature $= 20°C$

Outside temperature $= 0°C$

Thermal energy entering window per second $= 100 \text{ J}$

a. In 1 s, how much thermal energy flows out through the window with the curtain opened?
b. Repeat the calculation in (a) for a closed curtain.
c. How much thermal energy enters or leaves through the window in one second for the curtain opened? for the curtain closed?
d. Should you have the curtain open or closed?
e. If we have a storm window on the window, we add an R-value of 0.34 m² · s · °C/J to the total R-value. But the extra layer of glass reflects some of the sun's energy and only 90 J enter each second. Repeat the calculations and decide if the curtain on the window with the storm window should be open or closed.

C5. The ice in your picnic chest is melting quickly; you can slow down the melting process by wrapping the ice in an insulator.
a. Why does wrapping the ice slow down the melting?
b. What will happen to the temperature of the air in the chest?

C6. An old tale is that hot water will freeze more rapidly than cold water.
a. Based on the concepts presented in Chapter 10, would you expect hot water to freeze more rapidly than cold water?
b. Sometimes a tray of hot water will melt through the frost on a refrigerator shelf, while the cold water tray will sit on top of the frost. Why might the hot water freeze more quickly in this situation?

C7. On clear nights the temperature during the day will be much higher than the nighttime temperature. On cloudy days the daytime and nighttime temperatures will be only slightly different. To explain why, consider absorption and radiation.
a. On a clear night what happens to the energy radiated by the earth?
b. What happens to energy radiated by the earth on a cloudy night?
c. Use the answers to (a) and (b) to explain the temperature differences.

C8. The human body creates thermal energy, which must be removed. Some of this energy is removed by evaporation; some by conduction. The body's normal temperature is 37°C.
a. How does the rate at which thermal energy is conducted away from the body depend on the air temperature?
b. Why should you keep as much of your body as possible covered when the air temperature is very cold? (Explain this in terms of conduction.)
c. When the air temperature is above 37°C, can thermal energy be conducted away from the body?
d. Use the answers to (a)–(c) to describe why conduction can lead to health problems for some people.

C9. A bowl of hot soup will cool as it sits on a table.
a. How is the thermal energy transferred from the soup to the air immediately above it?
b. What happens to the air directly above the soup to aid in the cooling process?
c. Why will more than just the top layer of soup cool?

d. Use the answers to (a)-(c) to summarize the soup-cooling process.
   e. Would placing a lid on a soup bowl slow down the cooling? Explain your answer.

D. Activities

D1. While attending a football game on a cold, sunny day, you wish to have a supply of hot drinks. Design a solar drink heater. Describe how radiation-absorption, conduction, and convection are each important in your design.

D2. Design a window curtain that will be able to absorb solar radiation during winter days, not absorb during summer days, radiate on summer nights, and not radiate on winter nights. The curtain can have movable parts.

D3. Look at ads for devices that are supposed to decrease the energy loss up the chimney of fireplaces or increase the thermal energy supplied by the fireplace. Describe how each of the devices works.

D4. Using patterns of melted snow, analyze thermal energy losses from buildings. Explain why you think the energy is lost. (Reference: A. A. Bartlett, *The Physics Teacher,* January and February, 1976).

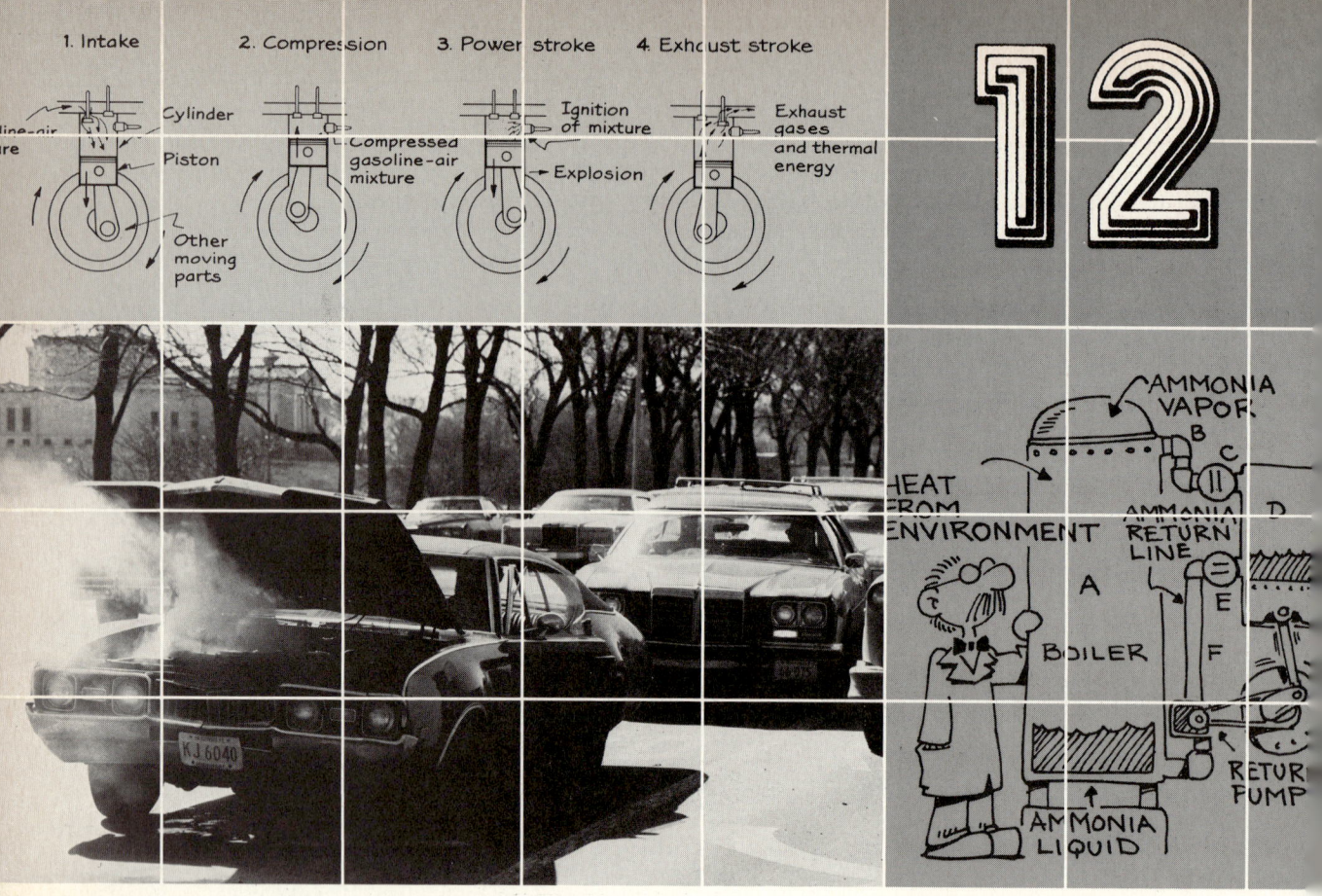

Freelance Photographer's Guild.

Thermodynamics

The energy crisis, the energy problem, the energy situation, decreasing energy reserves—these phrases appear over and over on politicians' lips, during TV newscasts, and in newspaper headlines. Behind these phrases lies the idea that we are somehow running out of energy. Yet energy is conserved. How can we run out of it?

Seen only from the physicist's concept of energy conservation, the notion that we are running out of energy seems absurd. When we drive an automobile, the energy stored in the gasoline does not disappear; it merely changes form. The chemical potential energy in the gasoline is converted into the kinetic energy needed to get us from place to place and the thermal energy that is exhausted to the environment. Nothing is lost—we still have the energy with which we started. All that we have lost is the original source, the gasoline.

If it is the gasoline that is in short supply, why not substitute a different source of energy? Better yet, why not recycle the energy we produced with the gasoline? We could build a "miracle machine" that sucks in the hot automobile exhaust, extracts the thermal energy from it, and exhausts cooler air to the environment. The thermal energy we extract could then be used to generate electricity or produce some other form of energy. If we used electric cars, it could even be used to run the car again. Such a machine would solve everyone's problems—we would have transportation, the environment would be cooler and we would have more electricity. Ah... but no such machine exists and, as our name for it implies, physicists are not betting that one will ever be invented. While energy is conserved, something irreplaceable is lost when the automobile converts gasoline into motion and heat. This something is the energy's usefulness, its availability to do work.

Thermodynamics is the study of the conversion of thermal energy into other forms of energy, and vice versa. This branch of physics originated in the practical study of heat engines such as the steam engine, but it has provided some of our most fundamental insights into the concept of energy. Using the *heat engine* as a basic model, this chapter explores the thermodynamic problems involved in building our miracle machine. As applied to heat engines, the law of conservation of energy is called the *first law of thermodynamics* and tells us we can never get more energy from a machine than we put into it. We introduce the concept of *entropy,* a measure of the disorder of a system, to distinguish between the chemical potential energy stored in gasoline and thermal energy found in the environment. The *second law of thermodynamics* limits the kinds of energy transformations that occur, according to the resulting changes in entropy. It also tells us that in every useful energy transformation, some energy must be wasted. This is the law that sounds the death knell of our miracle machine.

THE FIRST LAW OF THERMODYNAMICS: YOU CAN'T GET SOMETHING FOR NOTHING

Consider a machine even more fantastic than the miracle machine just described. Suppose someone proposed that you invest in the production of a machine that actually supplies more energy than it consumes. Such a machine could be used to drive two machines, which could be used to drive four machines, and so on. For the fuel used to drive one machine, we could drive hundreds of machines that would do useful work. Of course you would laugh off the would-be inventor. The law of conservation of energy tells us that we cannot end up with more energy than we had at the start.

Machines can never create more energy than we put into them, but they can make existing energy more useful to us. Most modern machines operate on the principle of the heat engine, a device which converts thermal energy into kinetic energy. In this section of the chapter we examine heat engines and use their operation to illustrate the first law of thermodynamics. To understand the operation of the heat engine, however, you must first know something about the behavior of gases.

Gases and Heat Engines

Heat engines convert the thermal energy stored in gases into the kinetic energy of moving pistons or turbines. In its simplest form, this conversion occurs through the two-step process shown in Figure 12-1. During the expansion

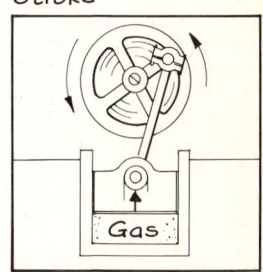
Expansion Stroke

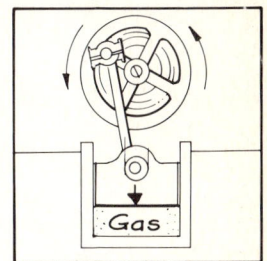
Compression Stroke

Figure 12-1

A simplified model of a heat engine. During the expansion stroke, the gas expands, doing work on the piston. During the compression stroke, the piston does work in compressing the gas.

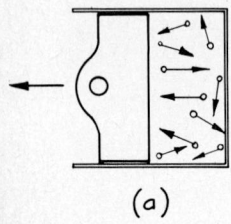

(a)

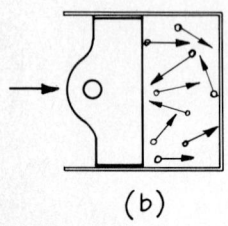

(b)

Figure 12-2

During expansion, the molecules push in the same direction the piston moves. During compression, the molecules push in the opposite direction, opposing the piston's motion.

stroke, a gas enters the cylinder and expands, pushing the cylinder to the left. If we connect the piston to a wheel as shown, then its motion can be used to rotate a fan, move a car, or do some other useful work. During the compression stroke, the piston moves back to the right, compressing the gas and eventually exhausting it from the cylinder. The entire process produces useful energy only when the two steps—expansion and compression—occur at different temperatures. To see why, we must look more closely at the interactions between individual gas molecules and the piston.

As we saw in Chapter 10, a gas is a collection of molecules moving about independently of one another. When these molecules strike an object, like the movable piston, they apply a force to it. These forces can either enhance or retard the motion of the piston, depending on the direction in which the piston is moving. During the expansion stroke (Figure 12-2(a)), the molecules exert forces that act in the same direction as the piston moves. The piston speeds up, gaining energy from the gas molecules with which it collides. As the piston moves back to the right during the compression stroke, however, the molecules exert forces that oppose its motion (Figure 12-2(b)). The piston slows down, losing energy to the individual gas molecules. In order for the piston to gain energy during the two-step process, the forces exerted by the molecules during expansion must be greater than those exerted during compression. A sample of gas molecules will exert a greater force when at a higher temperature. Consequently, the piston gains energy only if expansion occurs at a higher temperature than compression. Heat engines are designed with this in mind.

Heat Engines and the First Law

The first successful heat engine was the steam engine, invented by James Watt early in the nineteenth century. As its name implies, the steam engine uses the thermal energy stored in steam to create motion. A simple steam engine consists of three chambers connected to one another, as shown in Figure 12-3. Water is heated to produce steam in the boiler. A controlled quantity of high-pressure steam moves into the cylinder, where it pushes the piston to the right. The compression stroke then occurs at a lower temperature. The cooled steam flows into a condenser, where it is cooled and pumped back up to the boiler to be reheated and used again. The cycle continues as long as thermal energy is available to convert the water into steam in the boiler and the environment can handle the thermal energy released at the condenser.

If we trace energy through the steam engine, the original source of energy is the chemical potential energy of the fuel—wood or coal—used to produce the steam. This energy is converted into thermal energy, which is eventually stored in the steam. In the cylinder, part of the thermal energy stored in the steam is used to move the piston and the remaining thermal energy is exhausted to the environment. We have gone from chemical potential energy

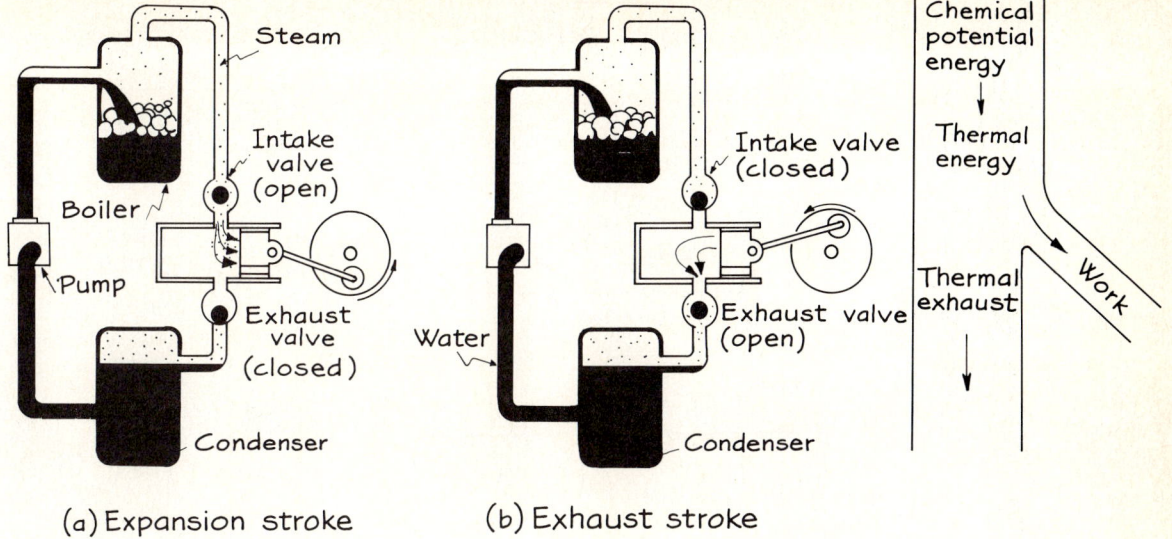

Figure 12-3 Wood or coal is burned to convert water into steam in the boiler. Steam flows into the cylinder, where it forces the piston to move to the right. The piston moves back to the left, forcing the steam into the condenser where it is cooled. The thermal energy in the steam is converted into the kinetic energy of the piston and the thermal exhaust dumped into the environment at the condenser.

to thermal energy to kinetic energy, with some thermal energy exhausted to the environment.

The steam engine provides us with a generalized model of a heat engine. A **heat engine** is any device that converts thermal energy into kinetic energy and thermal exhaust. The kinetic energy can be considered an end product, as in the case of steamships and steam locomotives, or it can be used to produce another useful form of energy, such as electricity in electric generators. Regardless of its use, this energy is called the **useful energy** produced by the process, the energy available to do work. The thermal energy released to the environment is often called the **thermal exhaust,** or waste energy.

Like other processes, heat engines must obey the law of conservation of energy. For each cycle, the thermal energy put into the system must equal the useful energy and waste energy released by the system.

Thermal energy input = useful energy + waste energy

For Watt's steam engine, the thermal energy stored in the steam must equal the sum of the kinetic energy of the piston and the thermal energy exhausted to the environment. As applied to heat engines, the law of conservation of

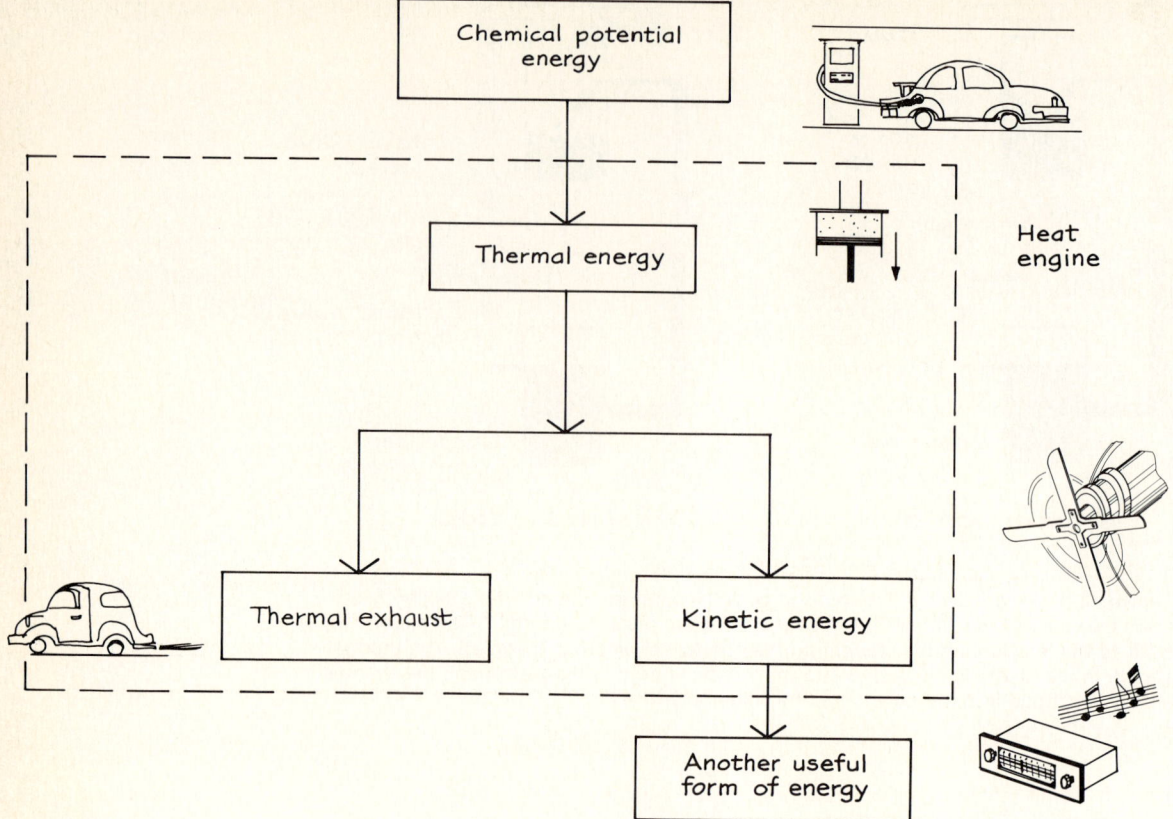

Figure 12-4
A heat engine performs one part of the complete process—converting thermal energy into kinetic energy and thermal exhaust.

energy is called the **first law of thermodynamics.** In this context the law of conservation of energy is often stated as: Energy is neither created nor destroyed, but only changed from one form to another. Or, as wits have long expressed it, you can't get something for nothing.

SELF-CHECK 12A

In 1 second (s), a typical power plant uses 3 billion joules (J) of thermal energy to produce 1 billion joules of electrical energy. How many billion joules of waste energy are exhausted to the environment?

Automobiles and Refrigerators

The generalized heat engine accurately describes the energy changes that occur in a variety of everyday devices. In addition to the steam engine, typical examples of heat engines include the moped motor, an electrical generating plant, and automobile engines. To see how some of these real engines can be described in terms of the generalized heat engine, we trace the form that

SOMETHING FOR NOTHING—ALMOST

What we need is a machine that does work, yet puts back into itself at least as much energy as it takes out. Commonly called perpetual motion machines, these devices have been the dreams of inventors for centuries. One perpetual motion machine was described by Robert Fludd in 1618. Water falling from a reservoir (A) drives a wheel (C) that turns a shaft (D). This motion turns the grinding wheel that enables the local smith to sharpen knives. At the same time a series of gears (E–L and R) turns a coiled pipe (Q) that brings the water back up to the reservoir. The water has regained its initial gravitational potential energy and has done useful work along the way. Energy comes out of the machine, but no net energy goes in. This type of machine violates the first law of thermodynamics (conservation of energy) and is called a *perpetual motion machine of the first kind*. It cannot work, but Robert Fludd had no way of knowing that. He devised it 220 years before the principle of conservation of energy was discovered.

The Bettman Archive, Inc.

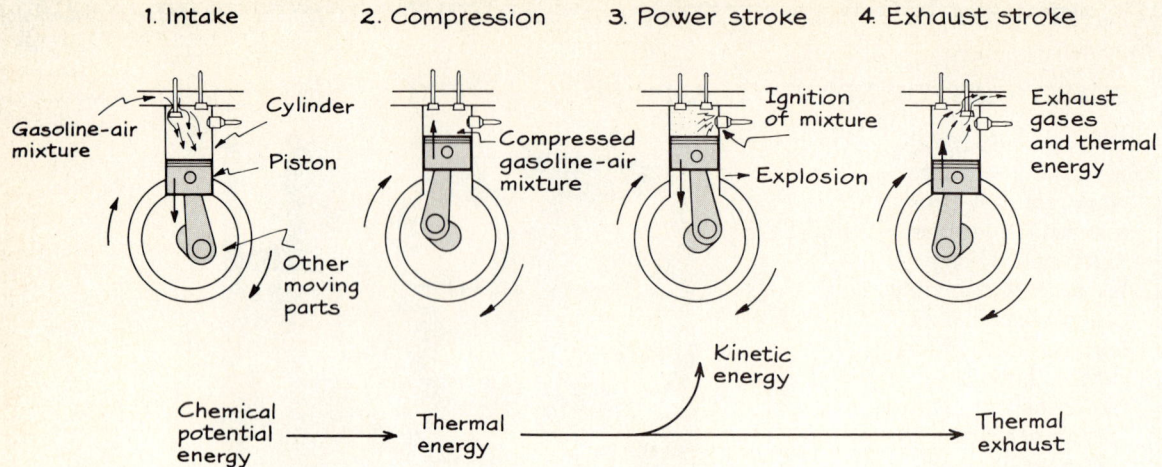

Figure 12-5

The generalized heat engine illustrates the operation of the four-cycle internal combustion engine. Thermal energy released in Step 3 is converted into the motion of the piston and the thermal energy exhausted to the environment.

energy takes throughout each step in the engine's operation. We consider one of the most well-known engines, the automobile engine, and a heat engine that runs backwards, the refrigerator.

Most automobiles use engines called *four-cycle internal combustion engines*. The four-step process, illustrated in Figure 12-5, is a slight variation on the steam engine cycle. In Step 1 the piston drops, drawing a mixture of air and gasoline (in the form of very fine droplets) into the cylinder. A valve closes and the piston moves upward to compress the gasoline-air mixture. In Step 3 the spark plug fires, igniting the gasoline-air mixture. The resulting explosion drives the piston downward. Finally, in Step 4 the piston moves upward, forcing the remaining gases out the exhaust system. Useful energy is derived primarily in Step 3. Energy must be supplied to move the piston in Steps 1, 2, and 4. Typically, an automobile has four to eight of these cylinders.

We can examine this four-step process in terms of the generalized heat engine. Instead of using coal or wood to convert thermal energy into kinetic energy, the automobile engine uses gasoline. The gasoline is burned directly in the cylinder—hence the name **internal combustion engine.** The chemical potential energy of the gasoline is converted into thermal energy, which causes the gases in the cylinder to expand. This expansion pushes the piston downward, converting thermal energy into kinetic energy. Thermal waste is released through engine cooling as well as through the exhaust system.

The refrigerator can also be described in terms of the generalized heat engine. As described in Chapter 10, liquid Freon is pumped through a series of coils inside the refrigerator. It absorbs thermal energy from the air inside the refrigerator and changes into a gas. Gaseous Freon is then pumped back outside the refrigerator by a compresser, where it is liquefied. In many respects a compresser is just a simple piston-cylinder arrangement like those used in steam engines and automobile engines. Gaseous Freon enters the cylinder. As the piston moves upward, it compresses the gaseous Freon, changing it back into a liquid. The thermal energy released by the Freon as it changes state is released to the environment, namely, your kitchen.

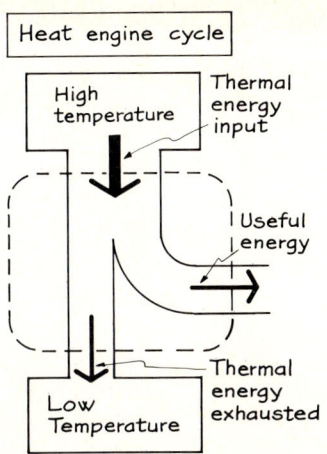 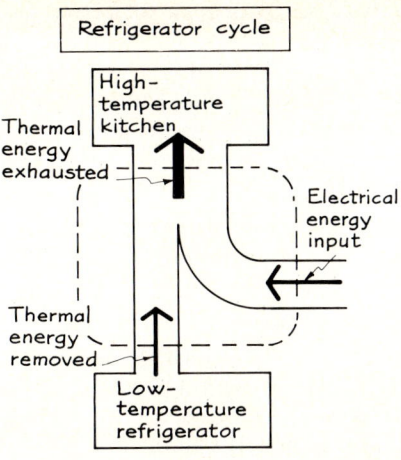

Figure 12-6
A conventional heat engine moves thermal energy from a high temperature reservoir to a low temperature reservoir, diverting some of it for useful work. A refrigerator reverses the process, using electrical energy to move thermal energy from the low temperature (refrigerator) and exhausting it at a higher temperature (the kitchen).

If we trace the various forms energy takes in the refrigerator cycle (Figure 12-6), we see that a refrigerator behaves like a heat engine running backwards. The original energy source is electricity, which is used to drive the compresser. The kinetic energy of the piston in the compresser is converted into thermal energy, which is then exhausted to the environment. Instead of converting thermal energy into kinetic energy, the refrigerator uses kinetic energy to move thermal energy.

THE SECOND LAW OF THERMODYNAMICS: YOU CAN'T EVEN BREAK EVEN

The first law of thermodynamics says we cannot get more energy out of a machine than we put into it. But the miracle machine we proposed at the beginning of the chapter does something more modest—it simply recycles thermal waste energy. Heat engines use thermal energy, and they also dump thermal exhaust into the environment. Why not use this thermal exhaust to run other heat engines?

What we are proposing is a machine that runs on thermal energy extracted from the environment. Actual heat engines, of course, rely upon some external source of energy to produce the thermal energy needed for their operation. Steam engines burn wood or coal, automobile engines burn gasoline, and the human body requires food. It is quite possible to build a heat engine that extracts thermal energy from the air if we use another form of energy to run the extraction process. But that would defeat the whole purpose of our miracle machine. What we need, in light of today's dwindling resources, is a machine that will extract thermal energy from the environment without using another form of energy to do so. Such a machine cannot be built. To see why, we need to investigate spontaneous processes, the concept of entropy, and the second law of thermodynamics.

Some Processes Are Not Spontaneous

Processes by which energy is changed from one form into another or transferred from one place to another can be separated into two categories: spontaneous and nonspontaneous.

Spontaneous processes are those processes that occur without the addition of energy. When a diver steps off a diving board, for example, her gravitational potential energy is converted into kinetic energy. When she hits the water, her kinetic energy is transferred to the water as thermal energy. Both processes are spontaneous because no additional energy is needed to make them occur.

Nonspontaneous processes require external sources of energy. We do not expect the diving-board scene to occur in reverse spontaneously. The thermal energy in the water will not collect itself and spontaneously convert itself into the diver's kinetic energy, lifting her out of the water. Nor does the diver's kinetic energy transform itself into gravitational potential energy, lifting her back up on to the diving board. Conceivably, we could build machines to accomplish both these tasks, but external sources of energy would be needed to run the machines. Nonspontaneous processes require external sources of energy.

The observation that we can separate processes into two categories—spontaneous and nonspontaneous—seems puzzling. The law of conservation of energy makes no distinction between the two. Energy conservation allows the diver to spontaneously rise from the water to the diving board. The thermal energy lost by the water would be gained by the diver as kinetic energy, then as gravitational potential energy. Energy would be conserved. Yet we know this never happens. Observation tells us that nature does make a distinction—energy transformations do have a preferred direction. In searching for an explanation, physicists compared the various forms of energy involved in energy transfer. The concept of entropy emerged to describe the differences they found.

Entropy

Entropy is a measure of the disorder of a system. The greater the disorder, the higher the entropy. To describe what we mean by this, consider the system of poker chips shown in Figure 12-7. We can arrange the poker chips neatly in piles according to color, as shown in (a), or we can scatter them about as shown in (b) and (c). While we cannot assign numerical values to the amount of disorder in each system, we can make qualitative comparisons. The system in (c) is more disordered than that in (b), and (b) is more disordered than (a). Using the concept of entropy, we say that (c) has the most entropy and (a) has the least entropy. As the entropy of a system increases, so does its disorder.

In a sense, entropy describes how easily we can locate something. In the poker-chip example, we might want to locate all the darkest chips. When the chips are scattered about over a large volume, they are distributed so randomly that it would take some time to locate all the darkest ones. When they are stacked neatly according to color, the location of each chip is very well

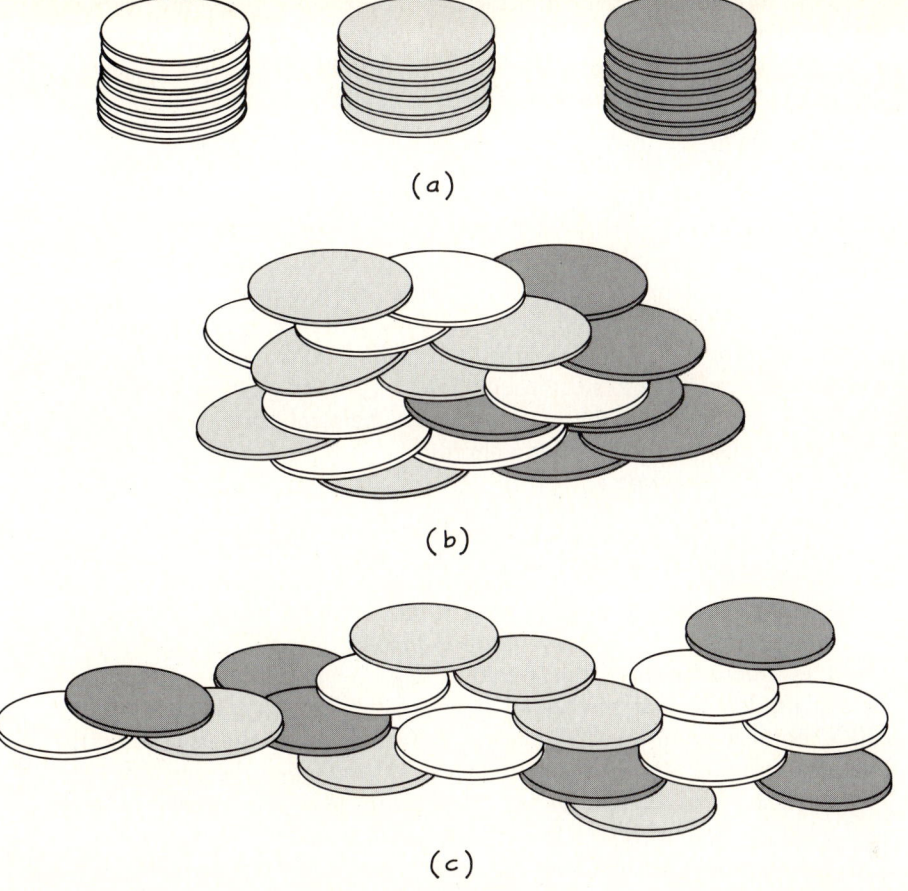

Figure 12-7
Entropy describes the disorder of a system; (a) has the least entropy and (c) has the most entropy.

defined. As the entropy of the system increases, we find it increasingly difficult to locate a specific group of chips.

SELF-CHECK 12B

Which system has the higher entropy: (a) two small glasses of water, one hot and the other cold, or (b) one large glass of lukewarm water?

Entropy and the Energy Stored in Matter

Entropy can be used to describe the manner in which energy is stored in a system—whether it is "stacked up neatly" or "strewn around randomly." In a sense entropy describes how easily we can locate and extract energy from the system. Two characteristics, the form in which energy is stored in the system and the distribution of that energy in the system, help describe differences in entropy. Let's consider examples of each.

Energy can be stored in matter as potential energy involved in molecular bonds and as kinetic energy stored in the motion of molecules. Either of these forms of energy can be converted into thermal energy. In gasoline, for example, energy stored in the electrical bonds among molecules is released when the gasoline burns. We have called this form of stored energy *chemical potential energy*. One liter (L) of gasoline contains roughly 32,000 kilojoules (kJ) of chemical potential energy, which we can extract by burning. We could extract this same amount of energy from the kinetic energy of the molecules—but to do this we would need to cool about 2.4×10^8 L of gasoline 100°C. The same amount of energy can be stored more compactly as chemical potential energy than as kinetic energy. In terms of the poker-chip example, we need to sift through only 1 L of gasoline to find 32,000 kJ of chemical potential energy, while we have to sift through 2.4×10^8 L of gasoline to find the same amount of kinetic energy in the molecules. Chemical potential energy is much more neatly stacked than kinetic energy.

Both the arrangement of energy and its form affect the entropy of a system. For example, consider a system that consists of 2 L of water at an average temperature of 50°C. We can arrange the energy of such a system in several ways. One might be to have a 1 L bottle of water at 100°C and a 1 L bottle at 0°C. A second arrangement would be to have a 2 L bottle at 50°C. The same amount of thermal energy is present in both arrangements, but it would be much easier to locate that energy in the first arrangement than in the second. In the first arrangement, more of the energy is distributed in the 100°C bottle than in the 0°C bottle. We would look in the 100°C bottle first. The first arrangement of energy has less entropy than the second.

Both the form of energy and the manner in which the energy is distributed within a system affect the entropy of the system. In general, potential forms of energy have less entropy associated with them than kinetic forms. And, given two systems with equal amounts of thermal energy stored in their molecules, the system in which the energy is distributed over fewer molecules has less entropy.

SELF-CHECK 12C

Consider a system that consists of the diver and the pool. Before hitting the water, the diver has all the kinetic energy. After entering the water, the diver transfers that kinetic energy to the water molecules. The diver stops moving and the water warms up slightly. Compare the distribution of energy before and after the diver enters the water. Which arrangement has more entropy?

Entropy Must Increase

In any spontaneous process, the entropy of the system increases. When the diver jumps off the diving board, her gravitational potential energy will be

© 1978 by Sidney Harris.

"HE WAS WORKING ON A THEORY OF ENTROPY, AND DEVELOPED A SEVERE CASE OF IT HIMSELF."

spontaneously converted into kinetic energy. Gravitational potential energy has less entropy associated with it than kinetic energy. When the diver enters the water, her kinetic energy is immediately shared with the surrounding water molecules. Kinetic energy located with the motion of a single object—the diver—has less entropy associated with it than the same amount of kinetic energy shared among thousands of separately moving water molecules. Because it is associated with the random motions of many objects, thermal energy has more entropy associated with it than any other form of energy. Thermal energy is the ultimate product of a series of spontaneous energy transformations.

Thermal energy itself flows spontaneously from warmer objects to cooler ones. If we place a glass of hot water in contact with a glass of cold water, both will be lukewarm a short time later. Initially, more thermal energy is distributed among molecules in the glass of hot water than among molecules in the glass of cold water. More of the thermal energy of the system is distributed among fewer molecules. When thermal energy moves from the glass of hot water to the glass of cold water, the thermal energy of the system is shared among more molecules. The entropy of the system increases. Were thermal energy to move from the glass of cold water to the glass of hot water, more thermal energy would be shared among fewer molecules and entropy would decrease. Such a process does not occur spontaneously.

What about nonspontaneous processes, like thermal energy flowing from a cooler object to a warmer object? When a refrigerator cools, thermal energy

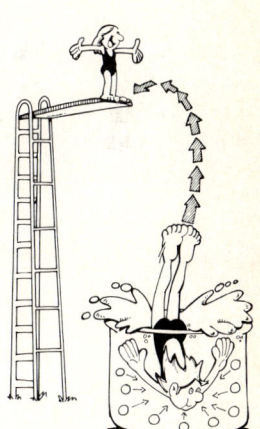

Figure 12-8

Thermal energy stored in the molecules cannot exert a net force upward on the diver, pushing her out of the water and back up to the top of the diving board. This is a nonspontaneous process.

moves from a cooler environment to a warmer environment—from the inside of the refrigerator to your kitchen. The refrigerator gets colder and the kitchen gets warmer. Like the two glasses of water, this arrangement has less entropy than one in which thermal energy is evenly distributed throughout the refrigerator and the kitchen. At first thought, we might say that the entropy of the system has decreased. However, electrical energy must be supplied in order for the heat extraction process to occur. Before drawing any conclusions about changes in entropy, we have to include electricity in our analysis.

Electrical energy is usually generated in power plants that burn oil, coal, or natural gas. To cool the refrigerator, the chemical potential energy of the fuel must be converted into electrical energy. This electrical energy is then converted into the kinetic energy of the piston, which is converted, in turn, into the thermal energy exhausted from the Freon gas. Entropy increases as low-entropy chemical potential energy is converted into higher-entropy thermal energy. The entropy increase in this conversion process far exceeds the entropy decrease that occurs when thermal energy moves from the cooler refrigerator into the warmer kitchen.

A nonspontaneous process does result in a local decrease in entropy, which is why it is nonspontaneous. Entropy does decrease in the refrigerator-kitchen system if we ignore the electrical generating plant. But we cannot ignore the plant. Without it, our nonspontaneous process would not occur. When we include the energy source required to initiate the nonspontaneous process, the entropy of the entire system increases.

These examples illustrate the way in which entropy can be used to explain our experiences with spontaneous and nonspontaneous processes. All processes result in an increase in entropy. Known as the **second law of thermodynamics,** this principle can be stated in a variety of ways, each specific to a particular application. Thermal energy never flows from a cooler object to a warmer object. Natural processes move toward greater disorder. No matter what the application or how it is worded, the second law of thermodynamics clearly reflects nature's preference in the direction of energy transformations.

SELF-CHECK 12D

The air around us contains lots of water vapor—water in the form of gas. Use the second law of thermodynamics to explain why the water vapor in air at room temperature (20°C) does not condense on our skin, transferring the latent heat of vaporization to our bodies at 37°C.

When applied to the heat engine, the second law of thermodynamics is often stated as: It is impossible to build an engine that transforms thermal energy into kinetic energy without exhausting some thermal energy in the

The Second Law of Thermodynamics: You Can't Even Break Even   269

© Beeldrect, Amsterdam V.A.G.A., New York, Collection Haags Gemeentemuseum—The Hague, 1981.

Figure 12-9
While you might not know why, experience tells you that something is wrong with Escher's drawing. What laws of thermodynamics does it violate?

process. A heat engine converts thermal energy into kinetic energy. The thermal energy stored in steam, for example, becomes the kinetic energy of a piston. By itself, such a process involves a decrease in entropy—it will not occur spontaneously. The local decrease in entropy that occurs inside the cylinder is offset by the increase in entropy that occurs at the condenser, where thermal energy once restricted to a certain volume of steam is dispersed among molecules in the environment. The second law of thermodynamics demands that this waste energy be present.

A conventional heat engine, like the steam engine, uses the environment as the condenser. Consequently, the gas must be heated above the temperature of the environment in order for thermal energy to be exhausted at the condenser. Coal or some other form of chemical potential energy is needed to produce the higher-temperature steam. In our miracle machine we wanted to use the thermal energy exhausted to the environment to drive yet another heat engine. In order to do so, we would have to build a condenser that operates at temperatures below environmental temperatures—we would have to build a refrigerator. Electricity or some other form of energy is needed to operate the refrigerator, so we have not gained a thing!

Billions of heat engines are in operation today (Figure 12-10). Each takes some form of potential energy, converts it into kinetic energy, and dumps the thermal exhaust into the environment. We continually diminish the world's

270   Chapter 12.   Thermodynamics

Figure 12-10
Billions of heat engines convert highly organized energy—fossil fuels—into kinetic energy and thermal exhaust.

Freelance Photographer's Guild.

supply of chemical potential energy—to the detriment of our pocketbooks—and increase the thermal energy of the environment, to the dismay of the environmentalists. The second law of thermodynamics allows us no way out of this state of affairs—you can't even break even!

HOW WELL CAN WE DO?

In any process, the entropy of the system must increase. The real energy crisis lies in the fact that we have been rapidly consuming known stores of useful energy and increasing the useless thermal energy dumped into the environment. In a sense, our energy crisis is an entropy crisis—entropy is increasing too rapidly. While we cannot change the second law of thermodynamics, we can use our understanding of it to slow this rate of increase.

The analysis of heat engines tells us that the entropy increase occurs at two places: in the conversion of a fuel into thermal energy and in the release of thermal waste into the environment. Let's examine each of these processes with the goal of minimizing the increase in entropy that accompanies all energy conversions.

Efficiency

In any heat engine, thermal energy in a hot reservoir is converted into kinetic energy (useful energy) and thermal exhaust (waste energy) which is released to a cold reservoir. We use the concept of efficiency to describe how effectively a heat engine or any other energy conversion device produces useful

THE SECOND LAW GETS IN THE WAY, TOO

In the continual search for energy from nothing John Gamge, in the 1880s, designed the zeromotor. Thermal energy from the environment vaporizes liquid ammonia in a tank (A). The ammonia gas travels to a cylinder (D), where it pushes down a piston (H), driving a wheel (I). Having lost its energy, the ammonia spontaneously becomes a liquid again and is pumped (G) back to the tank. There is only one problem: the second law of thermodynamics. When the gas expands and does work at D, its temperature must drop. It is a gas at a temperature lower than the environment. To become a liquid it must give up even more energy. However, since its temperature is already below the environment's, it cannot give up energy spontaneously. That would cause a spontaneous decrease in the entropy of the system. This device, an example of a *perpetual motion machine of the second kind,* will not solve our problems.

Perhaps the only thing truly perpetual about perpetual motion machines is the stream of inventors who claim to have designed them. An inventor obtained a patent for one as recently as 1979. He said he got the idea while taking a college physics course.

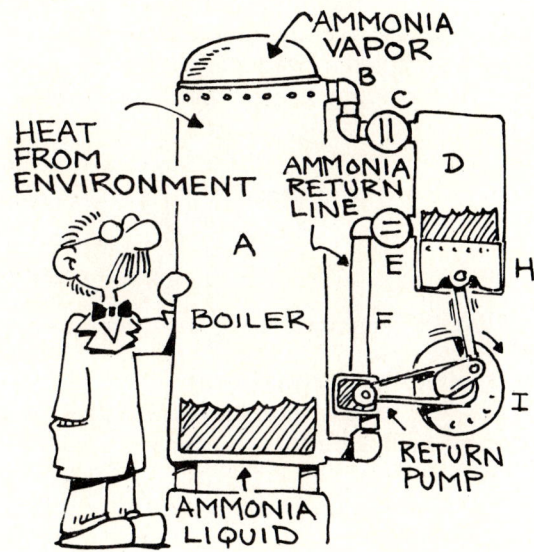

energy. **Efficiency** is defined as the ratio of the useful energy produced to the thermal energy with which we began. Since we often express efficiency as a percentage, this ratio is then multiplied by 100.

$$\text{Efficiency} = \frac{\text{useful energy}}{\text{thermal energy input}} \times 100\%$$

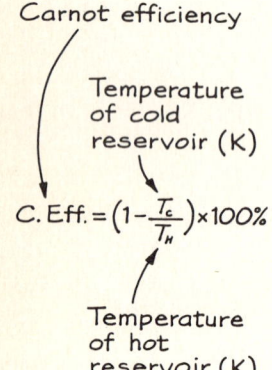

The second law of thermodynamics requires that some waste energy be produced. Consequently, the useful energy produced will never equal the thermal energy input. Heat engines can never be 100% efficient.

While heat engines are always less than 100% efficient, we can reasonably ask just how efficient they can be. An engineer, Sadi Carnot (pronounced Kar′-nō), examined this question; in 1824 he derived an expression for its efficiency.

Real heat engines lose thermal energy in the cylinder as well as in the condenser. Frictional and thermal energy lost to the environment are the usual culprits. While the second law of thermodynamics requires that thermal energy be released at the cold reservoir, losses that occur at other locations in the engine can be minimized. Carnot was able to show that an engine's efficiency ideally depends solely on the temperature difference between the hot and cold reservoirs—the greater the difference, the higher the proportion of thermal energy converted into kinetic energy. This ideal efficiency, called the **Carnot efficiency**, is given by the expression

$$\text{Carnot efficiency} = \left(1 - \frac{\text{absolute temperature of cold reservoir}}{\text{absolute temperature of hot reservoir}}\right) \times 100\%$$

(Absolute temperature (K) = temperature in Celsius (°C) + 273)

For example, an electrical power plant might operate between the temperatures of 700 K (hot reservoir) and 300 K (cold reservoir). The Carnot efficiency for such a plant is

$$\text{Carnot efficiency} = \left(1 - \frac{300 \text{ K}}{700 \text{ K}}\right) \times 100\% = 57\%$$

No matter how well designed it is, a power plant operating at these temperatures can never exceed 57% efficiency. Such a plant typically only reaches 30%–40% efficiency.

Carnot's ideal heat engine provides a goal toward which designers strive and a way with which to evaluate our consumption of useful forms of energy. The greater the temperature difference between the hot and cold reservoirs, the more efficient the engine. Practically speaking, this difference is limited by our use of the environment as the cold reservoir and by the maximum temperature at which most materials still function. Table 12-1 lists typical hot and cold reservoir temperatures, Carnot efficiencies, and usual operating efficiencies for several common heat engines. In most cases we can still improve.

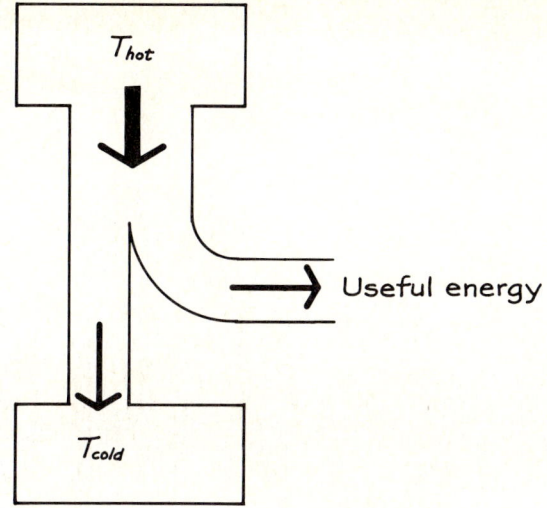

Figure 12-11
Carnot described the ideal efficiency of a heat engine in terms of the temperatures of the hot and cold reservoirs. As the difference in temperature increases, the efficiency of the engine becomes greater.

Table 12-1 Efficiencies of Several Heat Engines

Heat Engine	Temperature of Cold Reservoir (K)	Temperature of Hot Reservoir (K)	Carnot Efficiency (%)	Actual Efficiency (%)
Gasoline engine	300	450	33	25
Diesel engine	300	550	45	35
Piston steam engine	375	475	21	10
Copper-Bessemer engine*	240	700	65	42
Electrical generating plant	300	800	62.5	35

*Most efficient engine constructed so far.

SELF-CHECK 12E

Suppose your moped engine operates between the temperatures of 450 K and 290 K. What is the best efficiency it can reach?

Figure 12-12
A furnace and a heat pump differ in the way in which they produce thermal energy to heat your home. A furnace converts the chemical potential energy stored in a fuel into thermal energy. Some of this thermal energy heats your home, and the rest is exhausted up the chimney. A heat pump uses electrical energy to move thermal energy from outside to inside. The heat pump uses considerably less low-entropy energy to produce the same amount of thermal energy to heat your home.

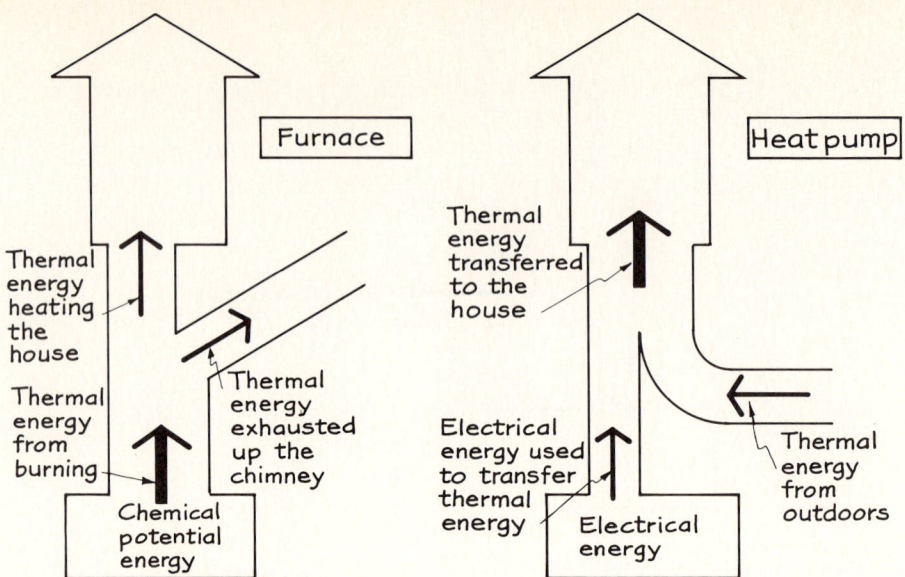

Selecting the Energy Source

The other way in which heat engines increase the entropy of a system is in the conversion of a fuel into the thermal energy needed to operate the heat engine. Usually, we select a low-entropy form of energy, the chemical potential energy stored in a fuel, and convert it into the highest-entropy form of energy, the thermal energy stored in the motion of gas molecules. We could decrease the rate of entropy increase by replacing the fuel with an energy source that has more entropy.

The heat pump uses this approach to decrease your home-heating costs. A conventional furnace takes a low-entropy form of energy (oil or gas) and converts it into thermal energy to be distributed throughout the house. A heat pump takes a somewhat higher entropy form of energy (electricity) and uses it to "pump" thermal energy from the outside air into the inside air. It works very much like a refrigerator, using electricity to extract thermal energy from a cooler environment and pump it into a warmer environment.

Heat pumps do require lower-entropy forms of energy, namely, electricity. Ultimately, however, they consume less of these lower-entropy forms of energy because they take advantage of the thermal energy already present in the environment. If we contrast a furnace with a heat pump (Figure 12-12), the two processes differ in how the thermal energy added to the air is obtained. In a furnace, all the thermal energy added to the air is provided by the chemical potential energy released when gas or oil is burned. By contrast, some of the thermal energy added to the air by a heat pump was thermal energy already present in the environment. The heat pump simply moves it to a different location.

Table 12-2 lists the amount of low-entropy energy needed to deliver 1,000,000 J of thermal energy to a house by four different devices. For the gas and oil furnace, the low entropy energy needed is measured directly in terms of the oil or gas consumed. Since all the thermal energy delivered by a furnace comes directly from the chemical potential energy of the fuel, a furnace consumes more energy than it produces. Real heat pumps and electrical furnaces use electricity, which ultimately has been produced by an electrical generating plant that burns fuel. Most electrical generating plants are about 33% efficient, so about three times the energy delivered by the electricity has been consumed from lower-entropy fuels. This makes an electrical furnace much less efficient than a gas or oil furnace that burns the fuel directly where the thermal energy is needed. The heat pump, however, delivers 1,000,000 J of thermal energy while consuming only 600,000 J of low-entropy energy. We have not gotten more energy out than we put in. Some of the energy delivered by the heat pump is present in the air that is being moved. This energy does not show up in a comparison of the low-entropy energy consumed. What we have done is move 1,000,000 J of thermal energy for one-third of the lower-entropy energy consumed by a furnace. Ideal heat pumps do even better.

This reasoning can be used to select appropriate methods of energy conversion for a variety of situations. The goal is to decrease, as much as possible, the amount of low-entropy energy consumed. When thermal energy is desired, using some of the thermal energy already present in the environment decreases the overall increase in entropy. The net effect—a warm house—is the same as when low-entropy energy is burned, but the entropy increase is much less. These considerations have led to the definition of a **second law efficiency.** Among those methods possible, the method that uses the least low-entropy energy has the greatest second law efficiency.

Why not use a heat pump as our miracle machine? It extracts thermal energy from the environment, so it could use the waste energy generated by other machines. However, a heat pump cannot do useful work. It merely moves thermal energy from one location to another; it does not transform thermal energy into kinetic energy. The second law of thermodynamics allows

Table 12-2 Comparison of Low Entropy Energy Needed to Produce 1,000,000 J of Thermal Energy

Method	Efficiency (%)	Low-Entropy Energy Needed (J)
Ideal heat pump	100	204,000
Real heat pump	33	600,000
Gas/oil furnace	60	1,667,000
Electric heat	33	3,000,000

us a few miracles, but not this ultimate one. The concept of using the thermal energy strewn throughout the environment is an intriguing one, however. And while physicists are not betting that our miracle machine will ever be invented, they also are not betting that people will stop trying. The idea is much too enticing!

ENTROPY AND THE UNIVERSE

The energy crisis we face is one of running out of lower-entropy forms of energy. Historians report past crises—water power shortages or wood scarcity. Our century seems to be the century of the oil crisis. The development of fusion and fission power plants may bring momentary relief. One day, their fuels, like the firewood, gas, and oil before them, will also be exhausted. We have no choice of endings. Our choice is only how quickly we reach the predictable end. Wise application of the laws of thermodynamics would allow future generations their chance at dealing with the problem.

The laws of thermodynamics apply beyond just our time and our planet. The largest closed system we have is the universe. The second law of thermodynamics tells us that the entropy of this system is continually increasing. Useful sources of energy, like stars, are transforming highly ordered nuclear energy into disordered, or degraded, thermal energy. Our own sun provides an enormous amount of thermal energy to us daily. Energy once localized will eventually be strewn about randomly, unavailable for the most useful task—maintaining life. When all forms of energy degrade to the thermal energy of molecules in space, we have reached what is called the *heat death* of the universe. If our understanding of thermodynamics is correct, the heat death cannot be avoided. Entropy must increase. But, must it increase forever? To answer this question, we will return to our molecular model of matter.

CHAPTER SUMMARY

Thermodynamics, the study of the transformation of thermal energy into other forms, and vice versa, originated largely in the study of heat engines. A *heat engine* is any device that converts thermal energy into kinetic energy and thermal exhaust. The kinetic energy can be an end product or can be used to produce another useful form of energy. Common examples of heat engines include the steam engine and the automobile engine. A refrigerator is a heat engine run in reverse. When applied to the heat engine, the law of conservation of energy states that for one complete cycle,

Thermal energy input = useful energy + waste energy

This is known as the *first law of thermodynamics*. The first law says that we can never get more energy out of a machine than we put into it.

Processes by which energy is changed from one form into another or transferred from one object to another can be separated into two categories: spontaneous and nonspontaneous. In both cases energy is conserved, but nonspontaneous processes do not occur without the addition of energy from external sources. In searching for a basis for this distinction, physicists have developed the concept of entropy. *Entropy* is a measure of the disorder of a system. The greater the disorder, the higher the entropy. *Spontaneous* processes are those in which entropy increases. *Nonspontaneous* processes are those in which the entropy of the system decreases, when the external energy sources are ignored. When the external energy source is included in the system, the entropy of the system always increases. *The second law of thermodynamics* states that in all processes, the entropy of a closed system increases. Applied to the heat engine, the second law of thermodynamics tells us that thermal energy cannot be used to do useful work without losing some of it to thermal exhaust.

The *efficiency* of an energy conversion device is defined as:

$$\text{Efficiency} = \frac{\text{useful energy}}{\text{thermal energy input}} \times 100\%$$

Since some of the thermal energy input must be exhausted to the environment, a heat engine can never be 100% efficient. The maximum efficiency attainable, called the *Carnot efficiency*, is determined by the temperature of the hot and cold reservoirs used in the heat engine.

$$\text{Carnot efficiency} = \left(1 - \frac{\text{absolute temperature of cold reservoir}}{\text{absolute temperature of hot reservoir}}\right) \times 100\%$$

Heat engines should be selected according to their operating efficiencies and the amounts of low-entropy energy consumed.

ANSWERS TO SELF-CHECKS

12A. Thermal energy input = useful energy + waste energy

3 billion J = 1 billion J + waste energy

Waste energy = 2 billion J

12B. (b) has the higher entropy because the thermal energy is spread among more water molecules. In (a) more thermal energy is located in the glass of hot water, so we would know to look there first.

12C. Before the diver enters the pool, the kinetic energy is located with her. After she enters the pool, the kinetic energy is shared between all the water molecules and herself. The arrangement in which the energy is shared with the water molecules has more entropy.

12D. In order for water to condense on our skin, thermal energy equal to the latent heat of vaporization of water would have to be transferred from the air to our skin. Our bodies are at a higher temperature (37°C) than the surrounding air (20°C). Thermal energy will not spontaneously move from a colder environment to a warmer environment, since the entropy of the system would have to decrease.

12E. Carnot efficiency $= 1 - \left(\dfrac{290 \text{ K}}{450 \text{ K}}\right) \times 100\% = 36\%$

PROBLEMS AND QUESTIONS

A. Review of Chapter Material

A1. Define the following terms:
  Heat engine
  First law of thermodynamics
  Spontaneous processes
  Nonspontaneous processes
  Entropy
  Second law of thermodynamics
  Efficiency
  Carnot efficiency
  Second law efficiency

A2. Describe the heat engine cycle and explain why it is useful.

A3. Describe what happens when a heat engine runs backwards.

A4. How is the first law of thermodynamics related to the law of conservation of energy?

A5. How is entropy related to disorder?

A6. How does the second law of thermodynamics distinguish between processes that occur spontaneously and those that do not?

A7. How do we calculate the efficiency of a device?

A8. What does the Carnot efficiency tell us about a heat engine?

A9. How is the second law of thermodynamics useful in selecting an energy conservation device?

A10. Why does a heat pump have a greater second law efficiency than a furnace?

B. Using the Chapter Material

B1. The Stanley Steamer was an automobile (circa 1900) that was propelled by a steam engine. Can the Stanley Steamer be described as a heat engine? If yes, identify the processes and forms of energy. If no, why not?

B2. An old tale is: On a hot summer day you can cool the kitchen by leaving the refrigerator door open. Why will this process make the kitchen warmer instead of cooler?

B3. Your friend decides that it is easier to operate his air conditioner under his bed rather than in a window. Use the concept of the heat engine to explain why his room gets warmer instead of cooler.

B4. An air conditioner uses 100 J of electrical energy to move 400 J of thermal energy from inside to outside. What is the total energy exhausted by the air conditioner?

B5. Hot water in a kitchen sink is allowed to cool. Why does entropy increase during this process?

B6. When a pan of water is heated on a stove, thermal energy is concentrated in the water. Why does entropy increase even though we can identify more energy as being in the water?

B7. A common practice is throwing empty beverage cans along roads rather than placing them in trash containers. Why do beverage cans strewn along the road represent a higher "beverage-can entropy" state than cans placed in a trash container?

B8. What is the Carnot efficiency of the following engines:

Hot Reservoir	Cold Reservoir
300 K	300 K
600 K	300 K
900 K	300 K
900 K	150 K
900 K	0 K

B9. A diesel engine operates at a hot temperature of 550 K, while a gasoline engine operates at 450 K. If everything else is equal, why is the diesel engine more efficient?

B10. What is the actual efficiency of a lawn mower that exhausts 70 J for every 100 J of chemical potential energy used?

C. Extensions to New Situations

C1. Bert and Ernie buy identical air conditioners and install them in identical houses. Both keep their homes at 25°C. Bert lives in a large city, where the temperature is higher than at Ernie's country home.
  a. Whose air conditioner will run more efficiently?
  b. Who will use more electrical energy?

C2. A freezer uses 100 kJ of electrical energy to move 330 kJ of thermal energy from inside the freezer to outside. We can determine how much the room temperature increases when 1 kg of ice freezes.
  a. How much thermal energy must be removed from 1 kg of 0°C water to turn it into 0°C ice? (The latent heat of fusion of water is 320 kJ/kg.)
  b. How much electrical energy is used to freeze the ice?
  c. What is the total energy, including waste heat, exhausted by the freezer into the kitchen?
  d. A kitchen contains 43 kg of air (specific heat capacity = 1.0 kJ/°C · kg). Before the ice freezes, the air temperature is 20°C. After the ice freezes, at which temperature will the air be—22°C, 25°C, or 30°C?

C3. An automobile engine uses 500 J of chemical potential energy to produce 100 J of kinetic energy. The operating temperature of the engine is 500 K and the exhaust goes to a 300 K environment.
  a. What is the highest efficiency this engine can reach?
  b. What is its actual efficiency?
  c. Could this engine be made three times as efficient as it is? Explain your answer.

C4. An inventor is looking for people to invest money in her great new engine. She claims that the engine has a hot reservoir temperature of 800 K and a cold reservoir temperature of 400 K. Further, the engine delivers 700 J of useful energy when the thermal energy input is 1000 J.
  a. What is the actual efficiency of the inventor's engine?
  b. What is the best possible efficiency of the engine?
  c. Does this engine violate the first law of thermodynamics? The second law of thermodynamics?
  d. Do you recommend investing in this engine?

C5. Dacia, whose father is a physicist, describes the concept of entropy as: If you clean up your room, the rest of the house gets messier. Is this a correct statement? Why or why not?

C6. You carefully hang up the clothes your friend has left scattered throughout the house.
  a. Has the entropy of the clothes increased or decreased?
  b. What is the entire system involved?
  c. Has the entropy of the system increased or decreased?

C7. A swamp cooler is a home-cooling device used in dry climates, like those found in the southwest part of the United States. Air is blown across water, causing the water to evaporate and the air to cool.
  a. Why is the air cooler after passing over the water?
  b. Will the swamp cooler have a greater second law efficiency than a conventional air conditioner? Why?

C8. Joe Engineer says he has solved an efficiency problem for his company. The factory has several engines operating in it. Joe argues that if they air-condition the factory, the cold reservoir to which the engines dump their exhaust will be cooler. Thus, Joe reasons, the engines will operate more efficiently.
  a. Will an engine's efficiency increase if the cold reservoir is cooled?
  b. Will the factory's overall energy consumption decrease if it is air-conditioned? Explain your answer.
  c. How will the rate of entropy increase change if air conditioning is installed?

C9. A refrigerator's efficiency is generally described by its coefficient of performance, which is defined as:

$$\frac{\text{energy removed from inside}}{\text{electrical energy used}}$$

The Carnot coefficient of performance is

$$\frac{T(K)\text{inside}}{T(K)\text{outside} - T(K)\text{inside}}$$

These numbers are usually greater than 1.
a. A refrigerator uses 400 J to remove 800 J of thermal energy that is inside. What is its actual coefficient of performance?
b. For the refrigerator in (a), how much energy is exhausted to the room?
c. If the inside temperature of this refrigerator is 275 K and the outside temperature is 300 K, what is the Carnot coefficient of performance?

C10. The Carnot coefficient of performance (See Problem C9) can be used to see how real performance changes with temperature. If the Carnot value decreases, so does the actual operating efficiency, and vice versa.
a. How does the Carnot coefficient change as the outside temperature increases? As the inside temperature increases?
b. On a hot night in the city, lots of air conditioners are operating. What happens to the outside air temperature of the city? Explain your answer.
c. How will the change described in (b) affect the performance of the air conditioner?

D. Activities

D1. Heat pumps and swamp coolers (see Problem C7) are two devices that slow down entropy increases. However, they do not work equally well in all climates. Investigate their usefulness to your area.

D2. Some science fiction stories are based on changes in entropy. You might enjoy *The Last Question* by Isaac Asimov.

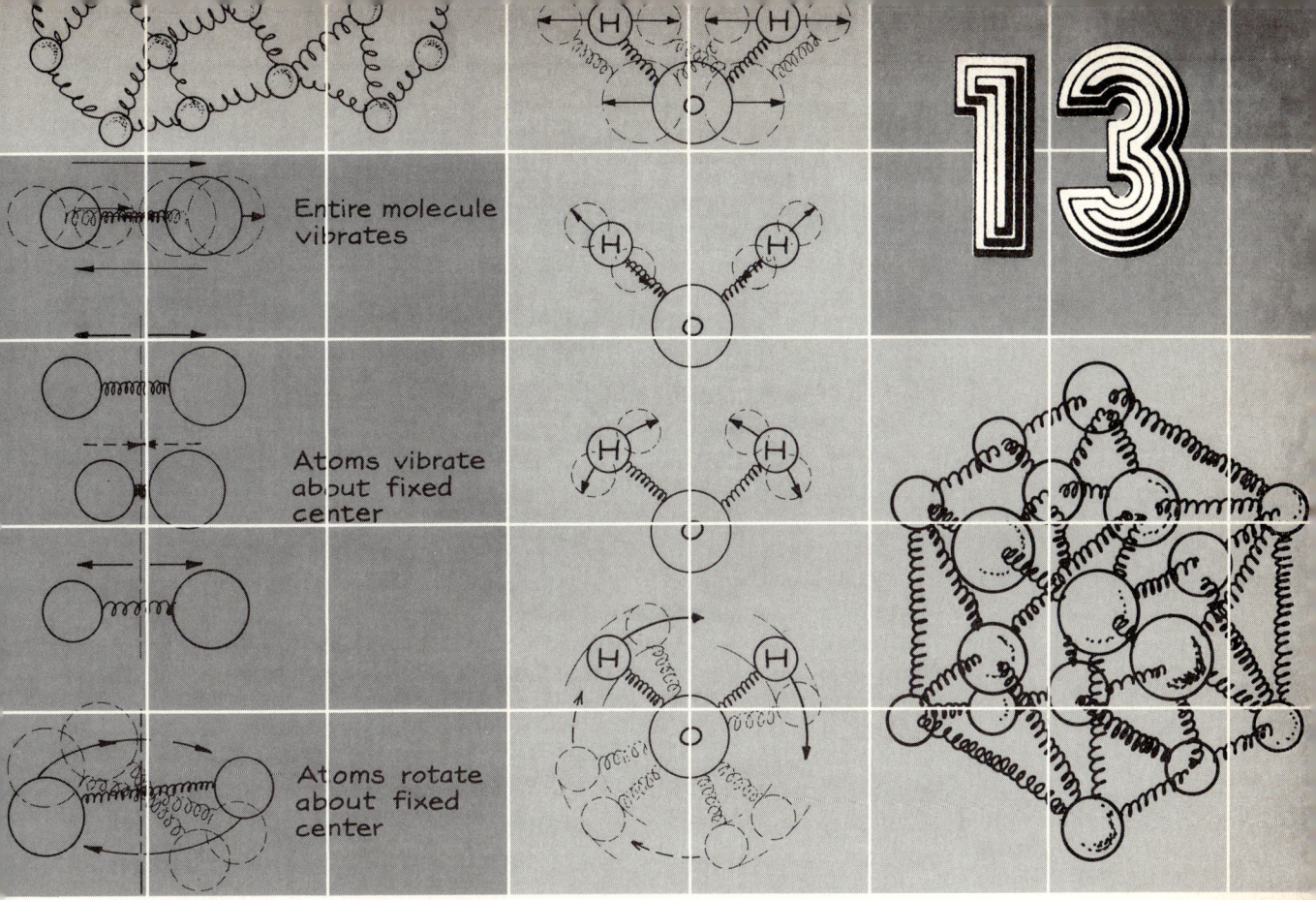

Atoms, Molecules, and Thermal Energy

So far we have treated thermal energy by observing matter—its temperature and its state—and then inventing concepts to describe what we see. Pizza sauce burns; crust at the same temperature does not. We invent the concept of specific heat capacity to describe differences in the amount of thermal energy transferred by the two materials. Ice at 0°C cools a glass of tea better than an equal amount of water at 0°C. We invent the concept of latent heats to describe the thermal energy transferred when substances undergo a change of state. Thermal energy can be transferred through empty space and through gases, liquids, and solids. Three processes, radiation-absorption, convection, and conduction, best summarize the differences we observe. Heat engines convert thermal energy into kinetic energy and thermal exhaust. Entropy explains the limits to the efficiencies of these devices. We could go on and on.

While you may have become pretty adept at using these concepts to describe a variety of interactions, lurking in the back of your mind may have been the questions: How does it all really happen? What goes on inside matter? Physicists use the molecular model of matter to provide a more complete image of thermal processes in matter. By applying Newton's laws and the laws of conservation of energy and momentum to individual molecules, physicists have built a fairly detailed picture of how thermal energy arises at the molecular level. Such a picture allows us to see the concepts of thermal energy from a broader perspective.

In this chapter we examine the concepts introduced in the past three chapters in terms of the molecular model of matter. We use the concept of *restoring forces* to describe the different latent heats of fusion and vaporization. The *motion of individual molecules* is used to describe differences in specific heat capacities and the three thermal energy-transfer processes. Because of the billions of molecules found in each speck of matter, we cannot have absolute knowledge of the behavior of each molecule. We introduce a *probabilistic view of matter* that gives rise to the concept of entropy and the second law of thermodynamics.

INSIDE MATTER

In Chapter 10 we briefly described the molecular model of matter. Matter consists of molecules that interact with one another through electrical forces. These forces allow individual molecules to move about, yet restrict liquids and solids to a definite volume. To picture how electrical forces can allow motion and still restrain molecules to a definite volume, we introduce an analogy with springs. We use this analogy to describe the different states of matter and the thermal energy associated with changes in state.

Restoring Forces and Springs

The electrical interaction among atoms and molecules is the glue that holds matter together. As discussed in Chapter 8, these interactions result in forces that can be either attractive or repulsive, depending on the types of electrical charge involved. In matter, large numbers of both positive and negative electrical charges are present. Consequently, any given molecule will feel many, many forces—some attractive and some repulsive. We can replace this complex set of forces with a single force, called a *restoring force*.

A **restoring force** is a net force that always restores the molecule to some central location. One way to picture restoring forces in matter is to imagine that the molecules in matter are connected to one another by springs. If you push two molecules toward one another, you compress the spring connecting them. The spring pushes back, trying to restore the molecules to their original separation. If you pull the two molecules apart, the spring again resists your force, this time pulling the molecules back toward one another. No matter which way you move the molecules, the spring tries to restore them to their original location. The complex set of electrical interactions that act on individual molecules in matter combine to act like restoring forces.

The strength of the restoring force affects how far the molecules can move about their central location. A very strong restoring force, much like a tight spring, holds the molecules together tightly. It allows very small displacements as the molecules vibrate about their central location. A weaker force, like a very flexible spring, allows the molecules to move farther away but still binds them together. No restoring force at all allows the molecules to move about freely, bumping into whatever gets into their way. These three types of restoring forces characterize the three states of matter.

Solids, Liquids, and Gases

The molecules of a solid are held together by strong restoring forces. Their motion is analogous to a bunch of balls connected by springs, like those shown in Figure 13-1. Each ball has some energy and vibrates back and forth about an equilibrium position. But its energy is not sufficient to allow the molecule to overcome the restoring force exerted by the springs. This restricted motion leads to the definite shape and volume characteristic of solids.

Liquid molecules do not experience as strong a restoring force as solids. A liquid is more like a collection of balls with loosely attached springs, like those shown in Figure 13-2. The balls move about, collide with one another, and temporarily attach to one another via relatively weak springs. Inside the liquid, molecules feel forces in all directions. At the surface of the liquid, molecules feel forces back toward the other molecules. The weaker restoring forces in a liquid allow the molecules to move about more freely. This leads to the lack of shape characteristic of a liquid. The restoring forces are strong enough, however, to maintain relatively fixed average distances between molecules. A liquid does maintain a definite volume.

The weakest restoring forces are those among molecules in a gas. Molecules in a gas behave much like billiard balls moving about on a billiard table. They move about freely, bump into each other, and collide with the edges of the table. Each ball moves independently of the others. In a gas, molecules behave as though they were independent of one another. They move about freely, assuming the shape of their container. When placed in a larger container, the molecules simply move about farther before colliding with one another or with the walls of the container. Since the restoring forces are extremely weak, the molecules will move as far apart as the container allows. Gases have neither definite shapes nor definite volumes.

Changes in State

We can use the different restoring forces in solids, liquids, and gases to explain the thermal energy involved when matter changes state. The restoring force among molecules is greatest when the molecules form a solid, smaller when the molecules form a liquid, and essentially zero when the molecules form a gas. When an object changes state, the strength of its restoring force must change.

The spring analogy provides us with a convenient way to imagine how the strength of these restoring forces can change. When we change a solid

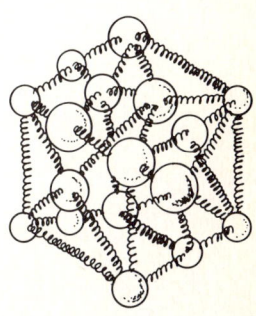

Figure 13-1

In solids, molecules are held in relatively fixed locations by strong restoring forces, represented as tight springs in this drawing.

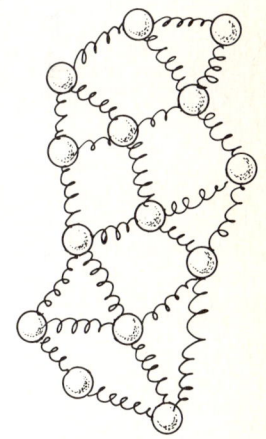

Figure 13-2

In liquids, molecules are free to move about more. Weaker restoring forces, represented here as loose springs, allow a liquid to change shape.

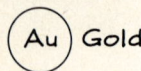
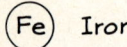
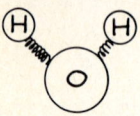
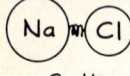
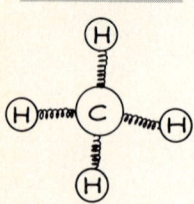

Figure 13-3
Molecules consisting of one, two, three, and four atoms.

into a liquid, we stretch the springs connecting the molecules until they are stretched out of shape. Longer and weaker, the springs allow the molecules to stay farther apart and move about more. Changing a liquid into a gas is like stretching the springs even farther—until they break completely. Broken springs allow the molecules to move freely about, much as the molecules in a gas move independently of one another.

Energy is required to stretch the springs and eventually break them. The **latent heat of fusion** and the **latent heat of vaporization** are the thermal equivalents of this energy. The latent heat of fusion describes the amount of energy needed to stretch the springs permanently. The latent heat of vaporization describes the amount of energy required to break the springs connecting the molecules. Since all substances differ in the strength of the restoring force holding their molecules together, each substance has unique latent heats of fusion and vaporization. The thermal energy needed to actually "break the springs" is much greater than that needed to "stretch the springs." Consequently, the latent heat of vaporization of a substance is always greater than its latent heat of fusion.

SELF-CHECK 13A

The latent heat of fusion of water (320 kilojoules per kilogram (kJ/kg)) is nearly three times that of alcohol (104 kJ/kg). Use the spring analogy to describe the difference in restoring forces present in the two substances.

INSIDE THE MOLECULE

In Chapter 10 we defined temperature as a measure of the average kinetic energy of the molecules in a material. Two substances at the same temperature have molecules with the same average kinetic energy. The pizza example suggests that different substances need different amounts of thermal energy to reach the same temperature. In order to understand both temperature and specific heat capacity at a molecular level, we need to look inside the molecule.

Atoms, Molecules, and Springs

A molecule can be composed of from one to hundreds of atoms. Like molecules bound together to form matter, atoms are glued together to form molecules by electrical restoring forces. Once again we can imagine the atoms within a molecule to be spherical balls held in place by springs. Figure 13-3 shows models for one-, two-, three-, and four-atom molecules. Since the restoring force holding atoms together in molecules is much greater than that hold-

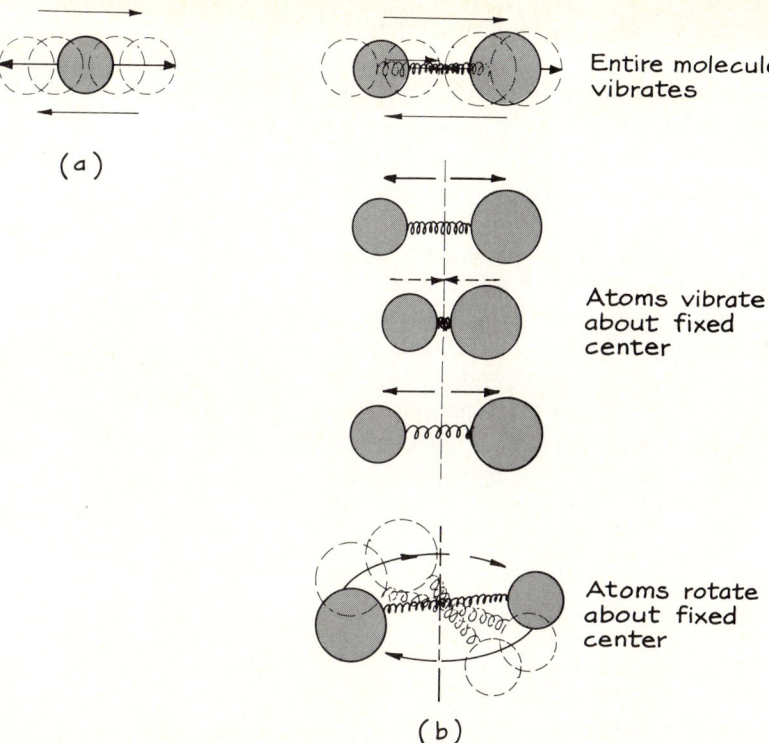

Figure 13-4
How a molecule moves depends on the number of atoms it has. **(a)** The one-atom molecule can only move back and forth. **(b)** The two-atom molecule can move back and forth or the atoms within it can vibrate and rotate while the center of the molecule remains stationary.

ing molecules in solids and liquids, we rarely see molecules fall apart during thermal interactions. Instead we see changes in state.

As shown in Figure 13-4, the number of different ways a molecule can move depends on the number of atoms from which it is built. The one-atom molecule in (a) can move about only in straight lines as it vibrates about its central location. Two-atom molecules (Figure 13-4(b)) can move in three possible ways. The entire molecule can move back and forth in a straight line, as did the one-atom molecule. Additionally, the two-atom molecule can remain in a fixed location while its atoms rotate or vibrate about its center. A three-atom molecule has even more possibilities. As the number of atoms in a molecule increases, the number of possible ways that the molecule can move increases.

The total kinetic energy of the molecule depends upon all of the different ways in which it can move. In the one-atom molecule, the total kinetic energy of the molecule depends simply on its back-and-forth motion. In the two-atom molecule, its kinetic energy depends on all three motions—the back-and-forth motion of the molecule, as well as the rotation and vibration of the atoms within the molecule. The more complex the molecule, the more complex the distribution of kinetic energy among the different ways in which the molecule can move.

Temperature and Kinetic Energy

The simplest way to measure the temperature of a substance is to place a thermometer in it. The molecules in the substance collide with the molecules in the bulb of the thermometer. These collisions are similar to the interactions

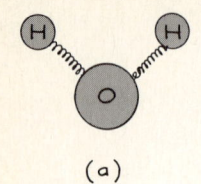

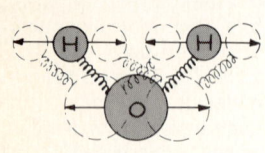

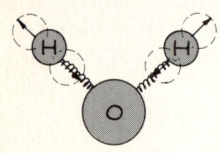

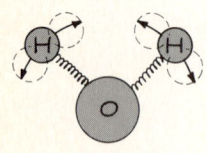

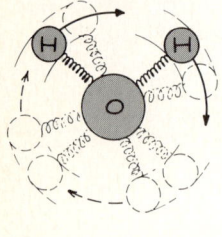

Figure 13-5
Water molecules consist of three atoms. Because of the number of different ways its atoms can move, water has a high specific heat capacity.

among billiard balls discussed in Chapter 5. Kinetic energy and momentum are transferred from the molecules in the substance to the molecules inside the thermometer. In collisions of this nature, the kinetic energy and momentum exchanged come primarily from the straight-line, or linear, motions of the atoms. When we say that two substances are at the same temperature, we are saying that the average kinetic energy associated with their linear motion is equal.

On any temperature scale, two different temperatures provide a qualitative comparison of the kinetic energy associated with the linear motions of molecules in a substance. The molecules of water at 60°C have more kinetic energy than those of water at 10°C. The Kelvin scale, however, provides a direct quantitative comparison of kinetic energies. The molecules of water at 60 K have six times as much kinetic energy as those at 10 K. If a material could actually be cooled to 0 K, the kinetic energy of its molecules would reach the smallest value it could possibly reach. This temperature, called absolute zero, is the temperature at which we imagine that the molecules reach minimum vibrations.

Specific Heat Capacity and Molecular Motion

Temperature does not provide a complete measure of the total kinetic energy of a substance. Different substances at the same temperature can have different amounts of total kinetic energy. While the average kinetic energy associated with their linear motions is the same, the energy associated with other motions can differ. **Specific heat capacities** describe these differences.

Complex molecules can do it all—move linearly, rotate, and vibrate—all simultaneously. Substances with high specific heat capacities consist of complex molecules that can vibrate and rotate in a variety of ways. Those with low specific heat capacities consist of much simpler molecules. For example, water consists of two hydrogen atoms and one oxygen atom, arranged as shown in Figure 13-5. Since its molecules can rotate and vibrate in a number of different ways, water has a high specific heat capacity. By contrast, helium consists of a one-atom molecule. All its thermal energy goes into the linear vibrations of the atom and consequently helium has an extremely low specific heat capacity.

In order to see how specific heat capacities vary with the complexity of the molecule, we need to compare the thermal energy absorbed per degree temperature change for samples consisting of equal numbers of molecules instead of equal masses. Table 13-1 provides such a comparison for one-, two-, and three-atom molecules. Each sample contains 6×10^{23} molecules of the specific gas. If we compare the amount of thermal energy absorbed per degree temperature change for the two one-atom gases, we find them to be equal. The two-atom gases absorb a little less than twice as much thermal energy per degree change as the one-atom gases. They vary from one another by no more than 1 J/K. Three-atom molecules require even more thermal energy per degree temperature change than either the two-atom or one-atom molecules. Here the energies become more diverse because the arrangement of the three atoms within the molecule can be more varied. Those with higher specific heat capacities, like sulfur dioxide, have more different ways in which their molecules can move. The specific heat capacity of a substance depends on the number of different ways the atoms within the molecules can move.

Table 13-1 Comparison of Specific Heat Capacities

Number of Atoms in Molecule	Name of Substance	Atoms	Thermal Energy Absorbed per 1 K (J)*
One	Helium	One helium	12.5
	Argon	One argon	12.5
Two	Hydrogen	Two hydrogen	20.4
	Nitrogen	Two nitrogen	20.8
	Oxygen	Two oxygen	21.1
	Carbon monoxide	One carbon One oxygen	20.9
Three	Carbon dioxide	One carbon Two oxygen	28.5
	Sulfur dioxide	One sulfur Two oxygen	31.4
	Hydrogen sulfide	Two hydrogen One sulfur	26.0
	Steam (water)	Two hydrogen One oxygen	28.8

*Each sample of gas contained 6×10^{23} molecules.

SELF-CHECK 13B

The thermal energy absorbed by 6×10^{23} molecules of air per degree Kelvin is approximately 21 joules (J). Use Table 13-1 to determine whether air consists mostly of one-atom, two-atom, or three-atom molecules.

THERMAL ENERGY IN TRANSIT

The thermal energy in matter is stored in the electrical bonds and in the motions of the individual molecules. While restoring forces are important in holding

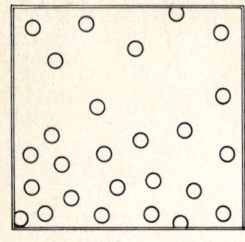

(a)

(b)

Figure 13-6
Gravitational forces produce an uneven distribution of molecules in the atmosphere. Faster-moving molecules will lose their energy more quickly near the bottom of a sample than near the top. Consequently, faster-moving molecules migrate upwards. Warmer air rises.

molecules together, the motion of the molecules is responsible for allowing thermal energy to be transferred from one substance to another or from one place to another. Let's consider the three transfer processes—convection, conduction, and radiation-absorption—in terms of the motions of individual molecules.

Convection

Convection is the primary process by which thermal energy is transferred in liquids and gases. Warmer samples of gas rise, taking thermal energy with them. The thermal energy is then transferred to other objects as the warmer samples cool. We can look at this process at the molecular level by considering the distribution of molecules in a fluid like air.

As shown in Figure 13-6, gravitational attraction between the earth and the air produces an uneven distribution of molecules. Each molecule is pulled downward toward the earth's surface. Collisions between molecules knock some of the molecules back upward, so that the air does not completely collapse. In spite of these collisions, more molecules are found near the bottom of the sample than near the top.

When we introduce thermal energy into this sample of gas, we are, in effect, adding rapidly moving molecules. These molecules will move in all directions, colliding with molecules in the air sample. Those that move downward run into a more dense layer of molecules than those that move upward. Consequently, the downward-moving molecules collide more frequently than the upward-moving molecules, losing energy with each collision. Within a relatively short time, only the energetic molecules that were initially moving upward are left. **Convection** occurs because the uneven molecular distribution created by gravitation favors upward rather than downward motion of the more energetic molecules.

SELF-CHECK 13C

In a gravity-free environment, such as Spacelab, the molecules in a sample of air will distribute themselves evenly throughout its volume. Use the molecular model to explain why convection will not occur in such an environment.

Conduction

The transfer of thermal energy by conduction occurs predominantly in a solid or at the interface between two substances—for example, air and the road. In solids, molecular motions are restricted by the large restoring forces between molecules. Consequently, thermal energy must be transferred within the average separation of molecules, rather than over the larger distances possible when molecules are actually free to move about.

If we heat one end of a solid, the average kinetic energy of the molecules near that end increases. Since the molecules in a solid are not free to move about, this increase in kinetic energy results in wider and more rapid vibrations of the molecules. Consequently, these molecules collide with their neighbors more often and transfer more thermal energy with each collision. The neighboring molecules then vibrate more rapidly, interacting with their neighboring molecules; and so on down the line. As with convection, the transfer of thermal energy by conduction is accomplished by molecular interactions. In conduction, however, the average position of each molecule remains fixed as the energy is transferred.

The **thermal resistivity** of a material, a measure of how effectively a material resists the transfer of thermal energy by conduction, depends on how easily the molecules transfer energy by interactions. This transfer, in turn, depends on the strength of the restoring force in the solid. Materials with a large resistivity have molecules with relatively weak restoring forces. The molecules tend to move about but interact only weakly with their neighbors. Materials with a small resistivity have molecules with large restoring forces. Since the molecules are so restricted in their motions, the transfer of thermal energy by interaction with their neighbors is much more efficient.

SELF-CHECK 13D

In Chapter 11 we discussed the excellent insulating ability of trapped air. Compare the restoring force of a gas (such as air) with that in a solid. Why would you expect still air to have a high thermal resistivity?

Radiation-Absorption

Thermal energy transfer by **radiation-absorption** is accomplished directly by electromagnetic interactions. The process involves the motion of atoms and molecules at the beginning (radiation) and end (absorption) of the process, but not during the actual transfer through space. The transfer process itself is accomplished by electromagnetic waves.

Atoms and molecules have electrical charges capable of exerting forces on other electrical charges. When a molecule vibrates, it exerts a vibrating force on other electrical charges. Through interaction at a distance, this force is capable of causing a molecule in a different material to vibrate. Consequently, the thermal energy present in the vibrations of molecules in the energy source is eventually transferred to the thermal energy in the vibrations of molecules in the energy receiver. Electrical forces are responsible for the

SECOND LAW OF THERMODYNAMICS

We know that thermal energy never spontaneously moves from a cold object to a warmer object, that heat engines must always exhaust some thermal energy, and that divers are never spontaneously thrown from the water back up to the diving board. We find that the preferred direction of energy transformations can be described in terms of increasing entropy, a measure of the disorder of the system. To understand entropy at the molecular level, we need to add the concept of probability to our molecular model of matter.

Thermal Energy and Chance

The molecules moving in a material do not prefer one direction over another; they move about randomly. When two of them happen to be at the same place at the same time, they collide. Since their motions are random, the collisions are accidental. However, some collisions are more likely to occur than others. For example, the probability for a collision is high when lots of molecules are around and low when few molecules are nearby. (More automobile collisions happen on congested freeways than on country roads.) Since collisions between molecules are responsible for the transfer of thermal energy, this **probability** argument can be used to explain why thermal energy is transferred from warmer objects to colder objects.

A warm substance and a cold substance can each be characterized as a collection of randomly moving molecules. The molecules in the warmer substance have a greater average kinetic energy than those in the cooler substance. When a container of warm water is brought into contact with a container of cold water, molecules in each strike the container walls. Each collision leads to the transfer of some thermal energy. Energy flows both ways—from cold to hot and from hot to cold. Since the molecules in the warm water are, on the average, moving faster, they collide with the walls more frequently (Figure 13-7). The probability of a warm-water molecule colliding with the container is greater than that of a cold-water molecule. Consequently, more thermal energy is transferred from hot to cold than from cold to hot. The net flow of thermal energy is from warmer objects to colder objects.

We can also apply the concept of probability to the high diver who converted her gravitational potential energy into thermal energy. Conservation of energy permits the reversal of this process—the transfer of thermal energy from the water back to the diver, pushing her out of the water and back to the top of the diving board. In order for this to occur, a huge number of water molecules would have to travel upward simultaneously, strike the diver, and transfer sufficient kinetic energy to her so that she flies out of the water. While not impossible, such an occurrence is extremely unlikely. Since molecules move about randomly, the probability that all the molecules in, for

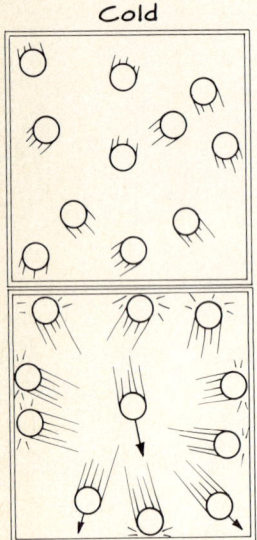

Figure 13-7
Because of their greater average speeds, molecules in warm water are more likely to hit the walls of their container than molecules in cold water. More thermal energy is transferred from warm to cold than from cold to warm.

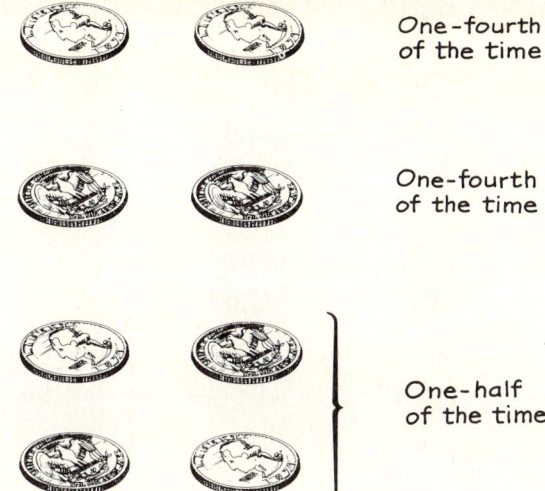

Figure 13-8
We expect to get two heads one-fourth of the time, two tails one-fourth of the time, and one head and one tail one-half of the time.

example, a cubic meter of water move upward at the same time is exceedingly tiny.

Entropy and Chance

We have seen that entropy is a measure of the disorder of a system. Stacked poker chips represent a system with a relatively low entropy. Poker chips strewn about randomly represent a system with a higher entropy. Applied to the way in which energy is found in a system, entropy describes how orderly the energy is distributed. We can now examine the concept of entropy in terms of the probabilities of various molecular motions.

Consider a common example. Suppose you have a system of two identical coins. If you flip the two coins simultaneously, the outcomes possible for any given flip are shown in Figure 13-8. Since flipping coins is a random process, each outcome (HH, HT, TH, or TT) is equally likely. Out of a large number of flips, each combination will occur one-fourth of the time. Since we cannot distinguish between identical coins, the combinations of one head with one tail (HT or TH) are lumped together. We expect to get two heads one-fourth of the time, two tails one-fourth of the time and one head and one tail one-half of the time.

We can use the concept of entropy to describe the relative order of the various outcomes. The most ordered arrangement of the two coins is the two-head or two-tail outcome. The least ordered is one head and one tail. Consequently, a system in which both coins are heads or both coins are tails has a low entropy. A system with one head and one tail has a higher entropy.

If we compare our description of the entropy of each outcome with its probability of occurring, we see a pattern. The two most ordered arrangements (HH or TT) each occur only one-fourth of the time. The least ordered arrangement (HT or TH) occurs one-half of the time. This pattern becomes even more striking when we increase the number of coins to three, as shown

292 Chapter 13. Atoms, Molecules, and Thermal Energy

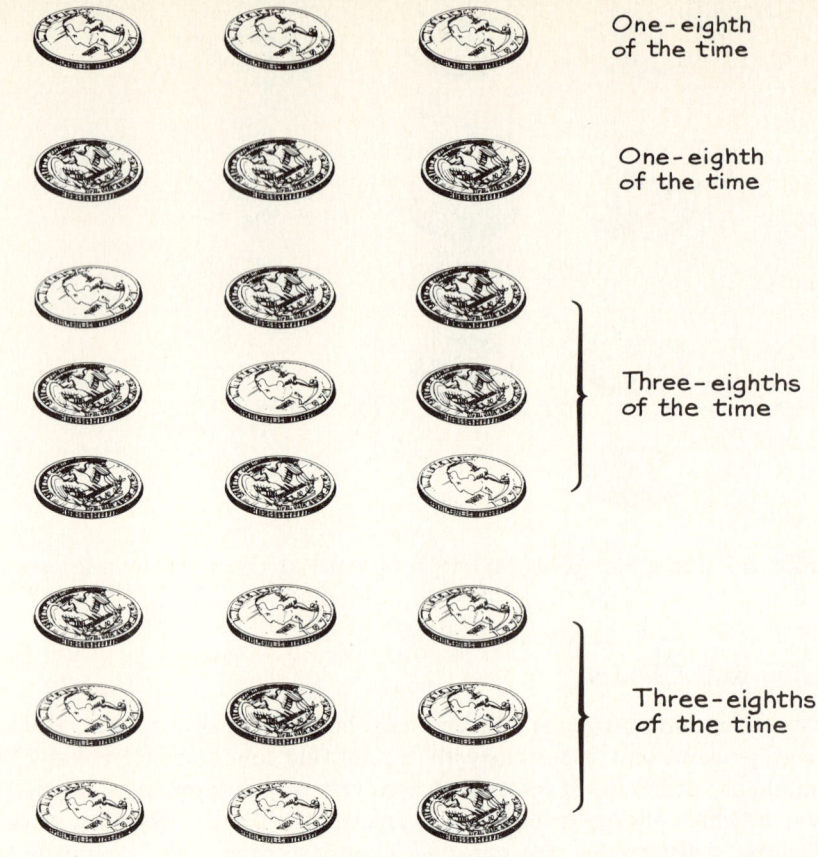

Figure 13-9
The arrangement with the most entropy (two heads and one tail or two tails and one head) are also the arrangements that are most likely to occur.

in Figure 13-9. The outcome with the most entropy is also the outcome most likely to occur.

SELF-CHECK 13E

Five outcomes are possible when we flip four coins: (a) all heads, (b) three heads and one tail, (c) two heads and two tails, (d) one head and three tails, or (e) all tails. Construct a figure like Figure 13-9 and use it to determine which outcome is the most probable one. Which outcome has the most entropy?

Second Law of Thermodynamics

The second law of thermodynamics states that in all processes the entropy of the system must increase. If we think of entropy in terms of high probabilities associated with high-entropy arrangements and low probabilities with low-entropy arrangements, the second law of thermodynamics states that systems move toward the most probable arrangement.

The distinction between spontaneous and nonspontaneous processes is one of probabilities. Spontaneous processes involve the transformation of a system from a less probable arrangement to a more probable arrangement. Nonspontaneous processes involve the reverse. While they can conceivably occur, nonspontaneous processes are highly improbable. The way in which we make them occur is to expand the system so that we include a spontaneous process. For example, thermal energy flows from hot to cold because a system in which thermal energy is evenly distributed is a more probable arrangement of molecules. We can make thermal energy flow from cold to hot but only by broadening our system to include entropy-increasing devices—electrical generating plants or refrigerators. Within a small section of the refrigerator, thermal energy flows from cold to hot. Within the entire kitchen, this energy is distributed over more molecules. The entropy of the entire system has increased.

A PHILOSOPHICAL CONCLUSION— OR IS IT A BEGINNING?

The most probable processes are those which move a system toward more disorder. Thermal energy represents the most disordered arrangement of energy within a system. The most probable state of any system, including the universe, is one in which all energy exists as thermal energy. The heat death of the universe discussed in Chapter 12 is our most probable end. Yet probabilities are not certainties; improbable events can and do occur.

Computer models based on our understanding of molecular motion enable us to investigate systems consisting of just a few molecules. Let's examine a simple system consisting of 10 molecules free to move about randomly in a room. We begin with a relatively ordered system—all 10 molecules are located in one-half of the room. As these molecules move about and collide with one another, they eventually distribute themselves throughout the entire room. However, as shown in Figure 13-10, eventually all the molecules once again find themselves in the same half of the room. While the system spends most of its time in the state in which the molecules are spread throughout the entire room, once every few minutes the molecules return briefly to a more ordered state. Given just 10 molecules, the system periodically experiences a spontaneous decrease in entropy.

If we increase the number of molecules in our room, the time during which the system remains in its disordered state increases to what seems to us to be eternity. One hundred molecules that begin on one side of the room spread out to fill the entire room in less than a minute. Then, for all practical purposes, they stay there! Very complicated calculations show that only once in every 1.5×10^{22} years will all 100 molecules return to one side of the room. Increasing the molecules in the room by a factor of 10 changes the frequency with which we observe the system decreasing its entropy from once every few minutes to once every 1.5×10^{22} years. If we calculated the probabilities for the actual number of molecules in the room (10^{30}), we would essentially never see the system spontaneously decrease its entropy.

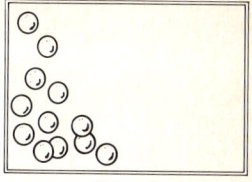
Start

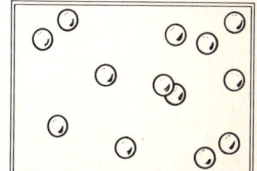
Less than 1 min. later

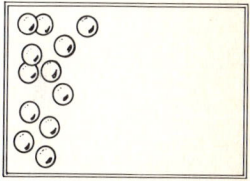
Still later

Figure 13-10
A computer simulation of ten molecules in a room shows that most of the time the molecules remain spread throughout the entire room. Periodically, however, they return briefly to just half the room.

One day our universe will reach its heat death. Then the bits and pieces of matter will bounce randomly about for billions and billions of years. During that time a probability—though exceedingly small—exists that these pieces of matter will again drift close enough to experience attractive forces and become more ordered. Atoms become molecules. Molecules become chemical elements. Chemical elements form the suns and the earths on which life can be sustained. A new universe comes into being. Will it happen after our universe has run down? Did it happen before our universe began?

CHAPTER SUMMARY

Matter is composed of atoms and molecules held together by restoring forces. A *restoring force* acts to return an atom or molecule to some central location. While the atom or molecule is free to move about to some extent, its average location appears fixed. Three types of restoring forces characterize the three states of matter. Solids experience the strongest restoring force, liquids have a somewhat weaker restoring force, and gases are characterized as having essentially no restoring force. The thermal energy required to break the restoring force characteristic of the state of matter is the *latent heat of fusion* (solids to liquids) and the *latent heat of vaporization* (liquids to gases).

Molecules can be composed of any number of atoms. The number of atoms in the molecule and the way in which these atoms are arranged determine the different ways in which a molecule can move. *Temperature* generally measures the average kinetic energy of a molecule as it moves along a straight line. Other molecular motions include vibrations and rotations of atoms within the molecule. *Specific heat capacity* depends upon these different motions. The higher the specific heat capacity of a substance, the more thermal energy is absorbed into motions other than the linear kinetic energy described by temperature.

Convection, conduction, and radiation-absorption can be described in terms of the motion of atoms and molecules. *Convection* arises from the uneven distribution of molecules in a fluid such as air or water, which is caused by the gravitational forces exerted by the earth. As thermal energy is added to the fluid, the more energetic molecules tend to drift upward because there are fewer molecules with which to collide. *Conduction* arises from collisions between neighboring atoms or molecules in the solid. The greater the restoring force, the lower the thermal resistivity of the substance. *Radiation-absorption* occurs when the vibration of electromagnetic waves is transferred to the vibration of atoms and molecules.

Entropy and the *second law of thermodynamics* arise from the application of the laws of probability to the motion and collisions between atoms and molecules. The most disorderly arrangement of energy within matter is also the most probable state of matter. Consequently, all processes lead to increasing disorder because disorderly states are the most probable states. Processes can proceed from disorder to order, but such processes are extremely rare.

ANSWERS TO SELF-CHECKS

13A. The stronger the restoring force between molecules, the more difficult it is to stretch the springs permanently, allowing the solid to become a liquid. Since water has a higher latent heat of fusion than alcohol, the restoring force between water molecules must be greater than the restoring force between alcohol molecules.

13B. A thermal energy of 21 J per 6×10^{23} molecules is approximately equal to those listed for two-atom molecules. Air must consist mostly of two-atom gases. (Air is roughly 78% nitrogen and 21% oxygen. Both exist as two-atom molecules.)

13C. If the molecules in a sample of air are spread about evenly, then a more energetic molecule has the same likelihood of colliding with molecules in all directions. Consequently, more energetic molecules will not rise and convection will not occur.

13D. Gases have extremely small restoring forces. Consequently, gas molecules will not transfer energy effectively through conduction.

13E. The most probable arrangement is two heads and two tails. It will occur, on the average, 6 out of 16 trials. This arrangement also has the most entropy.

PROBLEMS AND QUESTIONS

A. Review of Chapter Material

A1. What are the properties of a restoring force?

A2. Rank the three states of matter in order of the strengths of their restoring forces. Use the strength of the restoring force to describe why a liquid has no definite shape and a gas has neither a definite shape nor a definite volume.

A3. Use the spring analogy for molecular forces to describe why energy is needed to change states.

A4. How do the possible motions of a two-atom molecule differ from those of a one-atom molecule?

A5. To what molecular motion is temperature related?

A6. Why does the specific heat capacity of a substance depend on the number of atoms per molecule?

A7. Why is gravity important in convection?

A8. How is the thermal resistivity of a solid related to the strength of the restoring force?

A9. Use a probability argument to explain why entropy does not decrease.

A10. State the second law of thermodynamics in terms of molecular motions and probabilities.

B. Using the Chapter Material

B1. Use the molecular model to explain why more energy is needed to increase the temperature of 2 kg of water by 10°C than is needed to increase the temperature of 1 kg of water by 10°C.

B2. Use the molecular model of matter to describe why the total energy required to change a state depends on the mass of the substance.

B3. Listed below are the energies needed to increase the temperature of 6×10^{23} molecules of three gases by 1K. How many atoms would you expect to be present in each? How did you reach your conclusion?

> Mercury: 12.52 J
> Nitric oxide: 20.7 J
> Sodium: 12.5 J

B4. Smoke particles are more massive than air molecules, yet smoke rises. Use the molecular model of matter to explain why.

B5. Listed below are the thermal resistivities of three solids. Rank them in order from smallest to largest restoring force. Explain your ranking in terms of the molecular model of matter.

> Silver: 0.0023 m² · s · °C/J · m
> Gold: 0.0030 m² · s · °C/J · m
> Iron: 0.0125 m² · s · °C/J · m

B6. We have seen that the R-value depends on the thickness of a substance. Explain why in terms of the number of molecules between the energy source and the energy receiver.

B7. A neutron is a small particle that has no electric charge. Can it absorb thermal energy by the radiation-absorption process?

B8. Suppose you mix 1 kg of water at 300 K with 1 kg of water at 350 K. We know that the final temperature will be 325 K. Use the molecular model of matter to explain how this final temperature comes about.

B9. Since air molecules are moving about continuously, they could all move away from you right now and leave you with no air to breathe. Why have you never heard of this occurring before?

B10. Suppose you put a red marble and a green marble on a vibrating table. The table's vibrations keep the marbles constantly moving. Would you expect the arrangement of red on the right half, green on the left half to occur reasonably often? If you place 50 red and 50 green marbles on the table, would the arrangement of all red on one side and all green on the other side occur very often?

C. Extensions to New Situations

C1. In 1827 a Scottish botanist, Robert Brown, observed that small particles of pollen were always moving and the direction in which they were moving frequently changed. This motion is now called *Brownian motion* and was not explained adequately until the beginning of this century. How can Brownian motion be explained in terms of interactions between air molecules and pollen particles?

C2. Your brother walks into a room just after applying after-shave lotion. You notice it very quickly.
  a. Describe what must happen at the molecular level in order for you to smell the lotion.
  b. If the after-shave lotion were spread on a block of ice instead of your brother's face, would you smell it as rapidly?

C3. Evaporation occurs when molecules in a liquid break free from the liquid and move into the air.
  a. Of all molecules in a liquid, which are most likely to overcome the restoring force—those with high kinetic energy or those with low kinetic energy? Why?
  b. What will happen to the temperature of the remaining liquid?
  c. Alcohol evaporates more rapidly than water at the same temperature. What does that tell you about the strength of the restoring forces acting on water compared to alcohol?
  d. Use the molecular model to explain why evaporation of sweat cools your body.

C4. The temperature of a pot of boiling water remains at 100°C until all the water has turned to steam.
  a. Which molecules are likely to become steam—high-energy or low-energy ones?
  b. When a molecule has enough energy to become steam, what happens to it?
  c. As molecules leave, where does the thermal energy that has been added to the water go?
  d. In terms of molecular motion, why does the temperature remain constant during boiling?

C5. A metal rod has one end in a flame and the other end 3 meters (m) directly above the flame. If you place one hand at the top of the rod, you will feel heat conducted through the rod before you feel it transferred through the surrounding air by convection. Why?

C6. In general, solids expand as their temperatures increase. Explain why in terms of molecular motions.

C7. In this chapter we learned that thermal resistivity depends on molecular motions and interactions. However, we did not discuss how the molecular model of matter explains the other factors included in the thermal conduction equation.
   a. To describe the importance of temperature difference, consider a solid that has two ends at approximately the same temperature and one that has two ends at vastly different temperatures. Describe how this temperature difference affects the amount of thermal energy conducted across the solid.
   b. To explain the importance of distance between energy source and energy receiver, describe what might happen to vibrational energy as it moves through the molecules. (The analogy of a spring with friction will be useful.)
   c. Consider the number of molecules in small and large areas to explain the importance of area.

C8. In Chapter 11 we quoted one value for the thermal resistivity of various woods. Sometimes two values are given. One describes the thermal energy conducted across the grain; the other, the thermal energy conducted along the grain. For example, the across-the-grain thermal resistivity for hard maple is 5.5 $m^2 \cdot s \cdot °C/J \cdot m$, while the along-the-grain value is 2.3 $m^2 \cdot s \cdot °C/J \cdot m$. Use this result to create a molecular model of wood. Compare the strength of the restoring forces that act across the grain with those that act along the grain.

C9. Consider a sample of 10 identical molecules enclosed by a box. One molecule has a kinetic energy of 100 units, while the other 9 molecules are essentially stationary.
   a. What would be a rough estimate of the average kinetic energy of the 10 molecules?
   b. Is this situation one of high entropy or low entropy? Explain your answer.
   c. As the single molecule moves around the box, it eventually collides with another molecule. Use the concepts of conservation of momentum and energy to describe what the two molecules might do after such a collision.
   d. If the box is isolated so that no thermal energy enters or leaves, the temperature remains constant. If all 10 molecules are now moving, how has the motion of our molecule that originally had 100 units of kinetic energy changed?
   e. Describe the entropy of the system now that all 10 molecules are moving.

C10. To create a system in which entropy decreases, a nineteenth-century physicist, James Clerk Maxwell, invented a demon. Maxwell's demon could change the direction in which any molecule moved and could sort molecules by speed. How would this ability cause a decrease in entropy?

C11. "... if you could devise a system for beating the laws of chance, there are much more exciting things than winning money (at casinos) one could do with it. One could build cars that run without gasoline, factories that could be operated without coal and plenty of other fantastic things" (from *Mr. Tompkins in Paperback* by George Gamow). Explain this quote in terms of the molecular model.

D. Activities

D1. The restoring forces acting on an individual atom may not be the same in all directions. This leads to the observation that different materials break apart differently when struck. This is called *cleavage* when you are discussing minerals. Locate some mineral samples of mica, halite, pyrite, and sulphur and tap them lightly. Describe the way they break and speculate how the restoring forces in the three directions differ.

D2. Remember Mr. Tompkins? He is the guy who dreamed he was in a wonderland where the speed of light was 10 miles per hour. His wife, Maud, meets Maxwell's demon (See problem C10) in Chapter 10 of *Mr. Tompkins in Paperback* (George Gamow, Cambridge University Press, 1969). Read this delightful story.

INTERLUDE
Social Issues and the Energy Crisis

Our discussion of energy has been scientific. We have defined energy, its many forms, and the laws that summarize the transformation of one form of energy into another. We can do all this without mentioning gas lines, nuclear accidents, thermal pollution, windfall profits, or international oil cartels. Yet these problems face us daily. We pause here to examine briefly the social issues that arise from our use of energy.

Life itself is an entropy-increasing process. The remarkable order demonstrated in the evolution and maintenance of living organisms is accomplished at the expense of order elsewhere in the universe. We are not talking just about the refuse, the sewage, and the pollutants so abundantly obvious today. Entropy increases as plants transform radiant energy from the sun into the energy needed for their growth. Entropy increases as we transform the plants we eat into the thermal energy needed to sustain our lives. As we live and breathe, we have no choice but to increase the entropy of our environment. Our only choice lies in how quickly we make that increase occur.

Two variables—population and per capita energy consumption—dominate the rate at which we increase entropy. During the early 1970s, population in the United States was increasing at an average of about 1.5% per year. The amount of energy consumed per person during this same time interval was also increasing, roughly at a rate of 2.0% per year. Combining these two variables, we find that the total energy consumed in the United States was increasing at a rate of about 3.5% per year. While such a growth rate may sound modest, it can have disastrous results.

Exponential Growth

We commonly use a mathematical function, called the *exponential function,* to describe quantities that grow steadily at a fixed rate. Frequently the rates are stated as percentages for annual growth, like 3.5% per year. Another way to describe this rate of increase is the doubling time, the length of time for the quantity to double in value. For example, a growth rate of 3.5% per year corresponds to a doubling time of roughly 20 years. Given a fixed growth rate of total energy consumption, by the year 2005 we will be consuming twice as much energy as now; by 2025, four times as much; by 2045, eight times as much, and so forth. While the doubling time remains constant, the total energy consumed each year will soar out of sight. The impact of even modest growth on the consumption of a limited resource can be devastating.

To see how devastating, let's consider an analogy taken from biology. We begin with a single yeast cell in a petri dish filled with agar, a growing medium. The agar can supply enough food for a single cell to live a million days—a seemingly infinite food supply. Suppose the yeast cells divide, doubling in number every 72 minutes (min). We begin with one cell; 72 min later we have two cells; in another 72 min, four cells; and so forth. One day later we have more than a million cells. One cell could have lived a million days on the food supply. With a doubling time of 72 min, we have less than a day's supply of food in the second day. As the food supply is exhausted, exponential decay replaces exponential growth. Yeast cells begin to die. Instead of doubling time, we now describe the half-life of the cells—the time it takes half the remaining cells to die.

Our energy consumption can be compared to the growth of the yeast cells who doubled while gobbling their food. Unlike the yeast cells, however, we gobble more energy per person as the number of people grow. A doubling time of 20 years means that in 400 years we will have less than "a day's food" left from a resource people once thought would last 400 million years. One day our energy appetite will exceed available resources and exponential decay will replace exponential growth.

Limited Resources and a Limited Environment

Many argue that energy is not a limited resource—the only limitation we currently face is the technological development of new energy sources. Our knowledge of the heat engine tells us otherwise. Heat engines require some lower-entropy form of energy to operate and must exhaust waste energy. The current exponential growth in energy consumption poses two problems: (1) adequate supplies of lower-entropy sources of energy and (2) the ability of the environment to absorb thermal waste energy.

Our current crisis in lower-entropy sources of energy focuses on fossil fuels, which now provide nearly 97% of our energy needs. The exponential growth in energy consumption and the convenience of fossil fuels have conspired to consume in a matter of decades an energy resource that accumulated over hundreds of millions of years. History reports similar crises—water-power shortages in England and periods of wood scarcity in Europe. Each time, a shortage of existing energy resources stimulated the discovery and development of new sources. The current crisis is stimulating the use of coal and nuclear fuels and the recovery of smaller fields of natural gas and oil. Those who have learned the lesson of history now argue for the development of renewable rather than nonrenewable sources. Coal, natural gas, and oil are energy sources that cannot be replaced within a relatively short time interval, once consumed. Hence, they are called *nonrenewable energy sources*. Other energy sources, like the sun, the wind, and flowing water used in hydroelectric plants replace themselves, so to speak. As long as the sun continues to supply thermal energy to our planet, most of these sources will continue to exist. Consequently, they are called *renewable energy sources*. Solar energy, the recovery of methane from vegetable and animal wastes, and the use of wind and geothermal forms of energy head the lists of renewable sources. Regardless of the renewable sources developed or the new sources found, exponential growth will someday cause the demand for energy to exceed the supply.

Long before the energy sources diminish, the earth's environment will severely limit us because of its inability to absorb the thermal energy exhausted by the heat engines we run. Seventy percent of the thermal energy derived from fossil fuels is dumped into the lakes, the streams, and the atmosphere that surrounds us. Combined with pollutants, this thermal waste energy poses a serious environmental threat. Rather than starving to death like the yeast cells, we may well poison ourselves with pollution.

Energy and Societal Values

Our ultimate energy source, the sun, will one day dim, taking with it all the energy sources we now gobble so eagerly. We have no choice but to flow with the inevitable increase in entropy. The choice we have is in how long we allow ourselves to watch.

Exponential growth stands as a spectre against unlimited consumption of energy. Like the yeast cells in the petri dish of agar, we grow and multiply in a limited environment. Unlike the yeast cells, we can vary our energy consumption according to our wisdom or our greed. Whether we last another century or thousands of centuries depends largely on our willingness to impose limitations on our own consumption of energy.

The willingness to impose limits is largely a matter of values, not science. Science describes the limitations imposed by the environment; our choice of values describes the limitations we impose on ourselves. Consciously or unconsciously, our actions reflect our values. Do we value the present more than the future—our own comfort more than that of our children or grandchildren? Is the convenience provided by automobiles more valuable than the cleaner air that once surrounded us? Is the time saved by driving worth the obesity found in an increasingly sedentary society? Is our throw-away economy worth the cost to our environment and its resources? Is profit more valuable than the thermal energy lost by inefficient machines? Energy is like a candle in a dark night. The way in which we, individually and collectively, choose to burn that candle will determine how long and with how many the light may be shared.

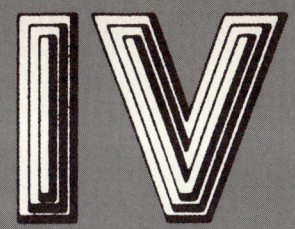

Waves and Particles

CHAPTER 14
Making Waves

CHAPTER 15
Waves: Sound and Electromagnetic

CHAPTER 16
Interference and Diffraction

CHAPTER 17
Wave-Particle Duality

CHAPTER 18
Light, Quanta, and Atoms

How is energy transferred? By the twentieth century, physicists felt they had observed two processes: particles and waves. Particles transfer energy when they make contact with one another, like a car colliding with a tree. Waves transfer energy through the motion of a disturbance through a medium, like a wave moving across the lake. Phenomena could be placed neatly in one category or another. Billiard balls are particles; sound is a wave. Electrons are particles; light is a wave.

As physicists began investigating the interactions between light and matter, the distinctions between the two models began to blur. Light, thought to be a wave, could display particlelike characteristics. Electrons, thought to be particles, could display wavelike characteristics. Mutually exclusive models suddenly stood side by side—dual participants in our explanations of phenomena. Current models of the atom provide insight into how physicists have reached peace with wave-particle duality.

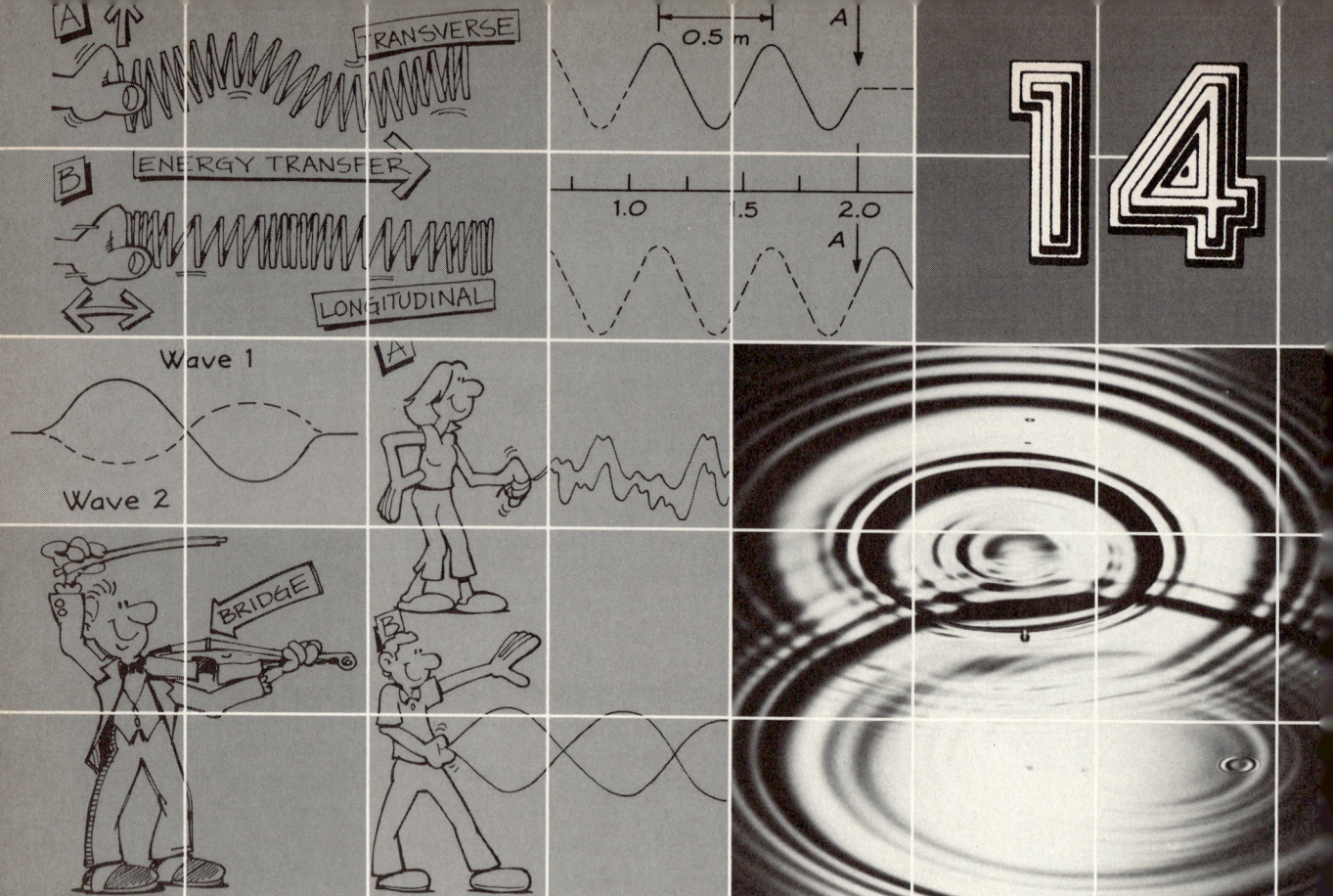

Making Waves

A friend lies asleep in a boat on the far side of a lake. A sound sleeper, he has not responded to any of your shouts. How can you get the message across? One way might be to toss a stone and try to hit him. The message travels with the stone. Another way would be to take a board and move it up and down at the water's surface, creating water waves that travel across and rock the boat. No object moves from you to your friend; instead the wave carries your message.

Stones and waves illustrate the ways we expect to contact or be contacted by life in outer space. While we have not yet received messages (either objects or waves) from extraterrestrial intelligence, science fiction writers freely imagine both methods of encounter. In the 1982 motion picture *E.T.*, the title character arrived on earth in a spaceship. Like the stone, E.T. carried his message with him. Finding that he was stranded on earth, E.T. contacted his ship with radio signals—sending waves, not himself.

Both methods—throwing stones and making waves—are used in real life as well. Objects have always carried messages. Using waves to carry messages is also nothing new—people have always communicated by means of sound. And for thousands of years people have used fires, lanterns, and lighthouses to signal each other by means of light. Light and sound, like radio signals, transmit messages in the form of waves.

Getting the message across requires the transfer of energy. Thus far, we have described how objects transfer energy. In this chapter we examine how waves transfer energy. Intuitively, our model of wave motion comes from experiences with the visible waves found in water and springs. Wave characteristics, like *amplitude, frequency, wavelength,* and *wave speed,* describe the differences we see among waves. How waves appear to us depend on whether we are moving relative to the wave source. Called *Doppler effects,* these differences remind us of how much our measurements depend on the reference frame we are in. Waves can interact with themselves and with matter. When waves meet waves, they combine according to the *principle of superposition.* When waves meet other matter, they can be *reflected, refracted,* or *absorbed.* In this chapter we use the visible waves found in water and springs to examine these and other wave concepts. In the next chapter we see how successfully the wave model explains the behavior of sound and light.

WAVE MOTION

Raindrops fall softly in a warm summer rain, causing ripples as they strike the surface of a puddle. You throw a stone into a pond. Plunk! A small wave moves outward from the point where the stone hits the water. As you stand on the beach on a windy day you see bigger waves—waves that gradually wear away the soil and rock along the shore. Deep in the ocean floor an earthquake occurs, giving rise to a giant wave called a tidal wave, or *tsunami.* Traveling at enormous speeds, the tsunami carries enough energy to destroy homes and lives.

While different in their destructive potential, these waves are similar in at least three respects. Each is created by a disturbance of the water. Each carries energy along the water's surface from the original disturbance to other locations. And, each transfers energy over distances much greater than those traveled by the individual water molecules. This last characteristic, the transfer of energy but not matter, makes wave motion an intriguing phenomenon.

Transferring Energy But Not Matter

Waves transfer energy but not matter. This can be hard to understand when we talk about something as continuous as water. An analogy with distinct objects makes the process clearer. Suppose you set up a line of dominoes from your finger to a bell (Figure 14-1). When you push one domino over, it pushes

Figure 14-1

Energy is transferred from your finger to the bell, but no single domino moves the entire distance.

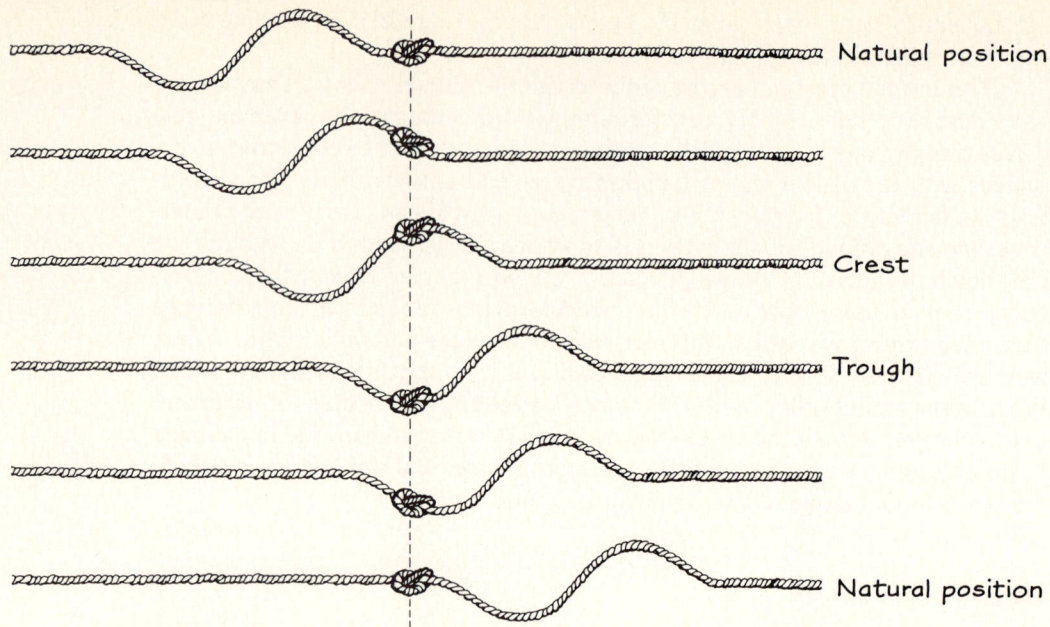

Figure 14-2
If we watch just one segment of the medium, it moves up and down about its central location as the disturbance passes through. One complete vibration occurs when the segment moves from its natural position to a crest, back to its natural position, to a trough, and back to its natural position again. The wave moves from left to right, but each segment of the medium returns to its original location.

the next one, which pushes the next, and so forth. Finally, the last domino falls over and rings the bell. Energy is transferred from your finger to the bell, but no single domino moves the entire distance. Likewise, when you drop a stone into a pond, energy is transferred from you to the shore, although individual water molecules do not travel the entire distance. A disturbance like that shown in Figure 14-2 travels from left to right while each section of the material simply vibrates up and down about a central location.

Like any energy-transfer process, wave motion requires an energy source and an energy receiver. The energy source creates a disturbance, which travels to the energy receiver. In the domino analogy, your finger acts as the energy source and the bell as the energy receiver. The raindrops, the stone, the wind, and the earthquake all act as energy sources in our examples of water waves. Soil and rock at the water's edge act as the energy receivers.

To our energy-source–energy-receiver model we must now add a third component. The disturbance created by the energy source must be transmitted through something, called the **medium.** The line of dominoes acts as the medium for the domino wave. Water is the medium for water waves. Matter transmits sound waves. Rock, soil, and water transmit the shock waves created by earthquakes. In each example, the medium transmits energy from a source to a receiver.

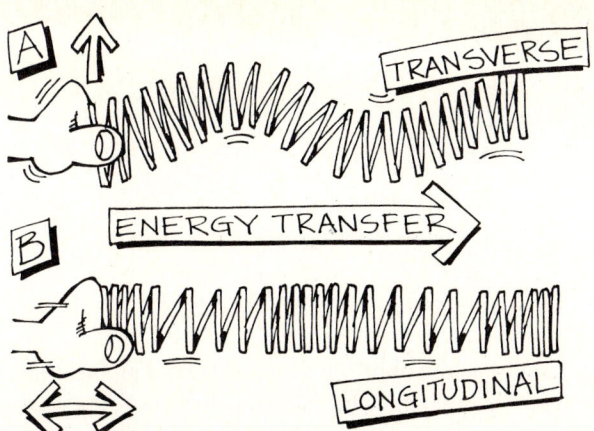

Figure 14-3
Both waves transfer energy from left to right.
(a) To produce transverse waves, you move your hand up and down.
(b) To produce longitudinal waves, you move the coils to the left and right.

Longitudinal and Transverse Waves

While the medium does not actually move from the energy source to the energy receiver, it does vibrate as it transmits the disturbance. The direction in which it vibrates leads to the definition of two different types of waves.

You can produce the two types of waves using a stretched spring attached to a wall, as shown in Figure 14-3. First shake your hand up and down, producing the wave shown in Figure 14-3(a). This wave distorts the linear shape of the spring as it moves from left to right. Now, while stretching the spring with one hand, use the other hand to compress several coils, then let go. Figure 14-3(b) shows the wave that results. The spring's linear shape is maintained, but its coils are compressed or pulled apart as the wave moves along its length.

We can describe the differences between these two kinds of waves by comparing the direction in which the energy travels with the direction in which the coils in the spring vibrate. In Figure 14-3(a), the energy moves from left to right, but the coils vibrate up and down. A wave in which the medium vibrates perpendicular to the direction of energy transfer is called a **transverse wave.** Waves along the surface of water are transverse. In Figure 14-3(b), the energy moves from left to right and the coils also vibrate from left to right. In **longitudinal waves,** the medium vibrates parallel to the direction of energy transfer. Sound waves are longitudinal waves.

While a spring can transmit both transverse and longitudinal waves, many media cannot. In order to transmit a wave, a medium must be able to mimic the motion of the energy source and then return to its original shape. Matter in any state—solid, liquid, or gas—will transmit longitudinal waves. Sound waves, for example, can be transmitted by air, water, even solid walls—as you have no doubt noticed if your neighbors enjoy loud parties. Solids and surfaces of liquids transmit transverse waves because both have strong molecular bonds defining a shape that can be restored once the disturbance moves on. If you place your finger near a plucked guitar string, you can feel the transverse vibrations of the stretched string. Liquids and gases, however, have no definite shape and cannot transmit transverse waves through their interior.

Figure 14-4
People transmit waves too! Is this "people wave" transverse or longitudinal?

CHARACTERISTICS OF WAVES

Ripples and tsunamis are both transverse waves, but the similarities end there. Compared to ripples, tsunamis transfer enormous amounts of energy. In a more modest example, think about the amount of energy you transfer when you shake the end of a spring. The distance you move the spring up and down, the rate at which you shake the end of the spring, and even the type of spring itself affect the nature of the transverse wave and the energy it transfers. To describe the differences among waves and the energy they transfer, we need to introduce some terms.

Amplitude, Frequency, and Wavelength

Varying the *distance* we move the spring up and down produces the differences in height between waves A and B in Figure 14-5. The **amplitude** of a wave is the maximum distance the parts of the medium move from their natural positions (Figure 14-6). Since we move the end of the spring up and down farther in B than in A, wave B has a greater amplitude than wave A. More energy is required to move the end of the spring farther. Tsunamis transfer more energy than ripples.

Varying the *rate* at which we shake the end of the spring produces the differences between waves A and C (Figure 14-5). These two waves have the same amplitude, but those in A seem squeezed together more than those in C. Two terms—*frequency* and *wavelength*—are used to describe this difference. Frequency describes the difference in terms of time; wavelength describes the difference in terms of space.

Frequency describes the rate at which we shake the end of the spring and, consequently, the rate at which each segment of the spring vibrates up and down. We measure this rate by watching one segment of the medium and counting the number of vibrations or cycles completed during a convenient time interval. As shown in Figure 14-2, a segment completes one cycle when it moves from its normal position to a crest, back to the normal position, to a trough, and finally back to its normal position. The **frequency** of a wave is the number of complete cycles divided by the time interval in which they occur.

$$\text{Frequency} = \frac{\text{number of complete cycles}}{\text{time interval}}$$

For example, if a segment completes 10 cycles in 5 seconds (s), then the frequency is 10 cycles divided by 5 s or 2 cycles per second. The units cycles per

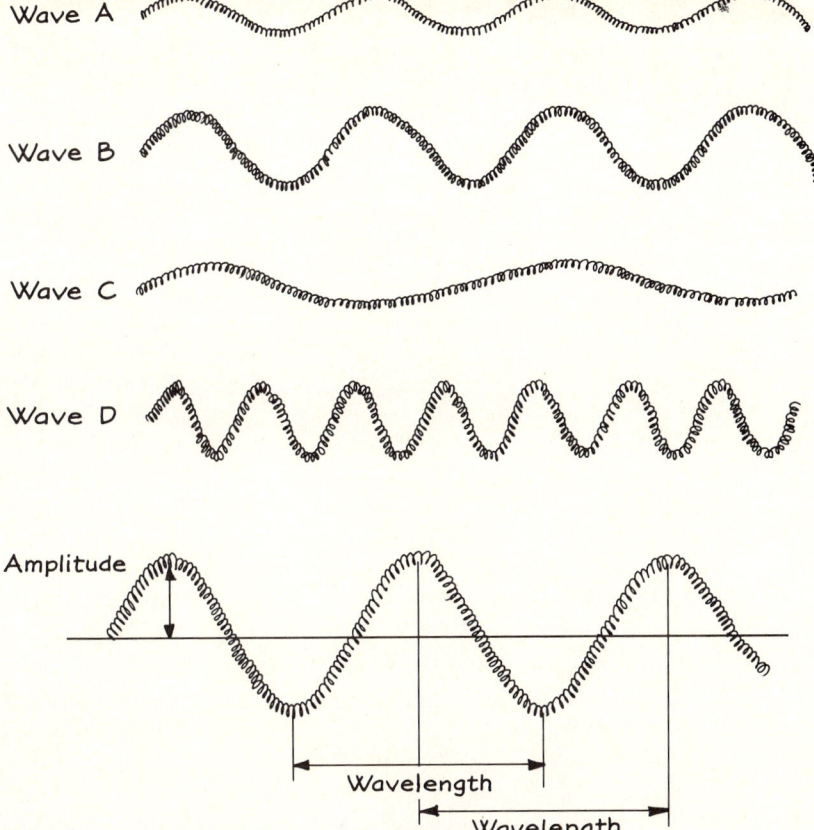

Figure 14-5
Waves A and B differ in amplitude. Waves A and C differ in frequency and wavelength. How does wave D differ from the others?

Figure 14-6
The distance covered by each unit is called the wavelength of the wave.

second are called hertz (Hz). A higher frequency means that pulses are occurring more frequently and results in waves that are more closely spaced. The frequency of the waves in A is greater than those in C. Since each cycle transfers energy along the spring, increasing the frequency increases the rate at which you transfer energy.

The frequency at which a wave medium vibrates is the same as the rate at which the energy source vibrates. Sound waves vibrate at rates from a few hertz to tens of thousands of hertz. Our ears respond only to waves in the range of 20 Hz to around 20,000 Hz. Tides, ocean waves, and earthquake waves are generally a few hertz. Light waves vibrate at much higher rates, about 10^{14} Hz.

Frequency describes differences between waves A and C in terms of time. These differences can be described in terms of space as well. If you look closely at the two wave patterns, you can see that they are composed of repetitions of a single unit (Figure 14-6). Each unit consists of a crest, the maximum displacement above normal, and a trough, the maximum displacement below normal. The length of this basic unit is called the **wavelength** of a wave. The wavelength of A is less than that of C; consequently, the waves in A are more closely spaced. Wavelength is measured in meters (m). The wavelength of visible light is extraordinarily tiny—10^{-7} m. The wavelength of tsunamis can be enormous—5×10^5 m.

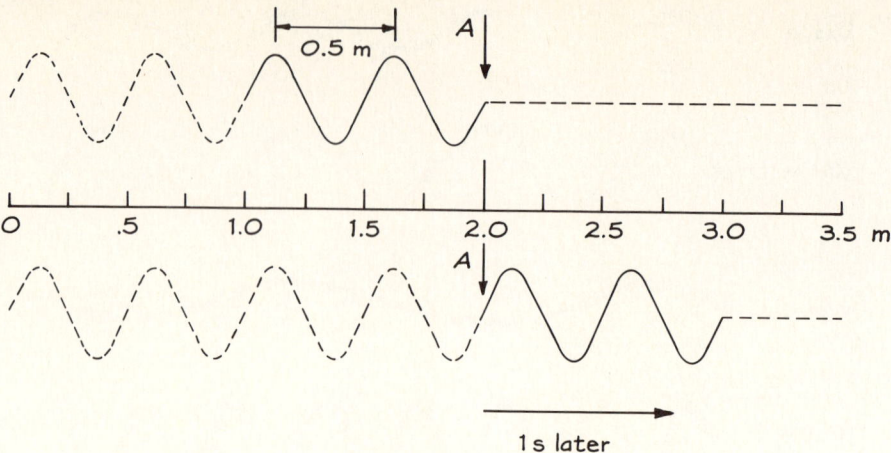

Figure 14-7
When 1 second has elapsed, a wave at A will have moved two complete wavelengths to the right. The wave speed is two wavelengths per second, or 1.0 meters per second.

SELF-CHECK 14A

Use the terms *amplitude, frequency,* and *wavelength* to describe how waves A, B, and C differ from wave D in Figure 14-5.

Wave Speed

Moving the end of the spring up and down farther increases the amplitude of the waves produced. Moving the end of the spring faster increases their frequency. Stretching the spring or getting an entirely new spring will affect the speed at which the waves move. Wave speed depends on the medium chosen.

Wave speed is the distance traveled by a disturbance divided by the time required for it to travel that distance. This speed can be described in terms of a wave's frequency and wavelength. To see how, we focus our attention on one segment of the wave medium, point A in Figure 14-7. A frequency of 2 Hz tells us that two complete waves pass point A each second. When one second has elapsed, a wave at point A will have moved a distance of two wavelengths. Its speed, then, is two wavelengths per second. If each wavelength is 0.5 m, then the wave speed is two wavelengths times 0.5 m per wavelength, or 1.0 m/s. The wave speed of a wave can be determined from the product of its wave frequency and wavelength:

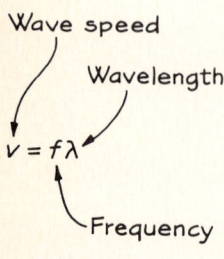

Wave speed = frequency × wavelength

In general, the speed at which waves travel depends on the medium through which the disturbance moves. Sound waves, for example, travel at speeds of about 330 m/s in air. If we compare the frequencies and wavelengths produced by a variety of musical instruments (Table 14-1), the product of the two quantities is always the same—330 m/s. You do not hear a

Table 14-1	Sound Waves	
Frequency (Hz)	Wavelength (m)	Wave Speed (m/s)
162	2.04	330
262	1.26	330
440	0.75	330
524	0.63	330

piccolo before you hear a trombone. All the sounds of the orchestra reach you at the same time. For a given medium, frequency and wavelength vary inversely so that the wave speed remains constant.

In a sense, wave speed describes how quickly the energy supplied by the energy source can be handed off from one part of the medium to another, from one coil in the spring to the next. The speed with which this occurs depends on the physical characteristics of the medium (how tightly the spring is stretched or whether the medium is a solid, liquid, or gas), on the substance from which the medium is made, and on the type of wave (transverse or longitudinal) being transmitted. The range of speeds is enormous. Sound waves travel through air at about 330 m/s, while light waves travel through space more than a million times faster, 3×10^8 m/s. This difference in speed produces the time delay between seeing lightning and hearing the thunder produced in distant storms.

SELF-CHECK 14B

A tuning fork vibrates at a frequency of 440 Hz. If the waves it produces have a wavelength of 3 m underwater, what is the speed of sound in water? How does this compare to the speed of sound in air?

Seeing the Earth's Interior

Wave speed has provided geologists with a valuable tool for building models of the earth's interior. Earthquakes, dynamite blasts, and nuclear tests all produce transverse and longitudinal waves that are transmitted through the earth's interior. By measuring the travel time of these waves to reporting stations located around the world, geologists have been able to estimate average speeds at various depths below the earth's surface. Since the speed of

310   Chapter 14.   Making Waves

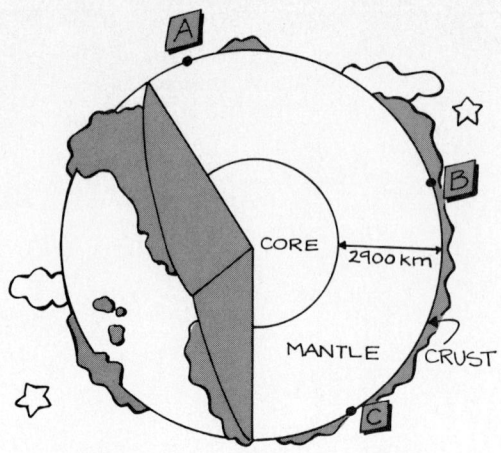

Figure 14-8
Geologists believe that the abrupt changes in wave speed identify the boundary between the earth's crust and mantle and between the mantle and a liquid or semi-liquid core.

both transverse and longitudinal waves depends on the medium through which they travel, changes in speed tell us that there are changes in the materials found in the earth's interior.

As shown in Table 14-2, the speed of longitudinal waves increases gradually with increasing depth. Two exceptions stand out. An abrupt increase in speed occurs at a depth of 35 kilometers (km). An abrupt decrease in speed occurs at a depth of 2900 km. Geologists believe that these sudden changes in speed locate boundaries between distinct regions of the earth's interior.

Our model of the earth's interior now includes concentric layers called the crust, mantle, and core (Figure 14-8). The abrupt increase in speed shown in Table 14-2 marks the boundary between the earth's crust and mantle. Similar increases in speed have been measured throughout the world, though

Table 14-2   Speeds of Longitudinal Waves Below the Earth's Surface

Depth (km)	Wave Speed (m/s)	Depth (km)	Wave Speed (m/s)
0	5,000	2,900	13,500
16	6,000		8,000
29	6,720	3,000	8,100
35	7,200	4,000	9,500
	8,320	5,000	10,500
1,000	12,000	6,000	11,500
2,000	13,000		

at depths that range from 5 km beneath the ocean floor to 60 km in mountainous regions. The earth's crust does not appear to be uniformly thick. Geologists believe that the increase in speed that occurs as waves leave the earth's crust and enter the mantle is due to a sudden increase in the presence of iron compounds in the mantle's rock layers. The abrupt decrease in speed that occurs at 2900 km marks the boundary between the earth's mantle and core. Since longitudinal waves travel more slowly in liquids than in solids, geologists believe that the earth's core is either liquid or semiliquid. We will never see the core or mantle directly, yet each longitudinal or transverse wave that passes through the earth's interior gives us a glimpse of what might lie beneath.

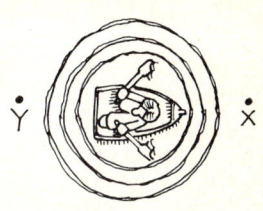

Figure 14-9

If you tap the surface of the water at a constant rate, a pattern of concentric circles moves outward from the boat.

MOVING SOURCES AND RECEIVERS

When an earthquake occurs, the wave source and receiver are stationary relative to one another. The frequency and wavelength of the wave emitted by the source are the same as that received by the receiver. Observers at both locations would describe the waves as being the same. However, in other examples of wave phenomena, either the source or receiver can be moving relative to the other. While the wave speed remains constant regardless of this motion, the frequency and wavelength of the waves change. The extent of these changes describes the relative motion between the source and the receiver.

A Moving Boat

Suppose you are sitting in a motionless boat on a lake. You tap the surface of the water and a wave moves outward in all directions. If you tap your finger repeatedly, say once per second, you will see the pattern of concentric circles shown in Figure 14-9. As the waves travel away from you, they pass people sitting in boats at points X and Y. By measuring the time between consecutive crests—1 s—these people can determine the rate at which you are tapping the water.

Now suppose your boat is moving toward X at a constant speed as you tap the water. At the end of the first second, the wave produced by the first tap will have spread out as shown in Figure 14-10(a). The boat moves during this time, so the wave produced by the second tap originates from a different place than the wave produced by the first tap. At the end of 2 s, the pattern we see is that shown in Figure 14-10(b). While the wave produced by each tap is circular, the waves are no longer concentric. The crests crowd together on one side and spread apart on the other side. As more waves are produced, the pattern in Figure 14-10(c) is created.

Observers at X and Y now report different time intervals between crests. Though you are still producing wave crests at a rate of one per second, observers at X report that crests arrive more often than once per second. Observers at Y, on the other hand, see waves arriving less often than once

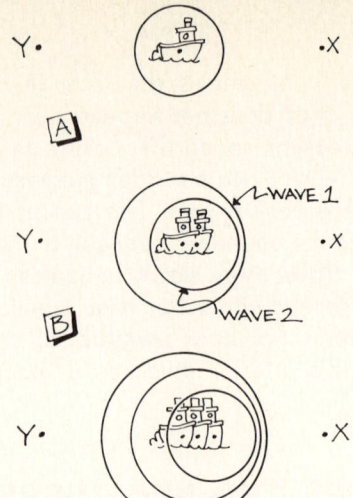

Figure 14-10
As the wave source moves, the waves no longer remain concentric. The crests crowd together on the right and spread apart on the left.

per second. The frequency at which the waves arrive at X and Y no longer matches the frequency at which the waves are produced, nor do they agree with each other. These differences arise from the motion of your boat—the source of the waves.

The Doppler Effect

A change in the observed frequency of waves caused by the motion of the source is called the **Doppler effect.** The frequency of waves received by the observers increases as the boat moves toward them and decreases as the boat moves away from them. The extent of the frequency shift depends upon the speed of the boat.

The Doppler effect is usually described in terms of the change in frequency. In our example, this change is the difference between the frequency of the waves emitted at the boat and the frequency of the waves received by observers at X or Y. As the relative speed between the wave source and receiver increases, the change in frequency increases. Police radar uses this relationship to measure the speeds at which cars travel.

While we have described the Doppler effect in terms of a moving wave source, the same changes occur when a wave source is stationary and the receiver is moving. In fact, both the source and the receiver can be moving simultaneously. The relative speed between the source and the receiver determines the extent of the change in frequency.

―――――――――― **SELF-CHECK 14C** ――――――――――

Suppose we stop moving the boat and observers at Y begin moving toward us. If we continue to tap the water at a frequency of 1 Hz, what do the observers at Y report? What do the observers at X report?

A STEP FURTHER—MATH

USING DOPPLER SHIFTS TO MEASURE SPEED

In most applications of the Doppler effect, the change in frequency, called the **Doppler shift,** is used to measure the speed of a moving object. Police determine the speed of a car by measuring the Doppler shift in reflected radar waves. Medical technicians use the Doppler shift in ultrasonic waves reflected by red blood cells to determine the speed at which blood flows. A Doppler shift in the light emitted by moving galaxies enables astronomers to estimate the speeds at which these galaxies move away from us. All these measurements require a quantitative description of the relationship between the Doppler shift and the relative speed between the wave source and receiver.

Two speeds are required to describe the Doppler shift—wave speed (the speed at which the waves travel from source to receiver) and the relative speed between the wave source and receiver. When the wave speed is much greater than the relative speed between source and receiver, the change in frequency due to the Doppler effect is given by:

$$\text{Change in frequency} = \frac{\text{relative speed}}{\text{wave speed}} \times (\text{frequency of source})$$

When the relative speed and wave speed are both expressed in the same units, such as meters per second, the change in frequency is expressed in the same units as the frequency of the wave source, usually Hz. The change in frequency is positive when the wave source and receiver move toward one another; negative when they move apart from one another.

To see how to apply this equation, let's return to the boat example. While sitting in the boat you produce water waves with a frequency of 1.00 Hz, one wave each second. Typically, water waves move at a wave speed of 0.100 m/s. If your boat moves at a speed of 0.005 m/s toward X, a stationary observer at X will report a Doppler shift of

$$\text{Change in frequency} = \frac{\text{relative speed}}{\text{wave speed}} \times (\text{frequency of source})$$

$$= \frac{0.005 \text{ m/s}}{0.100 \text{ m/s}} \times 1 \text{ Hz}$$

$$= 0.05 \text{ Hz}$$

Since the wave source is moving toward the wave receiver, the change in frequency is positive. An observer at X reports a change in frequency of +0.05 Hz. Under the same circumstances, the observer at Y reports a change in frequency of −0.05 Hz.

The police officer, medical technician, and astronomer rearrange the Doppler relationship to determine the relative speed in terms of the change in frequency, wave speed, and frequency of the source:

$$\text{Relative speed} = \frac{\text{change in frequency}}{\text{frequency of source}} \times (\text{wave speed})$$

$\Delta f = \dfrac{V_r}{V_w} f_s$

Change in frequency
Relative speed between source and receiver
Frequency emitted by source
Wave speed

$V_r = \dfrac{\Delta f}{f_s} V_w$

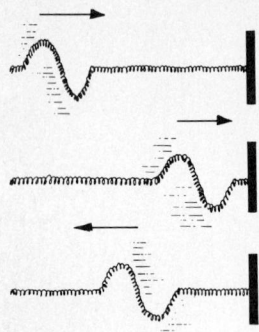

Figure 14-11
A wave can be reflected when it reaches a rigid boundary, such as the wall.

> Suppose you are driving along an interstate highway (speed limit of about 25 m/s.) A stationary police radar unit emits radar waves that move at a wave speed of 3×10^8 m/s and have a frequency of 1×10^{10} Hz. When these waves are reflected by your car, their Doppler shift is 100 (1×10^2) Hz. Will you get a ticket?

WAVES MEET MATTER

Waves do not move through a single medium. Earthquake waves, for example, move through a variety of rock layers, each of which has a slightly different composition and is under different pressures from the rock above it. When the change in the medium is gradual, waves gradually change speed and direction. When the change in the medium is abrupt, waves can also change direction or stop traveling altogether. They can be reflected, refracted, or absorbed.

Reflection

Light waves striking the surface of a mirror do not travel on through. They travel back in the general direction from which they came. Sound waves striking the smooth walls of a long corridor travel back to you as an echo. A transverse wave transmitted along a spring reverses direction when it reaches the wall (Figure 14-11). In many situations, waves remain in some initial medium rather than enter a new medium. These waves are said to be **reflected.**

Waves are reflected when the new medium offers a significantly more difficult path for the energy to travel. Since there are many possible media, the extent to which waves are reflected varies. A wall, for example, provides a very rigid medium compared to a spring. Consequently, the wave energy is transmitted back along the spring rather than into the wall. These waves are said to be *totally reflected.* If we were to replace the wall with a less rigid

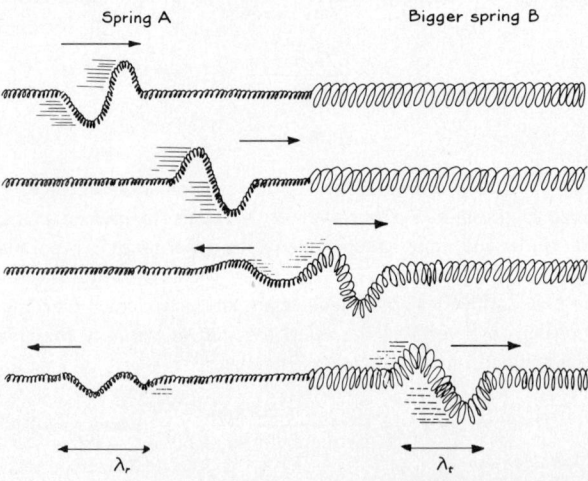

Figure 14-12
A wave can be both reflected and transmitted when it reaches the boundary between two media.

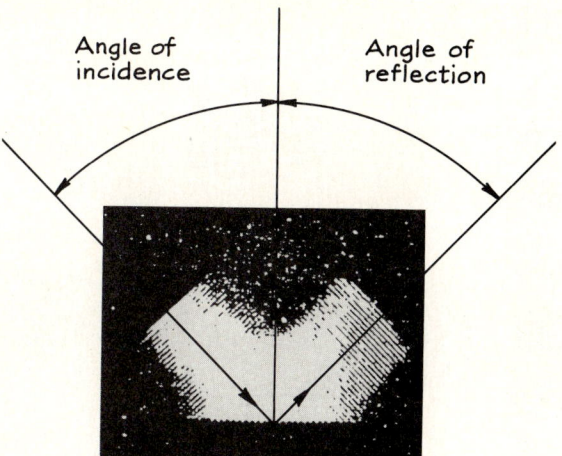

Figure 14-13
Waves are reflected such that the angle of reflection equals the angle of incidence.

medium, like the spring shown in Figure 14-12, some energy would be transmitted into the new medium. Some would still be reflected. These waves are said to be *partially reflected*.

In one dimension, reflected waves simply travel back in the direction from which they came. In two dimensions, waves are reflected from a boundary in a slightly more complex manner. If we compare the direction of motion for incoming and reflected waves, we see a pattern. Rays describe the direction in which incoming and reflected waves move. As shown in Figure 14-13, incoming rays and reflected rays make equal angles with a line perpendicular to the boundary. The angle made by the incoming ray and the perpendicular, called the **angle of incidence,** is equal to the angle made by the reflected ray and the perpendicular, called the **angle of reflection.** This relationship is called the **law of reflection:**

Angle of incidence = angle of reflection

Waves reflected from a boundary are always reflected such that the angle of incidence equals the angle of reflection. The law of reflection describes the behavior of both partially reflected and totally reflected waves.

SELF-CHECK 14D

Transverse waves are reflected at the boundary between the earth's mantle and core, shown in Figure 14-8. Use the law of reflection to determine whether seismic stations at B or C would report transverse waves received from an earthquake at A.

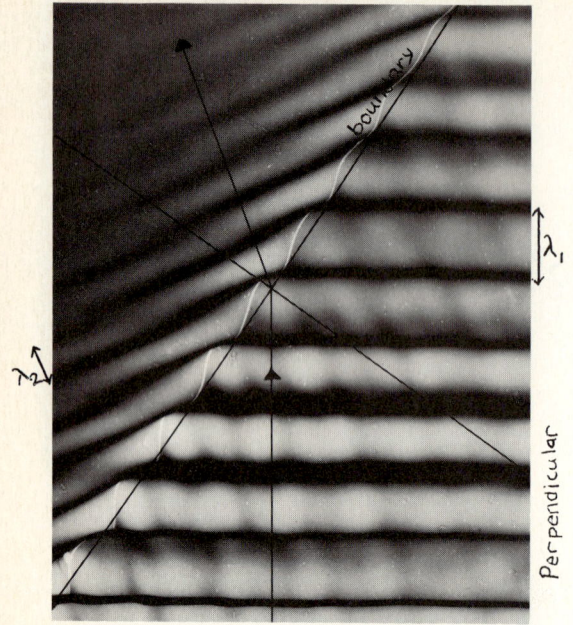

Figure 14-14
Water waves are refracted as they move from deep to shallow water. The wavelength decreases and the angle of refraction is less than the angle of incidence.

Figure 14-15
Waves are refracted in much the same fashion that a cart turns as it crosses the boundary between a paved surface and mud.

Refraction

Waves that strike a boundary between two media are sometimes transmitted rather than reflected. Light travels through water. Sound travels through walls. The transverse wave shown in Figure 14-12 is partially transmitted into the new medium. The waves transmitted into the new medium are not identical to the waves in the original medium. The change in speed that occurs as the waves change media affects both the wavelength and direction of the transmitted waves.

The change in wavelength that occurs as waves are transmitted into new media is particularly noticeable in Figure 14-12. The wavelength of the transmitted wave is shorter than the wavelength of the original wave. This change can be explained in terms of the change in speed. Waves travel more quickly in medium A than in medium B. Since the frequency of the wave motion is always constant, the wavelength must change when the speed of the wave changes. If the wave speeds up, the wavelength increases. If the wave slows down, the wavelength decreases. In Figure 14-12, the wavelength of the transmitted wave is shorter because the wave slows down as it enters medium B.

Water waves offer a striking example of the change in direction that often accompanies the change in wavelength when waves enter a new me-

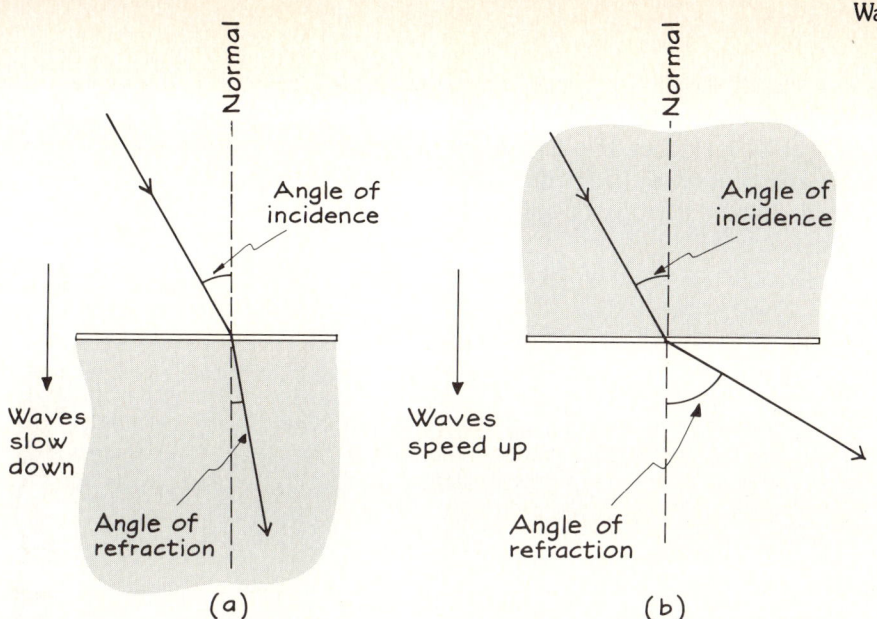

Figure 14-16
(a) When waves slow down, the angle of refraction is less than the angle of incidence.
(b) When waves speed up, the angle of refraction is greater than the angle of incidence.

dium. Water waves travel at different speeds in water at different depths. A new depth constitutes, in effect, a new medium. When waves move from deeper water to shallower water, we find that they change direction as well as wavelength (Figure 14-14). This change in direction can be explained by the change in speed that occurs as the waves enter the shallower water.

We can understand how a change in wave speed leads to a change in direction with the help of an analogy. Figure 14-15 shows a cart moving from a paved surface to a muddy surface. Naturally, the cart moves at a slower speed in the mud than on the pavement. When the cart moves straight toward the boundary, the wheels on the left and right sides enter the mud at the same time and the cart simply slows down. When the cart moves toward the boundary at an angle, however, one side enters the mud before the other. In Figure 14-15(b) the wheels on the left enter the mud before the wheels on the right. Consequently, the left side of the cart moves more slowly than the right side and the cart turns. Its direction of motion has changed. The water waves in Figure 14-14 behaved in much the same fashion—one side of the wave entered the shallower water before the other side did. Therefore, the wave turned.

Refraction is the term for the change in direction and change in wavelength that occur as waves are transmitted from one medium into another. To measure the change in wavelength, we compare the wavelengths in the new and old media. To measure the change in direction, we compare angles relative to a line perpendicular to the boundary separating the two media. The **angle of incidence** is the angle between the incoming ray and the perpendicular. The **angle of refraction** is the angle between the outgoing, or transmitted, ray and the perpendicular. Waves that travel along the perpendicular, so that the angle of incidence is 0°, do not change direction. The angle of refraction is also 0°. At any other angle of incidence, however, waves are refracted such that the angle of incidence does not equal the angle of refraction. If the waves move into a medium in which they slow down, the angle of refraction is less than the angle of incidence (Figure 14-16(a)). Light waves that move into a glass prism are bent toward the normal. If the waves move into a medium in which they speed up, the angle of refraction is greater than the angle of incidence (Figure 14-16(b)). Light waves that move from water to air are bent away from the normal. In both cases, the change in angle depends on the change in speed that occurs as the waves enter the new medium.

SELF-CHECK 14E

Longitudinal waves are transmitted from the earth's mantle into its core, but the waves slow down substantially (see Table 14-2). Sketch the path of the longitudinal waves produced by an earthquake at point A in Figure 14-8.

Absorption

Light can be completely absorbed by a black surface. Sound can be absorbed by the panels added to soundproof an office or room. In some situations, waves are neither reflected nor refracted. They simply appear to fade away. This is called **absorption.**

While the wave itself can disappear, the energy the wave transmits cannot. Conservation of energy does not allow it. Generally, the energy carried by the wave is converted into the thermal energy of the absorbing medium. The black surface becomes hot. Sound panels become warmer than the surrounding air. The wave energy is converted into the thermal energy of a medium's atoms and molecules.

WAVES MEET WAVES

Drop two stones in a pond at the same time. Each stone creates a series of circular waves. As the waves created by one stone move outward, they meet those created by the other stone. The familiar pattern shown in Figure 14-17 arises when wave meets wave.

Wave Interference

Figure 14-18 shows a series of drawings made of two wave crests as they move toward one another, meet in the center, and finally move away from one another. If we compare (a) and (e), the two crests seem to have passed through one another essentially unchanged. The drawing in (c), however, shows a completely new wave, which is a combination of the two crests. When waves interact, they combine according to the superposition principle.

Waves Meet Waves 319

Figure 14-17
Waves produced by each stone travel outward, overlapping to produce complex patterns.

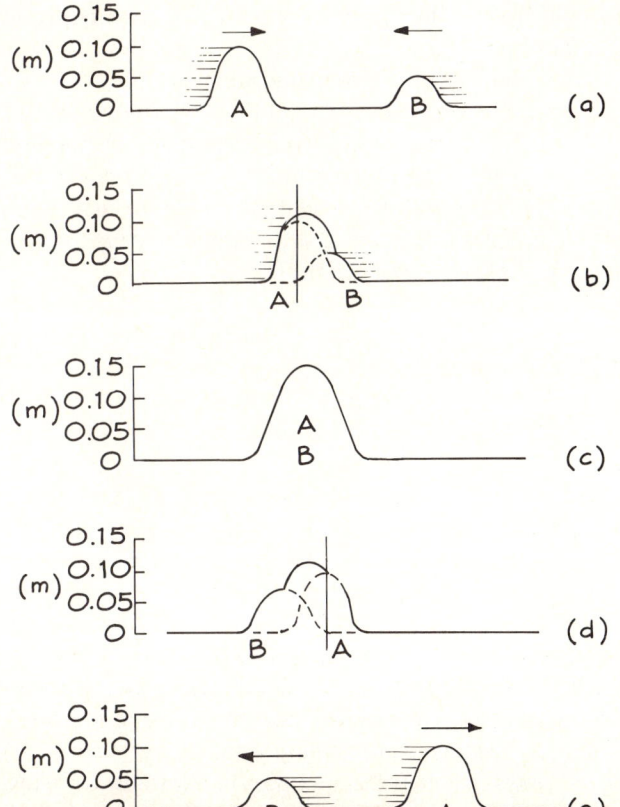

Figure 14-18
The two waves meet, combine briefly to form a new wave and then move on essentially unchanged.

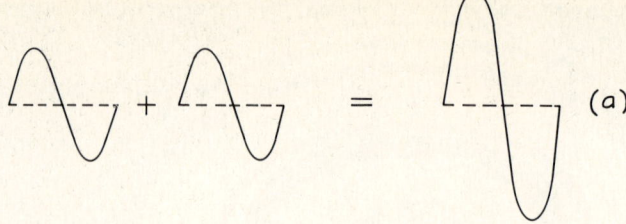

Figure 14-19
(a) Constructive interference. **(b)** Destructive interference.

The **superposition principle** states that when waves meet in space, their effects on the medium add. The total displacement of the medium is the sum of the displacements caused by each wave separately. We can apply this principle to the interaction sketched in Figure 14-18. Wave A has an amplitude of 0.10 m. Wave B has an amplitude of 0.05 m. When the two crests combine in (c), the amplitude is 0.10 m + 0.05 m, or 0.15 m. Had wave B been a wave trough rather than a wave crest, the combined wave would have had an amplitude of 0.10 m − 0.05 m, or 0.05 m. Amplitudes in the same direction add; those in opposite directions subtract.

When waves combine, they are said to **interfere** with one another. Waves can interfere with one another constructively or destructively. When two identical waves meet in such a way that each wave crest matches another wave crest and each wave trough matches another wave trough (Figure 14-19(a)), their combination is a wave with twice the amplitude of each single wave. This is called **constructive interference.** In Figure 14-19(b), the two identical waves meet so that the trough of one wave meets a crest of the other wave, and vice versa. The two waves cancel one another. This is called **destructive interference.** Constructive and destructive interference demonstrate the two extremes of wave addition. The amplitude of a combined wave can never be greater than the sum of the amplitudes of the waves which interact.

The superposition principle can be applied to all situations in which waves interact. When the waves do not have the same shape or the same wavelength, the combined wave can have a complex shape that bears little resemblance to the original waves that interact. You can imagine how complex the combined wave produced by a 100-piece orchestra might be. Still the principle governing wave interactions is the same. The complex wave that reaches our ears is the sum of a series of individual waves produced by each instrument. Each separate wave exists independently of the others and keeps its individual characteristics. Remarkably enough, we learn to separate out these individual waves against the complex background, following only the melodic line or picking out the parts played by the woodwinds.

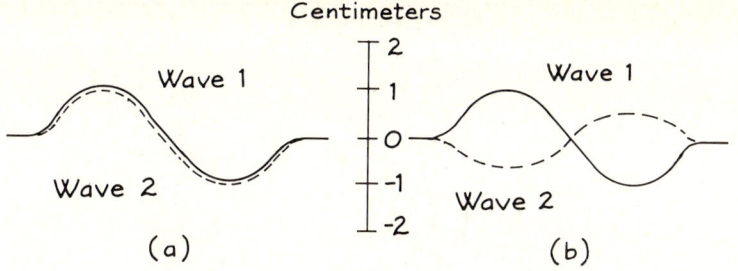

Figure 14-20

Figure 14-21

(a) At most frequencies, we see a jumble. **(b)** At a few frequencies we see a stationary wave pattern called a standing wave.

SELF-CHECK 14F

Sketch the shape of the combined waves shown in Figure 14-20.

Standing Waves

Suppose you are vibrating a spring that is attached to a wall. You produce continuous and identical waves by moving the spring up and down at a constant rate. As they travel back along the spring, the reflected waves interact continuously with identical incoming waves. At most frequencies, these interactions produce a jumble like that shown in Figure 14-21(a). However, at a few frequencies they produce patterns like those in Figure 14-21(b)—waves that move up and down but do not appear to travel along the spring. These stationary wave patterns are called **standing waves**.

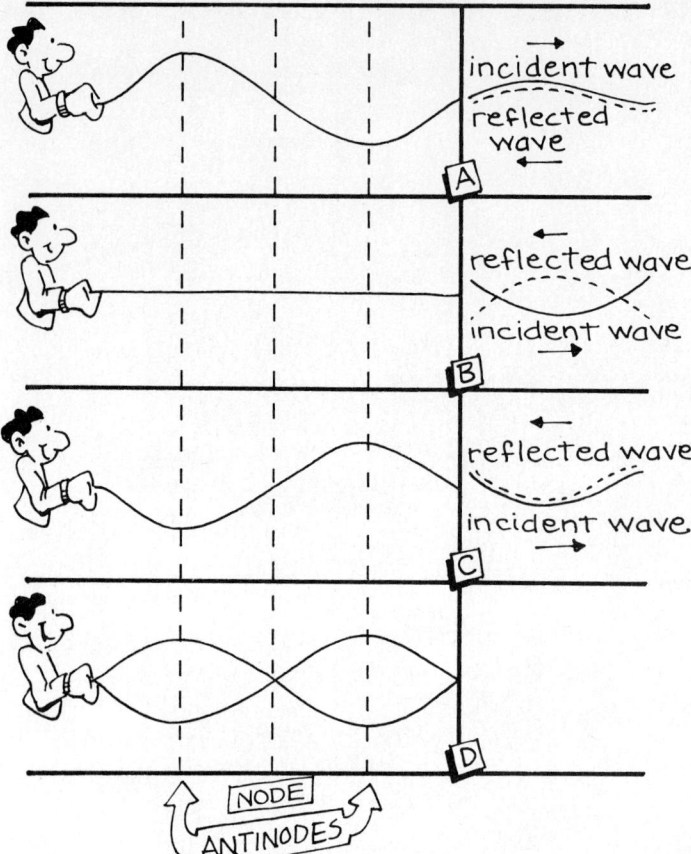

Figure 14-22
A standing wave is produced as the reflected and incoming waves alternate between constructive and destructive interference.

    A standing wave pattern is created by constructive and destructive interference of the incoming and reflected waves, as shown in Figure 14-22. When the incoming and reflected waves match (Figure 14-22(a)), they interfere constructively. As the incoming wave continues toward the wall and the reflected wave continues back, the crests of one eventually match the troughs of the other, and vice versa (Figure 14-22(b)). They then interfere destructively, and the entire spring is flat. Later the two waves again match (Figure 14-21(c)) and constructive interference occurs. This continual interference between incoming and reflected waves creates a wave pattern in the spring that stands still. Certain regions of the spring, called **nodes,** do not move at all. The incoming and reflected waves always interfere destructively at these points. Other regions of the spring, called **antinodes,** move up and down constantly. Here incoming and reflected waves alternately interfere constructively, destructively, constructively, and so forth.

    Vibrations at one frequency create standing waves, while those at another do not. If you shake the end of a spring you can actually feel the difference. A lot of energy is needed to keep a wave moving at most frequencies. But when you move the spring at a standing wave frequency—one at

Figure 14-23
The Tacoma Narrows Bridge provided a spectacular example of resonance shortly before its collapse on November 7, 1940.

which outgoing and returning waves match—the pattern is maintained with very little energy. Since the reflected waves constantly reinforce the outgoing waves, the system loses very little energy to the surrounding air or to the thermal energy of the spring itself. The small pushes you provide at just the right times replace this lost energy; more substantial pushes can increase the amplitude of the standing wave.

Standing waves occur in many objects—springs, ropes, metal rods, drum heads, and violin strings. Even buildings and bridges vibrate in standing wave patterns. Each object has unique frequencies, called **natural frequencies,** at which standing waves occur. In musical instruments, these standing waves are a means of creation. In buildings and bridges, they can be a means of destruction.

Resonance

On November 7, 1940, the bridge spanning the narrows at Tacoma, Washington, collapsed. From the day it was built, the concrete-and-steel suspension bridge had vibrated noticeably in the wind. Drivers on the bridge sometimes saw cars in front of them vanish and reappear as the bridge oscillated. On the fateful day these oscillations grew steadily until they became violent (Figure 14-23). The bridge looked very much like our spring rather than a semirigid structure made from solid iron girders 2 m thick. Within a few hours, the bridge had collapsed. The culprit was resonance.

Resonance occurs when the frequency of the energy source matches a natural frequency of the vibrating medium. The spring in Figure 14-21(b) is resonating; so is the Tacoma Narrows Bridge in Figure 14-23. In the case of the bridge, the amplitude of the standing waves increased to the point where the structure could no longer stand the strain and collapsed. The same phenomenon can cause a building to collapse during an earthquake. When the

Figure 14-24
The standing waves produced in the string are transmitted to the body of the violin through the bridge. Standing waves in the body of the violin create the musical sounds we enjoy.

ground shakes at a certain frequency, standing waves will be set up in a building that has that particular natural frequency. The building may crack or collapse under the strain, while an adjacent building with a different natural frequency remains undamaged.

The Tacoma Narrows Bridge had solid steel beams, which played a key role in its collapse. When a flat object such as a beam blocks the wind, it creates oscillating turbulence patterns on the downwind side of the object. The oscillating gusts of wind push alternately upward and downward at a rate that depends on the dimensions of the object blocking the wind. In the case of the Tacoma Narrows Bridge, the choice of solid steel beams led to wind oscillations on the downwind side that happened to match a natural frequency of the bridge. Each oscillation of the wind caused the bridge to oscillate higher, until the bridge finally tore itself apart.

Not all applications of resonance are destructive. The resonance that architects and engineers desperately try to avoid, instrument makers eagerly try to discover. A violin is designed so that the standing waves produced in the string can be transmitted to the wooden body of the violin (Figure 14-24). For centuries violin makers have refined the art of selecting the best woods and sculpting them to resonate in ways we find pleasurable.

Harmonic Series

Most vibrating objects produce standing waves at several different natural frequencies. The spring shown in Figure 14-25 vibrates in a standing wave pattern at frequencies of 2 Hz, 4 Hz, 6 Hz, 8 Hz, and so on. The lowest natural frequency of an object, in this example 2 Hz, is called the object's **fundamental frequency.** Other natural frequencies of the spring are integer multiples of its fundamental frequency. Collectively, all the resonant frequencies of an object form a **harmonic series.**

To understand how the frequencies in a harmonic series are established, consider the ways in which waves can fit along the spring. Once the standing wave has been established, your hand and the wall provide fixed points of contact for the spring. (While you do move your hand slightly to replace energy lost to the surrounding air, ideally the wave energy would just bounce back and forth between the wall and your hand.) Any standing wave established in the spring must have nodes located at these two end points. The standing wave produced at the fundamental frequency has only these two nodes. Consequently, one-half of a wavelength fits along the length of the spring. The next resonant frequency, 4 Hz, produces a wave with three nodes—one at each end and one in the middle. One full wavelength now fits along the spring. The next resonant frequency, 6 Hz, has four nodes—one and one-half wavelengths—and so forth. A series of resonant frequencies, such as that shown in Figure 14-25, consists of those frequencies for which one-half, one, one and one-half, two (and so on) wavelengths fit along the spring.

Since standing waves are produced when outgoing waves match the arrival of reflected waves, the fundamental frequency of a medium depends on

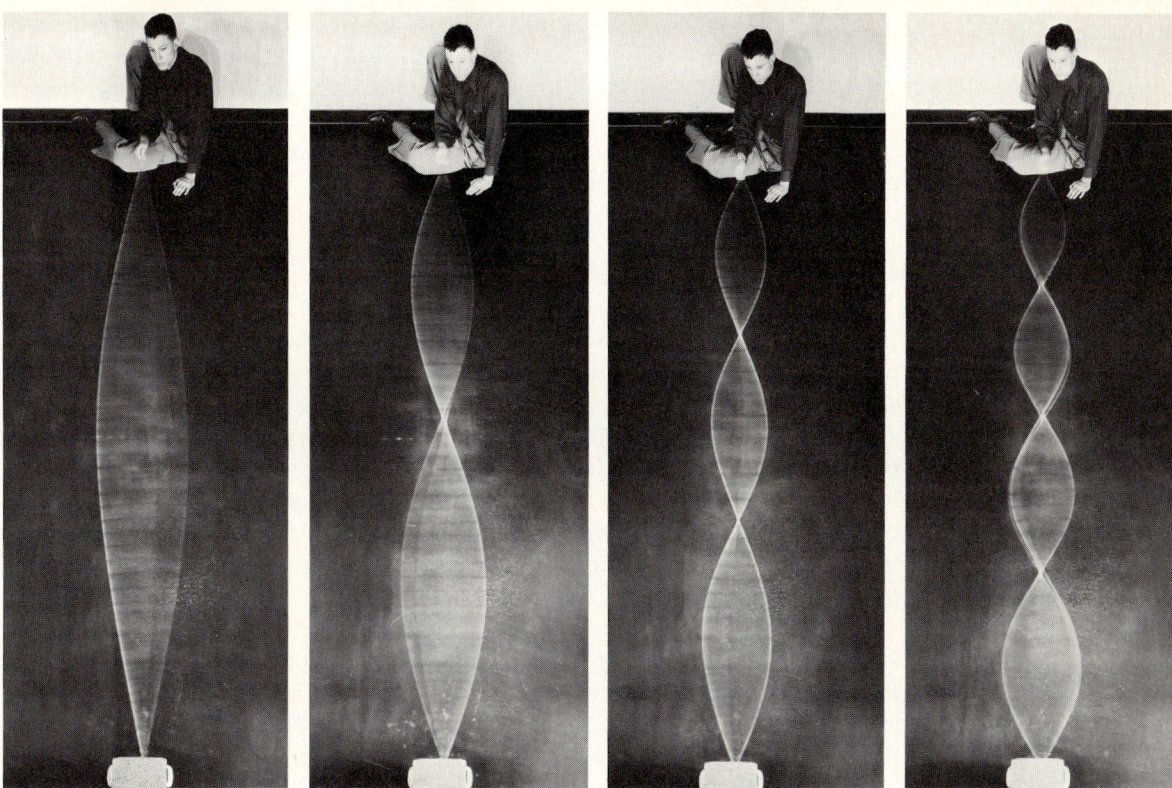

Figure 14-25
Most wave media produce standing waves at several different frequencies, called harmonics.

two variables: how fast the wave travels and how far it travels before it returns to the starting point. The first variable, wave speed, depends on physical properties of the medium. With two wave media of equal length, the fundamental frequency will be higher in the medium in which the waves travel faster. String instruments have tuning pins that tighten and loosen their strings. The tighter the string, the faster the waves travel along the string. Consequently, tightening the tuning pin increases the fundamental frequency produced by the string. The second variable, distance traveled, depends on the length of the medium. When two wave media have the same wave speed, the fundamental frequency will be lower in the longer medium. The longer strings used in bass viols produce lower fundamental frequencies than the shorter strings used in violins.

SELF-CHECK 14G

You have a spring twice as long as that in Figure 14-25. The wave speed is the same in the two springs. How will the fundamental frequency of the longer spring compare with that in Figure 14-25?

One goal in science is to simplify—to draw together the diverse phenomena we observe into as few categories as possible. The wave concepts that have emerged from observations of water waves, springs, and vibrating strings prove equally powerful in describing sound and light. Chapter 15 offers you a closer look at the wave model and its effectiveness in drawing together remarkably diverse phenomena.

CHAPTER SUMMARY

Wave motion is the propagation of a disturbance through a medium. It is a process by which energy is transferred from a source to a receiver without a simultaneous transfer of mass. Wave motion requires an energy source, an energy receiver, and a medium through which the energy is transferred. Comparing the direction in which the medium vibrates with the direction in which the energy travels leads to the identification of two kinds of waves. In a *transverse wave,* the medium vibrates perpendicular to the direction of energy transfer. A *longitudinal wave* is a wave in which the medium vibrates parallel to the direction of energy transfer.

A wave is described in terms of its amplitude, frequency, and wavelength. The *amplitude* of a wave describes its height, the maximum distance the parts of the medium move from their natural position. Amplitude is related to the amount of energy carried by the wave. The *frequency* of a wave is the number of complete cycles executed by a segment of the medium in 1 s. It is related to the rate at which energy is transmitted by the wave. *Wavelength* is the distance taken up along the medium by one complete cycle. The frequency and wavelength of a wave are related to one another by the concept of wave speed:

$$\text{Wave speed} = \text{frequency} \times \text{wavelength}$$

Wave speed describes how fast the disturbance moves through the medium and consequently the rate of energy transfer.

When either the wave source or receiver moves relative to the other, the frequency of the waves received is different from the frequency of the waves emitted. This difference in frequency is called the *Doppler effect*. The greater the relative speed between the source and receiver, the larger the difference in frequency reported by the receiver.

As waves move from one medium to another, they can be reflected, refracted, or absorbed. Waves that are *reflected* remain in the original medium. Reflected waves obey the *law of reflection:* The angle of incidence equals the angle of reflection. Waves that are transmitted into a new medium change speed. This change in speed leads to a change in wavelength and a change in direction called wave *refraction*. If the wave slows down, the angle of refraction is less than the angle of incidence. If the wave increases its speed, the angle of refraction is greater than the angle of incidence. Waves that are *absorbed* by the new medium convert their energy into some other form, usually thermal energy.

GETTING THE MESSAGE ACROSS

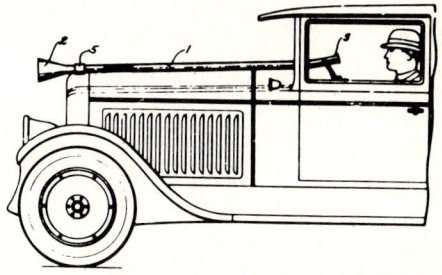

Today when we see a pedestrian in front of our car, we either stop or honk the horn. Eugene Baker apparently felt that the horn was too impersonal. He wanted a more direct communication between the driver and the pedestrian. To convey messages, Mr. Baker applied the ideas of wave reflection and resonance. If a driver saw someone standing in his way, he would speak into the tube (3). The sound would be reflected off the inside of the narrow tube (1) and thus travel to the megaphone (2) at the end. The shape of the megaphone assured that all of the sound was directed—via reflections—toward the pedestrian. The length of the tube affected the frequencies that would be transmitted. By selecting a rather long tube, Mr. Baker created a situation in which the low frequencies have standing waves in the tube. The driver's voice would seem to be lower to the pedestrian than it actually was because of these standing waves.

Waves that meet in space combine according to the *superposition principle*. The combined wave has a displacement at each point in the medium that is the sum of the displacements caused by each wave separately. Two identical waves that meet so that crest matches crest and trough matches trough combine to produce a wave with twice the amplitude of the two separate waves. This is called *constructive interference*. Two identical waves that meet so that crest matches trough annihilate one another. This is called *destructive interference*.

Under certain circumstances, an incident wave and its reflection will interfere with one another to produce a standing wave. A *standing wave* is a wave pattern that appears stationary in the medium. Points in the medium that do not move are called *nodes*. Points of maximum displacement are called *antinodes*. The motion of the medium in a standing wave pattern arises from alternating periods of constructive and destructive interference between the incoming and reflected waves. Standing waves are created when an energy source vibrates at one of the *natural frequencies* of the medium. The natural frequency of the medium depends on the wave speed and length of the medium. Vibration at a natural frequency, called *resonance,* can destroy structures like bridges or produce pleasurable sounds in musical instruments.

ANSWERS TO SELF-CHECKS

14A. Waves in A have a smaller amplitude, a greater wavelength, and a smaller frequency than those in D.

Waves in B have the same amplitude but a greater wavelength and a smaller frequency than those in D.

Waves in C have a smaller amplitude, a greater wavelength, and a smaller frequency than those in D.

14B. Wave speed = (440 Hz)(3 m) = 1320 m/s. Sound travels more rapidly in water than in air.

14C. It does not matter whether the wave source or the wave receiver is moving. As Y moves toward the boat, the waves will be pushed together. Y reports receiving a frequency higher than 1 Hz. X is not moving relative to the boat. X reports receiving a frequency of 1 Hz, the same frequency emitted at the boat.

14D. Seismic stations at B will report receiving transverse waves, but those at C will not.

14E. Since the waves slow down substantially when they enter the earth's core, the angle of refraction will be less than the angle of incidence, as shown below. The longitudinal waves will have a shorter wavelength in the core than in the mantle.

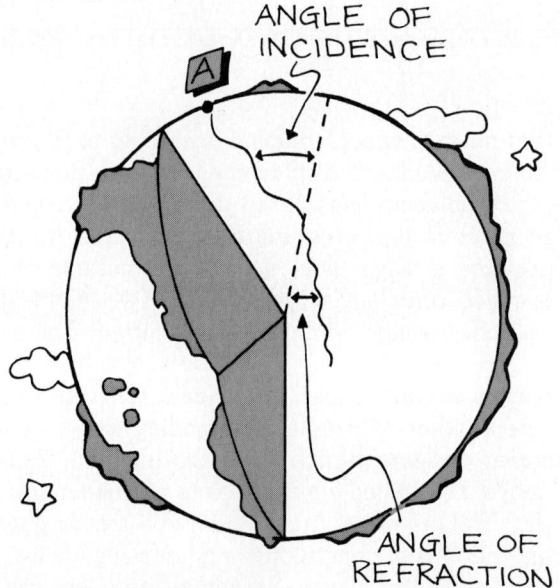

14F. The waves add as shown below.

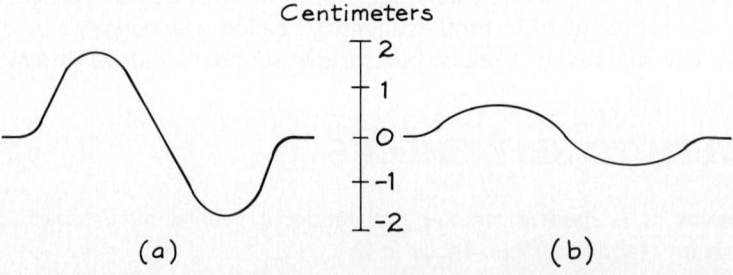

14G. Since the spring is now longer, the wavelength of the fundamental frequency will be longer. A longer wavelength corresponds to a lower fundamental frequency than that shown in Figure 14-25.

PROBLEMS AND QUESTIONS

A. Review of Chapter Material

A1. Define each of the terms listed below:
Wave motion
Transverse wave
Longitudinal wave
Amplitude
Frequency
Wavelength
Wave speed
Reflection
Refraction
Doppler effect
Absorption
Superposition principle
Constructive interference
Destructive interference
Standing wave
Nodes
Antinodes
Resonance
Fundamental frequency

A2. How is wave motion different from the other processes by which energy is transferred?

A3. How are longitudinal waves different from transverse waves?

A4. What affects the speed at which waves travel?

A5. How do geologists use earthquake waves to build a model of the interior of the earth?

A6. Why does a moving source or receiver cause a change in the frequency of waves reported by the receiver?

A7. Sketch the path of the refracted waves for each case.

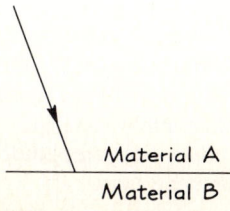

a. The wave speed in A is greater than in B.
b. The wave speed in B is greater than in A.

A8. Sketch the path of the reflected wave in each situation shown below.

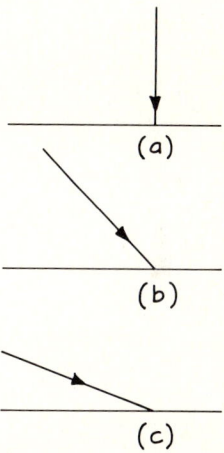

A9. When waves are transmitted into a new medium, they change speed. How does this change in speed lead to a change in wavelength? A change in direction?

A10. What happens to the energy transmitted by a wave when the wave is absorbed by the new medium?

A11. Under what circumstances will constructive and destructive interference occur?

A12. Describe how constructive and destructive interference explain the production of standing waves.

A13. Why are standing waves produced only at selected frequencies?

A14. What two variables determine the fundamental frequency of a medium?

B. Using the Chapter Material

B1. A spring is attached to the end of a swinging pendulum. Draw a diagram showing how the pendulum must move to produce longitudinal waves in the spring. How must it be attached to produce transverse waves?

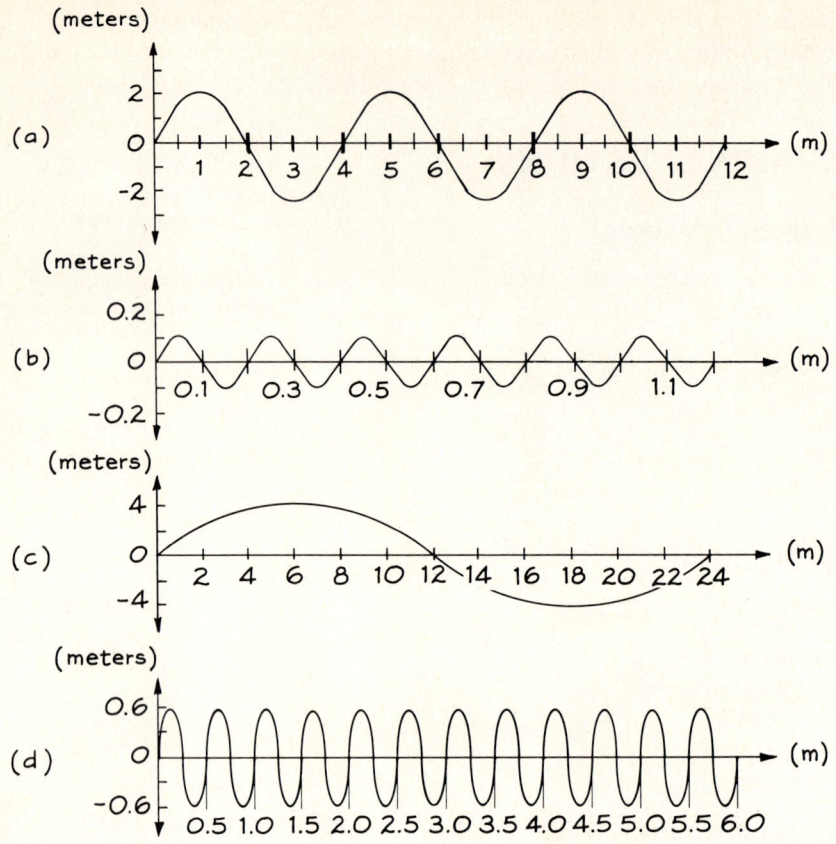

B2. What are the amplitudes and wavelengths of the waves shown above?

B3. In one of its vibrations, the Tacoma Narrows Bridge vibrated 6 times in 60 s. What was the bridge's frequency?

B4. Complete the following table.

Frequency	Wavelength	Speed
400 Hz	0.75 m	?
?	1.25 m	300 m/s
150 Hz	?	300 m/s
500 Hz	?	500 m/s
?	2.00 m	500 m/s
1000 Hz	0.50 m	?
100 Hz	0.40 m	?
100 Hz	?	80 m/s
?	1.60 m	160 m/s

B5. Use the results of the table completed in Question B4 to determine:
  a. what happens to the frequency if the wavelength doubles and the speed does not change
  b. what happens to the wavelength if the speed decreases by one-half and the frequency does not change
  c. what happens to the speed if the frequency decreases by one-half and the wavelength doubles

B6. The wave shown below is heading toward a point where the rope changes to thicker rope. The wave will travel more slowly in the new rope. Sketch the wave as it is transmitted into the new medium.

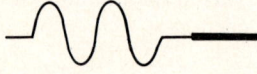

B7. The drawing that follows shows a wave at one instant in time. Draw waves that, if they interfered with this wave, would:
  a. completely cancel it
  b. reduce its amplitude to one-half its present value

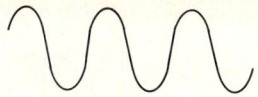

   c. increase its amplitude to twice its present value
   d. increase its amplitude to four times its present value
B8. Standing waves in an oboe have the same type of harmonic series as those in a spring. If the oboe has a fundamental frequency of 440 Hz, what will be some of the other resonant frequencies?
B9. The water strider is an insect that walks on water. As each of its feet strikes the water, a wave is sent out. Suppose the frequency at which its feet strike the water is 1 Hz; water waves travel at 0.1 m/s and the water strider moves east at 0.004 m/s. What frequencies of water waves are reported by a motionless receiver directly east of the insect? Directly west of the insect?
B10. If the Tacoma Narrows Bridge vibrated at 0.1 Hz, what must have been the frequency of the force applied by the wind?

C. Extensions to New Situations

C1. A transverse wave traveling on a spring strikes a wall. The reflected wave's amplitude is in a direction opposite to the incident wave. This result can be explained by applying Newton's laws to the situation.
   a. Draw a wave that is striking a wall.
   b. In which direction does the force applied by the rope act on the wall?
   c. What is the direction of the force applied on the rope by the wall?
   d. Which object will move more easily? Why?
   e. What will be the direction of the motion of the object mentioned in part (d)?
   f. Use the answers to parts (a)–(e) to explain why the amplitude reverses upon reflection from a wall.
C2. The concept of resonance can be used to create very large waves in a swimming pool.
   a. Suppose you and several friends stand at one end of a pool and jump into it simultaneously. Describe the wave that is created and describe how it travels through the pool.
   b. What happens when the wave reaches the other end of the pool?
   c. If the water waves travel at 0.5 m/s and the pool is 50 m long, how much time will pass before the wave returns to the end where you jumped in?
   d. Suppose you and your friends get out of the pool and then jump back in just as the wave returns to your end. How will the amplitude of the next wave compare with the first?
   e. If you keep repeating the process described in (d), what will happen to the amplitude of the wave?
   f. Do you think you could empty a swimming pool this way?
C3. Some evil people have been known to use the concept of resonance to destroy a waterbed. Use the ideas presented in Question C2 to explain how you could produce large-amplitude waves in a waterbed. (*Warning:* These waves can become large enough to split the plastic in a waterbed.)
C4. The Tacoma Narrows Bridge was not the first bridge to collapse because it resonated with the wind. A British bridge designer, Sir Samuel Brown, built three bridges during the nineteenth century, all of which collapsed after oscillating in the wind. He also built a footbridge that collapsed when soldiers marched across it in step.
   a. Explain why marching in step could set up large oscillations in a bridge.
   b. Today, soldiers break step and walk across suspended footbridges. Why does this procedure eliminate the creation of standing waves?
C5. By adding selected waves together, we can obtain waves that look quite different from the simple waves from which they were built.
   a. Sketch the addition of the two waves (a) shown on the following page.
   b. To the wave you sketched in (a), add the wave shown in (b) and sketch the result.
   c. To the wave you sketched in (b), add the wave shown in (c) and sketch the result.

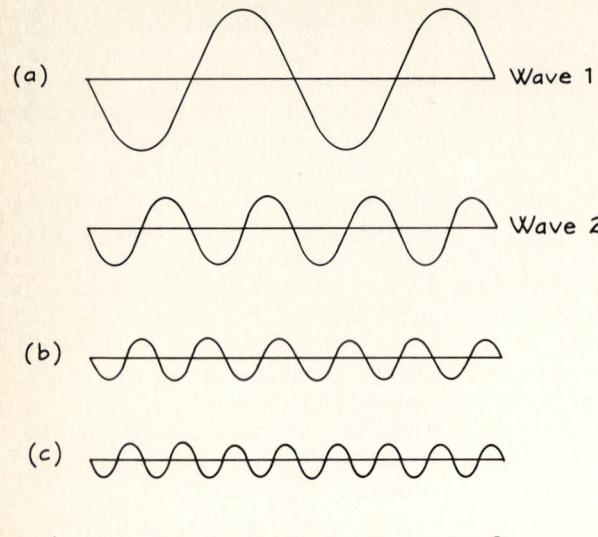

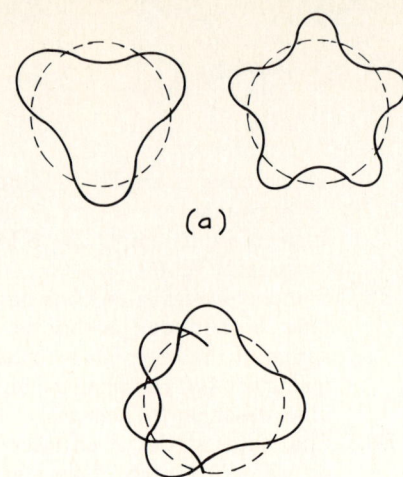

d. Do one more. To the wave you sketched in (c), add the wave shown in (d) and sketch the result.

e. Does your result in (d) look similar to any of the waves you added? (We discuss this idea more in the next chapter.)

C6. Earthquake waves travel through the earth and are received by stations on the other side. Some stations receive both longitudinal and transverse waves, while others receive only longitudinal waves. Explain why this supports the theory that the core of the earth is liquid.

C7. When you strike a water glass, you can establish standing waves on the rim of the glass. As you see it from above, the glass forms a circle. The standing waves are similar to those shown in the next column.
 a. Identify the nodes and antinodes for the wave shown in (a).
 b. Suppose a wave on a circular glass looks like that in (b). Explain why standing waves cannot occur.
 c. Sketch a standing wave that has fewer nodes than the one in the circle.
 d. Use your results to state a general rule relating the wavelength of circular standing waves to the circumference of of a circle.

C8. The earth is believed to have three quite different layers, shown in Figure 14-8. Use the information on wave speed given in Table 14-2 to discuss changes in the direction of motion as a wave travels from a point on the earth, through the earth to a point on the opposite side.

D. Activities

D1. Try the swimming pool experiment described in Problem C2.

D2. Try to find the nodes on a guitar string. Place small pieces of paper on the string and pluck it. Move the paper around until it undergoes the least vibration. Why is that location near a node?

D3. Watch electric wires, telephone lines, and clotheslines on windy days. Sometimes you will see standing waves.

Waves: Sound and Electromagnetic

After sitting in the sun for several hours, you have a sunburn. So, you decide to get a cold drink and relax in front of the television. As you sit down, you hear your roommate shout and the refrigerator door slam. (You left it open again!) In spite of the jackhammer vibrating the house and everything in it as it breaks up the concrete on the street outside, you try to relax while watching the evening news. Following a report on a new design for police radar comes the details of the destruction from yesterday's earthquake and tsunami. Whew! Perhaps you could relax better if you popped some corn in the microwave oven.

Sunburn, television, sound, jackhammer vibrations, radar, earthquakes, tidal waves, and microwaves. Each of these phenomena involves the transfer of energy from a source to a receiver without a transfer of matter. Each can be described in terms of the wave model developed in Chapter 14. Frequency and wavelength distinguish sound from earthquakes, radar from sunlight. Reflection, refraction, superposition, and resonance describe the behavior of sound and light as effectively as they described our observations of springs and water waves. The photographs and sketches of springs and water waves serve as guides as we construct mental images of phenomena we can sense but not necessarily *see*.

We begin by looking at the two major categories of wave phenomena: *mechanical waves* and *electromagnetic waves*. By ordering waves according to frequency, the *wave spectrum* organizes the different types of waves within each of these categories. For each wave category, we will describe phenomena due to reflection, refraction, absorption, standing waves, and the Doppler effect. *Sound* will be used to illustrate mechanical waves. *Light, microwaves,* and *radar* will illustrate electromagnetic waves.

WAVE CLASSIFICATION

Sunburn, television, sound, jackhammer vibrations, radar, earthquakes, tsunami, and microwaves—some of these phenomena are easier to imagine than others. Jackhammers and earthquakes shake the ground and the ground shakes you. You can watch tsunami waves move across the ocean. Even sound seems quite real as you strum a guitar and feel the vibrating strings. By contrast, other phenomena seem more abstract. It is hard to imagine how energy travels from the sun to our bodies, from the television tower in the next city to that box in the living room, from the police car to your car. Light, television signals, radar, and microwaves seem almost magical. The two categories of waves, mechanical and electromagnetic, reflect our sense of the concrete and the abstract.

Mechanical Waves

Sound, jackhammer vibrations, earthquakes, and tsunamis are all examples of mechanical waves, waves created when chunks of matter—vocal chords, jackhammers, rock, or water—vibrate. This matter collides with other matter, which collides with other matter, and so forth, as the disturbance is transmitted through the surrounding material. **Mechanical waves** can be produced and transmitted by matter in any form: solid, liquid, or gas. Table 15-1 lists several examples of mechanical waves, common sources of each, and whether these waves are transverse, longitudinal, or both.

The various kinds of mechanical waves differ in the frequencies at which they commonly occur. If you count water waves as they strike the beach, you find that they rarely exceed a frequency of one per second, or 1 hertz (Hz). Earthquakes and jackhammers produce waves in the ground that range from 1 to about 50 Hz. Vibrations that produce sound audible to the human ear range from 20 to 20,000 Hz. Dogs hear sounds at still higher frequencies.

We can arrange these frequencies in order from low to high, in a classification scheme called a **wave spectrum**. Figure 15-1 shows the wave spectrum for mechanical waves. Mechanical waves with frequencies below those of audible sound are called *infrasonic waves* and include the water waves and earthquakes discussed in the last chapter. Mechanical waves at frequencies

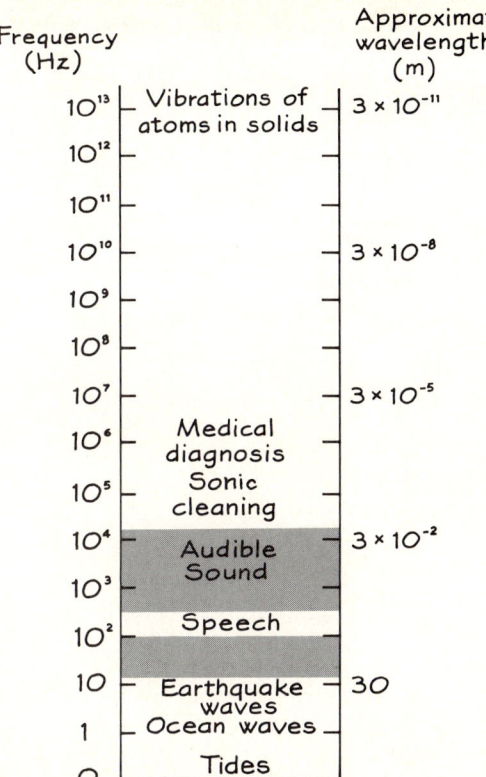

Figure 15-1
Mechanical wave spectrum.

Table 15-1 Examples of Mechanical Waves

Type of Wave	Source	Longitudinal/Transverse
Water waves	Objects thrown in water; wind-water interaction	Transverse on surface; longitudinal below surface
Earthquake waves	Sudden fault movement in earth	Transverse and longitudinal
Sound	Vibrations of certain objects, including musical instruments and vocal chords	Longitudinal
Ultrasound	Vibrations of certain objects, such as bats' vocal chords and ship sonar	Longitudinal

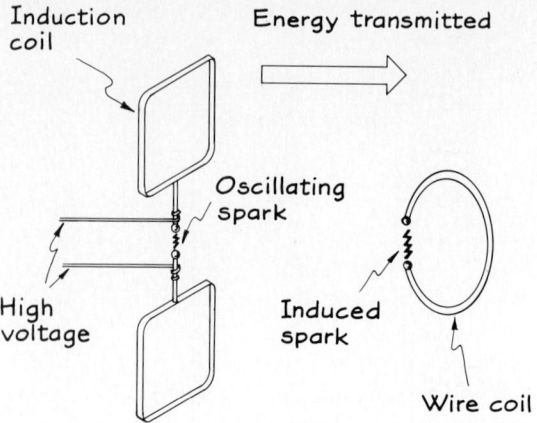

Figure 15-2
An oscillating spark at the left induces a similar spark on the right. Oscillating electrical charges produce electromagnetic waves.

larger than audible sound are called *ultrasonic waves*. These waves have found wide application in industry—welding, cleaning, and detecting flaws in a variety of materials. Ultrasonic imaging is used in medicine, primarily in situations where X rays are ineffective or dangerous. (For example, ultrasound is used to examine the development of a fetus.) Wavelengths, estimated from the media that usually transmit each kind of wave, are included in Figure 15-1.

Electromagnetic Waves

In the 1880s Heinrich Hertz performed a series of experiments with electricity and magnetism. Using a device called an induction coil, he became adept at producing vibrating electric charges, commonly called sparks, across an air gap. Near this oscillating spark Hertz placed a coil of wire, which also had a

Table 15-2 Examples of Electromagnetic Waves

Type of Wave	Vibrating Charged Source	Longitudinal/Transverse
Radio, television	Charges in an antenna	Transverse
Microwave	Molecules	Transverse
Infrared	Atoms	Transverse
Visible light	Atoms	Transverse
Ultraviolet	Atoms	Transverse
X rays	Electrons	Transverse
Gamma rays	Atomic nuclei	Transverse

small air gap (Figure 15-2). Each time a spark occurred across the induction coil, a spark was produced across the wire coil as well. Energy had been transmitted from one coil to the other, although no mass had been transferred. Our interpretation of this phenomenon is that oscillating electrical charges produce waves.

Light, television signals, radar, and microwaves are all examples of **electromagnetic waves,** waves created by vibrating electric charges. These charges can be a single charged particle like an electron or a chunk of electrically charged matter like an atom or molecule that has a net electric charge. One electric charge can affect another at a distance, even when there is no matter between them. Consequently, electromagnetic waves are transmitted through empty space as well as through matter. Light waves from the sun, for example, cross millions of kilometers of emptiness before reaching us. Table 15-2 lists several examples of electromagnetic waves and common sources of each type. As shown in the table, all electromagnetic waves are transverse.

Electromagnetic waves range in frequency from less than 100 Hz (radio waves) to more than a million million million times this amount (gamma rays). Figure 15-3 shows the complete electromagnetic wave spectrum. Visible light, perhaps the most famous member of the spectrum, actually occupies an extremely narrow range of frequencies. It extends from the lower frequencies of red light (4.5×10^{14} Hz) to the higher frequencies of violet light (7.5×10^{14} Hz). Differences we perceive as **color** are, in fact, differences in frequency. Like mechanical waves, electromagnetic waves travel at different speeds in different materials. However, in a vacuum all electromagnetic waves travel at 3×10^8 meters per second (m/s) (commonly called the speed of light). Wavelengths for electromagnetic waves in a vacuum are included in Figure 15-3.

SELF-CHECK 15A

Use Figures 15-1 and 15-3 to estimate the frequencies of human speech and of microwaves. What, besides frequency, is different about these two kinds of waves?

Wave Receivers

All waves transfer energy from a source to a receiver, but some receivers are more sensitive than others. Our eyes and photographic film detect visible light but seem unaffected by the shorter wavelengths in microwaves. A television antenna absorbs the longer waves emitted by a television station but virtually ignores the shorter wavelengths of visible light. Microwaves can pop corn, while gamma rays cannot.

To understand why receivers are more sensitive to some wavelengths than others, consider an example using water waves. Suppose you are sitting in a fishing boat that is 2 meters (m) long. While you are fishing, a water wave

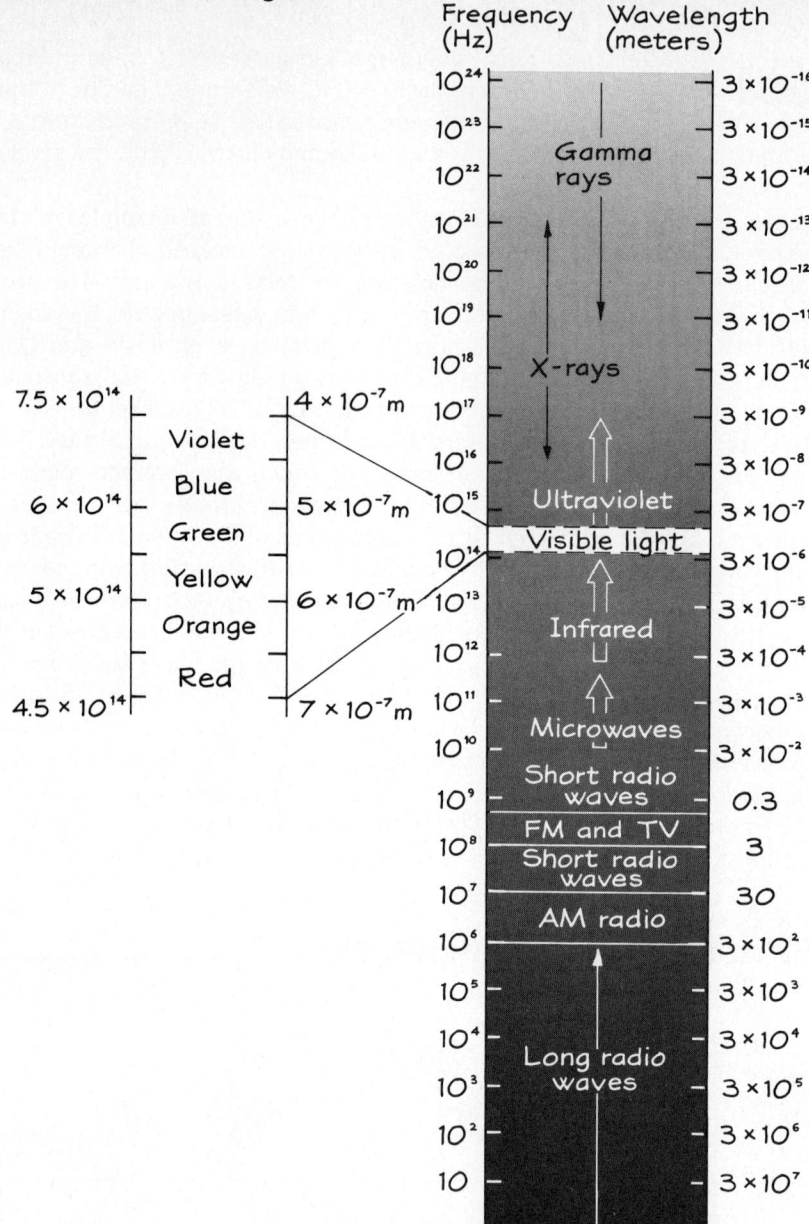

Figure 15-3
Electromagnetic wave spectrum.

with a wavelength of 20 m passes beneath you (Figure 15-4(a)). Your boat slowly rises and falls with the wave. Since the wavelength of the water waves is much longer than your boat, chances are your fishing will not be interrupted much. A wave with a 10-m wavelength will be only slightly more noticeable, while waves that are 4 to 5 m long will become apparent (Figure 15-4(b)). When the wavelength of the wave is approximately equal to the length of the fishing boat, fishing becomes virtually impossible (Figure 15-4(c)). The bow of the boat oscillates as every wave passes by. If the water waves become

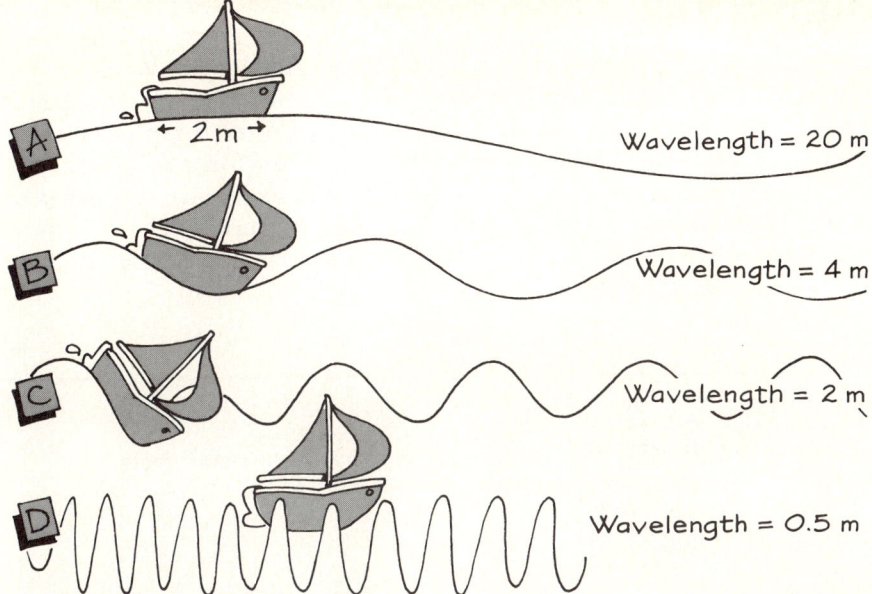

Figure 15-4
The closer the length of the boat is to the wavelength of the waves, the more easily it detects the waves.

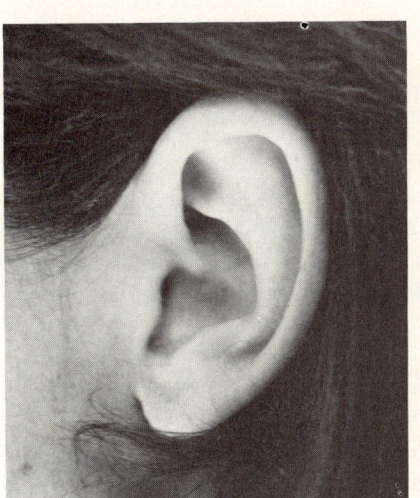

Figure 15-5

considerably shorter than the boat, the boat floats along the top of the crests (Figure 15-4(d)) and the waves again pass unnoticed. You notice only the waves whose wavelength is close to the length of the boat.

If we think of the fishing boat as an energy receiver, its behavior can be explained in terms of resonance. The bow of the boat oscillates more when the water waves push in time with its natural frequency. This natural frequency depends, in part, on the length of the boat. Consequently, waves that are about the same length as the energy receiver oscillate at frequencies near the natural frequency of the receiver. To be most effective, a wave detector needs to be about the same size as the wavelength of the waves it is to detect. Several examples are shown in Figure 15-5.

Radio antennae are about 1 m long and detect waves that range from about 1 m to several meters long. Microwave detectors are at most a few centimeters long and detect waves of about the same length. When we proceed to electromagnetic waves of very short wavelengths, we cannot build detectors small enough. Fortunately, nature does! Infrared radiation is best detected by molecules. Visible light, with wavelengths of about 8×10^{-7} meters, is best detected by atoms. A similar analysis applies to mechanical waves. For example, the size of the eardrum is appropriate for the range of frequencies we call audible sound. The size of detectors is closely related to the size of the wavelength being received.

SELF-CHECK 15B

The directions for installing automobile radio antennae frequently instruct the owner to set the antenna differently for FM reception than for AM reception. Why? Use Figure 15-3 to determine which would need the longer antenna, FM or AM.

SOUND

Pleasant music soothes us, unexpected noises frighten us, the roar of the crowd excites us. Friends recognize us by our voices. Sound waves may well be the most familiar type of wave we experience. While sound is not the only way to communicate, it carries a large fraction of our thoughts, feelings, and opinions. We need only compare the social barriers faced by the deaf with those encountered by the blind to realize how much human beings depend upon this form of wave motion.

Historically, sound was one of the first phenomena to which the wave model was successfully applied. The spread of sound in all directions from a source was likened to the spread of circular water waves from a pebble dropped into a quiet pond. Early interest in musical instruments contributed enormously to our understanding of standing waves and resonance. The concepts and images used to describe water and mechanical waves in a spring prove valuable in understanding the behavior of sound.

Describing Sound Waves

A bell is placed inside a jar to which a vacuum pump has been attached. Initially the jar is filled with air and we hear the bell ringing. The jar is then sealed and the pump begins removing the air inside it. Gradually, the sound dies out. Although we can still see the clapper as it pounds away on the bell, we can no longer hear any sound. Sound requires a medium.

Sound waves are longitudinal waves, capable of traveling through solids, liquids, and gases. We imagine atoms and molecules to be evenly spaced in matter, much like the evenly spaced coils in a spring. When a bell rings, its motion alternately compresses (squeezes together) and rarefies (pulls apart) the molecules in the surrounding air. A series of compressions and rarefactions travels through the air (Figure 15-6), much as compressions traveled in the spring. Each molecule vibrates parallel to the direction in which the sound energy is transmitted. When the compressions and rarefactions reach your ear, they establish vibrations in the eardrum that are eventually transformed into nerve impulses sent on to the brain.

The speed of sound has been measured in a variety of materials, some of which are listed in Table 15-3. Since wave speed depends on the medium, the speed of sound varies considerably in different materials. The speed of sound in air increases as the temperature of the air increases. At high temperatures the air molecules move rapidly and transfer the wave energy to their neighbors more quickly than at low temperatures. Sound travels progressively faster as we move from gases to liquids to solids because of the stronger restoring forces. The stronger the molecular bond, the more rapidly the motion of one molecule influences the next. Sound waves travel most rapidly in solids.

We describe sounds in terms of loudness and pitch. **Loudness** refers to our perception of the amount of energy carried by the sound wave. Consequently, it is related to the amplitude of the wave. As the energy carried by the sound wave increases, so does the amount by which molecules are compressed in each disturbance. **Pitch** describes how we perceive frequency or

Figure 15-6
A vibrating bell produces a series of compressions and rarefactions that travel to your ears.

Table 15-3 Speed of Sound in Various Materials

Material	Speed of Sound (m/s)
Gases (0°C)	
Carbon dioxide	259
Oxygen	316
Air	331
Helium	965
Liquids (25°C)	
Mercury	1450
Water	1498
Seawater	1531
Solids	
Rubber	1800
Lead	2100
Gold	3000
Glass	5000–6000
Granite	6000

Figure 15-7

The sound in **(c)** has a larger amplitude than the sound in **(b)**. The sound in **(a)** has a higher frequency and shorter wavelength than the sound in **(b)**.

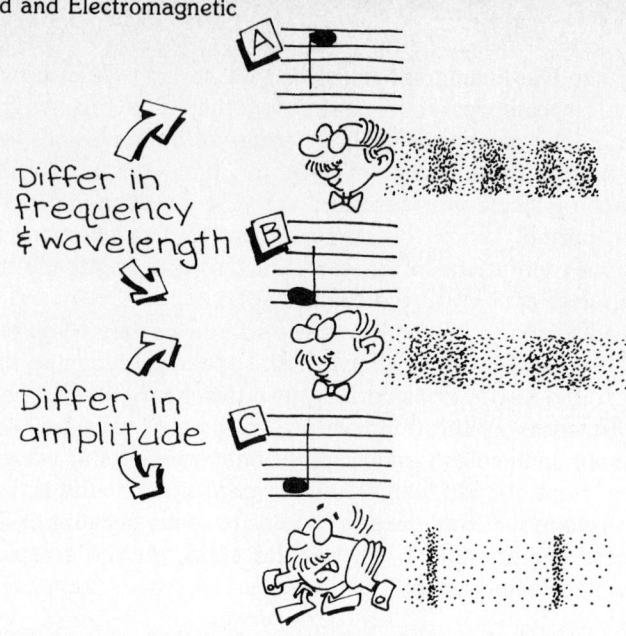

wavelength. A higher-pitched sound has a higher frequency and shorter wavelength than a lower-pitched sound. (Sound waves of differing amplitude, frequency, and wavelength are shown in Figure 15-7.)

Sound Waves in Matter

"Hello" you shout as you stare up at the surrounding canyon walls. "Hello" comes the echo back to you. "Hey, turn down that radio!" comes a familiar yell—funny how your neighbors always complain more in the evening than during the day. "I can't hear you through that ski mask" you tell a friend as you get on the lift. Like water waves, sound waves can be reflected, refracted, and absorbed as they move from one material to another.

Reflected waves have the same wavelengths as incident waves, but they travel in the opposite direction—back toward the energy source. Echoes reveal these characteristics. Your echo sounds similar to your voice; its energy is simply traveling back toward you rather than away from you. Sound can be reflected from any hard surface, from the walls in a hallway as well as the walls in a canyon. However, your ear can distinguish the echo as being separate from the original sound only when the original and reflected waves reach your ears at least 0.1 second (s) apart. In most buildings, the reflected waves reach your ear more quickly. Consequently, the reflections mix with the original waves, contributing to what we call the *brightness* of a sound instead of producing an echo.

A less well known application of sound reflection is the use of curved surfaces to focus sound. Figure 15-8 shows waves reflected by a curved barrier. Energy carried by the entire wave front is reflected to a single point, called the **focal point.** Sound waves behave in much the same fashion (Figure 15-9). Curved surfaces used in a directional microphone focus the incoming sound waves, allowing the microphone to pick up more sound. Ears, both

Sound 343

Figure 15-8
Waves reflected from a curved surface can be focused to a single point, called the focal point.

Figure 15-9
Directional microphones and ears take advantage of curved reflectors.

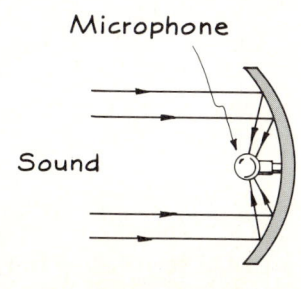

Figure 15-10
Sound waves are refracted **(a)** upward during the day and **(b)** downward at night.

human and animal, are shaped to focus as much sound as possible on the eardrum. In outdoor amphitheaters orchestras are often placed in front of a curved reflecting surface to direct more sound back toward the audience.

Since the speed of sound in air varies with the air temperature, sound is often refracted by the atmosphere. During the day, the air near the ground is warmed more than the upper levels. Since sound travels more quickly in warm air than in cool air, sound waves are gradually refracted upward (Figure 15-10(a)). At night, the reverse occurs. Air near the ground cools more rapidly and sound waves are refracted downward (Figure 15-10(b)). Refraction enables sound to travel farther along the ground at night than during the day. There is an explanation for your neighbor's complaints!

Waves are absorbed when the energy stored in the disturbance is dissipated into thermal energy found in the motion of molecules within matter. Though sound waves are easily transmitted through air, pockets of air within solids act as extremely effective sound absorbers. Sound waves enter the pockets of air, only to be reflected back and forth by the solid boundaries enclosing the air pockets. Eventually, the organized vibrations of the wave become the disorganized motions of the air molecules. Sound has been absorbed. Acoustic tiles designed to muffle room noise are made from porous foam or loosely woven fibers that create many such air pockets. Rugs and foam rubber are also excellent absorbers of sound. Architects take the absorbing properties of materials into account in designing buildings and concert halls.

Music and Standing Waves

A small-necked bottle stands under a dripping water faucet. Each time a drop hits the bottle, it makes a sound. When the bottle is empty, the pitch of the sound is low; but as the bottle fills up, the pitch becomes higher. As the air space above the water shortens, the pitch of the sound rises.

Figure 15-11
Which instrument produces lower frequency sounds?

Figure 15-12
As the string gets shorter, the wavelength associated with the fundamental frequency decreases. The shorter the string, the higher the pitch produced.

Musical instruments display the same relationship between length and pitch. Without ever hearing them, you know that the instrument on the left in Figure 15-11 can play lower notes than the one on the right. Within a family of instruments like the string family, the longer bass viols have a range that extends much lower than the shorter violins. This relationship between the size of the instrument and pitch can be explained in terms of standing waves and resonance.

Consider how different pitches are produced on a single string. To vary the pitch of a string, a guitarist presses the string against the guitar neck at different points, thereby changing the length of the string that is free to vibrate when plucked. A guitar string vibrates in much the same way as the spring we discussed in Chapter 14. Both ends of the string are fixed, one at the bottom where the string is tied off and the other wherever the guitarist places his or her finger. When the guitarist plucks the string, standing waves are produced at the resonant frequencies characteristic of that length string. These standing waves cause the wooden body of the guitar to vibrate, causing the air within it to vibrate, and eventually causing a sound wave to be transmitted to your ears. The sound we hear can ultimately be traced back to the standing waves possible in the guitar string.

The fundamental frequency produced in a guitar string varies with the length of the string (Figure 15-12). Since one-half of a standing wave is the minimum wavelength that can fit in each length of string, the wavelength associated with the fundamental frequency is twice the length of the string. As the string gets shorter, this wavelength gets shorter. Since frequencies increase as wavelengths decrease, short strings must have higher fundamental frequencies than longer strings. As the guitarist moves his or her fingers along the neck of the guitar, he or she lengthens and shortens the string to produce the desired pitch.

SELF-CHECK 15C

The standing waves produced in a bottle are somewhat different than those in a guitar string. Instead of nodes at each end, the bottle allows a node at the water level and an antinode at the open end, as shown in Figure 15-13.

a. How is the wavelength of the fundamental frequency related to the length of the air column above the water?

b. How does the fundamental frequency change as the bottle fills up with water? Does this match your experience?

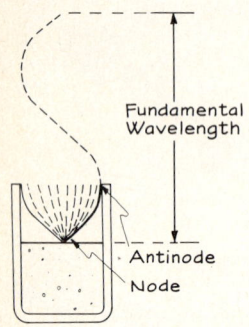

Figure 15-13

The analysis can be extended to wind instruments. Trumpets and sousaphones, for example, are long tubes that for convenience have been folded up about themselves. Air inside the tubes can be made to vibrate in a standing wave pattern with a node at the mouthpiece end and an antinode at the bell. Valves are used to change the lengths of the air columns. Since trumpets are made from much shorter tubes than sousaphones, their range of frequencies extends higher.

While the fundamental frequency in a violin string or air column is the dominant pitch we hear, other pitches are often present simultaneously. In a violin string, for example, the only constraint on the standing waves is that they have nodes at each end. A number of different frequencies satisfy this requirement. When a string vibrates, many frequencies are produced simultaneously but at different amplitudes. These vibrations are picked up by the wooden sound box, which further augments or diminishes their amplitudes to produce a sound wave in air that is relatively complex. Figure 15-14 contrasts the contributions of the lowest two frequencies for a violin and piano when each is playing the A above middle C. These different contributions lead to significantly different waveforms (Figure 15-15). While pianos and violins can play the same pitch, they do not produce the same sound waves. Differences in the higher frequencies establish the quality of the sound, the characteristic that enables us to distinguish one instrument from another.

The Sound of a Siren

WHEEE...OOOH—As the ambulance speeds by, its siren goes from a higher pitch to a lower one. Our most common experience with the Doppler

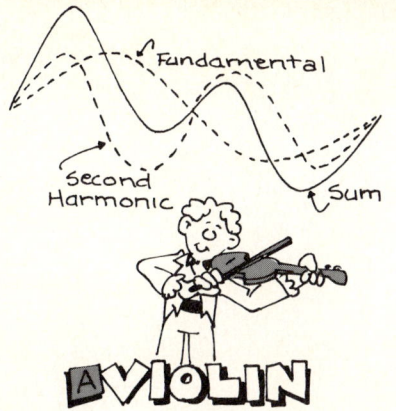

Figure 15-14
The sound we hear depends on the amplitude of the harmonics. **(a)** For a violin, the amplitude of the second harmonic is the same as the fundamental. **(b)** For a piano, the amplitude of the second harmonic is tiny compared to the amplitude of the fundamental. These differences produce two very different looking waves.

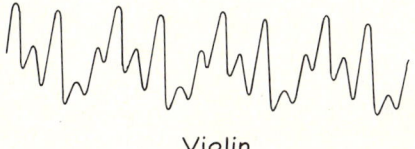

Violin

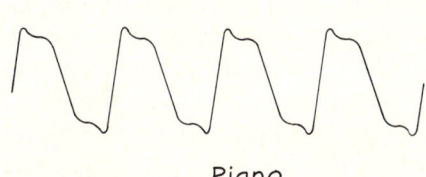

Piano

Figure 15-15
Two very different looking waves lead to two very different sounding sounds. Here the piano and violin play the same pitch!

effect occurs with sound. A drawn-out note like an ambulance siren or train whistle changes pitch as it travels past a stationary observer. Compared to the single note it produces when the ambulance is stationary, the siren sounds higher as the ambulance approaches the observer and lower as it moves away from the observer.

Like the water waves described in Chapter 14, sound waves are affected by the motion of the source or receiver. As the source moves toward the receiver, the sound wave compressions become more closely spaced and the receiver hears a higher pitch than that being emitted by the source. As the source recedes, the compressions are more widely spaced, and the receiver hears a lower pitch (Figure 15-16). The frequency of the sound we hear depends on the relative velocity between the source and ourselves.

Figure 15-16
The frequency of the sound we hear depends on the velocity at which the ambulance travels.

"I LOVE HEARING THAT LONESOME WAIL OF THE TRAIN WHISTLE AS THE MAGNITUDE OF THE FREQUENCY OF THE WAVE CHANGES DUE TO THE DOPPLER EFFECT."

© 1980 by Sidney Harris.

ELECTROMAGNETIC WAVES

All wave motion is defined as the transfer of energy without the transfer of matter. Nevertheless, mechanical waves do depend on the presence of matter. A bell clanging in a vacuum produces no sound. Electromagnetic waves are different—they do not depend on matter for their existence. Light is transferred continuously between the sun and the earth. Radio and television signals provide astronauts a link with home. X rays, gamma rays, visible light, and radio waves continually bring us messages from the stars, crossing an emptiness more vast than we can imagine.

Describing Electromagnetic Waves

Electromagnetic waves are created by vibrating electric charges. A vibrating electron can induce vibrations in another electron, which in turn induces vibrations in yet another electron, and so forth. Energy, but no mass, is transferred from one electron to another. Because electrical interactions occur at a distance, this energy can be transferred across empty space (as well as across matter). Electromagnetic waves are very hard to imagine.

To provide a medium analogous to water for water waves or air for sound waves, physicists use the concepts of electric and magnetic fields. Electrically charged objects create an electric field around themselves. This field describes their ability to influence other charged objects across empty space. When the object vibrates, it produces a disturbance that moves through this electric field, much as a water wave moves across water. When the disturbance in the field reaches another charged object, it can induce that object to vibrate. In a sense, electric fields provide us with a way of picturing the empty space surrounding the charged objects and the way in which energy can be transferred between them.

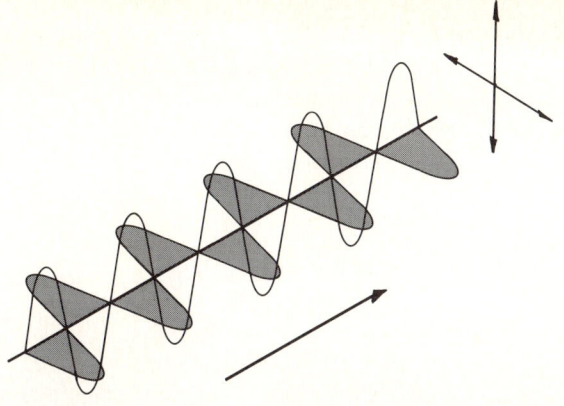

Figure 15-17
Two fields, one electric and the other magnetic, vibrate at right angles to one another as an electromagnetic wave is transmitted in the third direction. All electromagnetic waves are transverse.

Electromagnetic waves are actually a result of both electric and magnetic fields. A moving electrical charge produces a magnetic field and an electric field. This magnetic field can influence another moving electrical charge in yet a different way than the electrical field. Consequently, two disturbances are actually propagated—one through the electric field and a second through the magnetic field. Physicists picture electromagnetic waves in terms of changes in these two fields, as illustrated in Figure 15-17. The energy transferred by the wave travels perpendicular to the direction in which either field vibrates. All electromagnetic waves are transverse.

In a vacuum, all electromagnetic waves travel at the same speed—3×10^8 m/s, or 186,000 miles per second. In matter, electromagnetic waves travel at somewhat lower speeds. Like sound, light travels at different speeds in air at different temperatures. Representative speeds are included in Table 15-4. For distances on earth, these speeds are enormous. Consequently, electromagnetic waves seem to us to travel instantaneously from one point to

Table 15-4 Speed of Light in Various Materials

Material	Speed of Light (m/s)
Vacuum	2.998×10^8
Air	2.997×10^8
Liquids (20°C)	
Water	2.250×10^8
Glycerine	2.040×10^8
Carbon disulfide	1.840×10^8
Solids	
Ice	2.290×10^8
Quartz	1.940×10^8
Crown glass	1.970×10^8
Diamond	1.240×10^8

another. We can measure the time it takes for a thunderclap to reach us, but the lightning flash appears to take no time at all. Only when we look at distances as large as those between planets, stars, and galaxies does the finite speed of electromagnetic waves begin to take on meaning. Light from the moon reaches us in a little over a second; light from the sun, in about 8 minutes.

SELF-CHECK 15D

The nearest star, Alpha Centauri, is 4.2×10^{16} m away. If we received a radio message from Alpha Centauri today, how long ago would it have been sent?

Light in Matter

Our young son has just realized that there is not another baby in the mirror. He reaches out and finds the cold hard surface of the glass rather than the soft skin he expected. The hiker has just realized that the lake in the distance is a mirage. With each step he takes, the lake fades farther into the distance. You left your bicycle in the sun again—its black seat is hot. Reflection, refraction, and absorption describe the behavior of electromagnetic waves as effectively as they describe waves in sound and water.

We see most objects by virtue of reflected light. Light from a lamp strikes the table, for example, and is reflected to our eyes. Light waves moving parallel to one another before striking the table are reflected in all directions (Figure 15-18). Within this apparent chaos lies the information our eye-brain system uses to tell us that we are looking at a table. Light reflected by each point along the table's surface obeys the law of reflection—the angle of reflection equals the angle of incidence. The uneven, bumpy surface of the table offers a different angle of incidence at each point. Consequently, the pattern of reflected waves carries information about each and every point on the table's surface—information we learn to interpret in a fraction of a second.

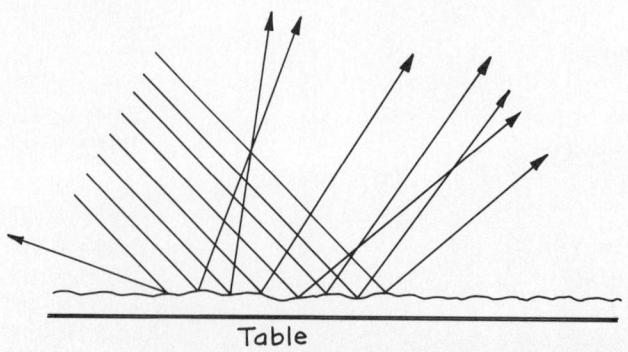

Figure 15-18

The bumps in the reflecting surface affect the way in which parallel rays of light are reflected. Each surface creates a unique pattern of reflected waves.

Electromagnetic Waves

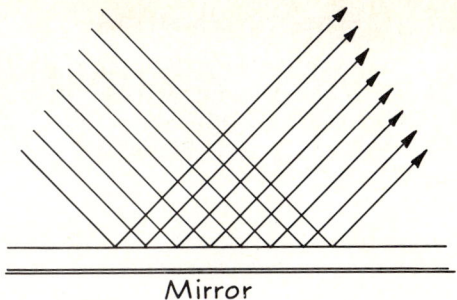

Figure 15-19
Mirror surfaces are so smooth that parallel rays of light remain parallel to one another after reflection. The mirror imposes no pattern of its own on the reflected light.

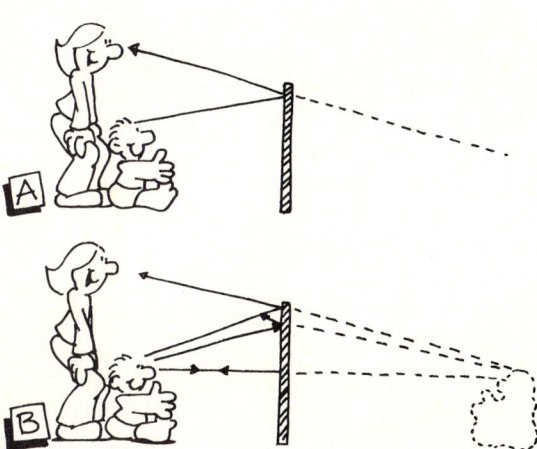

Figure 15-20
Our eyes trace light along straight lines. A light ray reflected by a mirror appears to have come from behind the mirror, as shown by the dotted line. We see an image that appears to be behind the mirror.

Mirrors allow us to see the baby instead of the mirror because they make little impression of their own on the light waves. Light waves moving parallel to one another before striking the mirror are still parallel once they are reflected (Figure 15-19). The mirror surface is so smooth that it imposes little, if any, pattern on the light it reflects. Consequently, light that has first been reflected by a baby will still carry information about the baby when it is reflected by the mirror. We see the baby instead of the mirror.

The information reflected by the mirror creates an image of the baby that appears to be located behind the mirror. This arises from the way we interpret light's motion. Experience with light sources has led us to expect light to travel along straight lines. Figure 15-20(a) shows one path the light waves travel as they strike the baby's head and are then reflected by the mirror. As we gaze into the mirror, the reflected light seems to travel along a straight-line path that originates behind the mirror, as shown by the dotted line. Our eyes trace that straight-line path, not the real path traveled by the light. Light waves traveling along paths that diverge from a single point on the baby's head appear to be diverging from a point behind the mirror (Figure 15-20(b)). The image we see is formed at that point. The same is true for light emerging from all other points along the baby's body. Thus, a complete image is formed behind the mirror. Only with experience do we come to realize that we are seeing just an image and not the real object.

Figure 15-21
Light is refracted as it strikes the glass surface of a prism.

Figure 15-22
Light is refracted toward the normal as it enters the medium.

SELF-CHECK 15E

Advertisements describe the effectiveness of furniture waxes in terms of the clarity of the images you see. Explain how a wax can help turn a table into a mirror.

Figure 15-21 shows several narrow beams of light as they strike a prism. A small part of the light is reflected at the glass surface, but most is transmitted into the prism. Light that is perpendicular to the surface continues to travel along the same path. Light incident at other angles, however, bends. Light, like sound and water waves, is refracted as it moves from one medium to another. You have seen this phenomenon often in swimming pools: The submerged part of your body looks shortened, due to the bending of light as it emerges from the water into the air.

Like other wave phenomena, light waves are refracted due to the change in wave speed that occurs as waves enter new media. Light travels more slowly in glass than in air. If we trace a single beam of light (Figure 15-22), we see the light refracted toward the normal as it enters the medium and away from the normal as it moves back into the air. Light waves, however, display a characteristic called **dispersion.** The speeds at which light waves move in matter depend on their wavelengths. The differences are small, but sufficient to refract each wavelength of light at a slightly different angle. White light, a mixture of all wavelengths, is separated into its colors. Red light

is refracted the least; violet light, the most. If we replace the prism with a drop of water, we have a rainbow!

Like sound, light waves experience atmospheric refraction due to the fact that the speed of light in air increases as the temperature of air increases. Mirages are caused by this refraction. Air near the surface of hot sand is warmer than the air above it. Light traveling downward from the sky is refracted as it encounters warmer layers, until its path is actually bent back upward (Figure 15-23). Hikers see this reflected light and trace its path backward along a straight line to the ground. Seeing an image of the sky on the ground, they naturally assume it is a reflection from the surface of a lake. Drivers see the same effect as they drive along asphalt highways on hot, sunny days.

Electromagnetic waves are absorbed when the energy stored in the electromagnetic disturbance is dissipated into thermal energy. The hot bicycle seat provides us evidence of how effectively this process can occur. Infrared radiation and visible light have wavelengths that correspond roughly to the sizes of atoms and molecules. Consequently, the energy stored in and transmitted by these electromagnetic waves is easily transformed into the thermal energy associated with the motions of atoms and molecules. In Chapter 10 we called this process radiation-absorption.

Figure 15-23

Mirages are formed when light from the sky is refracted upward by warmer layers of air. Our eyes trace the refracted light along straight lines, creating an image that appears to be located on the ground in front of us.

"SEEING" THE BAIT

In 1916 William Zeigler thought of a way to use the law of reflection to catch a fish. His artificial fish bait looks similar to many others except for the mirror (5) in the middle. A fish would swim near the bait and see its own reflection. Thinking that another fish would grab the bait before it could, the fish would bite, thus becoming someone's dinner. The law of reflection tells us that the angle of incidence equals the angle of reflection, so fish approaching the mirror from any angle other than straight on would not see their own reflections. One must wonder, then, about why Mr. Zeigler put the hooks on the back and bottom of his bait.

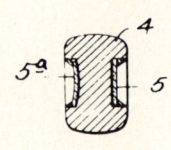

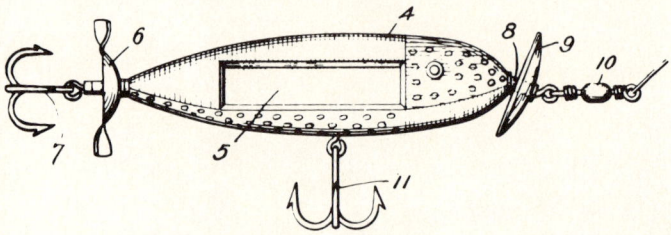

Standing Waves in Microwave Ovens

Oops . . . you forgot to take the casserole out of the freezer. Guess you will have to use the microwave oven. You shove the casserole in, close the door, set the timer, and push the button. Twenty minutes later it is done—steam rises as you lift the casserole out of the oven. You take a huge helping from the center and take your first bite. Ugh! It is still cold. A quick check shows that the casserole is plenty hot around the outside but still cold in several spots near the center. What happened?

Microwave cooking, one of the more recent additions to kitchen technology, is based on the principle of wave absorption. Microwaves, with a frequency of about 2×10^9 Hz, stimulate the water and fat molecules in foods to vibrate. The thermal energy generated by these vibrations cooks the food, often in a fraction of the time required in a conventional oven. Glass and plastic containers, as well as the surrounding air, absorb very little of the energy carried by the microwaves, so most of the energy goes directly into the food being cooked. One of the major disadvantages, however, has been the uneven cooking typical of many oven designs.

You can understand this uneven cooking in terms of standing waves. Microwaves generated by most microwave ovens have a wavelength of about 13 centimeters. Consequently, three to four complete wavelengths can fit into the oven cavity. To prevent leakage of microwaves to the outside, the oven's interior is lined with metal, which is an excellent reflector. The waves emitted by the microwave source interfere with the waves reflected from the sides, and a standing wave pattern is produced. Energy is distributed unevenly

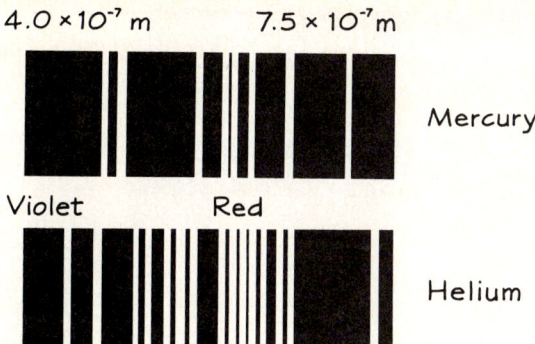

Figure 15-24
Light emitted by each chemical element has a light spectrum characteristic of that element.

about the oven's interior, with most of the energy concentrated at antinodes and little energy located at nodes. Your casserole's cold spot was located at a node.

Because of safety concerns, metal interiors are likely to remain an essential feature in microwave oven design. It would therefore be impractical to try to eliminate standing waves. Most recipes suggest rotating the food once or twice during cooking, thus rotating areas from nodes to antinodes for more even energy absorption. Some manufacturers have added rotating platforms that move the food continuously as it is being cooked. A second solution is to move the standing wave pattern itself. A rotating microwave source produces a standing wave pattern that moves about the food, moving nodes to different sections of the dish. Rotating fans have also been used. These fans reflect microwaves in continually changing directions, so that the standing wave patterns change as well.

The Expanding Universe and Police Radar

Electromagnetic waves, like mechanical waves, experience a shift in apparent frequency when either the source or the receiver is moving. This electromagnetic Doppler effect is used to measure the speeds of objects that are approaching or receding from us—objects as diverse as galaxies and automobiles.

For more than a century, astronomers have analyzed the light emitted by stars. The visible light emitted by a glowing gas, such as hydrogen or helium, is a mixture of different wavelengths. Each gas has its own characteristic mixture. When the light is passed through a prism, the different wavelengths are separated, creating a pattern characteristic of that gas, called its **line spectrum** (Figure 15-24). By comparing light from stars and galaxies with the spectra of known gases, astronomers can tell what chemical elements are present.

Astronomers found perplexing data when they analyzed light from very distant stars and galaxies. A spectrum would show all the lines characteristic

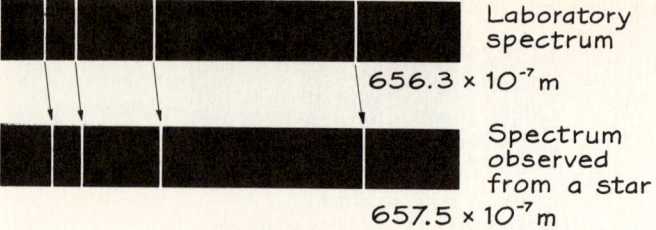

Figure 15-25
Light emitted by distant stars and galaxies shows shifts in frequencies characteristic of the Doppler effect. Astronomers believe that these shifts arise from the motion of stars and galaxies outward from the origin of the big bang.

of hydrogen, for example, but the lines were uniformly shifted to lower frequencies (Figure 15-25). Described in terms of wavelength, the lines were shifted toward the red end of the spectrum; thus the effect was called the **red shift.** The red shift is consistent with a model of the universe in which all galaxies are moving outward, away from one another. Like the sound of a train whistle as the train moves away from you, light is shifted toward lower frequencies as galaxies recede from us.

A more down-to-earth illustration of the electromagnetic Doppler effect is police radar. A police officer who wants to measure the speed of a car uses the shift in frequency of radar waves reflected by the moving car. Because of the car's motion, the reflected waves are shifted compared to the frequency emitted by the radar unit (Figure 15-26). By comparing the frequency of the radar emitted by the antenna with the frequency of the reflected waves, a computer built directly into the radar system can calculate the car's speed. As the car's speed increases, so does the magnitude of the shift in frequency.

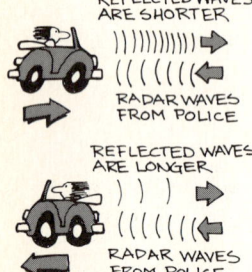

Figure 15-26
Police radar measures the speed of cars by looking at the shift in frequency of radar waves reflected by the moving car.

When the police car is stationary, the frequency change provides all the required information with which to calculate the car's speed relative to the ground. When both cars are moving, however, the frequency shift allows calculation only of the relative speed between the two cars. In order to determine the car's speed relative to the ground, moving radar units emit two signals. One is reflected by the moving car, allowing the radar unit to calculate the relative speed between the two cars. The second signal is reflected by the road and allows the radar unit to calculate the police car's speed relative to the ground. Given the two speeds, the computer can calculate the car's speed relative to the ground. From galaxies to automobiles, the Doppler effect allows us to measure the relative speed between objects.

The success of the wave model in describing a wide range of phenomena is indeed impressive. Starting with waves in springs and waves in water, we have been able to build mental pictures of sound and light waves. We cannot see the vibrations of atoms and molecules that combine to produce sound; but we can understand how sound behaves in terms of these vibrations. Wave reflection, refraction, and superposition provide an effective way of understanding echoes, sound transmission at night, and the design of musical instruments. We cannot see the electric and magnetic fields whose motions create electromagnetic waves, but we can understand how light, radar, and microwaves behave in terms of waves. Chapter 16 examines two additional wave phenomena—diffraction and interference—and their role in convincing physicists of the wave nature of light.

Right Off the End of the Piano!

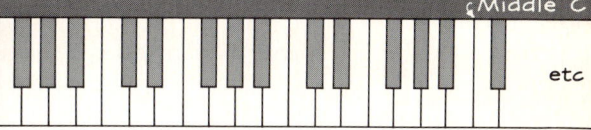

If electromagnetic radiation could be emitted by playing a piano, the visible spectrum would occupy about one full octave. The Doppler shifts observed by astronomers can be huge—literally taking us right off the end of the piano!

Red shifts have proven enormously useful to astronomers, providing what seems to be profound evidence in support of the big-bang model of the universe (see Chapter 8). As we look deep into space, the extent of the red shift becomes staggering. It is hard to appreciate the size of these shifts, however, without a benchmark—a reference frame for comparison. Fortunately, the piano provides a useful analogy.

Suppose that a train whistle emits a pitch corresponding to middle C on the piano. As the train moves away from us at 25 m/s (90 kilometers per hour), its whistle sounds about one half-step (one note) lower to us than the sound emitted by the train. We hear B below middle C. For sound this shift is one of the larger Doppler shifts we encounter. Most are far less. To hear a shift of a full octave, for example, the train would have to move about 500 kilometers per hour. Most ordinary objects do not move at such high speeds. For astronomers, however, Doppler shifts of light can be this large or even larger. Let's compare their observations with the more familiar shifts observed for sound.

We do not normally think of electromagnetic radiation in terms of octaves. However, for comparisons involving the Doppler shift, such a construct is quite useful. A musical octave ranges from one frequency to twice that frequency. Middle C, for example, has a frequency of 256 Hz. The C below middle C has a frequency of 128 Hz. In the electromagnetic spectrum, visible light has a range equivalent to about one octave (4×10^{14} Hz to 8×10^{14} Hz). Above and below this octave, we would find the other forms of electromagnetic radiation.

Consider the red shifts observed by astronomers. Beginning close to home, nearby galaxies emit light that is red-shifted about one step or so. The light drops in frequency by one note on our electromagnetic piano. As we progress to more distant galaxies, the red shift becomes larger. The farthest galaxies that have been observed so far have a red shift equivalent to the change of an octave. These shifts indicate that distant galaxies are moving away from us at speeds of about 6×10^7 m/s. If we look still deeper in space, the most distant quasars have red shifts equivalent to a couple of octaves. They are moving away from us at enormous speeds—up to 90% of the speed of light.

In 1965 Arno Penzias and Robert Wilson detected radio waves that seemed to come from even deeper in space. This radiation is coming to us from all directions, so we call it *background radiation*. Its red shift is so large that it falls right off the end of our piano! Presumably released at the time of the big bang itself, this radiation and its red shift hold clues to events that occurred an estimated 20 billion years ago.

CHAPTER SUMMARY

Wave phenomena can be separated into two categories: mechanical waves and electromagnetic waves. *Mechanical waves* are produced in matter in any form—solid, liquid, or gas. One piece of matter is displaced and the disturbance is transmitted throughout all parts of the surrounding matter. Electromagnetic waves are produced in empty space as well as in matter. Vibrating electrical charges produce disturbances that are transmitted across space. By ordering waves according to frequency, the *wave spectrum* organizes the different types of waves within each category.

Mechanical waves include water waves, earthquake waves, and sound waves. *Sound waves* are longitudinal waves. Displacing one atom or molecule produces a series of compressions and rarefactions that are transmitted by the surrounding matter. We use the terms *loudness* and *pitch* to describe our perceptions of the amplitudes and frequencies of sound waves. Sound waves are reflected in echoes, refracted by the atmosphere, and absorbed by trapped pockets of air. All musical instruments are designed to produce standing waves that establish the pitch and character of musical sounds. The lengths of the vibrating strings or air columns determine the fundamental frequencies. The fundamental frequency establishes the pitch; higher harmonics contribute to the character of the sound. Sirens and train whistles provide daily examples of the Doppler effect.

Electromagnetic waves can be described as disturbances in the electric and magnetic fields that surround a moving electric charge. Light, radar, and microwaves are three of the many kinds of electromagnetic waves. They differ from one another in frequency. Different *colors* of light have different frequencies as well. Light can be reflected, refracted, and absorbed by matter. The light reflected from objects to our eyes allows us to see the objects. Reflections from very smooth surfaces give rise to mirror images. Atmospheric refraction leads to the formation of mirages. Absorption of light and infrared radiation leads to the process of heat transfer called radiation-absorption. Absorption of microwaves is used to heat food in microwave ovens. Reflected waves in microwave ovens produce standing waves, resulting in uneven cooking. The electromagnetic Doppler effect is used extensively in measuring speeds—speeds of galaxies as well as speeds of cars.

ANSWERS TO SELF-CHECKS

15A. Human speech involves frequencies around 200–400 Hz. Microwaves include frequencies from 10^{10}–10^{12} Hz. Human speech involves mechanical waves. Microwaves are electromagnetic waves.

15B. AM waves have a lower frequency, and hence a longer wavelength, than FM waves. To be effective, an energy receiver needs to be about the same size as the waves it is to detect. Since AM and FM waves have different wavelengths, the radio antenna will need to be adjusted differently for best reception of each range. The antenna will have to be longer for AM reception.

15C. a. The fundamental frequency will have a wavelength that is four times the length of the air column.

b. As the bottle fills up, the length of the air column shortens. Consequently, the fundamental frequency will have a shorter wavelength and hence a higher frequency. The pitch of the sound goes up.

15D. $$\text{Time} = \frac{\text{distance}}{\text{speed of light}} = \frac{4.2 \times 10^{16} \text{ m}}{3 \times 10^8 \text{ m/s}} = 1.4 \times 10^8 \text{ s}$$

One year is equal to 3.2×10^7 s, so about 4.4 years have elapsed since the message was sent from Alpha Centauri.

15E. The furniture wax fills in the low spots in the wood grain, producing a much smoother surface. As the surface becomes smoother, light waves are reflected more regularly and an image is formed.

PROBLEMS AND QUESTIONS

A. Review of Chapter Material

A1. Define the following terms:
Mechanical waves
Wave spectrum
Electromagnetic waves

A2. Give examples of mechanical waves and electromagnetic waves.

A3. Describe how the size of a wave receiver is related to the wavelength of a wave. Explain the reason for this relationship.

A4. Describe the differences between mechanical and electromagnetic waves.

A5. Describe how a curved surface can focus waves.

A6. Use refraction to explain why sound travels further along the ground at night than during the day.

A7. Explain how the frequency of a sound wave is related to the size of the wave source.

A8. Explain how the Doppler effect is related to the changing pitch of a siren as it passes at high speed.

A9. How are electric and magnetic fields used to describe the motion of electromagnetic waves?

A10. Use the law of reflection to show how an image is formed by a mirror.

A11. Why do standing waves affect microwave cooking?

A12. Describe how the Doppler effect allows us to measure the speed of distant galaxies and passing automobiles.

B. Using the Chapter Material

B1. In Chapter 14 we studied waves on springs. Are these waves mechanical or electromagnetic? Explain how you reached your conclusion.

B2. Cyclotron radiation is a type of wave produced by electrons that are traveling in a circle. Is cyclotron radiation a mechanical or electromagnetic wave? Explain how you reached your answer.

B3. One way in which astronomers study the universe is to detect radio waves emitted by the hydrogen molecule. The wavelength of this wave is 0.21 m. Would you expect the astronomers' detectors to be larger or smaller than AM radio antennae? Why?

B4. A microphone is a detector of sound. Would you expect the size of a microphone to be greater, smaller, or approximately the same size as the human ear? Explain.

B5. A supernova is the explosion of a star. If a supernova occurs somewhere in space, will sound from it be detected on earth? Can light from the supernova be detected here?

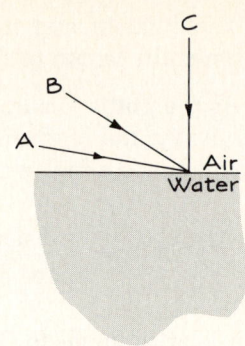

B6. Medieval musical instruments called krumhorns are shown above. One is an alto; the other is a higher-pitched soprano. Using physical principles, describe why you can determine which is the alto.

B7. Sound can be heard across a still lake much better than across a lake which has lots of waves. Apply reflection of sound to explain why.

B8. A car is traveling down the road with its horn blaring. What would you need to know so that you could measure the car's speed from the Doppler shift of the horn's sound?

B9. What frequency of electromagnetic wave would be produced by electrons vibrating at 60 Hz? 120 Hz? 240 Hz? What would be the speed, in a vacuum, of each of these electromagnetic waves?

B10. Each of the light waves shown below is striking a mirror as shown. Draw the direction of the reflected wave.

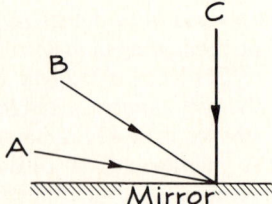

B11. Light travels more rapidly in air than in water. Draw the direction of the light waves in water for each of those shown in air in the next column.

C. Extensions to New Situations

C1. When lightning occurs, it creates two types of waves—sound and light. Because these waves travel at different speeds, their arrival times can be used to estimate the distance to the lightning flash. Light travels at 3×10^8 m/s and sound moves at about 300 m/s. Suppose we hear thunder about 2 s after we see the lightning flash.
   a. How far does sound travel in 2 s?
   b. How long would it take light to travel this same distance?
   c. Approximately how far away did the lightning flash occur?
   d. Can you make up a general rule to determine the approximate distance to a flash of lightning?

C2. The sitar, an Indian musical instrument, has a number of strings that are never plucked by the player. These strings are called *sympathetic strings* and are identical to strings that are plucked. Use the idea of resonance to explain why the sympathetic strings vibrate (and produce music) even though they are never touched by the sitarist.

C3. Could you ever sing and cause a stringed instrument to sound even though you do not touch it? Explain how and why.

C4. In a television commercial for a particular brand of audio tape, a recording of a singer is shown shattering a wine glass. Use the concept of resonance to describe how one can shatter glass with sound.

C5. The speed of electromagnetic waves is used to define astronomical distances. The distance that light, or any other electromagnetic wave, travels in 1 year through

© 1973 by
Sidney Harris.

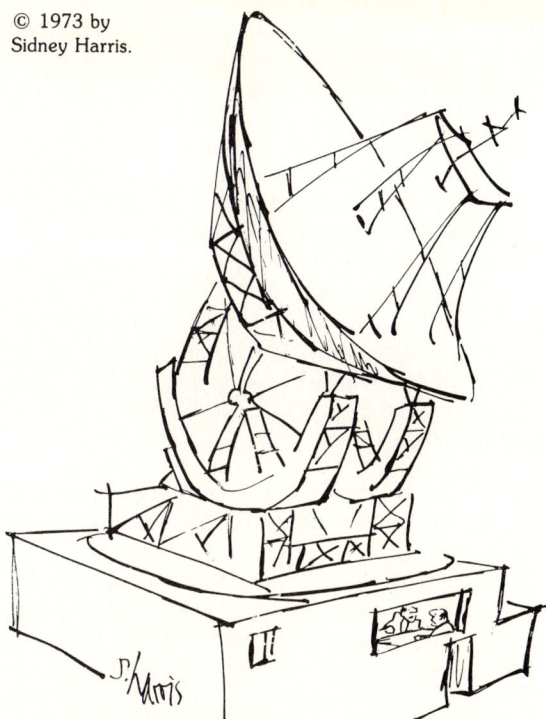

"AS I UNDERSTAND IT, THEY WANT AN IMMEDIATE ANSWER. ONLY TROUBLE IS, THE MESSAGE WAS SENT OUT 3 MILLION YEARS AGO."

empty space is called a *light-year*. The nearest star, Alpha Centauri, is 4 light-years away.
  a. How far away is Alpha Centauri in kilometers?
  b. In 1974 the radio telescope at Arecibo, Puerto Rico, sent a message toward a star cluster called M13, which is 27,000 light-years away. Within a day the following telegram arrived at Arecibo: "Message received. Help is on its way. M13." Why did the astronomers think this message was a prank?

C6. Suppose a movie set has a mirror on it. You wish to take a picture of the actor's image in the mirror but do not want images of the camera, lights, or crew. How should you arrange the mirror, actor, lights, and camera?

C7. Alternating current in the United States oscillates at 60 Hz.
  a. What frequency electromagnetic wave is produced around your lamp cord?
  b. This frequency must be all around us because of all the wires. Why do we not normally receive it on our radios and televisions?

C8. Each of our hearts contains a natural pacemaker, which stimulates our heart to beat about 60 to 80 times per minute. If this natural method fails, an electronic pacemaker can be inserted. The electronic device detects electromagnetic nerve signals in the heart. If no signal is detected, the electronic pacemaker fires. Why could it be dangerous or fatal for an electronic pacemaker to be near sources of strong electromagnetic waves, like microwave ovens?

C9. Your neighbor is playing music rather loudly, so you close your windows. Now you hear only the bass portion of the music.
  a. What size objects will vibrate in resonance with the bass?
  b. What objects of this size are between you and the music?
  c. Would you expect these objects to vibrate at high or low frequencies?
  d. Why do the bass pitches but not the higher pitches reach you when the window is closed?

C10. Astronomers have looked carefully at the spectrum of light emitted by the sun. They have discovered that rays of light from edges A and B have shifts in their frequencies. The frequencies emitted at A are greater than normal, while those from B are less.

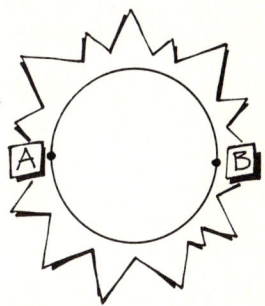

  a. Which side, A or B, is moving toward the earth?
  b. Which side is moving away from the earth?

c. What do these measurements tell you about the sun's motion?

C11. In Chapter 14 we stated an equation for the frequency shift of a wave emitted by a moving source. This equation is used to calculate the speed of a car from which police radar has been reflected. For this problem, we can rewrite this equation as:

$$\text{Speed of car} = \frac{(\text{speed of light}) \times (\text{change in frequency})}{\text{frequency of radar}}$$

Suppose a radar unit has a frequency of 10^{10} Hz.

a. What are the speeds of cars that cause frequency shifts of 100 Hz? 300 Hz? 700 Hz? 1000 Hz? (The interstate speed limit equivalent is about 25 m/s.)

b. When the radar unit is moving, the speed of the car is actually the relative speed between the car and the radar unit. Would the frequency shifts in part (a) increase or decrease if the car were traveling east and the radar unit were traveling west? Explain.

c. When moving radar is used, two frequency shifts are measured. One frequency shift is measured from radar waves reflected from the moving car. A second frequency shift is measured from radar waves reflected from the road. Suppose that the car and radar unit are moving in opposite directions. The frequency shift measured from radar waves reflected from the moving car is 750 Hz and the shift reflected from the road is 500 Hz. What is the speed of the car relative to the radar unit? The speed of the radar unit relative to the road?

d. Use the equations for relative velocities in Chapter 3 to determine if the driver of the car deserves to be cited for speeding.

C12. Singing in the shower has long been known to change the sound of one's voice compared to singing in an auditorium.

a. What range of frequencies would you expect to be enhanced by resonances in the shower room?

b. In terms of energy output, why is it particularly useful to enhance these frequencies?

D. Activities

D1. If you can borrow an oscilloscope, look at the waves produced by your voice or music.

D2. The next time you are in the bathroom, sing a scale. Can you identify the resonant frequencies of the bathroom?

D3. Visit a stereo store and learn how the designers of the store considered resonances in designing the speaker demonstration room.

D4. The next time you close your window to keep out your neighbor's loud music, place your hand on the wall or window. Can you feel it vibrate?

D5. Tune your radio to a distant station. Change the antenna length until you find the best reception. Is this length different for FM and AM reception?

D6. Investigate different brands of microwave ovens. Ask a salesperson to explain how the ovens avoid standing waves (usually called hot spots) and how they keep the energy inside the oven. Listen to the sales pitch carefully. Then evaluate the salesperson's knowledge of physics.

Interference and Diffraction

Despite its success in describing mirrors and mirages, the wave model of light met with formidable opposition during the seventeenth and eighteenth centuries. Although the wave model adequately explained the reflection, refraction, and absorption of light, a particle model of light offered what seemed to many physicists to be equally valid explanations of these phenomena. The ensuing wave-particle controversy over the nature of light offers a glimpse into how scientists deal with conflicting models.

Waves and particles offer two mechanisms for describing how energy gets from one place to another. Waves transfer energy but not matter. Particles transfer energy with matter. When phenomena are visible, distinguishing between the two models is relatively simple. Cars transfer energy like particles; springs transfer energy like waves. But when the phenomena are not visible, like sound, then distinguishing between the two models is more difficult. When we turn to light, we can see neither particles that carry the energy nor the medium responsible for transmitting the waves. The only way to distinguish between the two models is to find a behavior predicted by one model

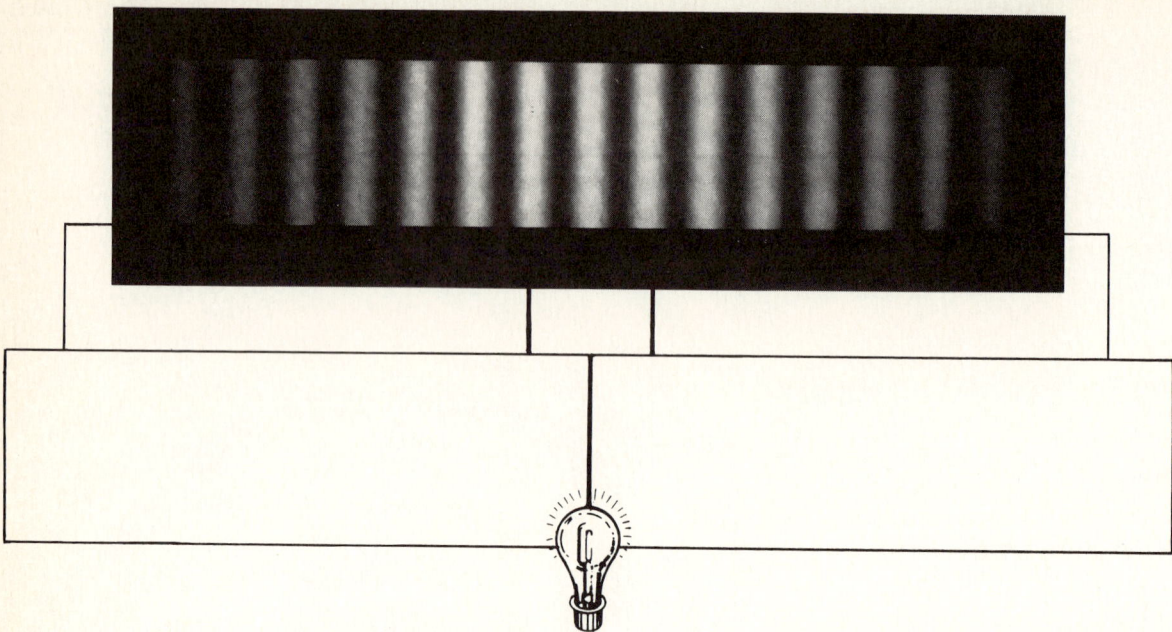

Figure 16-1

Young placed a barrier with a single slit and a second barrier with two slits between a light source and a screen. Instead of two bright slits, he saw a series of bright and dark bands spread across the screen.

but totally unexplained by the other. For the wave-particle controversy surrounding light, two phenomena—diffraction and interference—provided the critical test. At the turn of the nineteenth century these two phenomena could be explained only in terms of wave superposition.

In this chapter, we examine *interference* and *diffraction* of light in detail. Interference and diffraction effects were observed as early as the sixteenth century, but it was not until 1801 that Thomas Young explained light interference in terms of wave superposition. Later, diffraction was also explained in terms of the spreading and superposition of waves. You can see the effect of interference in the rainbow of colors reflected by oil patches and in the recently developed technology of *holography*. Diffraction of light plays an important role in photography and microscopy. Diffraction and interference were significant in temporarily resolving the wave-particle controversy, a controversy that would reappear as physicists began looking more closely at the structure of matter.

INTERFERENCE PATTERNS

In 1801 Thomas Young conducted what has become known as the Young double-slit experiment. As shown in Figure 16-1, Young placed two barriers between a light source and a screen. The first barrier had a single opening through which light could pass; the second had two such openings arranged side by side. The light transmitted by these two barriers produced a surprising pattern on the screen. Instead of two bright slits, Young saw a series of bright and dark slits spread across the screen. One way to understand these results is to look for similar patterns in other phenomena.

Figure 16-2
Circular waves produced by two sources spread out and combine to form a two-dimensional pattern. If we label the regions where nodes and antinodes occur along one dimension, we see Young's interference pattern.

Interference Patterns Occur with Waves

The pattern of bright and dark regions in Figure 16-1 is, in some ways, analogous to the antinodes and nodes formed by standing waves. The energy in standing waves is distributed unevenly. Nodes are areas in which no energy is found; antinodes are areas in which most of the energy is concentrated. If we think of these areas on the screen in terms of standing waves, the dark areas are nodes and the bright areas are antinodes. However, instead of being restricted to one dimension, like a standing wave on a rope, the bright and dark regions observed by Young are spread over two dimensions.

We can pursue this analogy further by looking for interference patterns in water waves. One way to create interference patterns in water is to attach identical beads to a piece of wood which can be vibrated at a constant rate. If this device is placed so that the two beads just touch the surface of the water with each vibration, each bead produces circular waves that spread continuously across the surface of the water. These waves overlap to produce a two-dimensional interference pattern (Figure 16-2). If we draw a line across the water's surface and label the regions where nodes and antinodes occur, we see the same pattern that Young observed with light. Nodes and antinodes alternate across the entire screen.

Use of Superposition of Waves to Explain These Patterns

Superposition of waves explains the patterns of nodes and antinodes shown in Figure 16-2. The two beads produce identical circular waves. As the two waves spread out and overlap with one another, constructive interference occurs wherever two crests or two troughs meet. Destructive interference occurs where a crest from one wave meets a trough from the other.

We can locate regions of constructive and destructive interference by comparing the distances the waves travel. Figure 16-3 gives an overhead view of the interference pattern. The most obvious region in which constructive interference occurs is the center line AB between the two wave sources S_1 and S_2. If we compare the distance traveled by one wave, S_1C, with the distance traveled by the second, S_2C, we find them to be equal. As the two waves arrive at C, a crest from one meets a crest from the other, and constructive interference occurs. The same result is true for any point along the

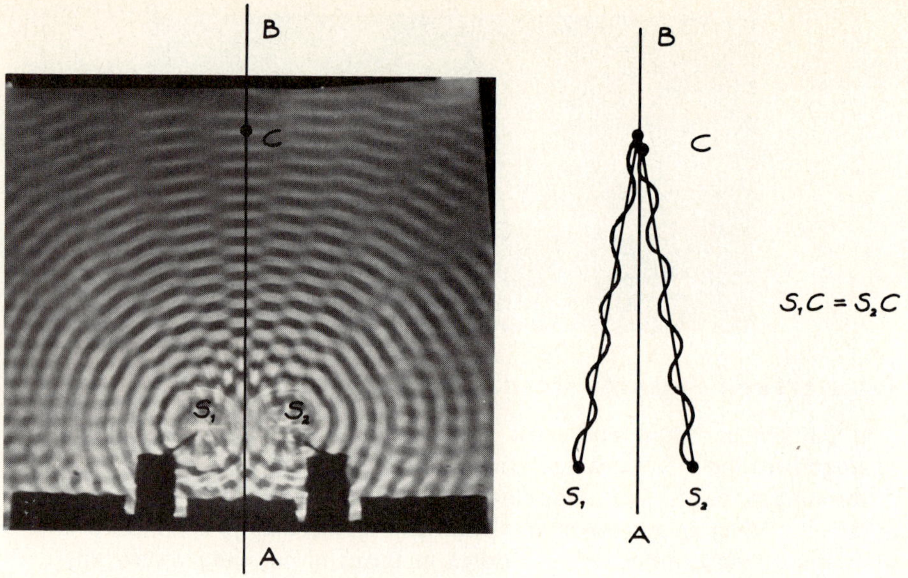

366 Chapter 16. Interference and Diffraction

Figure 16-3
Since waves from each source travel the same distance, constructive interference occurs along the center line, AB.

Figure 16-4
The type of interference, constructive or destructive, depends on the difference in the distances traveled by waves from S_1 and S_2. Destructive interference occurs at point D. Constructive interference occurs at point E.

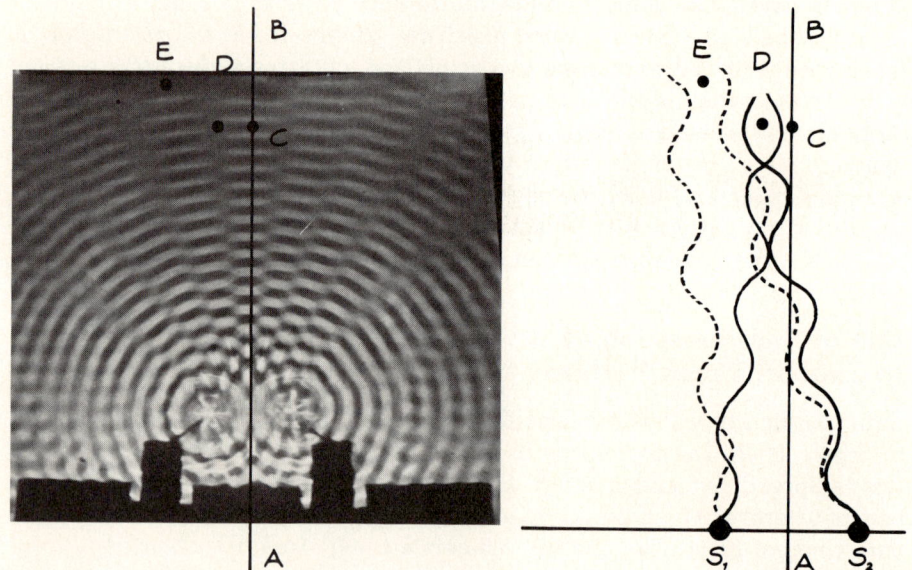

center line, AB. The center line locates a series of double-amplitude crests and double-amplitude troughs.

To reach a point slightly to the left or right of the center line, one wave has to travel farther than the other. The magnitude of this extra distance determines whether the two waves interfere constructively or destructively. At point D in Figure 16-4, the wave emitted at S_2 has traveled a distance that is one-half a wavelength farther than the wave emitted at S_1. If the wave from

Figure 16-5
Water waves help us create a mental picture of how light waves spread out and overlap with one another to produce the pattern of bright and dark bands on the screen.

S_1 arrives as a crest, the wave from S_2 arrives as a trough, and destructive interference occurs. At E the wave emitted at S_2 has traveled a distance that is one complete wavelength farther than the wave emitted at S_1. If the wave from S_2 arrives as a crest, the wave from S_1 also arrives as a crest, and constructive interference occurs. We can continue to apply this analysis to each point along the surface of the water.

Because they arise from constructive and destructive interference of waves, patterns like the one in Figure 16-2 are called **interference patterns.** The regions of constructive and destructive interference are called **interference bands.** A band of constructive interference occurs where the distances traveled by the two waves are equal or where they differ by a whole number of wavelengths. Bands of destructive interference occur where the distances traveled by the two waves differ by an odd number of half-wavelengths. Thus

Constructive interference: Extra distance $= n$ wavelengths

where $n = 0, 1, 2, 3 \ldots$

Destructive interference: Extra distance $= \dfrac{n}{2}$ wavelengths

where $n = 1, 3, 5, 7 \ldots$

Applied to each point along the surface of the water, superposition of waves correctly predicts the observed interference patterns.

Photographs like Figure 16-2 give us a way of visualizing what happens in Young's experiment with light. Light waves transmitted through the two slits spread out like the circular waves produced by the two beads. The waves overlap, interfering constructively to produce bright bands and destructively to produce dark bands (Figure 16-5). Applied to each point along the screen, superposition of light waves correctly predicts the bright and dark bands Young observed.

SELF-CHECK 16A

Assume that water waves have a wavelength of 0.6 cm. The distance from S_1 to point I is 3.6 cm and from S_2 to I is 4.5 cm. Predict whether the waves interfere constructively or destructively at point I.

Seeing Interference of Light

When Young conducted his experiments, interference patterns like that in Figure 16-1 had never been observed. You have probably never seen one when you have looked through two narrow openings. Young saw them because he chose narrow slits that were very closely spaced.

The separation between interference bands depends on three variables: how far away from the wave sources we observe them, the separation between the two wave sources, and the wavelengths of the waves emitted by the two sources. Figure 16-6 shows the effect of the distance from the wave sources. The farther away from the wave sources you observe the pattern, the more the interference pattern spreads out. Figure 16-7(a) and (b) shows

Figure 16-6
Interference bands spread out as we move away from the sources.

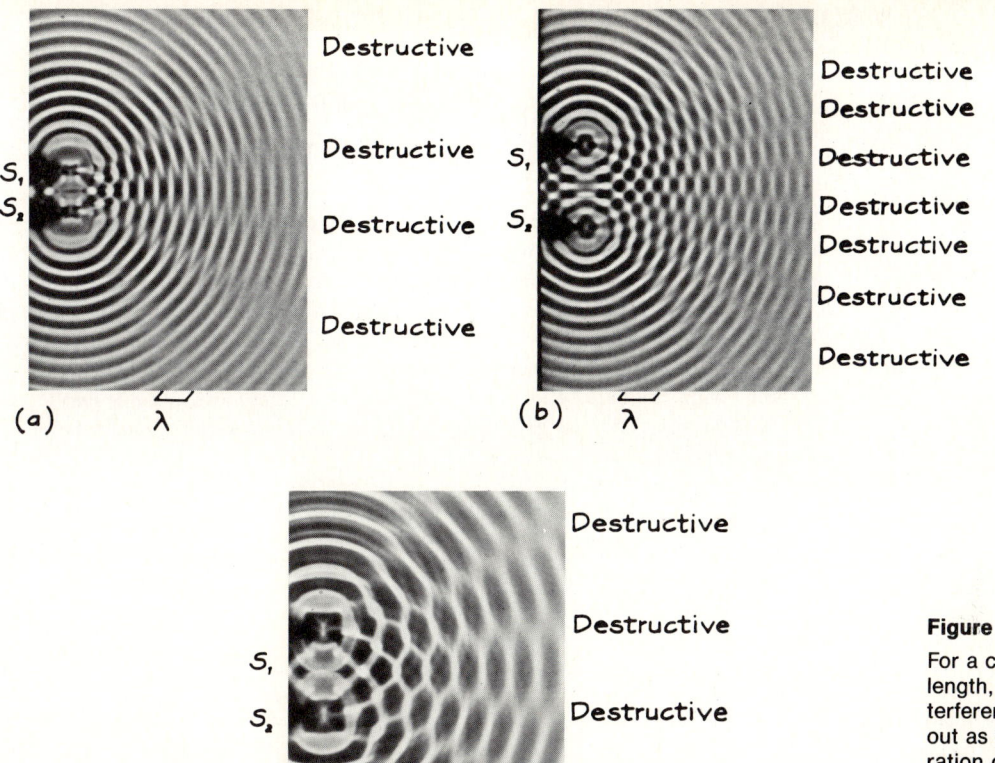

Figure 16-7

For a constant wavelength, **(a)** and **(b)**, interference bands spread out as the source separation decreases. For a fixed source separation, **(b)** and **(c)**, interference bands spread out as the wavelength increases.

the effect of the separation between wave sources. The wavelength is the same, but the wave sources are more separated in (b) than in (a). As the separation between sources increases, the interference bands become more compressed, more closely spaced. Finally, Figure 16-7(b) and (c) shows the effect of wavelength. As the wavelength increases, the interference bands spread out.

As mentioned in Chapter 15, the wavelength of visible light is extremely small—about 5×10^{-7} meters (m) long. The interference bands produced by waves of such small wavelengths are so closely spaced that our eyes cannot resolve them. Slits spaced about a centimeter apart will produce interference bands that are separated by a distance less than the thickness of a strand of hair. The only way to spread the pattern out is to place the slits as close together as possible and observe the pattern as far from the slits as possible. Two slits less than a millimeter apart will produce interference bands separated by several millimeters on a screen 1 m from the slits. Our eyes are able to resolve these separations. Young succeeded where others had failed partly because he experimented with slits that were extremely closely spaced.

A STEP FURTHER—MATH

DESCRIBING INTERFERENCE PATTERNS QUANTITATIVELY

We can describe interference patterns quantitatively in terms of the location of each interference band. As shown in the figure, the location of each interference band can be described in terms of its distance from the center line, AB. X_1, X_2, X_3, ... refer to the distances from the center of the pattern to the first band of constructive interference, second band, third band, and so on. $X_{1/2}$, $X_{3/2}$, $X_{5/2}$, ... refer to the distances from the center to the first band of destructive interference, the second band, the third band, and so on. Since the interference pattern is symmetrical about the center line AB, these distances can be measured to either the left or the right of center.

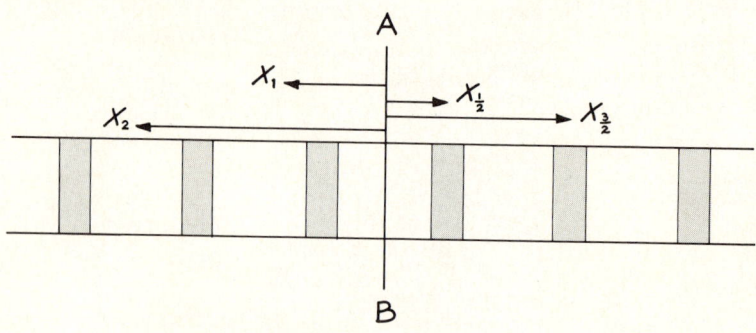

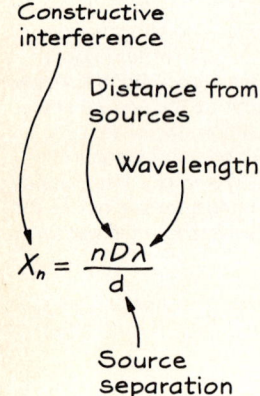

As described in the text, these distances depend on three variables: how far away from the wave sources we measure them, the separation between the two wave sources, and the wavelength of the waves emitted by the sources. We can combine these relationships into a single expression for the distance from the center line to the center of each interference band:

Constructive interference:

$$X_n = \frac{(n)(\text{distance from source})(\text{wavelength})}{(\text{separation of sources})} \qquad n = 1, 2, 3, \ldots$$

Destructive interference:

$$X_{n-1/2} = \frac{(n - \tfrac{1}{2})(\text{distance from source})(\text{wavelength})}{(\text{separation of sources})} \qquad n = 1, 2, 3, \ldots$$

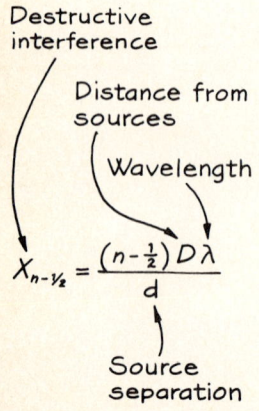

To see how these relationships can be applied to interference patterns, we examine the pattern produced by two slits 1 cm apart.

Let's calculate the distance from the center line to the middle of the *first* band of constructive interference. We need to know the distance from the source, the wavelength of light used to produce the interference bands, and the separation between the two slits. We assume that we are using red light with a wavelength of 6.5×10^{-7} m. The two slits are separated by 1 cm, a little less than half an inch. If we place a screen 1 m from the two slits, then the distance from the source is just 1 m. Substituting this information into our equation we have:

$$X_n = \frac{(n)(\text{distance from source})(\text{wavelength})}{(\text{separation of sources})} \quad \text{for } n = 1$$

$$X_1 = \frac{(1)(1 \text{ m})(6.5 \times 10^{-7} \text{ m})}{(1 \times 10^{-2} \text{ m})} = 6.5 \times 10^{-5} \text{ m}$$

6.5×10^{-5} m is equivalent to 0.065 mm. This distance is less than the width of a strand of hair! A separation of 1 mm produces a band of constructive interference 0.65 mm away from the center line. A separation of 0.1 mm produces a band 6.5 mm away—a distance easily noticed.

INTERFERENCE PHENOMENA

Young's double-slit experiment was important to nineteenth-century physics because it was the first phenomenon explained wholly in terms of light waves. Once understood in terms of wave superposition, Young's patterns provided the basis upon which a variety of everyday phenomena could be explained. More recently, they have provided the basis for a new technology—holography.

Interference Colors

The interference pattern in Figure 16-1 was created using a single color of light and captured on black-and-white film. Had we seen the actual pattern created by white light, we would have seen a central white region with bands of yellow, magenta, and blue-green to the right and left. You may have seen colors like these when looking at light through a lace curtain. Called **interference colors,** these bands arise from the different wavelengths of visible light that combine to form white light.

As shown in Figure 16-7(b) and (c), the spacing of interference bands depends on the wavelengths of the waves emitted by the two sources. As the wavelengths increase, the interference bands spread out. With its longer wavelength, red light (6.5×10^{-7} m) produces interference bands that are more spread out than those produced by blue-violet light (4.5×10^{-7} m). Although slight, this difference is enough to produce interference colors.

One way to explain interference colors is to look at the position on the screen where the various colors interfere destructively. Because of its longer wavelength, red light will interfere destructively farther from the center line than blue light. As shown in Figure 16-8, blue light interferes destructively at point A, green light at point B, and red light at point C. At point A we see white light minus blue light, which leaves yellow. At B we see white light minus green light, or magenta. Finally, at C we see white light minus red light, which leaves blue-green light. Bands of yellow, magenta, and blue-green are

Figure 16-8
Each color interferes destructively at a different location. The color we see is white light minus the color that interferes destructively.

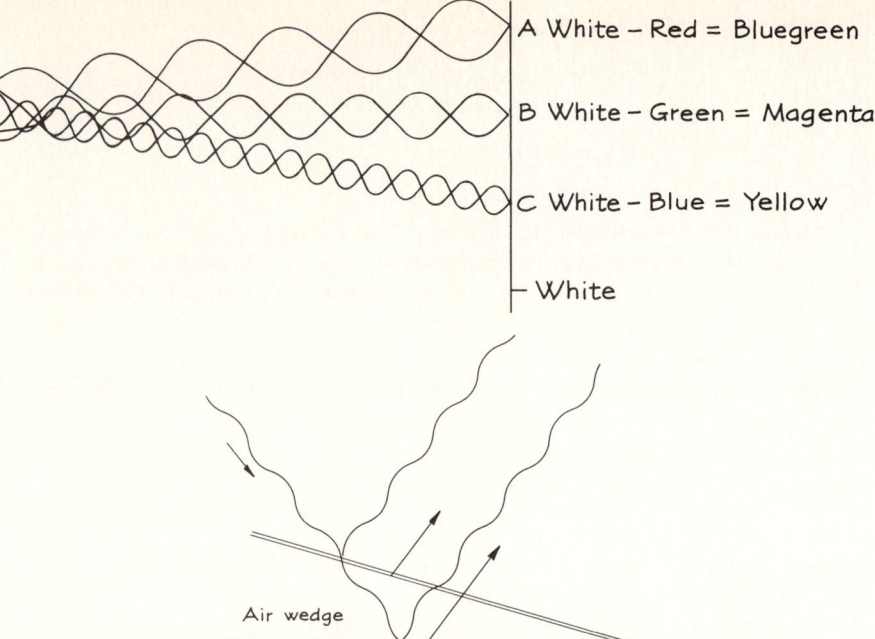

Figure 16-9
Light reflected by the bottom slide interferes with light reflected by the top slide.

formed on either side of a central white band, each the result of the absence of a particular wavelength due to destructive interference.

A lace curtain produces a similar effect because each opening in the lace acts as a slit. Rather than just two small openings, the curtain has many. The resulting pattern of colors is a little more complex. It is, however, produced by the same phenomenon—interference of white light as its waves pass through a series of narrow openings.

Interference Colors by Reflection

A puddle of water is covered by a thin layer of oil. As you walk by, you see a swirl of colors. Soap films left on dishes produce the same effect. You can see bands of color on soap bubbles as they float through the air. Water, soap films, and bubbles produce interference colors by reflecting light from two surfaces.

To understand how interference colors are produced by reflection, consider two pieces of glass arranged to form a V, as shown in Figure 16-9. Since glass both reflects and transmits light, the top and bottom piece will each reflect some light. However, light reflected by the bottom piece will have traveled farther than light reflected by the top piece. When this extra distance is an integer number of wavelengths (λ, 2λ, 3λ, . . .), light reflected by the top and bottom glass pieces interferes constructively. When it is an odd number of half-wavelengths ($\lambda/2$, $3\lambda/2$, $5\lambda/2$, . . .), it interferes destructively.

The pattern of colors we see depends on the wedge of air separating the two glass pieces. If the wedge of air is uniformly thick, we see a single color. If the air wedge varies in thickness, we see bands of different colors similar to the pattern in Young's double-slit experiment.

The bands of color that occur on soap bubbles and thin films of oil are produced by the same process. Light reflected from the top of the surface interferes with light reflected from the bottom surface. Since the thickness of the layers varies, swirls of color are produced.

SELF-CHECK 16B

Most camera lenses have a thin layer of material placed on their surfaces. When you look at these surfaces, they have a bluish appearance. What does this tell you about the thin layer of material? Does it vary in thickness?

Holography Records Images Using Interference Bands

One of the more exciting applications of interference by reflection is **holography**—a three-dimensional version of photography. In order to understand how this process produces a three-dimensional photograph, we need to contrast it with conventional two-dimensional photography. When you take a conventional photograph, you use sunlight or a flash attachment to provide a source of light. Light reflected by the object enters a camera lens (Figure 16-10(a)), which in turn directs it to the film. Photographic film responds to the amount of brightness of the light striking it. It simply adds together all the light that reaches it while the shutter is open, forming a two-dimensional image of the object.

Figure 16-10
(a) Light reflected by the object is recorded on the photographic film in an ordinary camera.
(b) Light from two paths interfere to record a hologram. Along path A, light travels directly from the laser to the photographic plate. Along path B, light travels from the laser to the object and then to the photographic plate.

Figure 16-11

(a) A single three-dimensional image is produced with light of a single wavelength.
(b) A mercury-arc lamp, which emits a mixture of five wavelengths, produces five three-dimensional images. Can you find them?

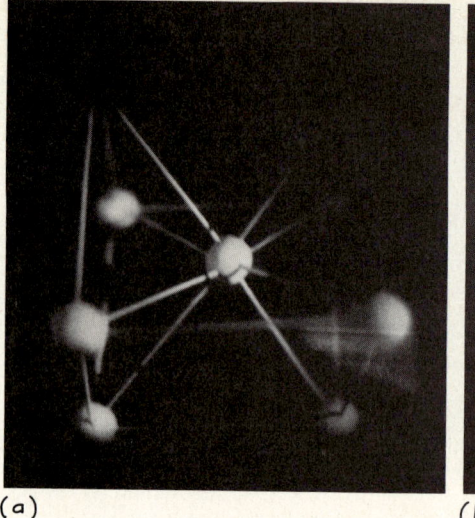
(a)

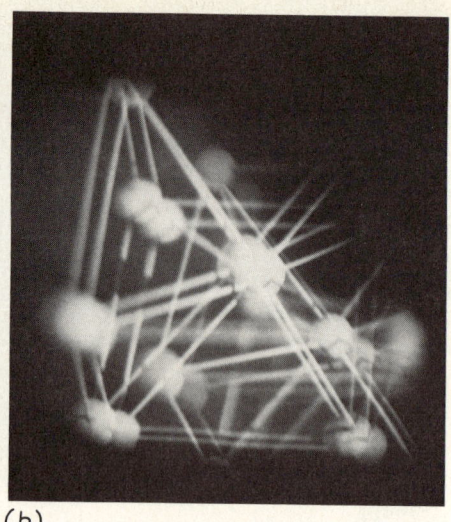
(b)

Holography differs from conventional photography in its ability to add information about the third dimension—the depth of the object. A hologram is produced by the interference between two laser light beams—one that is reflected from the object, as in conventional photography (path B in Figure 16-10(b)), and a second that travels directly from the laser to the photographic film (path A in Figure 16-10(b)). This second beam is often called the **reference beam.** When the reflected light interacts with the reference beam, they interfere. Since light reflected from a person's nose will have traveled a shorter distance than light reflected from his or her neck, the interference patterns for the two will be slightly different. Holography allows us to record information about the depth of an object because of the addition of a reference beam against which to measure the distance traveled by the reflected light.

The pattern captured on the holographic film looks like a series of bright and dark bands—similar to but more complex than the simple interference patterns Young produced with his double-slit experiment. Stored in these interference bands is the information needed to reconstruct a three-dimensional image of the original object. If you hold the holographic film in front of a light source, light transmitted by the bright bands and blocked by the dark bands creates the image. The image is truly three-dimensional. As we move our viewing positions, we see parts of the object that were not visible from our earlier positions.

Interference patterns depend on wavelength as well as on the distance the light has traveled. Consequently, holograms are generally made with light of a single wavelength and then viewed with this same light. A hologram viewed with a mercury-arc lamp, for example, produces four separate images—one for each wavelength present in the light (Figure 16-11). You can imagine how confused a hologram would look when made by the complete spectrum of wavelengths found in white light. The information about the depth of the object becomes lost in the interference patterns produced by each separate wavelength. Lasers, which produce light of a single wavelength, offer the most convenient light source with which to record holograms.

Because the interference pattern recorded is specific to the wavelength of light with which it was made, we usually view the hologram with this same wavelength. If you hold a hologram in front of an ordinary light source, you will not see an image.

In conventional photography, a set of lenses is used to focus the light reflected from the object. As the reflected light spreads out from a point on the object, the lenses intercept a portion of the light and direct it to a point on the photographic film. Each point on the film corresponds to a point on the scene being photographed. In making a hologram, lenses are not needed.

Interference and Acoustics

All physicists seemed to agree that sound was a wave. Unlike light, observations of interference phenomena were not required to resolve any controversy. However, interference of sound waves has frequently been the critical point in a different kind of problem—the quality of sound in an auditorium.

A singer on stage emits sound waves that spread out and travel throughout the auditorium. When the sound strikes a hard surface, it is reflected. These reflections can travel back toward the stage and throughout the auditorium, interacting with the sounds later emitted by the singer. The law of superposition applies to these interactions. These results of these interactions affect what we hear. A particularly interesting effect occurs when the singer holds a note long enough that the frequencies of the reflected and emitted sounds are equal. Then, both constructive and destructive interference are possible. If the interference is constructive, the audience hears a note that is louder than the original. Destructive interference, however, means that they hear little or no sound—not very satisfying to the patron. One of many possible interactions is shown in the figure. The black semicircles locate the crests of each wave; the white semicircles identify the troughs. When the crest from the incident wave meets the crest from the reflected wave, constructive interference occurs. Destructive interference occurs at the solid dark areas, where a crest from one wave meets a trough from the other. As shown in the figure, reflections from the surface are capable of interacting destructively with incoming waves, leaving a wide swath of silence across the audience.

In an actual auditorium, the situation is much more complex. Sound reflects off all objects, resulting in many reflections that interfere constructively and destructively with each other, as well as with the original sound. These complications give rise to the difficulties architects face in designing concert halls. Even a good design can be less than ideal depending on the behavior of the audience. In one symphony hall, the designer assumed that the patrons would leave their coats at the checkroom. He included the possibility that sound would be reflected by the people in the audience and corrected for interference patterns accordingly. When the audience brought their coats into the hall, the reflections were slightly different and changed the interference patterns, albeit slightly.

Figure 16-12

Light transmitted through a single slit produces a broad central band with bright and dark bands on either side.

Both the reflected and the nonreflected beams spread out across the photographic film. The pattern recorded on the film is made up of the interference of light reflected from every point on the object. If we cut the hologram into small segments, each segment contains a full interference pattern and, thus, the information necessary to reconstruct the entire image. Try that with a conventional photograph!

DIFFRACTION PATTERNS

In designing his double-slit experiment, Thomas Young was preoccupied with a search for evidence that light, like water and sound, experienced interference. The two slits in the second barrier acted as wave sources, providing waves that interfered with each other to produce the pattern shown in Figure 16-1. Had he removed the second barrier entirely, Young would have found additional evidence supporting the wave model of light. Instead of seeing a single band about the same width as the slit, he would have seen a much broader band with bright and dark bands on either side (Figure 16-12). Such a pattern is called a **diffraction pattern.** Light seems to spread out and interfere with itself as it passes through a single slit.

Waves Spread and Interfere

When straight water waves pass through an opening in a barrier, they spread out rather than producing a band of waves the same width as the opening (Figure 16-13(c)). At some distance from the barrier, the region across which energy has been transmitted can be considerably wider than the width of the opening through which the waves traveled. This spreading of waves after passing through a narrow opening is called **diffraction.**

The extent to which waves are diffracted depends on the width of the opening compared to the wavelength of the waves. The figure shows a series of experiments in which water waves of the same wavelength pass through openings of different widths. When the width of the opening is considerably larger than the wavelength (Figure 16-13(a)), very little diffraction occurs. As the opening narrows (Figure 16-13(b)), diffraction becomes more noticeable. When the opening becomes about the same size as the wavelength (Figure 16-13(c)), diffraction is substantial.

An additional characteristic of diffraction can be seen in Figure 16-13(b). If you look to either side of the broad central region in the diffraction pattern, you will notice that the wave crests seem to fade out, reappear, and then fade out again. A side view of the water's surface would show that the water re-

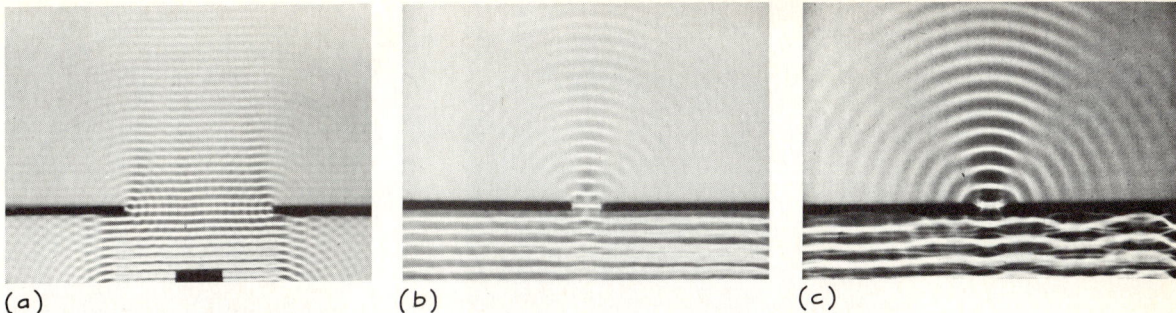

(a)   (b)   (c)

mains undisturbed in the regions where the pattern fades. As they are diffracted, water waves interfere to produce regions of constructive and destructive interference on each side of the central band.

Figure 16-13

Diffraction becomes noticeable when the width of the opening and the wavelength of the water waves are about equal.

SELF-CHECK 16C

Sound waves have wavelengths that range from a centimeter to several meters. Most voices produce sounds with wavelengths of about a meter. Light waves have wavelengths that are much smaller—averaging 5×10^{-7} m. Explain why sound is noticeably diffracted through doorways but light is not.

Superposition of Circular Waves Explains Diffraction

We can understand diffraction patterns if we imagine a straight-line wave as consisting of a series of individual circular waves. As each point in the water vibrates, it produces a circular wave that transmits energy in all directions. When several of these point sources vibrate together along a straight line, the individual circular waves combine to form the straight-line wave we actually see (Figure 16-14). We say that a straight-line wave is the sum of a series of waves from point sources, each producing circular waves that spread out and combine according to the principle of superposition.

This model for a straight-line wave explains why waves diffract as they pass through a narrow opening. Sources in the middle of the opening produce waves that combine to produce a straight-line wave. Sources near the edge of

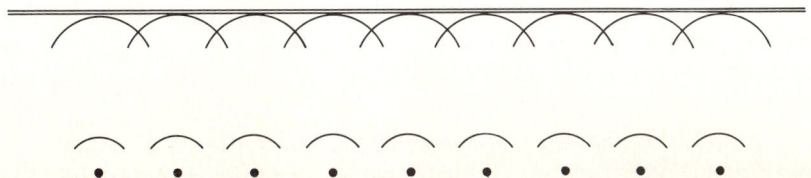

Figure 16-14

A straight-line wavefront can be constructed from a series of circular waves that have spread out from a line of point sources.

378 Chapter 16. Interference and Diffraction

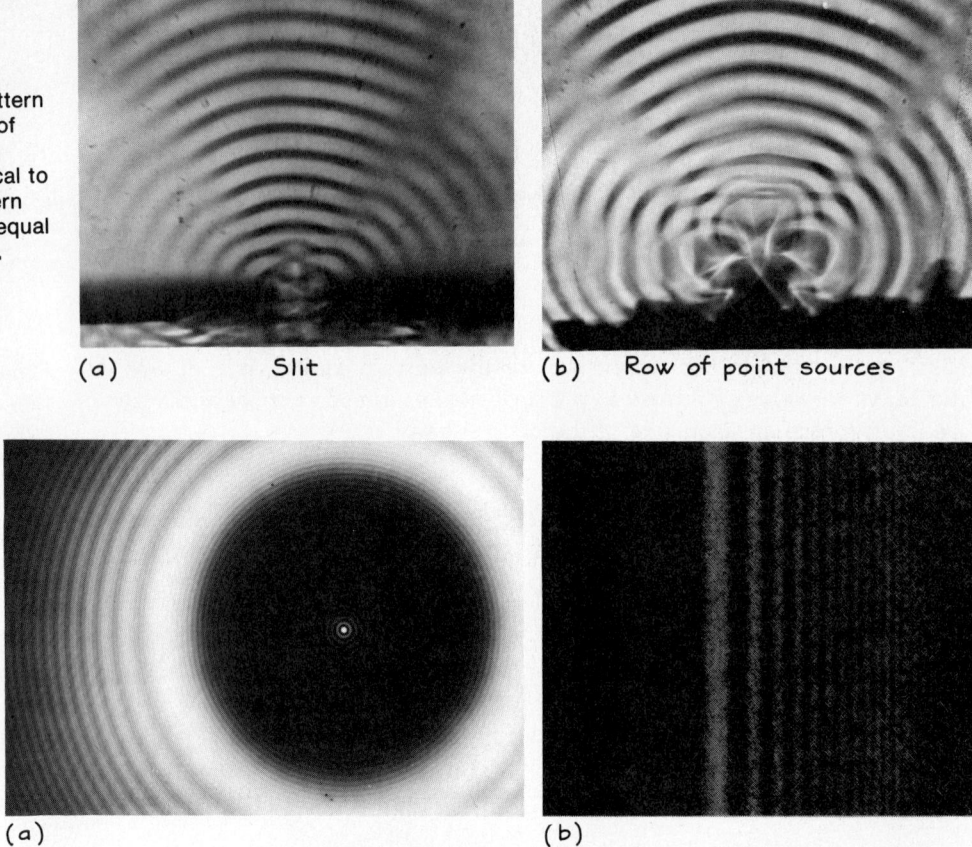

Figure 16-15
The interference pattern produced by a line of point sources in **(b)** looks almost identical to the diffraction pattern formed by a slit of equal width, shown in **(a)**.

(a) Slit (b) Row of point sources

(a) (b)

Figure 16-16
Diffraction patterns are formed as light passes through narrow openings or bends around small objects.
(a) Shadow of a penny.
(b) Shadow cast by a straight barrier.

the opening have nothing with which to combine, so their wavefronts remain circular. This bending at the edge of the pattern is what we call diffraction.

This model is equally effective in describing the bright and dark bands that accompany diffraction. Figure 16-15 shows a diffraction pattern formed by straight-line waves passing through a narrow opening (Figure 16-15(a)) and an interference pattern formed by a line of point sources (Figure 16-15(b)). The length of the row of point sources is the same as the width of the opening used to produce the diffraction pattern. Near the sources we see some differences in the two patterns. Far from the sources, however, the two patterns look identical. Circular waves passing through the slit combine to produce regions of constructive and destructive interference just like those due to waves from individual point sources.

Once again, the behavior of water waves helps explain the patterns produced by light passing through narrow slits. Before Thomas Young completed his work, physicists were not able to demonstrate diffraction consistently. Once they understood the need to provide openings as small as the wavelength of light, they found an abundance of diffraction phenomena to study.

Figure 16-16 shows a sample of diffraction patterns produced as light passes around obstacles or through openings of various shapes. Light is dif-

Figure 16-17

fracted, producing shadows larger than the obstacle or bright regions broader than the opening. Interference bands surround both the shadows and the central bright regions. One of the crowning achievements of the wave model of light was its success in predicting the small bright spot in the shadow formed by a penny (Figure 16-16(a)). Such a spot can arise only from the diffraction and interference of waves striking the edge of the coin.

SELF-CHECK 16D

For artistic purposes, photographers often like to capture the starlike effect shown in Figure 16-17. You can create similar patterns by squinting your eyes while looking at a light source. Use diffraction to explain how these patterns are produced.

DIFFRACTION AND RESOLUTION

The diffraction and interference of light waves produce striking and attractive patterns. The appeal of these phenomena diminishes, however, when we turn to the problems of designing optical instruments that produce faithful images. Instruments that use light to "see" must pass that light through some type of narrow opening. Because diffraction is created at each opening, cameras, microscopes, telescopes, and even the human eye face limitations due to diffraction.

Resolution of Closely Spaced Objects

Driving late at night, you see a distant car coming toward you. When it is very far away, its headlights all seem to blend together—you see only one light. As the car comes closer, you see headlights shining from both sides of the car.

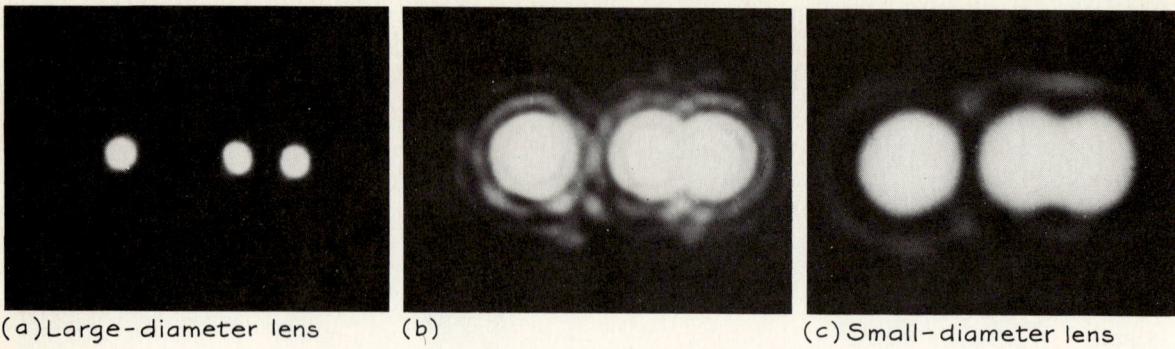

(a) Large-diameter lens (b) (c) Small-diameter lens

Figure 16-18

As the diameter of the lens decreases, we're less able to resolve the three point sources of light.

But, you still cannot tell whether each headlight is single or double. Finally, as the car comes still closer, you see the dual headlights on each side. Your difficulty in separating the four headlights is related to the resolving ability of the human eye.

The **resolution** of an optical instrument describes its ability to produce separated images of objects that are closely spaced. Two sources of light are resolved when we see separate images of them. Figure 16-18 shows a series of photographs of three closely spaced light sources. The pictures were taken through lenses identical to one another except in diameter. Each lens acts like a pinhole, providing a circular aperture through which light waves must travel. Because light is diffracted as it passes through small apertures, the images produced are slightly enlarged and fuzzy. When viewed through smaller and smaller apertures, they become so large and fuzzy that we are no longer able to resolve the three light sources.

Figure 16-18(a) was made with a relatively large-diameter lens, while Figure 16-18(c) was made with a small-diameter lens. As the diameter of the lens decreases, the diffraction of each image becomes more noticeable, just as the diffraction of the water waves increases as the passageway becomes narrower. We are able to distinguish three separate point sources for a large-diameter lens but are less able to do so as the diameter becomes progressively smaller. Because of diffraction, the image of each point source overlaps with the next. We say that the images cannot be resolved. The larger the lens, the better its resolution—that is, the more closely spaced the objects it can resolve.

One reason you are unable to resolve the distant set of headlights is related to the size of the pupil in your eye. Light must pass through this small opening in order to reach the retina. Diffraction occurs, causing the images produced to be fuzzy and slightly enlarged. When the images overlap, as they did photographically in Figure 16-18(c), you see two images instead of three.

Optical Instruments

Diffraction places some limitations on photographers. Cameras use adjustable apertures to control the amount of light that enters them. In a camera, the aperture is adjusted in a series of stages called *f*-stops—the larger the *f*-stop number, the smaller the aperture. Apertures usually range in diameter from a few millimeters to 20 or 30 millimeters. One of the most common problems with diffraction in a camera is that a small dot or point of light takes on fuzzy edges. While the pattern is not very noticeable on the film itself, it becomes obvious if the photograph is enlarged. Photographers who routinely take pictures intended for posters must be keenly aware of these effects.

Diffraction places a slightly different limitation on microscopes. Microscopes are designed to enlarge very small objects. The aperture size of the lens affects the resolving power of the instrument, just as it does for the camera and the eye. An additional problem is posed by the size of the objects the microscope is magnifying. Diffraction patterns become significant when the object being illuminated is the same size as the wavelength of light. Consequently, light transmitted or reflected by matter will not reveal the structure of objects smaller than 10^{-7} m. Microscopists have had to turn to electromagnetic waves of shorter wavelengths to investigate objects as small as those found within the cell.

SELF-CHECK 16E

The diameter of the pupil in the human eye averages 4 to 5 millimeters. The diameter of a large research telescope can be as much as 5 m. Use the problems posed by diffraction to explain why a large telescope should be able to resolve objects better than the naked eye.

WAVES VERSUS PARTICLES

The nature of light puzzled scientists for centuries. Most light phenomena—reflection, refraction, and absorption—could be explained as well by assuming light was a particle as by assuming it was a wave. For example, light is reflected so that the angle of reflection equals the angle of incidence. Balls are also reflected off walls so that the angle of reflection equals the angle of incidence. Light is refracted as it crosses a boundary between two media; balls, too, change speed and direction as they move into regions where different forces act. Light is absorbed by some forms of matter; objects stick together after certain kinds of collisions. True, physicists could not see the particles thought to carry light energy, but then, neither could they see a medium through which light waves could travel. One description seemed as reasonable as the other. However, when physicists observed interference and diffraction phenomena, the scales tipped in favor of the wave model of light. Diffraction and interference could not be explained by particle motion. They could only be explained in terms of wave superposition.

The wave-particle controversy surrounding the nature of light provides one of the most fascinating confrontations in science. Many people think that science concerns itself with facts—with measured observations about which there can be little disagreement. They are surprised to hear the word *controversy* used in this context. In reality, science deals with explanations and reasons for facts as much as with the facts themselves. Controversy about these explanations can and does arise quite often. Physicists agreed about the behavior of light as it was reflected, refracted, and absorbed by matter. The underlying reasons for these observations, described by some in terms of particles and by others in terms of waves, were what sparked the controversy. Only when they made additional observations were physicists able to resolve the controversy—temporarily.

The acceptance of the wave model of light provided two models for two types of phenomena. When energy is transmitted with an object, the particle model applies. When energy is transferred from a source to a receiver but matter is not, the wave model applies. At the beginning of this century, the dichotomy between the two models was well understood. Waves explained some phenomena; particles, other phenomena. This neat and clean separation between the two models was short-lived, however. Investigations into the structure of the atom blurred the distinctions and led to a renewal of the wave-particle controversy. The renewal of the wave-particle dilemma is the subject of the next chapter.

CHAPTER SUMMARY

Light from two slits or two point sources interacts to form a series of bright and dark bands on a distant screen. The resulting pattern, called an *interference pattern,* can be explained in terms of wave superposition. Constructive interference, producing the bright bands, occurs where the distances traveled by the two waves are equal or differ by a whole number of wavelengths. Destructive interference, producing the dark bands, occurs where these distances differ by an odd number of half-wavelengths. Interference bands are most noticeable when the separation between the light sources is about the same as the wavelength of light. We see interference patterns daily in the swirls of color reflected from oil and soap films. *Holography* uses interference bands to produce three-dimensional photographs.

Light from a single slit spreads out and interacts to form a broad band of light with light and dark bands on either side. The resulting pattern, called a *diffraction pattern,* can be explained in terms of circular wavelets. Each point in a wavefront acts as a point source for a circular wave. These individual wavelets combine to form the wavefronts we actually see. In a narrow opening, wavelets near each edge have nothing with which to combine, so they contribute to the spreading of the wavefront. This spreading is called *diffraction.* Within the opening, wavelets produce circular waves that can interfere constructively and destructively with one another. This interference produces the light and dark bands on either side of the central band. Diffraction patterns are most noticeable when the opening through which the waves travel is about the same size as the wavelength of light. Diffraction places limits on the ability of optical instruments to *resolve* images of separate objects that are closely spaced. Microscopes that use light to illuminate matter can only resolve structures larger than the wavelength of light.

ANSWERS TO SELF-CHECKS

16A. If we subtract the two distances, we find that waves from S_2 have traveled 0.9 cm farther than waves from S_1. This extra distance is $\frac{3}{2}$ times the wavelength of the water waves. The waves will interfere destructively at point I.

16B. Since we see just one color of light reflected from the lens, the thin layer must be uniformly thick. The thickness was chosen to allow blue wavelengths of light to interfere constructively while other colors interfere destructively. (This layer helps prevent unwanted reflection during photography.)

16C. Diffraction becomes most noticeable when the size of the opening is about the same size as the wavelength of the waves. A door is about 1 m wide. Sound has about the same wavelength, so sound waves are diffracted. Light has a much smaller wavelength, so we notice little diffraction.

16D. Diffraction of light as the sun is viewed through a narrow opening leads to the starlike effect. Light that passes near the edge of the opening spreads out, producing the streams that move outward from the source. When you squint your eyes, you create a narrow opening through which the light is diffracted.

16E. Light will be diffracted more by the pupil in the human eye than by the telescope lens because the lens is larger than the pupil. Since they diffract light less, telescopes should be able to resolve images better than the human eye.

PROBLEMS AND QUESTIONS

A. Review of Chapter Material

A1. Define each of the following terms:
Interference
Interference pattern
Interference colors
Holography
Diffraction
Diffraction pattern
Resolution

A2. In what ways are the bright and dark bands in Figure 16-1 analogous to nodes and antinodes in standing waves?

A3. Two identical waves travel from two different sources to a single point in the medium. What do you need to know in order to predict whether constructive or destructive interference occurs at that point?

A4. How does the interference pattern change if you change the distance from the source at which you observe it? If you change the wavelength of the waves? If you change the separation between sources?

A5. Why is white light separated into a series of interference colors after it passes through two closely spaced slits?

A6. How can interference patterns be produced by reflected light?

A7. How does the interference between the reference beam and the reflected beam contain information about the depth of an object?

A8. How does the diffraction pattern change when you change the width of the opening?

A9. How are individual point sources of circular waves used to explain the diffraction of a straight-line wave as it passes through a narrow opening?

A10. How are individual point sources of circular waves used to explain the interference bands in Figure 16-13(b)?

A11. How does diffraction affect the resolution of an optical instrument?

A12. Why were observations of interference and diffraction phenomena so crucial to the resolution of the wave-particle controversy in light?

B. Using the Chapter Material

B1. Two sources, S_1 and S_2, produce identical waves with wavelengths of 0.06 m. For each point described below, describe the type of interference that occurs—constructive, destructive, or in between.

Point Along Water's Surface	Distance from S_1	Distance from S_2
A	0.06 m	0.06 m
B	0.03 m	0.06 m
C	0.12 m	0.09 m
D	0.18 m	0.12 m
E	0.18 m	0.14 m

B2. Calculate the distance to the center of the first band of constructive interference for each situation described below. Use these distances to describe the way in which the interference pattern changes if we:
a. change the distance to the source
b. change the wavelength of the waves
c. change the separation between sources

Distance from Source	Wavelength (λ)	Source Separation
0.5 cm	6 cm	2 cm
1.0 cm	6 cm	2 cm
2.0 cm	6 cm	2 cm
1.0 cm	6 cm	30 cm
1.0 cm	6 cm	6 cm
1.0 cm	12 cm	2 cm
1.0 cm	2 cm	2 cm

B3. Red light with a wavelength of 7×10^{-7} m strikes a layer of gasoline that is 28×10^{-7} m thick. Will you see any red light reflected from the gasoline? Will you see any blue light with a wavelength of 4.3×10^{-7} m?

B4. Two slits are placed 1 mm (1×10^{-3} m) apart. A screen is placed 1 m from the two slits. Calculate the distance from the center line to the first band of destructive interference, $X_{1/2}$, for red light (6.5×10^{-7} m), green light (5.4×10^{-7} m) and violet light (4.0×10^{-7} m).

B5. Suppose you performed Young's double-slit experiment with red light, with green light, and with violet light. How would the interference patterns change as the color of light is changed?

B6. Seashells, butterfly wings, and bird wings often change color as you change the position from which you look at them. Use light interference to explain this phenomenon.

B7. When your car has a thin layer of oil on the windshield, you can see a rainbow of colors. Use light interference to explain why.

B8. Sound from two stereo speakers behaves like light traveling through two slits. If both speakers are emitting the same frequency of sound, would you expect to find regions where the sound is loud or soft? Why? Why do we not normally notice interference of sound with stereo speakers?

B9. Any point along a wave can itself be regarded as a point source of circular waves.
a. Draw 10 points, each separated from the next by 0.5 cm. Then draw circular arcs representing circular wave crests that have traveled 2.0 cm from their source. Show how these circular arcs combine to produce another straight-line wave.
b. Now imagine that these 10 point sources are equally spaced along a single slit. What happens to the circular wave produced by each point at the edge of the slit? Use this construction to explain diffraction of light by a single slit.

B10. Which would result in better resolution, a microscope that uses blue light or a microscope that uses red light? Why?

C. Extensions to New Situations

C1. Automobile headlights provide a common example of two sources of light. Yet we see no interference effects like those that occur in water. To understand why, let's begin with an analogy.

a. Suppose two friends are marching in step on the floor above you. Describe what you would hear.
b. Now suppose your two friends are just randomly walking about. What will you now hear? How is it different from what you described in (a)?
c. In both situations, (a) and (b), sound waves add according to the superposition principle. How does the pattern we hear depend upon whether the two people are marching in step or not?
d. Most light sources, including automobile headlights, emit a steady stream of independent bursts of light that are generally out of step with those emitted by other sources. Use the analogy provided by (a)–(c) to explain why we do not see interference patterns like those in Figure 16-1.

C2. Ordinary light sources do produce interference patterns, but the bands constantly change because the light emitted by one source is out of step with the light emitted by a second source. Thomas Young and his contemporaries were familiar with the problem. Young's solution was to use a single slit in the first barrier that acts like a single point source of light. He then placed the second barrier such that the two slits, S_1 and S_2, were equidistant from the single slit.
a. How does this procedure ensure that the light rays passing through the two slits remain in step with one another?
b. Light passing through the single slit will be diffracted. In order to avoid problems posed by diffraction, should the second barrier be placed near or far from the first barrier? Explain your answer.

C3. Light from a single source can be reflected by two different surfaces, leading to an interference pattern. The same thing often occurs with sound waves. In this case, the sound wave emitted by a musical instrument often interferes with some of its reflection. As shown in the next column, the sound that traveled directly (along the path AX) interferes with the sound that was reflected from the floor (along the path $AB + BX$).
a. How would we decide whether point X

was a region of constructive or destructive interference?
b. If the wavelength of sound is 1 m and the distances along AX, AB, and BX are 7 m, 6 m, and $5\frac{1}{2}$ m, respectively, will point X be a region of constructive or destructive interference?
c. What will a listener hear if he or she is located in a region of constructive interference? In a region of destructive interference?
d. This problem is a common one encountered by people who design concert halls. How might they solve this problem?

C4. As you walk by a puddle covered with gasoline, you see a changing rainbow of colors. This effect occurs because the difference in distance between the top and bottom layers depends on the actual path light has taken. The figure below shows four possible paths and the distances light travels to reach the bottom reflecting surface.

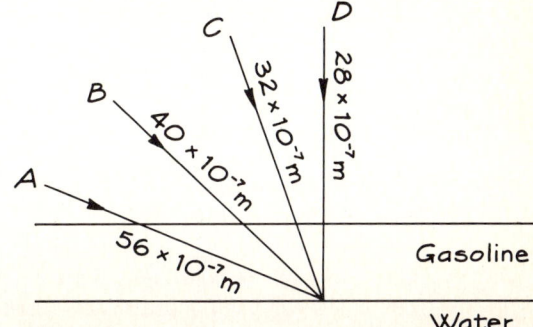

a. For each path, determine which wavelength interferes constructively and which interferes destructively.
b. Describe what you see as you walk from point A to point B to point C and finally to point D.
c. Does this explain the changing rainbow of colors you see?

C5. One way to take a photograph is with a pinhole camera. As its name implies, a pinhole camera is simply a box with a pinhole in one end and a piece of photographic film on the other end. When we open the pinhole, light travels into the box, exposing the film like a conventional camera. If the pinhole is too small, the image on the film will be fuzzy. Why?

C6. An interference pattern produced by two slits, the figure shows both the interference pattern and the diffraction pattern produced by the slits.

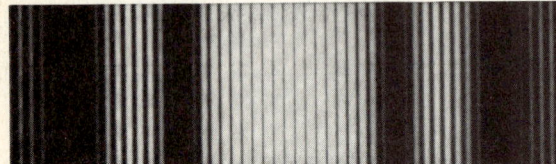

a. The narrow bands of bright and dark regions result from interference between light from the two slits. Measure the distances to the centers of the first two bright and dark bands.
b. What happens to the bright and dark bands produced by interference as we move farther away from the center line?
c. The fading of interference bands arises from the diffraction pattern produced by each slit. How would the location of this fading change if the slits were made wider?

C7. You are casting a shadow on a wall. Near the wall, you see a shadow with sharp edges. As you move away from the wall, the edges become fuzzy. Why?

C8. Frequently scientists need to measure the wavelength of light emitted by a light source. We can use the spacing of interference bands produced by the double-slit experiment to provide such a measurement. Two slits are separated by 2×10^{-3} m. On a screen placed 1.5 m away, the distance from the center line to the first bright band was measured to be 4.4175×10^{-4} m. What was the wavelength of light emitted by the source?

C9. All electromagnetic waves are capable of producing interference patterns. Two radio towers broadcasting in all directions produce interference patterns that completely surround them. As shown in the figure below, two towers are separated by a distance X. Both towers emit identical waves of wavelength 500 m.

a. If the distance X is 500 m, predict what type of interference (constructive or destructive) will occur at points A and B.
b. If the two towers are placed closer together, X = 250 m, what type of interference occurs at A and B?
c. Use your answers to parts (a) and (b) to describe how the placement of radio broadcasting towers can be used to control the direction in which the signals travel.
d. How would you arrange two radio towers to broadcast 200 m waves to the city but not to the ocean or mountains (Figure 16-C9(b))?

D. Activities

D1. Look at lights through a very fine mesh such as fine lace curtains. Explain what you see in terms of interference.

D2. With a straight pin, make a number of different-sized holes in a piece of heavy paper. Look at light through each hole. Explain what you see.

Figure 17-1 R. L. Gregory, The Intelligent Eye, 1970, McGraw Hill.

Wave-Particle Duality

Which do you see—the young lady or the old woman? Designed by an American psychologist, E. G. Boring, the sketch that opens this chapter illustrates a phenomenon called object ambiguity. We can perceive the same sketch to be two very different objects. The black band at her neck, the rear profile of her chin, the dainty eyelashes—all combine to produce an image of a beautiful young woman. Then, before our eyes, the image dissolves. The black band becomes a cruel mouth; the dainty chin turns into a hideous nose. We now see an ugly old woman. These two very different perceptions can emerge as we shift our gaze from one part of the sketch to another. They also emerge spontaneously as we keep our gaze fixed. Perceptions arise from the way we think about what we see as well as from what we actually see.

This chapter deals with an ambiguity in science—what we might call *model ambiguity*. Like the young woman and old hag in Boring's sketch, the wave model and the particle model provide us two very different perceptions of nature. Both models emerge from our attempts to explain how energy gets from one place to another. The particle model associates energy with mass. Energy is transferred from a source to a receiver during collisions—relatively

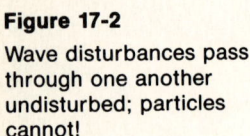

Figure 17-2
Wave disturbances pass through one another undisturbed; particles cannot!

abrupt events. By contrast, the wave model associates energy with the amplitude of a disturbance. The transfer of energy from source to receiver is more gradual and continuous. Phenomena can be categorized as behaving like particles or waves. When ambiguities arise, superposition phenomena such as interference and diffraction offer the critical test. Wave disturbances pass through one another undisturbed—particles cannot!

Early in this century, physicists began to realize the extent to which both models describe the same phenomena. Waves can be described as particles; particles, as waves. Diffraction and interference phenomena reveal a wavelike character in light, but the *photoelectric effect* shows light to have a particlelike character as well. Electrons have mass but can be diffracted like waves. All particles turn out to have a wavelike character described by *de Broglie wavelengths*. Nature reveals a *wave-particle duality*—an ambiguity uncharacteristic of science. While the meaning of this wave-particle duality remains a subject of intense debate, many physicists now accept the *Bohr complementarity principle*. The two models exclude one another, yet both are necessary for a complete description of nature.

THE PARTICLE NATURE OF WAVES

Early in the nineteenth century, Thomas Young and his contemporaries *appeared* to resolve the intense debate over whether light was a particle or a wave. The diffraction and interference effects were so striking and Young's mathematical explanation of the phenomena in terms of waves so compelling that few scientists challenged the conclusion that light is composed of waves. Ironically, just as the controversy was being resolved by diffraction and interference experiments, investigations into the photoelectric effect revealed a completely new facet of light. These experiments not only reopened the wave-particle controversy surrounding the nature of light; they eventually shook the very foundations of physics.

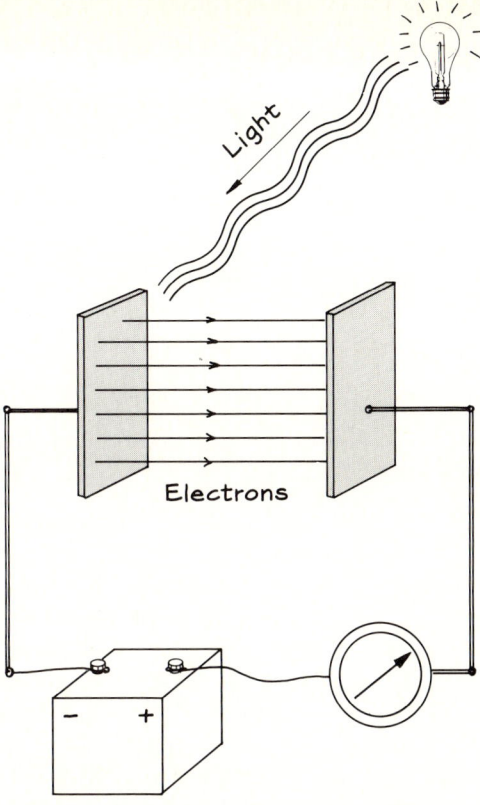

Figure 17-3
Two oppositely charged metal plates are placed inside an evacuated glass tube. When illuminated with visible light, the negative plate ejects electrons, which are then drawn across to the positively charged plate.

The Photoelectric Effect

During the latter part of the nineteenth century, a series of accidental discoveries showed that light could knock electrons loose from metal surfaces, a phenomenon called the **photoelectric effect.** These observations led to a series of controlled experiments similar to that illustrated in Figure 17-3. Two metal plates, one positively charged and the second negatively charged, were placed inside a glass tube. Light was directed toward the negative plate. Electrons ejected from the negative plate were attracted to the positive plate. A meter detected the motion of these electrons.

We can think about the photoelectric effect in terms of our two models, the particle model and the wave model. Light, the energy source, transfers the energy stored in its electromagnetic waves to electrons, the energy receivers. In turn, the electrons transform this wave energy into the electrical potential energy required to release them from the metal plate and the kinetic energy needed to move across to the positive plate. Light delivers wave energy; electrons transform it into particle energy. This hardly seems startling—wave energy is transformed into particle energy daily as the earth absorbs the thermal radiation transmitted to us from the sun. What is startling, however, is that the energy delivered by the light cannot be explained in terms of the wave model. To see why, let's examine what we expect to happen when we vary the intensity and frequency of the light source used to illuminate the metal plates.

Figure 17-4
(a) When we think of light as an electromagnetic wave, its changing electric fields cause the electrons to vibrate. Some vibrate far enough to break free from the metal surface.
(b) When we think of light as a particle, photons collide with individual electrons, knocking them free from the metal surface.

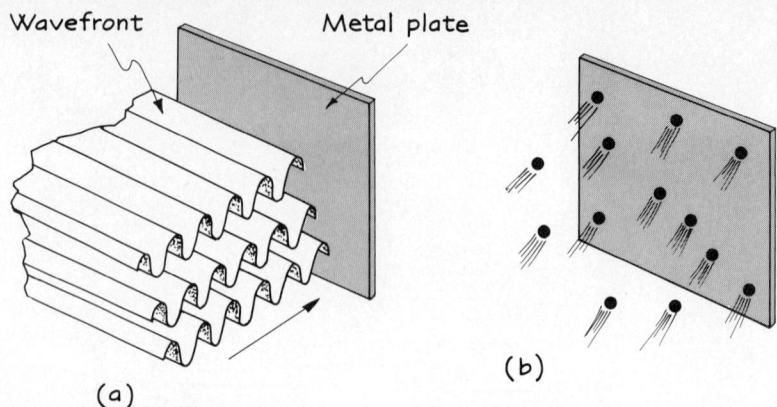

Light Waves Can't Do It

As you might expect, the interaction between a light wave and an electron is complex. Figure 17-4(a), however, shows one way of imagining what happens when light waves reach the metal surface. Light waves consist of vibrating electric and magnetic fields. As the light wave enters the metal, electrons (which are negatively charged) begin to vibrate in response to these changing electric and magnetic fields. The electrical bonds holding the electrons to their atoms are a little like rubber bands. Small vibrations do not enable the electron to escape—the rubber band always pulls it back. Larger vibrations, however, could eventually stretch the rubber band far enough that it breaks. The electron escapes from the atom with whatever energy is left over after the bond has been broken. This extra energy becomes the kinetic energy of the ejected electron.

We can use this model to predict what might happen if we vary the intensity (brightness) or frequency (color) of the light source. The intensity of the light source describes the amplitude of the individual waves. Dim light consists of waves with small amplitudes; bright light consists of waves with large amplitudes. We could imagine metal surfaces for which the amplitude of the waves in dim light is too small to break the electrical bonds holding the electrons. Increasing the intensity of this light should increase the vibration of the electrons, eventually to the point that the electrons break free. Increasing the intensity even further should increase the kinetic energy of the ejected electrons. On the other hand, the frequency of the light should have little effect on either the ejection of electrons or their kinetic energy. Changing the frequency of the light source simply changes the rate at which wave disturbances reach the metal surface. If the amplitudes of the waves are not sufficient to help the electrons break the bonds, changing the rate at which these waves reach the metal surface should not make much difference. Similarly, changing the frequency of the incident light should not affect the kinetic energy the ejected electrons have.

Much to the bewilderment of early experimenters, the ejection of electrons and the kinetic energy they had once ejected turned out to depend on the *frequency* of the incident light, not the intensity. The intensity of the light affected the number of electrons ejected but little else. Table 17-1 summarizes the results typical of experiments performed with cesium, sodium, and tungsten plates. At some frequencies of incident light, no electrons are ejected regardless of the intensity of light used. Tungsten, for example, will not eject electrons when illuminated with visible light. Only ultraviolet frequencies (10×10^{14} hertz (Hz)) work. Exposing cesium plates to red light (4.5×10^{14} Hz) has no effect, no matter how long or with what intensity of light you illuminate them. Orange light (5×10^{14} Hz) causes the immediate ejection of electrons. Each metal surface seems to require some minimum frequency of light before it ejects electrons. Once ejected, the electrons have kinetic energies that depend on the frequency, not the intensity, of the light used. At frequencies above the minimum frequency needed to eject the electrons, increasing the frequency of light increases the kinetic energy of the ejected electrons.

We cannot explain these results in terms of the wave model of light. The energy supplied to each electron depends on the frequency of the light, not the intensity. A wave picture like that in Figure 17-4(a) no longer gives us any way of picturing the transfer of energy from light to the electrons. Changing the intensity of the light and thus the amplitude of the waves does not have any effect until the right frequency of light is used. The electrons just seem to wait for the right frequency to come along. Once we have the minimum frequency for a given metal surface, increasing the intensity of the light still does not affect the energy of the ejected electrons. Individual electrons absorb energy associated only with the frequency of the light. The wave model, so beautifully demonstrated by interference and diffraction, so well accepted at the end of the last century, seems inadequate after all.

Packets of Energy

The results of the photoelectric experiments led Albert Einstein to propose that in its interaction with electrons, light could be best described as a stream of small particles, or "energy packets." Each packet, called a **photon,** acts as

Table 17-1 Results of Photoelectric Experiments*

Frequency of Incident Light (Hz)	Kinetic Energy of Released Electrons (J)		
	Cesium	Sodium	Tungsten
(Red) 4.5×10^{14}	None emitted	None emitted	None emitted
(Orange) 5.0×10^{14}	0.109×10^{-19}	None emitted	None emitted
(Green) 5.5×10^{14}	0.440×10^{-19}	0.120×10^{-19}	None emitted
(Blue) 6.0×10^{14}	0.772×10^{-19}	0.451×10^{-19}	None emitted

*These results are the same for all intensities of light illuminating the metal surface.

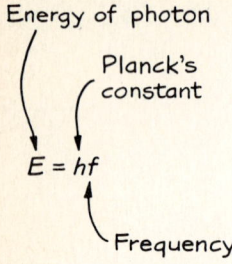

a unit when it interacts with an electron. Using an idea introduced originally by Max Planck, Einstein suggested that the energy carried by each photon is defined by:

$$\text{Energy} = (\text{Planck's constant}) \times (\text{frequency})$$

where Planck's constant is defined to be 6.63×10^{-34} joule · seconds (J · s). Planck's constant is a proportionality constant that converts frequencies in hertz to energies in joules. For example, the energy carried by a photon of orange light (frequency = 5.0×10^{14} Hz) is (6.63×10^{-34} J · s) × (5.0×10^{14} Hz), or 3.32×10^{-19} J. Higher frequencies of light have photons with larger energies.

SELF-CHECK 17A

Compared to a photon of orange light, how much energy is carried by 1 photon of green light? (The frequency of green light is 5.5×10^{14} Hz.)

Einstein's concept of the photon provides a solution to the problems posed by the photoelectric effect. Each photon acts as a unit when it interacts with matter. When a photon encounters an electron, it transfers all its energy to the electron and ceases to exist. By associating energy with frequency, the photon explains why electrons are ejected by light of one frequency but not of another. It also explains why the kinetic energy of the released electrons depends only on the frequency of the incident light. For example, the energy needed to remove an electron from a sodium atom is 3.52×10^{-19} J. As shown in an earlier example, a photon of orange light carries 3.32×10^{-19} J —not enough to free the electron. A photon of blue light, however, carries 3.97×10^{-19} J—more than enough to free the electron. The electron absorbs all the energy carried by the photon, uses 3.52×10^{-19} J to escape from the atom and has 0.45×10^{-19} J left over as kinetic energy. This matches the observations listed in Table 17-1.

The concept of the photon also explains how the intensity of the incident light is related to the number of electrons ejected by the metal surface. Einstein's picture of light is that it consists of a stream of photons. The greater the number of photons emitted by a light source is, the more intense the light appears. If we illuminate a metal surface with a frequency of light large enough to cause the emission of electrons, the number of electrons ejected depends on the number of photons striking the surface. More intense light means more photons strike the metal surface and thus more electrons are ejected. The total energy carried by light in a given time interval depends on the light intensity (number of photons per second) and the light frequency (energy carried per photon).

Einstein's solution to the problems posed by the photoelectric effect was based upon the concept of **quantization of energy,** which had been introduced a few years earlier by Max Planck. After studying the energy radiated by solids, Planck proposed that energy is not continuous. The energy emitted cannot be any value; it appears in chunks, or discrete quantities, called quanta. The photon is a light **quantum.** The energy carried by orange light, for example, appears in chunks of 3.32×10^{-19} J. When orange light with a frequency of 5.0×10^{14} Hz is absorbed by matter, the energy transferred is never some fraction of 3.32×10^{-19} J. It is always equal to some whole number times 3.32×10^{-19} J. Admittedly tiny, these quanta play an enormously important role in explaining phenomena like the photoelectric effect.

The idea that a quantity can be quantized is not new. We are familiar with many materials that appear continuous from a distance but are, in fact, built up of identical chunks. From a hundred yards or so, a brick wall looks rather continuous. But as you walk closer it becomes apparent that the wall is, in fact, made up of lots of discrete "quanta," called bricks. A piece of gold looks solid and continuous, but chemists will hasten to point out that it is built up from billions of identical "quanta" called atoms. The mass of the chunk of gold will be some whole number times the mass of an individual gold atom. Other quantities, like electrical charge, are also quantized. The concept that energy might be quantized is a little surprising but hardly impossible to accept. What is startling, however, is that something thought to be a wave, like light, transfers energy like particles.

Does it Really Work?

Step into the beam of light traveling across a closing elevator door; the door reopens. Hold the product code on a box of corn flakes in front of a small beam of light; the cash register adds the cost to your grocery bill. As the sun rises, the electric eye on a street lamp senses the coming daylight; the street lamp turns off. These and a variety of other devices take advantage of the photoelectric effect and, in so doing, assure us that photons are real.

Sometimes called an electric eye, the device used in elevator doors and street lamps is a **photoelectric cell** (Figure 17-5). Two separated metal plates are placed inside a glass tube. When light strikes the negative plate, electrons are released and attracted to the positive plate. The motion of the electrons can then act as a switch, turning the device on or off. In automatic doors, a light source is placed directly across from the photoelectric cell. As long as light strikes the photoelectric cell, the door remains closed. Once you step between the light source and the photoelectric cell, the motion of electrons in the cell stops and the door is opened. In street lamps, the photoelectric cell is used to detect sunlight. When daylight strikes the photoelectric cell, the street lights turn off. When night falls and no light strikes the cells, the street lights turn on.

Photoelectric cells have proven enormously useful in coding and decoding information. In laser-operated cash registers, for example, laser light is absorbed by the series of black lines that make up the universal product code

Figure 17-5

Photoelectric cells, often called *electric eyes,* can be found in a variety of everyday devices.

Figure 17-6
In laser-operated cash registers, a photoelectric cell is used to detect the pattern of light reflected from the universal product code.

(UPC) found on most grocery store products (Figure 17-6). A photoelectric cell detects the pattern of absorbed and reflected light, converting it into a pattern of electrical signals, which operate the cash register and change inventory records. Many libraries use similar systems to maintain lender records and book inventories.

SELF-CHECK 17B

Most photoelectric cells in street lamps and automatic doors respond to ordinary sunlight, which has an average frequency of 5.5×10^{14} Hz. Of the three metals listed in Table 17-1, which would be the most convenient to use? Why? Which metal could not be used for these applications? Why?

Seeing with Photons

Visual information, what we see and how we interpret it, reaches us in the form of light. Many aspects of vision, such as the need for corrective lenses, can be described best in terms of the wave model of light. Others are best treated with the particle nature of light. Two of the more interesting questions posed by those investigating the quantum nature of vision are:

What is the minimum number of photons needed before we see any light?

How many photons must be reflected from an object in order for us to recognize it?

These two questions have been the subject of extensive research.

To learn about the minimum number of photons required before a person sees any light, researchers asked subjects to sit in totally dark rooms until their eyes were completely adapted to the dark. (In one experiment the subjects sat in a totally dark room with their heads held in a fixed position for 45 minutes before the experiment began.) Then, flashes of light were emitted at very low intensities. The intensity was gradually increased until the subjects first reported that they saw the light. The energy and frequency of the light emitted was then recorded. As you might expect, the results of the experiment varied from one person to the next. At a frequency of 5.9×10^{14} Hz, people reported seeing light at energies that ranged from 3×10^{-17} J to 6×10^{-17} J.

In order to determine the number of photons emitted with each flash of light, experimenters used the relationship between energy and frequency, $E = hf$. To illustrate the process they used, we assume that the light used in the experiment had a frequency of 5.9×10^{14} Hz. At this frequency, each photon has an energy of:

$$\text{Energy} = (\text{Planck's constant}) \times \text{frequency}$$
$$= (6.63 \times 10^{-34} \text{ J} \cdot \text{s})(5.9 \times 10^{14} \text{ Hz})$$
$$= 3.9 \times 10^{-19} \text{ J}$$

If the average person reported seeing light at an energy of 4×10^{-17} J, we can determine the number of photons present by dividing the total amount of energy by the energy/photon:

$$\text{Number of photons} = \frac{4 \times 10^{-17} \text{ J}}{3.9 \times 10^{-19} \text{ J/photon}}$$
$$= 102$$

We can conclude that, on the average, about 102 photons are emitted by the lamp when the average person first reports seeing the flash. While the experimenters were careful to be sure that all these photons struck the eye, not all of them actually reached the retina. Some photons were reflected by the front of the eye; others were absorbed by the liquid in the eye. When these and other effects were taken into account, researchers concluded that some people can see light when as few as 2-5 photons strike the retina.

While 2-5 photons enable us to see light, they are not sufficient for us to identify the object emitting or reflecting them. To determine the number of photons needed to convey visual information about an object, Albert Rose, a researcher in vision, prepared the photographs shown. Each represents a photograph of the same object but with different numbers of photons present. As you would expect, the amount of information conveyed increases as more photons are used. How many photons do you need to see that the image is a face? That it is a woman's face?

Vision seems a remarkable gift—regardless of the model of light with which you investigate it. The incredibly small number of photons needed to initiate vision and the enormously large number of photons we process with each meaningful image speak eloquently of the sensitivity and intricacy of the human eye.

(a) 3,000

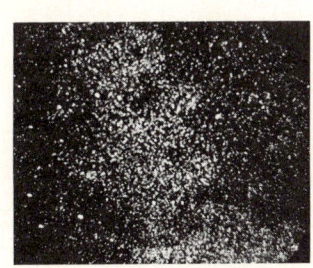

(b) 12,000

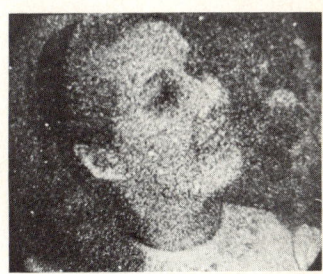

(c) 93,000

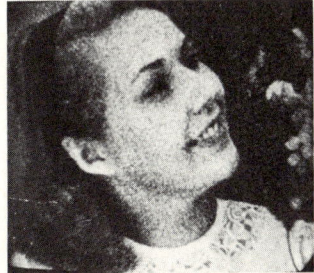

(d) 760,000

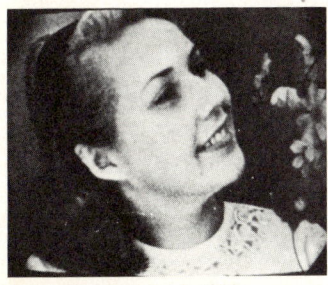

(e) 3,600,000

(f) 28

Albert Rose, Vision: Human and Electronic. Plenum Press (1973).

WAVE NATURE OF MATTER

Einstein's quantum picture of light is compelling, yet paradoxical. It presents a solution to the problems posed by the photoelectric effect, yet it simultaneously contradicts the wave model of light. The photoelectric effect tells us that in interactions with electrons, light energy is carried in discrete bundles called photons. Interference and diffraction phenomena tell us that in interactions with itself, light energy is spread continuously across space in the form of waves. Light seems to have a dual nature—particlelike in some situations, wavelike in others.

De Broglie Wavelengths

The discovery of the particlelike character of light waves led some physicists to wonder if particles might, under certain circumstances, display a wavelike character. Based strictly on an intuitive belief in the symmetry of nature, Louis de Broglie proposed that particles have a wavelength associated with their momentum. Called the **de Broglie wavelength,** this wavelength is defined as:

$$\text{de Broglie wavelength} = \frac{\text{Planck's constant}}{\text{momentum of particle}}$$

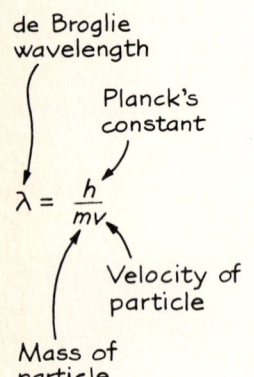

Using Planck's constant (h) as 6.63×10^{-34} J · s, mass in kilograms (kg), and speed in meters per second (m/s), the de Broglie wavelength is given in meters (m).

The size of the de Broglie wavelength associated with a particle depends on its momentum. An automobile traveling down the highway has a momentum of about 25,000 kg · m/s and a de Broglie wavelength of 10^{-38} m. A person sitting in the automobile has a larger wavelength—10^{-37} m—but not much larger! A bullet traveling at a speed of 500 m/s has a smaller momentum, 5 kg · m/s, but its de Broglie wavelength is still too small to detect. Smoke particles drifting about have a de Broglie wavelength of about 10^{-23} m, still too small to notice. Only when we enter the world of the atom, do we begin to notice de Broglie wavelengths. A nitrogen molecule drifting about in space has a de Broglie wavelength of about 10^{-11} m. Electrons have longer wavelengths—10^{-10} m. The smaller the momentum of the particle is, the larger its de Broglie wavelength will be.

The size of Planck's constant reflects the fact that we do not notice the wavelike character of ordinary objects. Ordinary masses moving at ordinary speeds have de Broglie wavelengths of about 10^{-30} m, extraordinarily tiny. When we move into the submicroscopic world of the atom, however, the tiny masses associated with particles like electrons result in de Broglie wavelengths that we can detect. We notice the wavelike character of electrons, but not the wavelike character of automobiles. George Gamow's delightful stories (Activity D1) transport you to an imaginary world in which Planck's constant is a much larger number. The wavelike character of objects becomes part of the ordinary world, enabling us to imagine the reality found within the atom.

A STEP FURTHER—MATH

DE BROGLIE WAVELENGTHS

An automobile with a mass of 1000 kilograms moves at a speed of about 25 m/s. Its de Broglie wavelength is:

$$\lambda = \frac{h}{mv}$$

$$= \frac{6.63 \times 10^{-34} \text{ J} \cdot \text{s}}{(1000 \text{ kg})(25 \text{ m/s})}$$

$$= 2.65 \times 10^{-38} \text{ m}$$

This wavelength is shorter than any we are capable of detecting.

An electron with a mass of 10^{-30} kg moves at a speed of about 6×10^6 m/s. Its de Broglie wavelength is:

$$\lambda = \frac{h}{mv}$$

$$= \frac{6.63 \times 10^{-34} \text{ J} \cdot \text{s}}{(10^{-30} \text{ kg})(6 \times 10^6 \text{ m/s})}$$

$$= 1.10 \times 10^{-10} \text{ m}$$

This wavelength is about 0.001 that of visible light. We commonly detect wavelengths of this size.

The size of Planck's constant makes the de Broglie wavelengths associated with ordinary objects too small to notice. Only on the atomic scale do we begin to see the wave nature of matter.

SELF-CHECK 17C

How does your de Broglie wavelength while walking at a speed of 2 m/s compare to the de Broglie wavelength of the automobile? Of the electron?

Electron Diffraction

De Broglie's concept that wavelengths can be associated with particles was originally based on a hunch. Before they could be accepted by the scientific community, these waves had to be detected. Electrons, for example, needed to demonstrate a wavelike behavior that could be explained only in terms of their de Broglie wavelengths. As you might expect, diffraction and interference phenomena provided the critical test.

As you saw in Chapter 16, interference and diffraction phenomena become noticeable only when waves pass through openings or around obstacles that are about the same size as the wavelength of the waves. Openings that are considerably larger than the wavelength of the waves simply cast ordinary

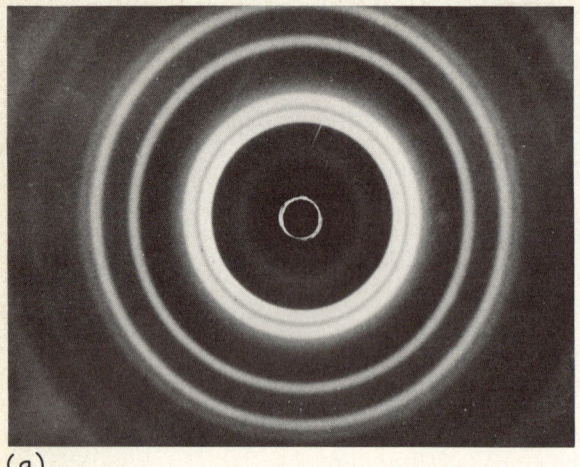 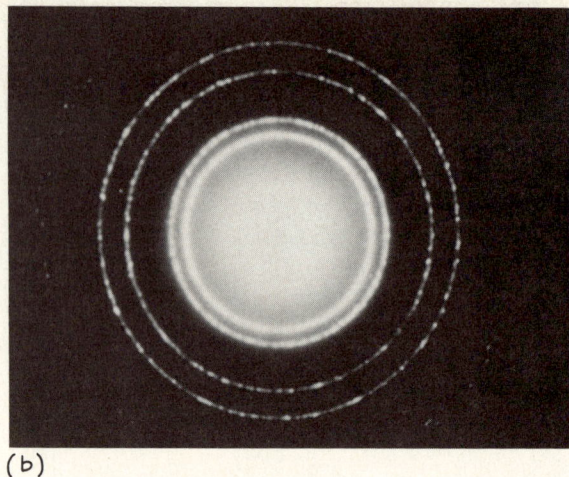

Figure 17-7

(a) X rays produce a diffraction pattern after passing through a thin layer of aluminum foil.
(b) Electrons produce a remarkably similar pattern when they pass through this same thin layer of foil.

shadows. The de Broglie wavelength associated with electrons is about 10^{-10} m, about the same size as X rays. Physicists expected that experimental arrangements that revealed diffraction and interference of X rays should reveal a similar behavior if moving electrons were used instead of X rays.

X rays are typically diffracted when they pass through thin layers of metal. The spacing between atoms in a solid metal is about the same distance as the wavelength of X rays. These spacings act like an array of slits that diffract the incident X rays, producing the pattern shown in Figure 17-7(a). If we repeat these experiments using electrons instead of X rays (Figure 17-7(b)), we see a pattern almost identical to that produced by the diffraction of X rays. Later experiments showed a resemblance between the patterns produced by electrons and those created by other wave phenomena. Electrons directed at the edge of a thin piece of magnesium oxide (Figure 17-8) produce a diffraction pattern similar to that created by visible light striking the edge of a barrier. The similarities are striking!

Once the de Broglie wavelengths associated with electrons had been detected, physicists eagerly turned to designing experiments that would enable them to observe diffraction and interference phenomena with other particles. In each case, de Broglie's relationship could be used to predict the wavelengths associated with particles of a certain size and speed. When barriers or obstacles could be created with these exceedingly small sizes, diffraction and interference effects were always observed. Protons and neutrons displayed exactly the same behavior. The de Broglie waves associated with particles are now commonly referred to as **matter waves**.

Electron Microscopes

Less than a decade after de Broglie proposed that particles have wavelengths associated with them, technology put electrons to work in building more powerful microscopes. In Chapter 16 we discussed the role that wave diffraction plays in limiting the resolution of the microscope. When the object being observed is about the same size as the wavelength of light, diffraction patterns blur the image. No matter how much you magnify an object, it will still be

Wave Nature of Matter 399

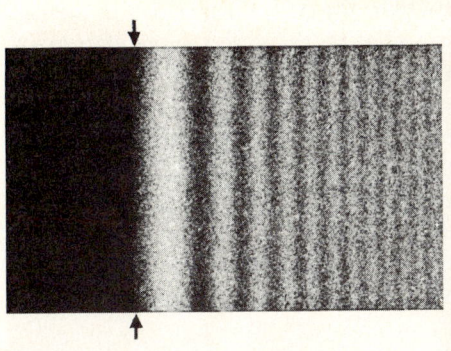

Visible light

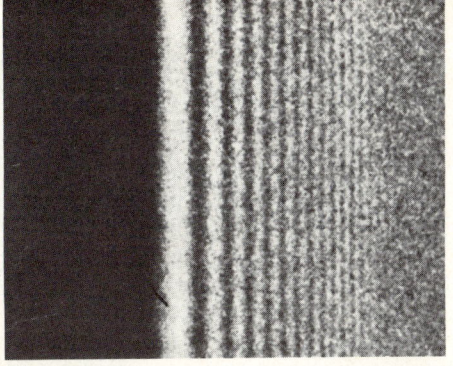

Electrons

Figure 17-8
When they strike the edge of a barrier, electrons produce diffraction patterns similar to those observed with visible light.

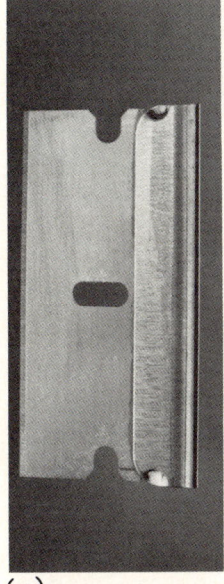
(a)

(b)

(c)

blurred. Structures smaller than the wavelength of visible light—10^{-7} m—cannot be resolved with conventional microscopes.

Electrons have much shorter wavelengths associated with them, typically about 10^{-10} m. Based strictly on a comparison of wavelengths, electron waves should be able to resolve structures that are a thousand times smaller than structures resolved with visible light. In practice, this turns out to be the case. A transmission electron microscope directs a beam of electrons toward a very thin specimen. The waves transmitted through the specimen are focused to form an image. Using electron waves, structures can be magnified more than 10,000 times their actual sizes. The major drawback seems to be the damage done to the specimen by the electrons. Live specimens often cannot be observed.

Figure 17-9 contrasts the images of a razor blade produced by a camera, a light microscope, and an electron microscope. Your eye is capable of resolving objects within 0.1 millimeter (mm) at a distance of 10 centimeters (cm).

Figure 17-9
A razor blade viewed **(a)** through an ordinary eye, **(b)** through a light microscope, and **(c)** through an electron microscope. How straight is straight?

While the image on film is slightly smaller than the actual razor blade, we can still resolve detail within about 0.1 mm. The width of the small line in (a) shows the region magnified by a light microscope and displayed in (b). A light microscope can often extend our abilities to resolve structures by a factor of 1000. Light microscopes can resolve structures separated by 10^{-7} m. Finally, the thick block in (b) shows the region magnified by a electron microscope and displayed in (c). An electron microscope, capable of resolving structures separated by 10^{-10} m, allows us an even closer glimpse of the razor blade. Our ability to resolve objects has been extended by yet another factor of 1000. If that's a new razor blade, think what an old one must look like!

SELF-CHECK 17D

Neutrons are about 2000 times more massive than electrons. If neutrons moved at the same speed as electrons, would their de Broglie wavelengths enable us to resolve structures smaller than those resolved with the electron microscope? Why or why not?

WAVE-PARTICLE DUALITY

The particlelike character of light waves and the wavelike character of particles present a curious ambiguity. Prior to this century, physicists had come to classify energy transfer processes into two categories: particle and wave. Energy is transferred between particles when a discrete event—one particle colliding with another—occurs. Billiard balls transfer energy when they collide; bats transfer energy when they hit a baseball. Energy is transferred by waves in a more continuous fashion. Water waves continuously transfer energy, gradually wearing away the rock and soil along the land's edge. The realization that light demonstrates characteristics associated with both categories was startling. The realization that electrons—real particles with measurable masses—also demonstrate both kinds of characteristics seemed almost catastrophic. **Wave-particle duality** seemed the only answer.

Duality is unusual in science. While controversy is a common occurrence, physicists expect to be able to design experiments that can differentiate one model from another—to decide whether light is a wave or a particle. Conflicts are resolved so that one model replaces another. For the first time, conflicting models could not be resolved. When physicists performed diffraction and interference experiments, light behaved like a wave. When they performed experiments on the photoelectric effect, light behaved like a particle. What was worse was that electrons behaved in much the same way. In recounting the events of the 1920s, Jacob Bronowski reports that the physicists in Göttingen, a center for research on wave-particle duality, took to saying that on Mondays, Wednesdays, and Fridays electrons behaved like particles and on Tues-

M. C. Escher—Sky and Water II, 1938. Collection Haags Gemeentemuseum—The Hague, 1981. © BEELDRECHT, Amsterdam/V.A.G.A., New York.

Figure 17-10
One picture merges into the next. Wave and particle models offer complementary views of the microscopic world. Neither by itself is sufficient.

days, Thursdays, and Saturdays they behaved like waves (*The Ascent of Man*. 1973. Boston: Little, Brown and Company, p. 364).

In one sense, wave-particle duality arises from the way we think. The sketch that opened the chapter demonstrates the extent to which we impose our own order on the things we observe. Boring drew a series of lines on a piece of paper. When we look at the lines one way, we form an image of a beautiful young lady. When we look at the lines yet another way, we see an ugly old hag. Both images arise from the same set of lines. What we see is the order that we impose on those lines. A person from an alien culture might not see either the young lady or the old hag. In science, the images we see are the models we build to describe phenomena. Our everyday world of billiard balls, baseball bats, and water waves seems best ordered by two models—waves and particles. When we look into the microscopic world, we impose these models on entirely new phenomena. To our question of whether you are a particle or a wave, the microscopic world replies both!

Niels Bohr proposed a resolution to wave-particle duality called the **complementarity principle.** He suggested that the wave model and the particle model provide complementary pictures of reality. Neither provides a complete view of reality; each is needed to explain some of our experiences in the microscopic world. Asking whether microscopic objects are particles or waves is no longer a meaningful question. We use whichever model is required to explain a particular experiment. Like Escher's *Sky and Water* (Figure 17-10), reality reveals clear models for some phenomena and ambiguous models for others. We see fish in the sea and birds in the air. Where they meet, however, we see whichever we wish to see.

CHAPTER SUMMARY

Light incident on a metal surface causes the emission of electrons. This process, called the *photoelectric effect,* converts light energy into electron energy and is used extensively in photoelectric cells. In looking at the way in which energy was exchanged in this transfer process, physicists found some surprising results. Contrary to results predicted by the wave model of light, the amount of energy transferred to the electrons depends on the frequency as well as on the intensity of the light used. These observations led Albert Einstein to propose that the energy carried by light is transferred to electrons in discrete packets, called light quanta, or *photons.* The energy carried by each photon is defined by:

$$\text{Energy} = \text{Planck's constant} \times \text{(frequency)}$$

Einstein's concept of the photon explained the results of the photoelectric effect but seemed to contradict the wave model of light established by diffraction and interference effects.

Since light waves behave like particles, physicists speculated that particles might also behave like waves. Experiments revealed that electrons transmitted through thin metallic foils produce diffraction patterns analogous to those produced by light passing through narrow slits. Particles appear to have wavelike characteristics. Their wavelength, called the *de Broglie wavelength,* is defined as:

$$\text{de Broglie wavelength} = \frac{\text{Planck's constant}}{\text{momentum of particle}}$$

The short wavelengths associated with electrons have proven useful in the design of electron microscopes capable of resolving structures as small as 10^{-10} m in size.

Wave-particle duality has led to fundamental questions regarding the models we build to explain our observations. In his *complementarity principle,* Niels Bohr suggested that the wave and particle models of reality provide complementary pictures of nature. Both are needed; neither can describe the microscopic world without the other.

ANSWERS TO SELF-CHECKS

17A. Green light has a higher frequency than orange light. One photon of green light will carry more energy than one photon of orange light: $E = hf = (6.63 \times 10^{-34} \text{ J} \cdot \text{s})(5.5 \times 10^{14} \text{ Hz}) = 3.65 \times 10^{-19}$ J.

17B. Cesium would be the most convenient metal to use because it emits photoelectrons over the broadest range of visible frequencies. Tungsten does not emit any photoelectrons when illuminated with visible light. It would be useless in these applications.

17C. The mass of an average person is about 80 kg. At an average walking speed of 2 m/s, a person's momentum would be 160 kg · m/s. $E = h/mv = (6.63 \times 10^{-34}$ J · s$)/(80$ kg$)(2$ m/s$) = 4.1 \times 10^{-36}$ m. The de Broglie wavelength associated with a person walking is longer than that associated with an automobile but much shorter than that associated with an electron.

17D. With a mass 2000 times greater, neutrons will have a momentum 2000 times larger than the electrons. The larger the momentum of a particle is, the smaller is its de Broglie wavelength. Neutrons should enable us to resolve still smaller structures.

PROBLEMS AND QUESTIONS

A. Review of Chapter Material

A1. Define each of the following terms:

Photoelectric effect
Photon
Quantum
Matter waves
Photoelectric cell
de Broglie wavelength
Bohr complementarity principle

A2. If light energy is transferred by waves, what variables should we expect to increase the energy supplied to the electrons in the photoelectric effect?

A3. In measurements conducted with the photoelectric effect what variable(s) affected: (a) the number of electrons ejected, (b) the kinetic energy of the electrons once released, and (c) whether the metal surface would or would not release electrons?

A4. How did Einstein's concept of the quantum explain the measurements described in Problem A3?

A5. How does the energy of an individual photon change when the frequency of light increases?

A6. Describe how a photoelectric cell is used in a street lamp.

A7. What led Louis de Broglie to propose that particles had wavelike characteristics associated with them?

A8. How does the de Broglie wavelength associated with a particle change as the particle's momentum increases?

A9. Why do we not notice the wavelike character of baseballs and automobiles?

A10. What experiments did physicists conduct to detect the wavelike character of particles proposed by de Broglie? Why were those experiments selected?

A11. How is wave-particle duality analogous to the object ambiguity described at the beginning of the chapter?

B. Using the Chapter Material

B1. Light with a frequency of 5.8×10^{14} Hz strikes metal surfaces made from each of the three metals listed in Table 17-1. Which surface(s) will emit electrons?

B2. Orange light (5.0×10^{14} Hz) causes cesium surfaces to emit electrons with kinetic energies of 0.109×10^{-19} J. Green light (5.5×10^{14} Hz) causes the electrons to be emitted with kinetic energies of 0.440×10^{-19} J.

a. What is the change in kinetic energy that results when we use green light instead of orange light?
b. Does the same change occur if we use blue light (6.0×10^{14} Hz) instead of green light?
c. Predict the kinetic energy of electrons ejected when cesium plates are illuminated by violet light (6.4×10^{14} Hz).

B3. Calculate the energy associated with a single quantum for each of the following kinds of electromagnetic waves.

Type of Electromagnetic Wave	Frequency (Hz)
Radio	10^6
Television	10^8
Microwave	10^{10}
Infrared	10^{13}
Ultraviolet	10^{16}
X ray	10^{18}

B4. Blue light (6.0×10^{14} Hz) and red light (4.0×10^{14} Hz) are both incident on a sodium plate. If 3.52×10^{-19} J are required to free the electron, determine the kinetic energy of the electron after it interacts with (a) a photon of blue light, (b) a photon of red light.

B5. Two light sources, A and B, emit light of identical frequencies but different intensities. A is more intense than B.
   a. Assuming that the light has sufficient energy to release electrons from the metal plate, which will cause the release of more electrons?
   b. Which source causes the release of electrons with more kinetic energy?

B6. Calculate the de Broglie wavelength associated with each of the following objects moving at a speed of 10 m/s. How does the mass of an object affect its de Broglie wavelength?

Object	Mass
Electron	10^{-30} kg
Proton	10^{-27} kg
Gold atom	10^{-25} kg
Glass of water	0.5 kg
Person	100 kg
Automobile	1000 kg

B7. What is the de Broglie wavelength associated with the gold atom in Problem B6 when it moves at a speed of 1 m/s? 100 m/s? 1000 m/s? 10^6 m/s? 10^8 m/s? How does an object's speed affect its de Broglie wavelength?

B8. The de Broglie wavelength of electrons (10^{-10} m) is considerably shorter than wavelengths of visible light (10^{-7} m). How would a diffraction pattern produced by visible light compare to that shown in Figure 17-7b?

C. Extensions to New Situations

C1. Some smoke alarms use photoelectric cells with light sources placed directly across from them. As long as light strikes the cell, the alarm will not sound. When no light reaches the cell, the alarm sounds.
   a. How does this arrangement allow the alarm to detect smoke?
   b. Do you need to worry about the light source burning out and making the smoke alarm ineffective?

C2. We often hear that certain types of electromagnetic radiation cause more serious damage to human tissue than others. This damage is a result of the energy transferred to the tissue. Use the results of Problem B3 to explain why each of the following is true.
   a. X rays and gamma rays are said to be the most dangerous forms of radiation.
   b. Ultraviolet radiation causes sunburn, while visible and infrared radiation does not.

C3. Electrons in metal surfaces must receive a characteristic amount of energy, called the *work function,* before they have enough energy to leave the metal. The work function is the same for all metal plates made of the same metal. We can use the data supplied in Table 17-1 to determine the work function for cesium and sodium plates.
   a. How much energy does a single photon of orange light (5.0×10^{14} Hz) supply to a cesium plate?
   b. If an electron released from the cesium plate had a kinetic energy of 0.109×10^{-19} J, how much of the energy supplied by a photon of orange light went into releasing the electron?
   c. Repeat (a) and (b) for the data given for green and blue light. What is the work function for cesium?
   d. Determine the work function for sodium.

C4. In photochemical reactions light absorbed by a molecule initiates a chemical reaction. Two of the more common reactions occur when light strikes photographic film or a green leaf.
   a. Photographic paper and photographic film involve similar photochemical reactions. However, film must be processed in total darkness, while paper can be processed in red light. What does this tell you about the amounts of energy required to initiate the photochemical reactions in each case?
   b. Two distinct frequencies initiate the photochemical reactions involved in photosynthesis. One lies in the red portion of the visible spectrum. The second lies in

the blue portion. Which reaction requires more energy?

C5. Plants convert carbon dioxide to oxygen when the leaves absorb red light (4×10^{14} Hz).
   a. What is the energy associated with a single photon of this light?
   b. A total of 8×10^{-19} J are required to convert a single molecule of carbon dioxide into a molecule of oxygen. What is the minimum number of photons needed to supply this amount of energy?
   c. Measurements show that 10 photons are actually required to initiate the reaction. What is the efficiency of the reaction?

C6. The momentum associated with a photon can be calculated by rearranging de Broglie's relationship:

$$\frac{\text{Momentum}}{\text{of photon}} = \frac{\text{(Planck's constant)}}{\text{wavelength}}$$

   a. What is the momentum of a photon of light that has a wavelength of 7×10^{-7} m?
   b. Suppose the photon in (a) is directed toward a stationary metal plate. What is the total momentum of the photon-metal plate system before the collision?
   c. What must be the total momentum of the system after the collision?
   d. Suppose the photon is absorbed by the metal plate. Describe the motion of the plate after the collision.
   e. Suppose the photon is reflected by the metal plate. Describe the motion of the plate after the collision.
   f. Will the momentum of the plate in (d) be higher or lower than that in (e)?
   g. A piece of metal with one side painted black, so light is absorbed, and the other side painted white, so light is reflected, has equal light shining on both sides. Which way will it move? (The toy shown in the next column does exist, but it moves in the opposite direction because of air currents.)

C7. Electron microscopes enable us to look more closely at matter than any visible

light microscope. Typically, electrons are directed toward the specimen at a speed of 1×10^8 m/s.
   a. What is the wavelength associated with each electron?
   b. Why does this wavelength allow us to see smaller objects than light microscopes?

C8. Planck's constant plays an important role in determining the situations in which we notice the particlelike characteristics of light and the wavelike characteristics of matter.
   a. $E = hf$ describes the size of the "chunks" of energy carried by waves. What would happen to these chunks if Planck's constant were a larger number?
   b. $\lambda = h/mv$ describes the wavelengths associated with matter. What would happen to these wavelengths if Planck's constant were a larger number?

D. Activities

D1. In *Mr. Tompkins in Paperback* (Cambridge University Press, 1969), George Gamow gives us the opportunity to imagine what life would be like if Planck's constant were a larger value. Read Chapter 7, "Quantum Billiards," and Chapter 8, "Quantum Jungle," of this delightful book.

D2. Obtain some pieces of transparent plastic that are different colors, such as red, green, and blue. Place each in front of the light source for a photoelectric cell of an elevator. How does each affect the operation of the door?

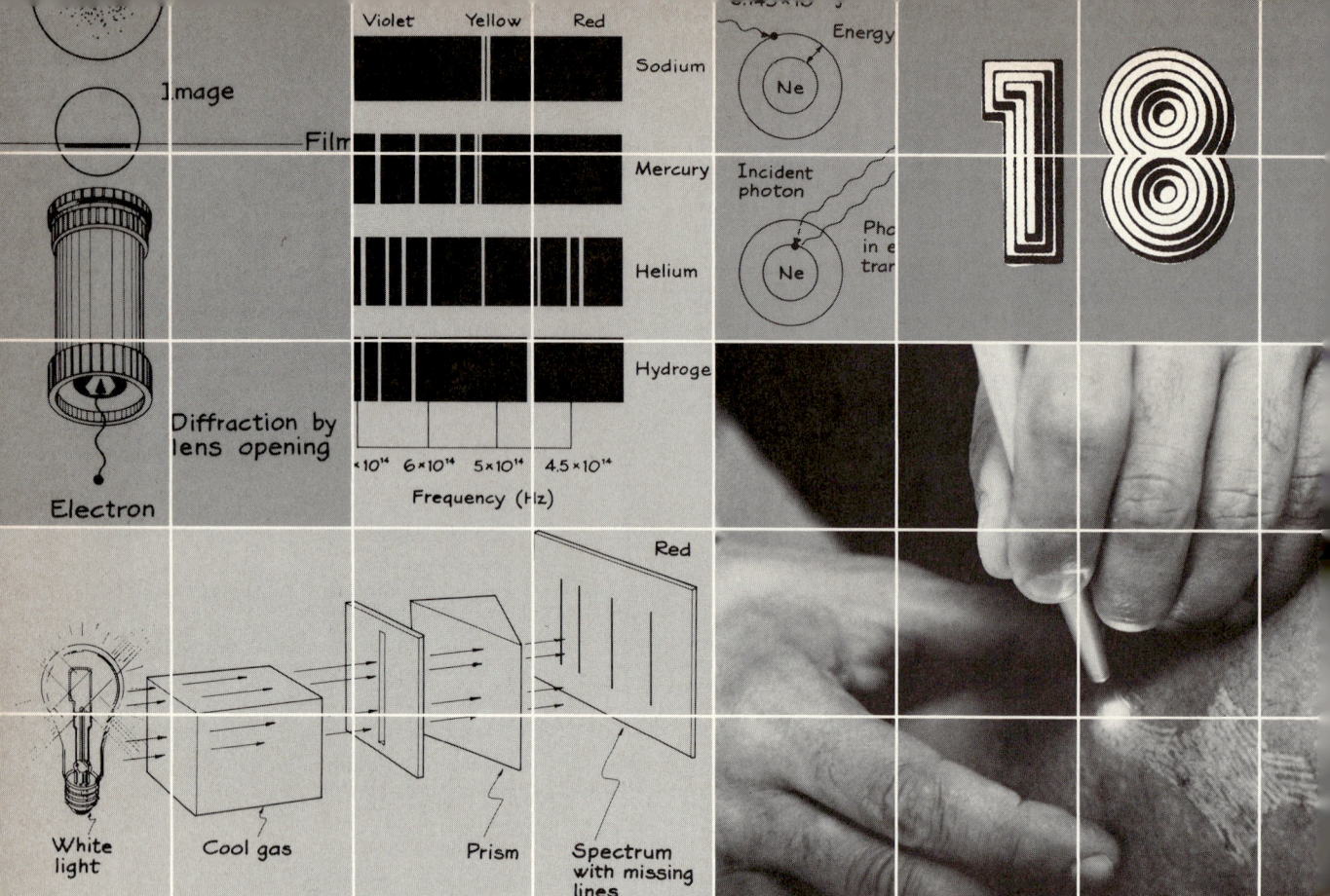

Light, Quanta, and Atoms

People create games and puzzles because it is fun to try to figure out the unknown. In Mastermind® (Pressman Toys) you guess patterns of colored pegs; in Battleship® (Milton Bradley) you figure out the location of enemy battleships. Some video games include hidden evils for you to discover before they discover you. These games differ in how clues are gathered, but all depend on our natural human enjoyment of figuring out something we cannot see.

Physicists do physics because they experience this same sense of enjoyment. Much of modern science has dealt with figuring out the nature of things we cannot see—stars and galaxies too far away to see in any detail, atoms and molecules too small to see even when viewed up close. Following up clues provided by nature, physicists build models in much the same fashion that we solve puzzles. While the models themselves often seem abstract, the processes used to develop them are familiar.

Within the context of solving puzzles, we can see how the current model of the atom evolved. *Emission* and *absorption spectra* provide information about how atoms absorb and release energy. Niels Bohr combined these observations, the quantum nature of light and Newtonian physics to build a model of the atom, called the *Bohr model*. By applying de Broglie's concept of matter waves to Bohr's model of the atom, electron orbits can be explained in terms of *circular standing waves. Quantum mechanics* now interprets these standing waves as *probability clouds,* which describe the probability of finding the electron at various locations in the atom. This quantum mechanical model, the result of many people working on different pieces of the puzzle, is the currently accepted picture of the atom.

LIGHT SPECTRA AND ATOMIC STRUCTURE

Some street lamps produce a yellowish light, while others have a bluish-white color. A campfire looks yellow until you throw in the color comics from last Sunday's paper, producing a multicolored display. From common experiences we know that the colors of burning gases depend on the material being burned. So consistent are these colors that nineteenth-century chemists were able to devise a series of flame tests to identify the elements in unknown materials by the colors of light they emit when they burn.

Just looking at the color of the light emitted has its limitations, though. Two colors that seem to be the same may actually be slightly different mixtures of light frequencies. Scientists eventually found that they could analyze substances far more accurately by passing the emitted light through a prism. The resulting spectra provided unmistakable signatures for each of the elements. And, though no one suspected it at the time, these spectra were also a major clue in the puzzle of atomic structure.

Emission and Absorption Spectra

Physicists study light spectra by means of a spectroscope, a device that separates or spreads light out into its component frequencies. The pattern can be captured on a screen or on photographic film. By using film sensitive to infrared or ultraviolet frequencies, we can extend the range of frequencies we can detect into regions below and above the visible range.

The spectrum of light emitted from a light source is called an **emission spectrum.** When we look at emission spectra from different sources, we see two different types. A **continuous spectrum** is a smear of colors, each merging with the next. These colors usually range from red (4.3×10^{14} Hz) on one end to violet (7.5×10^{14} Hz) on the other. The spectrum of light emitted by the sun—a rainbow—is an example of a continuous spectrum. Campfires and incandescent lamps also produce continuous spectra. In contrast to the continuous spectrum, a **discrete spectrum** reveals isolated colors with regions of black between them. Because the regions of color look like lines against a black background, a discrete spectrum is often called a *line spectrum.* Emission spectra produced by pure samples of chemical elements are always discrete. A few of these are shown in Figure 18-1.

A discrete spectrum is always unique to the chemical element that produced it. Hydrogen always produces exactly the same spectrum, and it produces a spectrum that is distinctly different from that of neon. The spectral lines combine to produce the colors characteristic of chemical elements used

408   Chapter 18. Light, Quanta, and Atoms

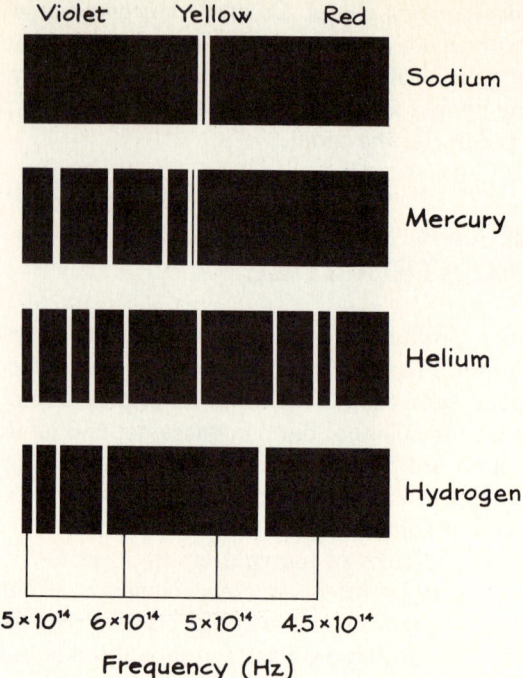

Figure 18-1
Emission spectra for selected chemical elements.

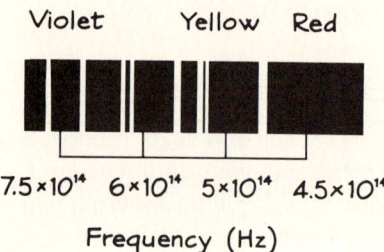

Figure 18-2

in the flame tests. Emission spectra can be used to identify all the elements in any substance.

SELF-CHECK 18A

The emission spectrum shown in Figure 18-2 was obtained from a mixture of two gases. Use the spectra in Figure 18-1 to identify the gases.

Absorption, the reverse of emission, occurs when light travels through a gas. Figure 18-3 shows an experiment in which white light (which contains all visible frequencies) passes through a sample of hydrogen gas and then through a prism. The resulting pattern looks like a continuous spectrum, ex-

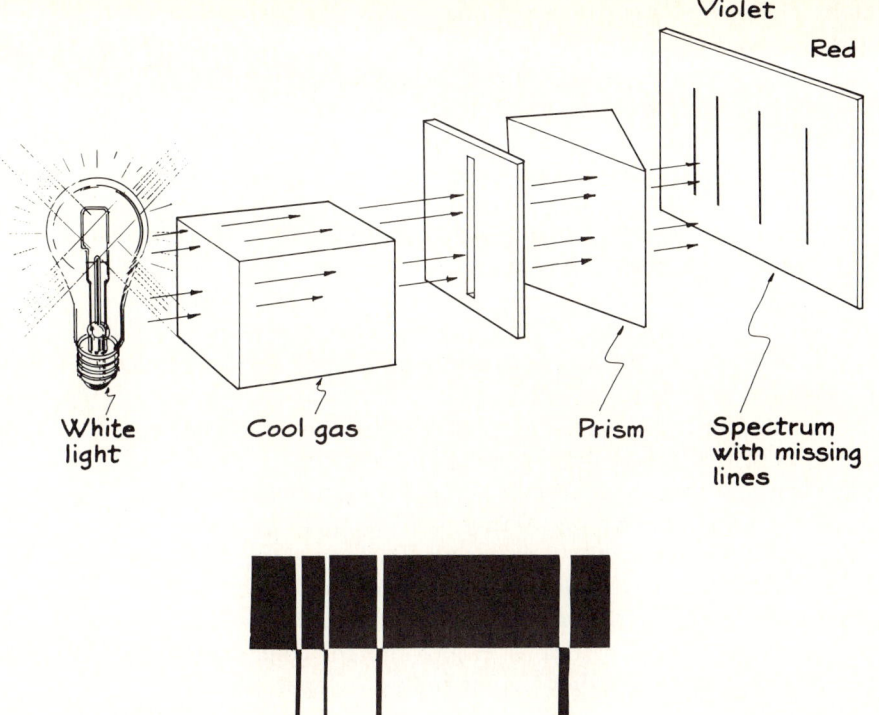

Figure 18-3
An absorption spectrum is produced when light emitted by an incandescent bulb passes through a sample of cool gas before striking the screen. The pattern of dark lines against a continuous spectrum shows which frequencies the gas has absorbed.

Figure 18-4
The emission and absorption spectrum of hydrogen.

cept that it has several dark lines in it. Some of the frequencies of light have been removed, absorbed as the light traveled through the gas. A continuous spectrum in which a few frequencies have been removed by absorption is called an **absorption spectrum.**

When we compare the emission and absorption spectra of hydrogen (Figure 18-4), we see a striking similarity between the two. The emission and absorption lines occur at precisely the *same* frequencies. The same is true for the emission and absorption spectra produced by other chemical elements. The frequencies of light absorbed by any chemical element match exactly the frequencies of light that it emits. Similar processes must be involved in light emission and light absorption. The changes in the atom that produce emission need only be reversed to produce absorption.

Discrete spectra baffled nineteenth-century scientists. It was not hard to understand how burning gases give off light; that is simply a matter of converting thermal energy into electromagnetic energy. But no one had any idea about why each chemical element produced or absorbed only a few of the frequencies of all those possible within the enormous range of the electromagnetic spectrum.

Discrete Spectra and Atomic Structure

As some physicists were puzzling over discrete emission and absorption spectra, others were working on a related problem—the internal structure of the atom. Rutherford's scattering experiments with alpha particles (Chapter 8) demonstrated that the atom contains a massive, positively charged nucleus with negatively charged electrons distributed around it. The nucleus is extraordinarily tiny, about one ten-thousandth the size of the atom itself. Since the positively charged nucleus attracts the negatively charged electrons, it seems curious that the electrons stay so far away. They should simply fall into the nucleus.

To explain the fact that the electrons remain apart from the nucleus, Rutherford suggested a "planetary" model of the atom. In our solar system, the sun's gravitational attraction provides a centripetal force that keeps planets moving in its orbit about it. Rutherford proposed that the electrical attraction exerted by the nucleus provides a centripetal force that keeps electrons circling it.

Rutherford's planetary model answered the immediate question of how atoms were structured, yet inadvertently raised other questions. As they orbit the nucleus, electrons accelerate. Accelerating electrons emit electromagnetic waves—that is how we get X rays, for example. In so doing, however, the electrons would constantly lose energy. Each emission of electromagnetic energy would decrease the electron's kinetic energy. Like the satellite that gradually loses energy to friction and falls back to earth, the electron *should* gradually spiral back into the nucleus. Rutherford still could not explain why atoms continue to exist—why they do not simply collapse.

The Bohr Model of the Atom

Early in the twentieth century, Niels Bohr examined the problems posed by Rutherford's planetary model of the atom in terms of the discrete emission and absorption spectra of chemical elements. Collections of atoms, like those found in a sample of hydrogen gas for example, do release energy in the form of discrete lines in the electromagnetic spectrum. They also absorb energy—precisely the same lines in the electromagnetic spectrum. If we assume that the electrons are responsible for these lines, then the details of emission and absorption spectra are clues to the internal structure of the atom.

Bohr linked the frequency of emission and absorption lines to a specific quantity of energy, using the relationship that energy is equal to Planck's constant times the frequency of the light. If electrons emitted electromagnetic energy continuously as they spiraled inward toward the nucleus, they would emit all possible frequencies of light (Figure 18-5(a)). We would expect to see a continuous spectrum of light, a rainbow of colors. The observation of discrete rather than continuous spectra suggested to Bohr that electrons release energy while making jumps from one orbit to another, as shown in Figure 18-5(b). However, even jumps from one orbit to another could produce continuous spectra. If all orbits were equally possible, then all possible jumps could occur. The electrons would emit all energies and we would again see a continuous spectrum.

Bohr needed a radical modification to Rutherford's planetary model of the atom. He suggested that electrons move only in certain allowed orbits at specific distances from the nucleus, like those shown for the hydrogen atom in Figure 18-6. While moving in these allowed orbits, electrons do not emit any

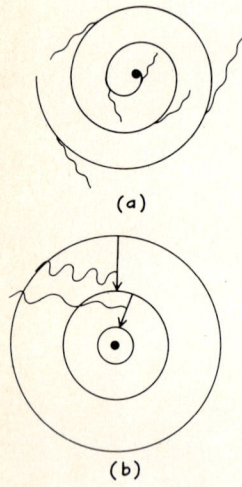

Figure 18-5
(a) Electrons that release energy as they spiral into the nucleus should produce a continuous spectrum.
(b) Bohr proposed that the electrons make jumps from one allowed orbit to another, producing a discrete emission spectrum.

electromagnetic energy. When moving from one orbit to another, however, the electrons absorb or release energy. If the electron moves to an orbit farther from the nucleus, it absorbs an amount of energy equal to the energy of a photon in one of the absorption lines. When the electron jumps to an orbit closer to the nucleus, it releases an amount of energy equal to the energy of a photon in one of the emission lines. Allowed orbits restrict the possible jumps to just a few. Consequently, the light spectrum emitted by atoms would be discrete. Since Bohr's model uses the same orbits to explain both emission and absorption of energy, it predicts emission and absorption lines that have identical frequencies.

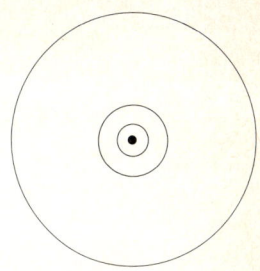

Figure 18-6
The allowed orbits for the hydrogen atom.

In proposing that electrons move only in allowed orbits, Bohr was suggesting that the energy of the electron is quantized. In any given orbit, an electron has a certain amount of kinetic energy associated with its motion and electrical potential energy associated with the electrical force of attraction exerted by the nucleus. If electrons move only in allowed orbits, their energy takes on selected or discrete values. Bohr saw the same quantization of energy in the atom that Planck saw in the radiation emitted by solids and Einstein saw in the photoelectric effect. At the atomic level, energy appears in discrete amounts.

The energies associated with each allowed orbit provide a useful way of describing what happens as atoms absorb or emit light. Figure 18-7 shows an

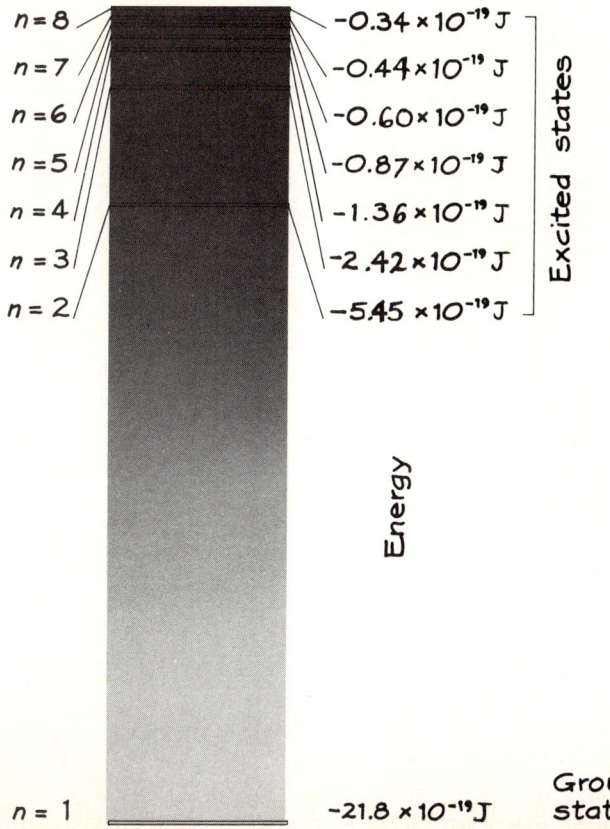

Figure 18-7
The energy level diagram for hydrogen.

energy-level diagram that illustrates the allowed energies for electrons in the hydrogen atom. The $n = 1$ line corresponds to the energy of an electron when it is moving in the smallest orbit, the orbit closest to the nucleus. This is often referred to as the **ground state** of the electron because it represents the smallest energy the electron can have. The $n = 2$ line corresponds to the energy the electron has while in the next orbit; $n = 3$ while in the third orbit; and so on. Collectively, the energy levels above ground state are called **excited states.** The energy of the electron increases as we move to higher excited states.

The energy-level diagram provides a convenient way to describe the energy emitted or absorbed by the atom. The distance between states is proportional to the energy difference between states. So, for example, the energy difference between $n = 3$ and $n = 2$ is smaller than that between $n = 2$ and $n = 1$. Electrons absorb and emit light as they move from one orbit to another. When they move from a lower state to a higher state, electrons absorb energy. When they drop from a higher state back down into a lower state, they release energy. The energies absorbed or released with each jump can be found from the energy level diagram.

In a sample of hydrogen gas at room temperature, each atom is most likely to have its electron in its ground state. An external energy source—heat, light, or electricity—can provide energy to these electrons. When we pass white light through a sample of hydrogen gas for example, electrons "select" photons of energy needed to move into excited states. Electrons might move from $n = 1$ to $n = 2$; from $n = 1$ to $n = 3$; from $n = 1$ to $n = 2$ and then to $n = 3$; and so forth. The photons with energies equivalent to each of these jumps are absorbed while the others are ignored. Consequently, the light that emerges is white light minus the frequencies used to move electrons into excited states. We see a discrete absorption spectrum. Once excited, electrons eventually drop back to the ground state. Again, several paths are possible. Electrons in the $n = 3$ state, for example, can drop from $n = 3$ to $n = 1$ in one jump or they can go from $n = 3$ to $n = 2$ and then from $n = 2$ to $n = 1$. With each jump, the electron emits a photon of light with an energy equal to the energy difference between the two states. Discrete energy levels lead to discrete emission spectra.

In addition to explaining the discrete spectra found for chemical elements, Bohr's model also resolves the problem of why atoms do not collapse. The electron can lose only specific amounts of energy equivalent to the jumps in the energy-level diagram. When it reaches some lowest energy, called the *ground state,* the electron can neither lose more energy nor move closer to the nucleus. The stability of the atom is assured. Bohr's model of the atom moved several pieces of the atomic puzzle into place. It did, however, leave some questions unanswered. Before turning to those questions, let's examine the design of the laser in terms of Bohr's model of the atom.

A STEP FURTHER—MATH

MATCHING THE EMISSION SPECTRUM

Bohr showed that the energy (E) of a given state (n) can be expressed by a simple relationship:

$$E_n = -\frac{E_{\text{ground state}}}{n^2} \quad \text{where } n = 1, 2, 3 \ldots$$

The $n = 1$ state is called the ground state of the electron. At $n = 2$, the first excited state, the electron's energy is one-fourth that at the ground state. At $n = 3$, the energy is one-ninth that of the ground state and so forth. (By convention, the total energy of the atom is expressed as a negative number and approaches zero as the electron moves farther from the nucleus.) For hydrogen, the measured ground state energy of the electron is 21.8×10^{-19} J. Using Bohr's relationship:

$$E_2 = -\frac{21.8 \times 10^{-19} \text{ J}}{(2)^2} = -5.45 \times 10^{-19} \text{ J}$$

$$E_3 = -\frac{21.8 \times 10^{-19} \text{ J}}{(3)^2} = -2.42 \times 10^{-19} \text{ J}$$

We could continue this process until the energy of the electron becomes zero. The energies associated with the first few states of the hydrogen atom are shown in its energy-level diagram (Figure 18-7).

We can use the energies associated with each state to predict the energy released when the electron jumps from a higher to a lower state. Consequently, we could check to see whether Bohr's model predicts the emission lines observed for hydrogen. When an electron moves from a higher state to a lower state, it emits energy equivalent to the energy difference between the two states. For example, an electron that moves from the $n = 3$ to the $n = 2$ state loses:

$$E_3 - E_2 = (-2.42 \times 10^{-19} \text{ J}) - (-5.45 \times 10^{-19} \text{ J})$$

$$= 3.03 \times 10^{-19} \text{ J}$$

We can determine the frequency of light to which this energy corresponds by using the relationship:

$$E = hf \quad f = \frac{E}{h}$$

Since the energy released by the electron is 3.03×10^{-19} J, we have:

$$f = \frac{E}{h} = \frac{3.03 \times 10^{-19} \text{ J}}{6.62 \times 10^{-34} \text{ J} \cdot \text{s}} = 4.58 \times 10^{14} \text{ Hz}$$

This corresponds to the frequency of the red line in the hydrogen spectrum!

Bohr's model of the atom predicted the frequencies of light found in the emission spectrum of hydrogen. Three other lines are visible. They have frequencies of 6.2×10^{14} Hz, 6.9×10^{14} Hz, and 7.3×10^{14} Hz. All occur when the electron jumps from a higher state down to the $n = 2$ state. Can you figure out to which jump each line corresponds?

> ### SELF-CHECK 18B
>
> Use the energy-level diagram for hydrogen shown in Figure 18-7 to describe:
>
> a. the possible jumps an electron could make in moving from the $n = 4$ level to the $n = 1$ level
>
> b. the number of different lines in the emission spectrum that could be produced by these jumps
>
> c. the order of jumps, from least to most energy released

LASERS

The laser is one of the most interesting practical results of our increased understanding of the internal structure of the atom. Invented in 1960, the **laser** (Light Amplification through Stimulated Emission of Radiation) has found many applications. The fundamental process of light emission is the same in a laser as in an ordinary light bulb. Electrons move from a higher energy level to a lower one, releasing energy in the form of light. However, our understanding of energy levels in the atom has enabled us to manipulate this process to produce coherent light.

Lasers and Coherent Light

Light emitted by a laser differs from light emitted by a light bulb because of a single property, called *coherence*. Laser light is coherent; light emitted by a light bulb is incoherent. **Coherent light** has two characteristics: (1) it has a single wavelength, and (2) the waves produced by individual atoms are in phase, or in step, with one another. We examine each of these characteristics in more detail.

Light emitted by most light sources does not consist of a single wavelength. While the light has a characteristic color, this color is usually a composite of at least several discrete lines in the emission spectrum. The whitish blue color of a mercury-vapor lamp, for example, comes from the violet lines in the mercury spectrum; but other lines (green and yellow) are also present. We can convert ordinary light into single-wavelength light by using a filter to block all wavelengths but one. The light is then a single wavelength, but it is not yet coherent.

The second characteristic of coherent light is that the waves produced by individual atoms are in step with one another. Ordinarily, excited atoms emit light spontaneously. The light waves produced by individual atoms move independently of one another (Figure 18-8(a)). The waves often interfere destructively, reducing the intensity of the light finally emitted. In a laser, atoms are stimulated to emit each photon so that its wave has a definite relation to all other waves. The crests of all waves line up (Figure 18-8(b)) and interfere

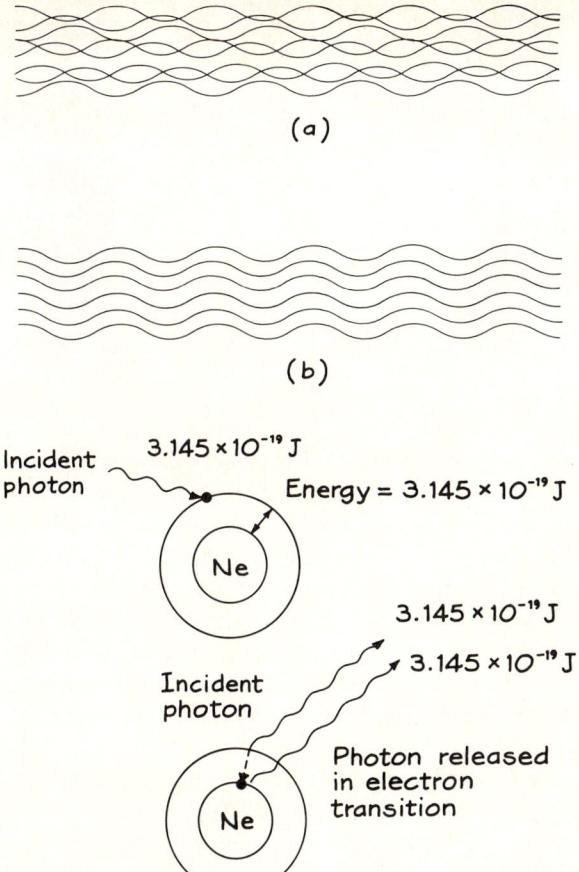

Figure 18-8
(a) Ordinary light sources produce waves that are independent of one another. (b) Coherent light consists of waves that are in step with one another.

Figure 18-9
Neon in its excited state can be stimulated to drop to a lower energy state by a photon equal in energy to the energy lost by the neon atom.

constructively with one another. Using an analogy drawn from rock concerts, coherent light is similar to applause that occurs when people keep time with the music. It is stimulated by the music, so everyone claps at the same time. Incoherent light is like the general applause that concludes a performance.

The process by which coherent light is produced needs to do two things. First of all, it needs to emphasize one specific electron transition in order to produce light of a single wavelength. This can be accomplished by moving a majority of the atoms into a single excited state. Secondly, each atom must be induced to emit its photon so that the crests of each light wave are lined up with the others. This occurs as a result of stimulated emission.

Stimulated Emission of Radiation

Stimulated emission of radiation occurs when an atom in an excited state interacts with a photon of precisely the energy that would be emitted by the atom as it drops to a lower energy state. This interaction stimulates the atom to make the transition and to release a photon that is coherent with the incident photon. For example, suppose one of the electrons of a neon atom is in an excited state, as illustrated in Figure 18-9. The energy emitted by the electron if it moves down to the next lowest orbit is 3.145×10^{-19} joules (J).

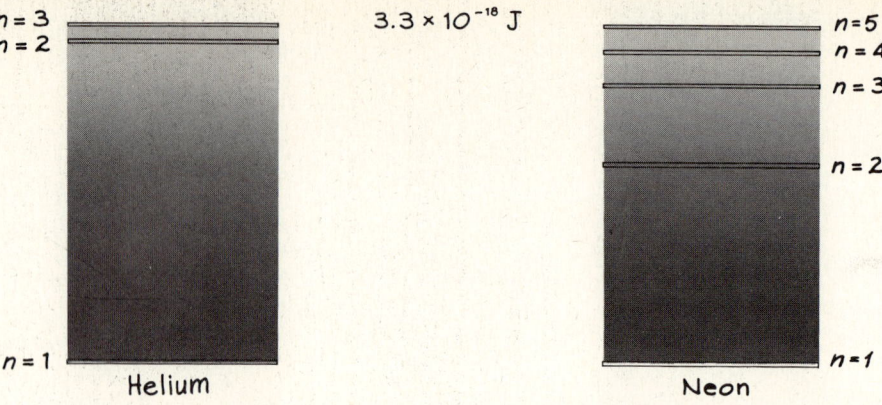

Figure 18-10

The $n = 3$ state of helium and the $n = 5$ state of neon are both 3.3×10^{-18} J above the ground state. Helium atoms in the $n = 3$ state can excite the neon atoms to move to the $n = 5$ state.

When a photon of 3.145×10^{-19} J interacts with this excited atom, it will stimulate the electron to drop to the lower orbit. A photon of 3.145×10^{-19} J is emitted. The stimulation does more, however. In effect, the incoming photon "shakes" the electron to cause it to change orbits. In the process, the electron emits a photon that matches crest and trough of the incident photon. The two photons are in step with one another. Each of these photons can, in turn, stimulate other excited atoms in the same state to emit identical photons. Since all the affected electrons make the same energy jump, the resulting radiation is of a single frequency. And since all the photons are emitted by stimulation, the radiation is coherent.

We might expect stimulated emission to be relatively simple to produce. In fact, however, it is somewhat complex. First of all, you have to move a great many atoms into a single excited energy state. Secondly, the atoms must remain in that state long enough to enable the photon to interact with the electron. Most atoms emit light approximately one-millionth of a second after having been excited, too short a time for the photon-electron interaction. However, a few excited states last a few thousandths to a few hundredths of a second. These states can be used for stimulated emission.

Helium-Neon Lasers

Lasers use the process of stimulated emission to produce the intense, coherent light for which they are famous. While several materials can be used to establish this process, helium and neon—used in the helium-neon laser—are the most common. Let's examine the role of each element in establishing stimulated emission of light.

Helium atoms are used to raise the neon atoms to the particular excited state required for stimulated emission. The neon atoms then emit the photons that we collectively call the laser light. Figure 18-10 shows the energy level diagrams of helium and neon. For convenience, we have called the ground-state energy zero and expressed the energy of higher levels relative to it. Helium and neon both have excited states at 3.3×10^{-18} J above the ground state—the $n = 3$ state for helium and the $n = 5$ state for neon. When a neon electron reaches the $n = 5$ state, it remains there for a relatively long time.

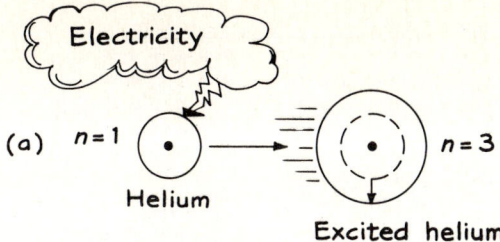
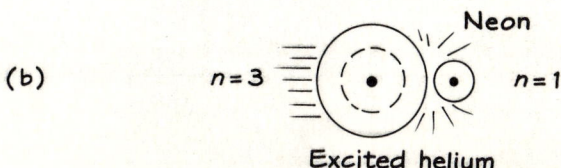
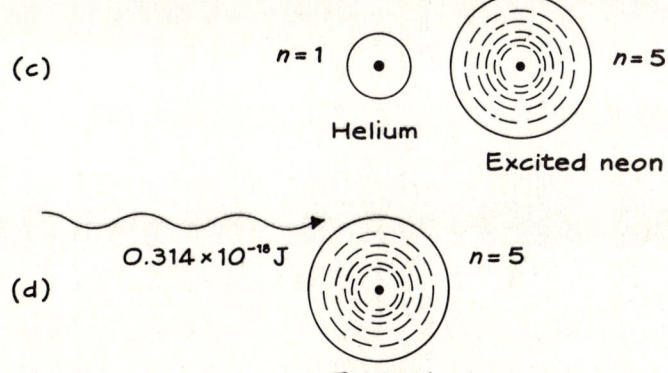
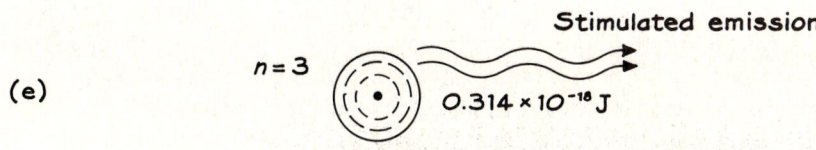

Figure 18-11
A high voltage is applied to a mixture of helium and neon gas, exciting some helium atoms to the $n = 3$ state. Collisions between excited helium atoms and neon atoms excite neon atoms to the $n = 5$ state. A single photon of the proper energy can then stimulate neon atoms to make a transition from the $n = 5$ state to the $n = 3$ state, emitting red light.

Thus, it is an excellent candidate for stimulated emission. When it does change states, the electron moves down to the $n = 3$ state and releases a photon of red light.

Figure 18-11 shows the complete process by which a helium-neon laser emits coherent, red light. A relatively high voltage (1000 volts) is applied to a tube containing a mixture of helium and neon gas. (Usually the mixture consists of approximately 85% helium and 15% neon.) The high voltage excites some of the helium atoms to the $n = 3$ state (Figure 18-11(a)). These excited helium atoms collide with neon atoms (Figure 18-11(b)), exciting the neon atoms to their $n = 5$ state. As the excited neon atoms are stimulated to drop down to the $n = 3$ state (Figure 18-11(d)), each of them emits a single photon of 0.314×10^{-18} J (Figure 18-11(e)). Each of these photons will stimulate the emission of other 0.314×10^{-18} J photons by other excited neon atoms.

418 Chapter 18. Light, Quanta, and Atoms

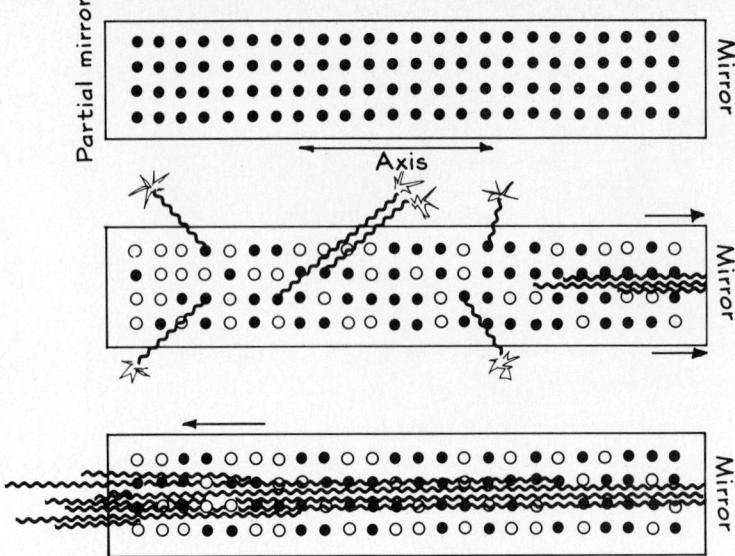

Figure 18-12
Mirrors reflect photons traveling along the axis of the tube. Those not traveling along the axis merely escape and do not contribute to further stimulated emissions.

The construction of the helium-neon laser is shown in Figure 18-12. Mirrors at both ends of the tube reflect most of the emitted photons in order to build up an intense beam of light by repeated reflection. A small amount of coherent light escapes after each complete trip up and down the laser. Photons not traveling along the axis of the tube merely escape and do not contribute to further stimulated emissions. Consequently, the light emitted by the laser is well focused into a narrow beam.

The helium-neon laser is a very inexpensive laser to produce and consequently has found wide application. Other lasers, using solids, liquids, and gases as the lasing material, are also used widely. Lasers find broad applications in fields like the construction industry, medicine, energy research, and even entertainment (Figure 18-13). Lasers are often used to align underground tunnels, the most famous application being the alignment of the Bay Area Rapid Transit (BART) tunnel under the San Francisco Bay. Surgeons use the intensity and small diameter of laser beams to assist them in performing microsurgery. Grocery store cash registers read product codes with laser light. Lasers are now used as a type of "record needle" for videodiscs. The puzzle of atomic spectra has lead to an enormously useful array of laser devices.

THE QUANTUM-MECHANICAL ATOM

Rarely do we solve a puzzle on our first attempt. Mastermind requires at least two tries and sometimes many, many more. We do crossword puzzles with pencils so we can erase. In a sense, each try is a temporary model. We take the observations we have, build a temporary model, and then test it with additional observations. Eventually we reach a final model, a solution to the puzzle. Building models in science is in some ways the same. We take the available observations, build a model, and then look for additional observa-

The Quantum-Mechanical Atom 419

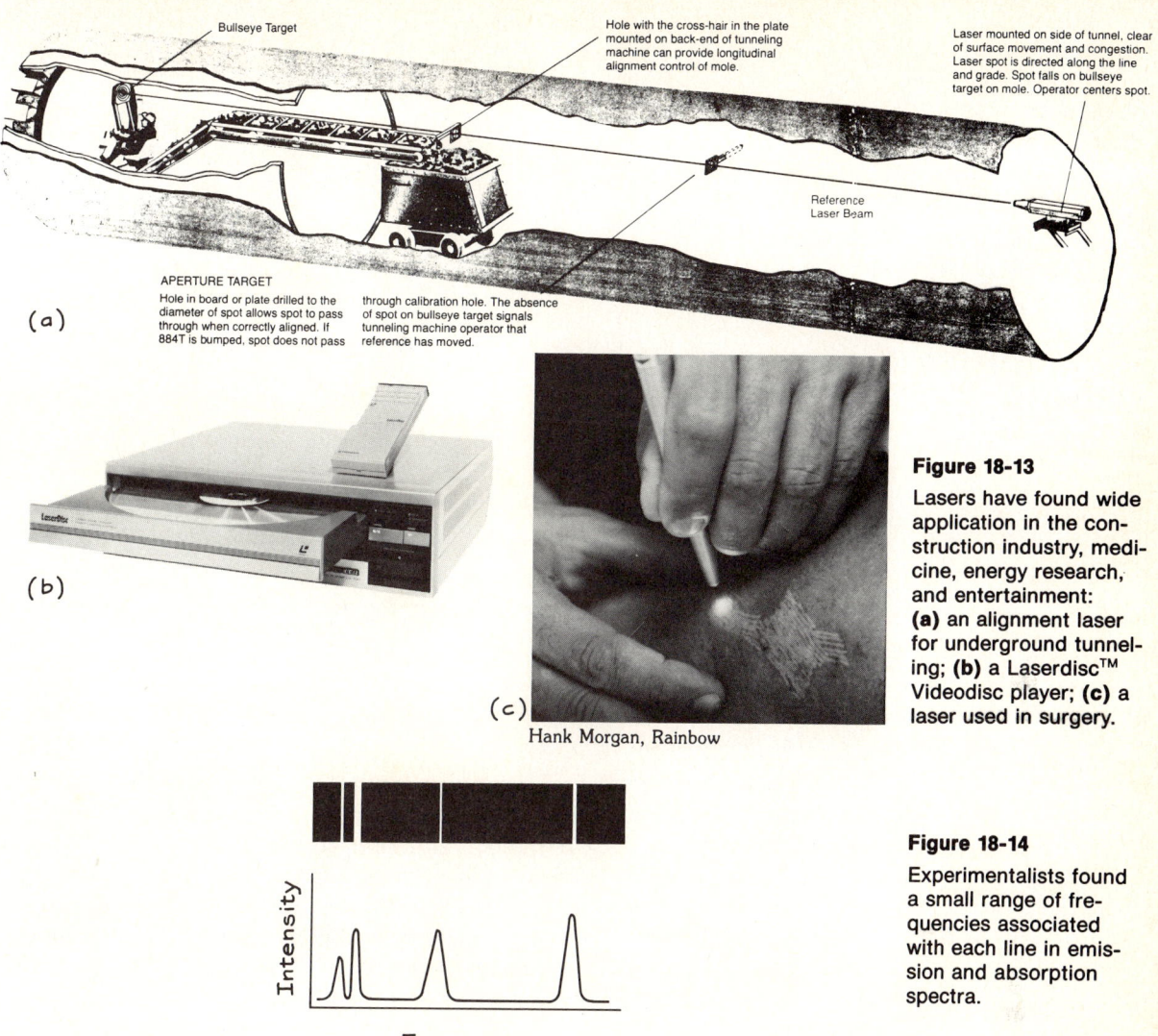

Figure 18-13
Lasers have found wide application in the construction industry, medicine, energy research, and entertainment: **(a)** an alignment laser for underground tunneling; **(b)** a Laserdisc™ Videodisc player; **(c)** a laser used in surgery.

Hank Morgan, Rainbow

Figure 18-14
Experimentalists found a small range of frequencies associated with each line in emission and absorption spectra.

tions that can test the model's validity. Sometimes the model survives; sometimes it does not.

Bohr's model of the atom was a temporary one. While the concept of allowed orbits explains both the stability of the atom and discrete emission and absorption spectra, it is not based on any known principles of physics. There seems to be no reason for electrons to jump instantaneously from one energy state to another, shunning all energies in between. Furthermore, additional observations revealed limitations to the model. While it was enormously accurate in describing the hydrogen spectrum, Bohr's model was far less successful in describing the spectra produced by more complex atoms. In addition, more detailed measurements showed that the so-called lines in atomic spectra are not confined to the frequencies predicted by Bohr's allowed orbits. Each line is really a narrow band centered at the predicted frequency (Figure 18-14). Moreover, the intensity of individual spectral lines varied.

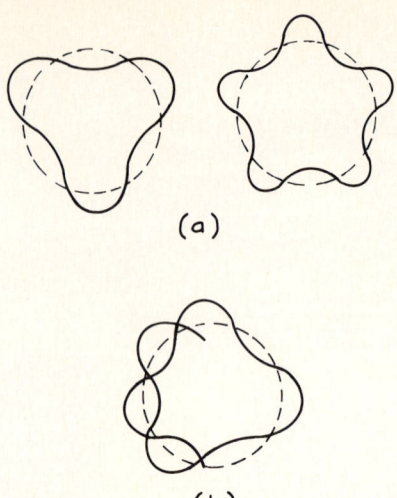

Figure 18-15
Because the electron moves in a circular orbit, the de Broglie wavelength must close upon itself. **(a)** Orbits exist only where a wave can close on itself in phase, so that crest meets crest or trough meets trough. **(b)** Other matches produce destructive interference and the wave dies out.

Some lines were brighter than others—an observation for which Bohr's model offered no explanation. Bohr's picture began to grow fuzzy. The resolution to these problems lay in yet another model of the atom—the quantum mechanical model.

De Broglie Wavelengths and Circular Standing Waves

As we saw in Chapter 17, de Broglie wavelengths explain the wavelike behavior of electrons as they move through a thin metal foil. Like light waves and X rays, electrons produce diffraction and interference patterns. If electrons orbiting a nucleus also have wavelike characteristics, each of them should have a de Broglie wavelength associated with its momentum. An electron's momentum varies with the distance of its orbit from the nucleus; consequently a unique wavelength can be associated with each possible orbit. Electrons in orbits near the nucleus have longer wavelengths than electrons in orbits far from the nucleus.

If we forget about electrons for a moment and think just in terms of waves, circular paths demand that a wave close in upon itself. Closing waves into circles presents circumstances similar to those that result in resonance and standing waves. If the wave closes so that a crest meets a crest or a trough meets a trough (Figure 18-15(a)), the wave reinforces itself. We would have a **circular standing wave.** If a wave closes so that a crest meets a trough, it cancels itself. Other matches (Figure 18-15(b)) create a jumble that gradually dies out.

In applying his concept of matter waves to the Bohr model of the atom, de Broglie showed that Bohr's allowed orbits were those orbits for which de Broglie wavelengths could establish circular standing waves. For a given orbit, an electron has a de Broglie wavelength that depends on its momentum in that orbit. When a whole number of these wavelengths fit into the circumference of the orbit ($n\lambda = 2\pi r$, where $n = 1,2,3,\ldots$), the orbit is stable. When a whole number of de Broglie wavelengths do not fit into the circumference of the orbit, the orbit is unstable and the electron moves on. Electrons select those orbits in which their de Broglie wavelengths can establish circular standing waves.

De Broglie's matter waves provided a much-needed mechanism to explain the discrete orbits found in atoms. Once a wavelike character can be associated with the electron, the identification of acceptable orbits becomes a mathematical puzzle in which the orbit's circumference (determined from the radius of the orbit) must match a whole number of de Broglie wavelengths (determined by the electron's velocity at that radius). Classical concepts of force, momentum, and circular waves are integrated with the modern concept of matter waves.

SELF-CHECK 18C

De Broglie wavelengths and circumferences have been calculated for a series of possible orbits in the Bohr atom. Select the orbits that satisfy de Broglie's criteria and indicate, in each case, how many de Broglie wavelengths fit into the orbit.

Orbit	de Broglie Wavelengths	Circumference
A	3.33×10^{-10} m	3.33×10^{-10} m
B	4.71×10^{-10} m	6.66×10^{-10} m
C	6.67×10^{-10} m	13.34×10^{-10} m
D	5.72×10^{-10} m	9.80×10^{-10} m
E	9.99×10^{-10} m	29.97×10^{-10} m

Waves and Probabilities

While de Broglie's circular standing waves give us a way of understanding why some orbits are allowed and others are not, they thrust us back into the ambiguous world of wave-particle duality. Added to the diffraction and interference phenomena discussed in Chapter 17, the wave nature of orbiting electrons provides impressive evidence that particles do behave like waves. But most of us end up wondering just what all this means. When we say that a moving ball or electron has a wavelength, what does this wavelength represent? A few years after de Broglie published his work, his entire concept of matter waves was incorporated into a much broader model called quantum mechanics. **Quantum mechanics** presents a mathematical description of phenomena in terms of wave equations—a complex description of a wave view of reality. It is extremely abstract and by itself offers us little additional insight. But the probabilistic interpretation given to its results can be very helpful and, in a sense, offers the only real bridge between wave and particle views of matter.

The **probabilistic interpretation** given to the wave view of matter proposes that the waves associated with electrons and other particles describe

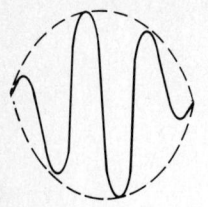

Figure 18-16
A wave packet describes the wave associated with a particle. The larger the amplitude of the wave within the packet, the greater the probability of locating the particle at that point.

the probability of locating that particle at each point in space. For an isolated particle, quantum mechanics predicts a wave like that shown in Figure 18-16, called a **wave packet**. The wavelength of the wave is the de Broglie wavelength, determined from the momentum of the particle. The amplitude of the wave varies from zero at the ends of the packet to a maximum in the middle. According to the probabilistic interpretation, the amplitude squared at each point in space describes the probability of finding the particle at that point. For the wave packet shown in Figure 18-16, the probability of finding the electron is the largest at the center of the packet and essentially zero near the ends and outside the packet.

SELF-CHECK 18D

The sketches in Figure 18-17 describe the wave packets associated with two electrons. Which has the greater momentum? Use the probabilistic interpretation of the wave amplitude to describe the probable locations of each electron. Which can be located within a narrower region of space?

Figure 18-17

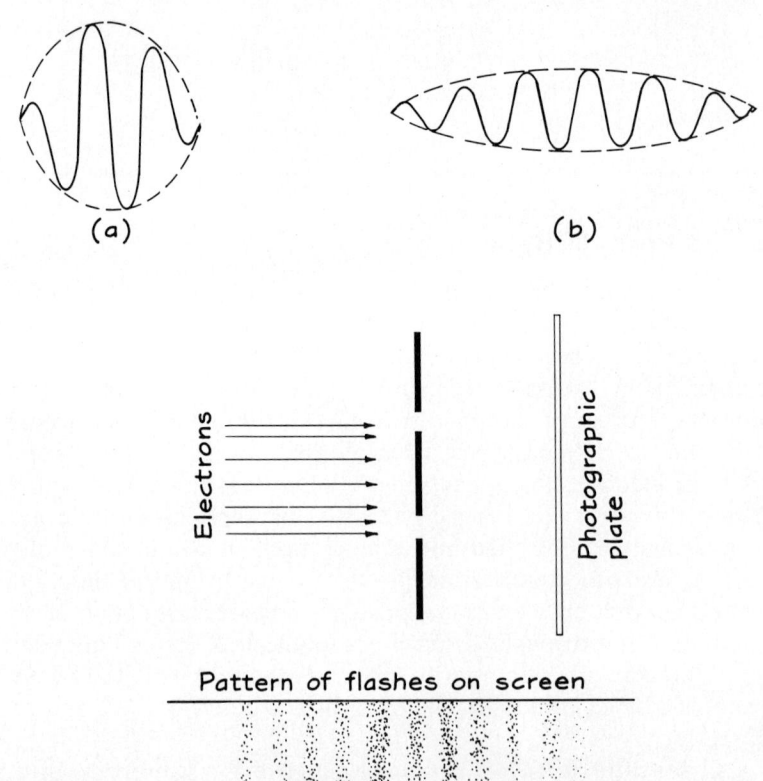

Figure 18-18
We can think of an interference pattern as describing the most probable paths of a collection of electrons. Most go to regions of constructive interference: few go to regions of destructive interference.

We can extend this probabilistic interpretation to the behavior of a group of electrons, like those that create diffraction and interference phenomena. Let's imagine that we fire a continuous stream of electrons at a barrier with two slits in it. The electrons that make it through the slits strike a photographic plate and leave a dot exposed on the plate. At first the dots seem to be spaced across the plate somewhat randomly. But gradually a pattern builds up (Figure 18-18). The electrons are not spread randomly across the plate—they are clustered in bands that are identical to interference bands produced by waves. Regions where a lot of electrons go alternate with regions where few electrons go. We can think of the interference pattern produced by waves as a description of the probability that an electron will reach each point of the screen.

Probabilities and the Atom

When applied to the atom, quantum mechanics presents a model of the atom that is more sophisticated than the Bohr model. Quantum mechanics replaces the discrete orbits in the Bohr atom with probabilities. Imagine for a moment that we measure the position of the electron in the hydrogen atom and record that position with a small dot. If we repeat this experiment hundreds of times, a pattern like that in Figure 18-19(a) builds up. Dots are spread about throughout the space outside the nucleus, but they are more dense in some regions than others. The region of space in which the density of the dots is the largest corresponds to the first Bohr orbit. In this sense quantum mechanics presents a model of the atom that corresponds to Bohr's model. However, Bohr's restriction that electrons exist only in discrete orbits is replaced with a description that says that electrons are most probably found in regions that correspond to Bohr's orbits. Theoretically, the electron can exist anywhere in the atom. Quantum mechanics replaces Bohr's discrete orbits with fuzzy **electron probability clouds.**

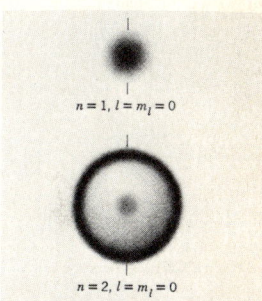

Changes in state can be described by sketches like those in Figure 18-19. Bohr's model of the atom leads us to imagine an electron spontaneously jumping from one orbit to another, from one energy level to another. Quantum mechanics replaces this image with one in which one picture dissolves into another. As an electron moves from the $n = 2$ state to the $n = 1$ state, the distribution in (b) fades into and becomes the distribution in (a). The transition occurs over an extremely short time interval, 10^{-8} seconds, during which time light is emitted. The reverse happens when light is absorbed. As the electron gains energy, the probability distribution in (a) fades into and becomes the distribution shown in (b). Our knowledge of the electron remains incomplete. The different probability distributions reflect the energy of the electron but provide only a guess at where the electron might actually be found.

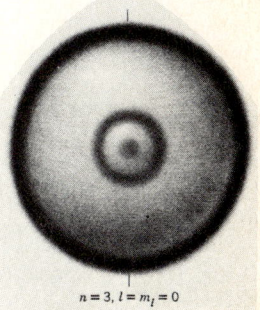

Figure 18-19

The electron probability clouds for $n = 1$, $n = 2$, and $n = 3$ states of the hydrogen atom.

Ultimately, the success of the quantum mechanical model of the atom rests with its ability to take up where Bohr's model left off. Bohr's model was unable to describe the spectra of the more complex atoms, the different intensities of spectral lines, and the broadening of individual spectral lines. The explanation of the broadened spectral lines lies in the probabilistic nature of electron orbits. Since the quantum mechanical atom predicts only probabilities, the electron can be in a variety of locations about the nucleus. When electrons change state, the most probable transition they will make is from one Bohr orbit to another. In reality, of course, they can make any number of jumps—some larger and some smaller than the discrete jumps predicted by Bohr's model of the atom. The spectral lines will be broadened slightly because electrons are making transitions slightly larger and slightly smaller than the transition predicted by Bohr orbits. The different intensities of the spectral lines reflect the different probabilities associated with each transition. Some transitions are very improbable and, thus, occur very seldom. Others occur more frequently, to the point that we see a dim spectral line. Finally, highly probable transitions occur frequently. These yield the most intense spectral lines. The different intensities of spectral lines arise from the different probabilities associated with each transition. While the mathematics becomes increasingly complex, quantum mechanics has been applied to complex atoms as well as to the collisions among atoms and detailed structures of materials. Quantum mechanics provides us with a significantly more powerful model of atomic structure.

Today the quantum mechanical model of the atom is regarded as the most complete picture of atoms and their interactions. It fits observations and experimental results extremely well. But, is it the correct solution to the puzzle of the atom? We do not know. Here our analogy between games and puzzles and model-building in science fails. Ultimately, there are no "correct solutions" in science. All scientific models and theories are temporary—reflections of the observations available at the time they were built. As we learn more, our theories change. Perhaps that is what keeps the puzzle so interesting—what keeps physicists doing physics.

PROBABILITY AND UNCERTAINTY

The concept that matter waves describe probabilities is at once comforting and disquieting. When applied to diffraction and interference effects where we are dealing with many electrons at once, a probabilistic interpretation seems reasonable. After all, a lot of electrons can travel a lot of different paths, and we can use probability to show the collective result, just as we can use it to show the collective results of tossing a coin a thousand times. When applied to a single electron, however, the concept seems disquieting. Quantum mechanics predicts the probability distribution for a single electron, not a collection of one thousand one-electron atoms. We have replaced the concept of knowing where an electron is with knowing the probability of its being at each point in space. This is rather like saying we can never know whether an individual coin has landed heads or tails up, only how likely it is to end up one way or the other. Such an idea is at odds with our everyday experience!

Limits to Measuring

Traditionally, physicists believed that they measured, and consequently described, phenomena from an objective viewpoint. According to the traditional

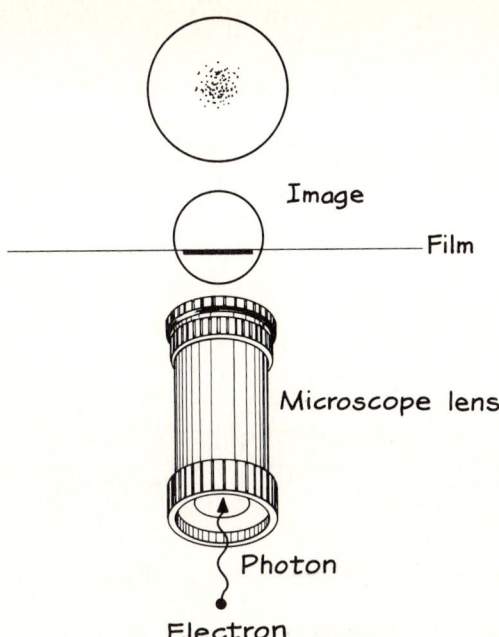

Figure 18-20
We "see" an electron by causing light to interact with it. Light strikes the electron, is reflected into a microscope lens and transmitted to a piece of photographic film.

view, it should be possible to measure the position and velocity of an object as precisely as the measuring instruments allow. The fact that these measurements do not tell us the exact position or velocity of a particle seemed to be a reflection of the measuring instruments, not of our ability to know these quantities.

Quantum mechanics and its probabilistic interpretation force us to conclude that we cannot know things as exactly as we once thought. To understand how this inherent uncertainty arises, we conduct a thought experiment. Suppose we want to measure the position of a stationary electron as precisely as possible. Measuring the position of something usually involves seeing it in some way, so we use light to help us detect the electron. Light strikes the electron and is reflected into a microscope lens and transmitted to a piece of photographic film (Figure 18-20). There, the magnified image of the electron's location is recorded. We generally trace backward from the film image to infer the original location of the electron.

Difficulties arise when we think about the interaction between the light and the electron. Scientists have always conceded that the process of measurement alters the thing being measured, but traditionally they have contended that this alteration is quite small. When light strikes an electron, the energy and momentum carried by its quanta can be transferred to the electron, thus altering the electron's position and momentum. We can minimize this effect by allowing only one quantum of light to strike the electron, but that single quantum will still affect the electron.

We can examine what happens to our measurements of position and momentum when we try to minimize the effect the experimenter has by looking at the electron with a single quantum of light. In order to restrict ourselves to this single quantum, we look through narrower and narrower microscope lenses. Because of diffraction, however, smaller and smaller lenses produce

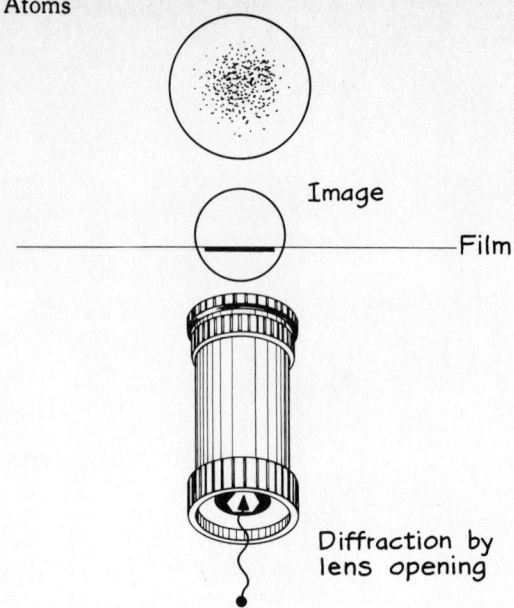

Figure 18-21
Because of diffraction by the microscope lens, a well-defined image of the electron is spread into a diffuse circle. The diffuse circle describes an uncertainty associated with locating the electron.

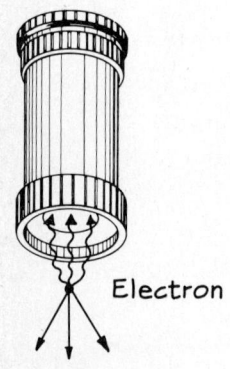

Figure 18-22
Once it has collided with the photon, the electron can move away in one of many possible directions.

images of the electron that are more diffuse (Figure 18-21). As we try to minimize the momentum transferred to the electron during the process of measuring, our knowledge of where the electron is becomes more and more vague. The uncertainty in locating the electron increases. Suppose we relax our one quantum restriction and allow more quanta to interact with the electron. Now our knowledge of the position of the electron improves, but our knowledge of the direction it will move once struck by the photons (Figure 18-22) deteriorates. What we have gained in being more able to locate the electron precisely, we have lost in knowing where it will be after we have found it. Large uncertainties in momentum make it more difficult for us to predict where the electron goes after the collisions with the quanta.

You might imagine that we could design an experimental apparatus that would reduce the uncertainties in position and in momentum. For example, we can reduce the uncertainty in momentum by using light quanta of lower and lower energies. Consequently, less energy is transferred to the electron. However, quanta of lower energy correspond to waves of longer wavelength. Since the spreading of light due to diffraction depends directly on the wavelength, the diffraction pattern spreads as the wavelength increases. The location of the electron would now become less certain. If we decrease the wavelength to locate the electron more precisely, we simultaneously increase the frequency and transmit more energy to it, making the momentum less certain. We cannot know both position and momentum with precision at the same time.

Heisenberg Uncertainty Principle

The lack of certainty inherent in physical measurement was first described by Werner Heisenberg. The **Heisenberg uncertainty principle** states that in any measurement, the uncertainty in momentum times the uncertainty in position must be greater than a fixed number, Planck's constant divided by 2π, or 1×10^{-34} J · s. In equation form:

$$\begin{pmatrix}\text{Uncertainty} \\ \text{in momentum}\end{pmatrix} \times \begin{pmatrix}\text{uncertainty} \\ \text{in position}\end{pmatrix} \geq \frac{\text{Planck's constant}}{2\pi}$$

This principle places a lower limit on the product of the uncertainty in the momentum and the uncertainty in the position. While the uncertainties can change, they must change in response to one another. If the uncertainty in position decreases, the uncertainty in momentum increases, and vice versa. Measuring one quantity precisely limits our knowledge of the other. Expressed in words, Heisenberg's principle says that we can never know both the exact position and the exact momentum of any object at the same time.

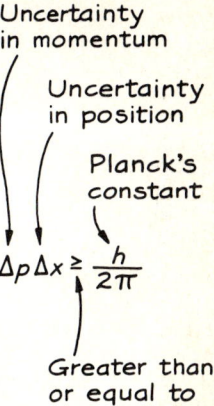

A STEP FURTHER—MATH

HEISENBERG'S UNCERTAINTY PRINCIPLE

An automobile with a mass of 1000 kg moves at a speed of about 25 m/s. Its momentum is 25,000 kg · m/s. A reasonable uncertainty in measuring this momentum might be 5%—1250 kg · m/s. The uncertainty in the position of the car would then be:

$$\Delta x \Delta p \geq \frac{h}{2\pi}$$

$$\Delta x (1250 \text{ kg} \cdot \text{m/s}) \geq \frac{6.63 \times 10^{-34} \text{ J} \cdot \text{s}}{2\pi}$$

$$\Delta x \geq 8.4 \times 10^{-38} \text{ m}$$

It is not hard to find the automobile within 10^{-38} meters. This uncertainty is too small to be noticeable.

An electron with a mass of 10^{-30} kg moves at a speed of about 6×10^6 m/s. Its momentum is 6×10^{-24} kg · m/s. Assume the same uncertainty in momentum as we did for the automobile—5%. The uncertainty in momentum would be 0.3×10^{-24} kg · m/s. The uncertainty in position would then be:

$$\Delta x \Delta p \geq \frac{h}{2\pi}$$

$$\Delta x (0.3 \times 10^{-24} \text{ kg} \cdot \text{m/s}) \geq \frac{6.63 \times 10^{-34} \text{ J} \cdot \text{s}}{2\pi}$$

$$\Delta x \geq 3.5 \times 10^{-10} \text{ m}$$

This uncertainty is about the size of th While it seems small to us, it is eno the electron. With such a large u would not be able to locate th its atom.

The size of Planck's constant makes the uncertainty i position too small to notice for automobiles but too ignore for electrons. At the atomic level we begi an inherent uncertainty or unknowability in na

"BUT YOU CAN'T GO THROUGH LIFE APPLYING HEISENBERG'S UNCERTAINTY PRINCIPLE TO EVERYTHING."

© 1979 by Sidney Harris.

The size of Planck's constant limits the situations in which the uncertainty principle is noticeable. If we calculate the uncertainties in position and momentum associated with objects larger than subatomic particles, we find them to be incredibly small. For example, we can locate a grain of wheat within about a millimeter, 10^{-3} meters (m). Heisenberg's uncertainty principle then predicts an uncertainty in momentum of no less than 10^{-31} kilogram · meters/second (kg · m/s). As it is being blown through the air, the wheat grain might have a momentum of 10^{-6} kg · m/s. The uncertainty, 10^{-31} kg · m/s, is so small compared to the momentum itself that we cannot possibly notice it. It is, therefore, quite easy to follow the motion of the wheat grain.

On a more philosophical level, Heisenberg's principle can be interpreted as either an uncertainty or an indeterminacy. When we say that position and momentum are uncertain, we mean simply that we cannot measure them as precisely as we might like. The principle describes limitations of the observer, not of nature. When we say that position and momenta are indeterminate, we mean that we will never be able to measure them precisely. The limitation is one of nature. If you think of position as a measure of an object's "present" and momentum as a measure of its "future," then the Heisenberg uncertainty principle describes the inherent unknowability of the future based on an accurate knowledge of the present.

CHAPTER SUMMARY

Atoms produce discrete spectra characteristic of the chemical element they represent. Light that has been emitted by an atom produces a series of colored lines against a black background, called an *emission spectrum*. Light that has been absorbed by an atom produces a series of black lines against a rainbowlike smear of color, called an *absorption spectrum*. The emission and absorption lines occur at identical frequencies for all atoms of a given chemical element. Emission and absorption spectra serve as fingerprints for the chemical element that produced them.

Light spectra produced a significant clue as to the internal structure of the atom. Niels Bohr combined Rutherford's nuclear model of the atom with the information provided by emission spectra to develop a model of the atom. The *Bohr model* proposes that electrons orbit atomic nuclei much like planets orbit the sun. Unlike planetary orbits, however, only *selected electron orbits* are permitted. As an atom emits or absorbs energy, the electron jumps into orbits nearer or farther from the nucleus. The energy of the atom associated with these permitted orbits is quantized. The smallest orbit corresponds to the *ground state* of the atom. Higher orbits, $n = 2$, $n = 3$, and so on, are called *excited states*. Bohr's model correctly predicts the centers of the emission and absorption lines for hydrogen.

Lasers can be described in terms of Bohr's discrete orbits. Lasers produce coherent light through the stimulated emission of radiation. *Coherent light* is light of a single wavelength that is in step. It is produced when atoms in an excited state are stimulated to emit a photon by a photon identical to it. Helium-neon lasers accomplish this by using excited helium atoms to excite a large number of neon atoms to the $n = 5$ state. A photon with the same energy as the $n = 5$ to $n = 3$ transition will stimulate other neon atoms to emit photons in phase with it.

Matter waves can explain Bohr's discrete orbits in terms of *circular standing waves* produced by de Broglie wavelengths associated with the electron. The orbits selected by Bohr were, in fact, the only orbits for which a whole number of de Broglie wavelengths would fit into the circumference of the orbit. The use of matter waves to justify Bohr's discrete orbits led to the development of the *quantum mechanical model* of the atom. This model replaces discrete orbits with *electron probability clouds* that describe the probability of finding the electron at each location in space. The regions of greatest probability correspond to the radii of Bohr orbits. The quantum mechanical model is regarded as the most complete model of the atom.

Quantum mechanics has led to fundamental questions regarding the extent to which we can know the physical world. The *Heisenberg uncertainty principle* places limits on the preciseness with which we can simultaneously measure the location and momentum of a particle. The uncertainty in momentum times the uncertainty in position must always exceed Planck's constant divided by 2π. In a sense, Heisenberg's uncertainty principle limits the extent to which we can know the present and future of an object simultaneously.

ANSWERS TO SELF-CHECKS

18A. The spectrum in Figure 18-2 shows emission lines characteristic of mercury and hydrogen.

18B. a. There are six possible jumps: 4 to 1; 4 to 2; 4 to 3; 3 to 1; 3 to 2; and 2 to 1.

b. Each jump produces a unique spectral line. Six spectral lines can be produced by these jumps.

c. We can estimate the relative energy released in terms of the space between the energy level. In order from least to most energy released, we have 4 to 3, 3 to 1, 4 to 2, 2 to 1, 3 to 1, and 4 to 1.

18C. Orbit A: One wavelength fits into the circumference; orbit C: Two wavelengths fit into the circumference; orbit E: Three wavelengths fit into the circumference.

18D. Since the two waves have the same wavelength, the two electrons have the same momentum. The extent of the wave packet in (a) is narrower than that in (b), telling us that the particle's location is restricted more in (a) than in (b). The variation in amplitude of the wave in (a) tells us that the greatest probability for locating the electron is within the center of the packet. There is little variation in amplitude of the wave in (b). Consequently, the probability of finding the particle is about the same for all locations within the packet.

PROBLEMS AND QUESTIONS

A. Review of Chapter Material

A1. Define the following terms:
Emission spectrum
Continuous spectrum
Discrete spectrum
Absorption spectrum
Bohr model
Ground state
Excited state
Coherent light
Stimulated emission
Laser
Wave packet
Quantum mechanical model
Probabilistic interpretation
Electron probability clouds
Heisenberg uncertainty principle
Energy level diagram

A2. In what ways are the processes we use in playing games and solving puzzles similar to the processes scientists use in building models? In what ways are they different?

A3. How are emission and absorption spectra different?

A4. How can emission or absorption spectra be used to identify chemical elements?

A5. How did Bohr's model of the atom draw upon:
a. Rutherford's nuclear model of the atom?
b. emission and absorption spectra?

A6. How are jumps from one Bohr orbit to another related to the discrete lines in emission and absorption spectra?

A7. Describe the role of helium atoms and neon atoms in the helium-neon laser.

A8. How are de Broglie wavelengths used to explain Bohr's discrete orbits?

A9. What characteristics of a wave packet describe:
a. the space within which the particle will be found?
b. the momentum of the particle?
c. the probability of finding the particle at each position inside the wave packet?

A10. In the quantum mechanical model of the atom, what do the Bohr orbits represent?

A11. In what ways do electron probability clouds explain emission and absorption spectra more completely than Bohr's orbits?

A12. According to the Heisenberg uncertainty principle, if we measure the position of an object more precisely, what happens to our knowledge of its momentum?

B. Using the Chapter Material

B1. A lithium atom has a ground state energy of -8.0×10^{-19} J. Use Bohr's relationship to calculate the energy of the first two excited states.

B2. An electron moves from one orbit, total energy of -0.137×10^{-19} J, to a second orbit, total energy of -0.550×10^{-19} J.
   a. What energy is associated with this transition?
   b. Will the energy be emitted or absorbed by the atom?

B3. An electron is initially in an orbit for which the atom has a total energy of -3.6×10^{-18} J. The atom absorbs a photon with an energy of 0.5×10^{-18} J.
   a. What is the total energy of the atom now?
   b. Will the electron be closer to or further from the nucleus?

B4. An electron is in the $n = 5$ orbit of a hydrogen atom. List all the possible ways that the electron can return to the $n = 1$ orbit. What is the total energy associated with each possible route?

B5. An electron orbit has a circumference of 3×10^{-10} m.
   a. List the series of de Broglie wavelengths that would fit into the circumference.
   b. If the orbit corresponds to the $n = 3$ state, which of the possible wavelengths in (a) represents the de Broglie wavelength for an electron in that orbit?

B6. The figure below shows the emission spectrum taken with a photographic plate above a graph showing the intensity associated with each frequency.
   a. In what ways are the two spectra similar?
   b. What additional information is provided by the lower spectrum?

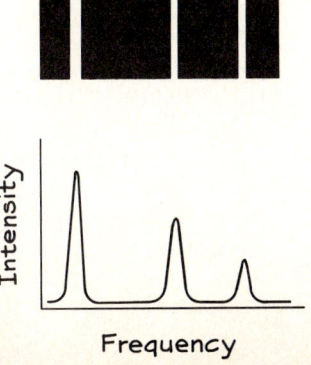

c. List the spectral lines in order from most probable to least probable transitions.

B7. An atom has a relatively long-lived state at -3.1×10^{-18} J. What energy photon is needed to stimulate emission to the -5.6×10^{-18} J state?

B8. Diffraction effects are the major limitations we face in locating a particle precisely. How is diffraction affected if we use light of longer wavelengths? Shorter wavelengths?

B9. The energy exchanged between the particle and the quantum is the major limitation we face in measuring the momentum of a particle precisely. How are momentum measurements affected if we use light of longer wavelengths (lower frequencies)? Shorter wavelengths (higher frequencies)?

C. Extensions to New Situations

C1. Helium was first identified in the absorption spectrum of the sun before its presence was detected on earth. Use the emission spectra in Figure 18-1 to identify the location of the helium absorption lines.

C2. Art historians are frequently faced with determining whether paintings are originals or forgeries. Suppose you have a painting that was supposedly painted in 1704. However, the colors look remarkably like pigments from cobalt blue, a paint mixture not used until 1804. How might you use the ideas presented in this chapter to determine whether the painting contains cobalt blue?

C3. Sodium has 10 inner electrons held in orbit close to the nucleus and a single outer electron that behaves much like the single-electron hydrogen atom. The figure below shows the first six energy levels for the outer electron.

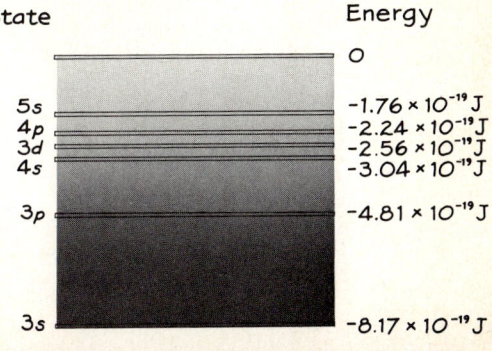

a. How much energy would be required to remove the outer electron from the atom?
b. Sodium produces a characteristic yellow color because of the intensity of a spectral emission line of 5.09×10^{14} Hz. To what energy does this correspond?
c. Use the figure to determine which transition corresponds to the yellow emission line.
d. List the possible energies of photons that could be absorbed or released as the outer electron moves from one orbit to another.

C4. Electrons can be released from gases much as they are released from metal plates in the photoelectric effect.
a. Use Bohr's energy levels for hydrogen to predict the amount of energy needed to release the electron from the atom if the electron were in the $n = 1$ orbit.
b. What would be the threshold frequency needed to release this electron?
c. If we illuminate hydrogen with light of 4×10^{15} Hz, how much kinetic energy will the released electrons have?
d. How would the threshold frequency change if we could somehow keep the hydrogen atoms in excited states?

C5. Black-light posters use the process of fluorescence to emit light. When illuminated with black light, the poster gives off light.
a. A black light emits frequencies in the ultraviolet region of the spectrum. Suppose photons with energy of 5.0×10^{-19} J are incident on the poster. If its atoms need only 4.0×10^{-19} J to move into an excited state, how much energy will be left over from each interaction?
b. This leftover energy is released. Based on its frequency, in what form will this energy be released? (An electromagnetic spectrum can be found in Chapter 15.)
c. The excited atom now returns to its ground state. What frequency of light is emitted? Is it visible?
d. Compare the frequency of the incident light (black-light source) with the frequency of the light emitted by the poster. Explain how the process of fluorescence works.

C6. Many fluorescent lamps are mercury-vapor lamps with a flourescent coating along the outside. The mercury-vapor lamp emits ultraviolet light, as well as visible frequencies of light.
a. Use the results in Problem C5 to explain how the ultraviolet frequencies can be converted into visible frequencies.
b. Does the fluorescent coating increase or decrease the amount of visible light available?
c. Many large mercury lamps place the fluorescent coating along the inside of the container and a mercury light source inside the container. What happens if the container breaks but the mercury light source continues to operate? In these situations, why do people frequently get sunburns? (Ultraviolet light is the primary cause of sunburns.)

C7. The first laser used a ruby crystal that contained chromium impurities. This crystal was surrounded by a flash tube, which emitted sudden flashes of bright green light.
a. The light emitted by the flash was absorbed by the chromium atoms. Based upon the frequency of green light, approximately how much energy did the chromium atoms absorb?
b. After excitation, the chromium electrons would spontaneously drop to a lower energy state but not to ground state. This intermediate state was long-lived, allowing it to serve as a state for stimulated emission. How could electrons be stimulated to move from this state?
c. The light emitted by the chromium atoms when they moved from the intermediate state to ground state was the light characteristically emitted by the laser. Will this light be in the red end or blue end of the spectrum? Why?
d. To get laser action to occur, the ends of the ruby tube reflect light, but the sides do not. Explain why.

C8. Another light-emission process is phosphorescence. In this case a material that has been exposed to light will continue to emit light long after the original light source has been removed. Use the model of the atom to explain how this could occur. (The discussion of long-lived states used in lasers will be helpful.)

C9. Standing waves can exist for particle waves. One example is a particle that is confined to a box. Since the particle cannot leave the box, the amplitude of its matter wave must be zero at the box walls and outside the box. The figure below shows three examples of wave functions for a particle confined to a box.
   a. Which of these wave functions represent the particle with the greatest momentum? With the least momentum?
   b. Why can wavelengths with values between those of A and B or between those of B and C not exist in this box?
   c. Why can momenta between those represented by the waves of A, B, and C not exist in the box?

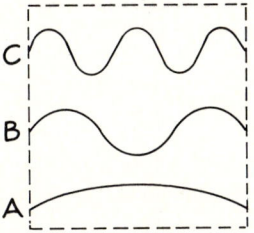

   d. What conclusions can you draw about the particle momenta that can exist in the box? How does this relate to the electron momenta that can exist in the atom?

D. Activities

D1. A diffraction grating will separate light into its separate wavelengths much like a prism. See if you can borrow an inexpensive diffraction grating. Take it outside and view the various lights you see. Describe the spectra and compare them to spectra in the book or emitted by laboratory sources. Identify the types of gases in the lamps.

D2. Tape a diffraction grating on the front of a camera. Use high-speed color film to take photographs of the spectra of interesting lamps.

D3. George Gamow's stories "Quantum Billiards" and "Quantum Jungle" (see Activity D1 in Chapter 17) give you some concept of how Heisenberg's uncertainty principle would affect our experience in the macroscopic world if Planck's constant were a larger number. You might want to read these stories again.

INTERLUDE
Newton to Heisenberg

More than 50 years have passed since the wave and particle models merged to become a new model of the physical world. In the early days of this century, physicists voiced strong arguments for and against wave-particle duality and its interpretations. Now, the arguments have become less emotional: the concepts, less unsettling. Passing years and new generations of physicists have a way of turning a revolutionary thought into a tradition; the new physics into the old physics. In the midst of this settled acceptance of modern physics, we must realize the enormous impact quantum mechanics and wave-particle duality have upon a physicist's view of "reality." We pause briefly to examine the remarkable transformation from the physics of Newton to that of Heisenberg.

When Isaac Newton introduced his three laws of motion, he provided a structure within which we could understand all motion—from the falling apple to the orbiting planet. Once we knew all the forces acting on an object, we could predict all future motions with complete accuracy. By placing certainty squarely within the grasp of human intelligence, Newton created an enormously comforting view of our universe. This feeling of certainty was stated well by the French mathematician Pierre LaPlace:

An intelligence which at a given instant knew all the forces acting in nature and the position of every object in the universe—if endowed with a brain sufficiently vast to make all necessary calculations—could describe with a single formula the motions of the largest astronomical bodies and those of the smallest atoms. To such an intelligence, nothing would be uncertain; the future, like the past, would be an open book.

Newton's model created an image of a rational world proceeding in a rational way—a world view eagerly embraced by philosophers, theologians, and physicists alike.

Beneath this world view lie two very important assumptions. The first is that all events are ordered, not random. To Newton and his contemporaries, all motion was completely determined by whomever or whatever started the universe. These motions obeyed and would continue to obey a series of orderly rules that could be discovered by the careful observer. The second assumption was that the physicist acts as an objective observer of events. Newton and his contemporaries believed that while the measurer does have some impact on the events he or she measures, this impact is minimal and predictable. Events continue, according to a system of ordered rules, with an existence independent of the observer. All that remained was for science to discover the rules.

During the eighteenth and nineteenth centuries, when Newton's laws were applied to objects as small as molecules, this world view prevailed. In principle, physicists believed, once they knew the momentum and position of each molecule, they could predict all future motions of all molecules. Completing these measurements and calculations for a gram of water, let alone the entirety of the universe, was not humanly possible, so statistical or probabilistic descriptions of nature were adopted. Consistent with Newton's world view, probabilities were needed only to compensate for an information overload, not because of the inherent unknowability of nature.

What does the new world view have to say to us about our knowledge? Implicit in the probabilistic interpretation now given to matter waves is the assumption that, on the microscopic level, events are random. Wave descriptions provide us information about the probabilities associated with this random behavior; particle measurements convert these probabilities into brief certainties. Further, objective observers have become active participants in the world that they are trying to describe. Physicists now acknowledge that the types of measurements they undertake affect the observations and models they subsequently construct. Words like *particle, position,* and *path* have no meaning apart from the way in which the experimenter measures them. These words describe our way of ordering the events we see, not a true underlying structure of nature. Newton's view of an orderly nature that exists independent of how we observe it exists no more.

For many physicists the radical departure from more traditional ideas was difficult to accept. Erwin Schrodinger, whose equations were the Newton's laws of quantum mechanics, remained uncomfortable with the probabilistic interpretation given to matter waves. Albert Einstein, whose quantum explanation of the photoelectric effect won a Nobel Prize, also remained unconvinced. He felt that quantum theory was only a stepping stone to a more complete understanding of matter. In this view, probabilities do not represent nature but rather, people's limited ability to comprehend nature. In a letter to Max Born in 1926, Einstein summarized his and perhaps many others' feelings:

Quantum theory is certainly imposing. But an inner voice tells me that it is not yet the real thing. The theory says a lot, but does not really bring us closer to the secrets of the "old one." I, at any rate, am convinced He is not playing at dice.

Only time will tell whether Einstein's inner voice was the voice of wisdom or the voice of a past, unwilling to give way to the future.

From Electricity to the Nucleus

CHAPTER 19
Turning on the Lights

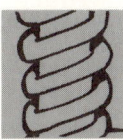
CHAPTER 20
Electromagnetism

CHAPTER 21
Radioactivity

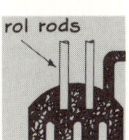
CHAPTER 22
Nuclear Energy

Explorer of nature . . . shaper of the landscape. Two strands of human nature are intricately woven along the path from electricity to the nucleus. Previously of interest only to the explorers of nature, electricity and magnetism remained a curiosity until the middle of the nineteenth century. The realization that vibrating electric and magnetic fields could provide an extraordinarily efficient means of energy transfer aroused the entrepreneur. A glance about our homes or at the network of power lines above remind us of the extent to which we have reshaped our landscape with this knowledge.

Similar strands of human nature can be found in the nuclear studies that have dominated the twentieth century. Seen initially as a curiosity, the radioactive emissions released by uranium aroused only the interest of scientists. By 1950, however, the immense energy released during fission and fusion events had aroused the interests of the entrepreneur and general alike. Once again, we reshape the landscape in response to the knowledge and understanding we have gained.

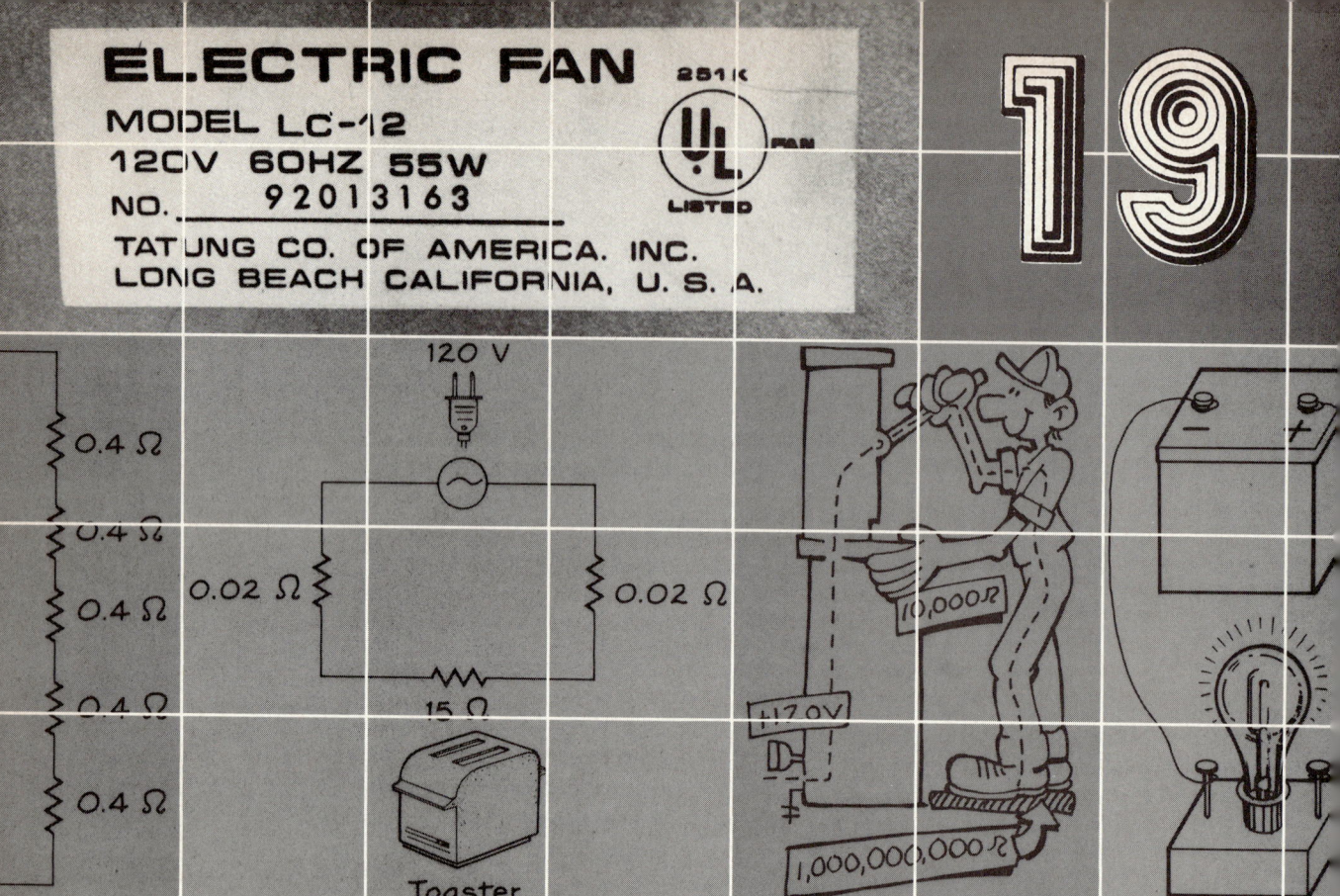

Turning on the Lights

"Live better electrically" was an advertising slogan of the electrical industry during the 1950s. The advantages were clear. Electrical energy could be transmitted over hundreds of kilometers through relatively small wires, was instantly available, and was controlled at the flip of a switch. Automatic devices ranging from washing machines to garage door openers offered more freedom and less drudgery to the average American family. As we used more and more electrically operated gadgets, we required larger amounts of electrical energy. By the mid-1970s our needs began to outstrip our supplies. Living better electrically began to threaten limited energy resources.

In this chapter we examine the process by which electrical energy is transferred from an energy source to an energy receiver. Lightning and sparks are uncontrolled transfers of energy, while *electric circuits* provide a means of controlling the transfer. The *voltage* and *current* in the circuit describe the rate of energy transfer. The *resistance* of any energy receiver states how that receiver resists or retards the motion of the current. Given two of the three circuit variables—voltage, current, and resistance—*Ohm's law* allows us to determine the third. Circuits that involve two energy receivers can be either *series circuits* or *parallel circuits*. We discuss the way in which each configuration affects the total power and current delivered by the energy source. Electrical energy is delivered to receivers by electrons that move through the components in the circuits. The motion of these electrons depends on the type of circuit, the voltage of the source, and the resistance of the receiver.

SPARKS AND LIGHTNING

The electrical interaction, one of the four fundamental interactions, has proven useful in explaining the structure of the atom and is one of the most common ways by which we transport energy from one place to another. As we have seen, electrical charges come in two varieties, positive and negative. In matter, the positive charges are carried by the protons in the nucleus of the atom, while the much less massive electrons carry the negative charge. The attraction between opposite charges and repulsion between like charges can be controlled to move energy about.

Before people learned to control the movement of electrical energy, they observed the uncontrolled transfer of electrical energy by sparks and lightning. These two phenomena are variations on the same general theme. In each, some type of motion or rubbing pulls electrons away from their atoms, building up concentrations of electrical charge. When the electrical forces become very great, the charges jump toward a place of opposite charge. This sudden motion of the electrical charges is what we call a spark or lightning.

During a thunderstorm, molecules and atoms move about rapidly in the clouds. As this motion occurs, electrons are stripped from the atoms. One result is a separation of charges within the clouds (Figure 19-1). Electrons with their negative charges tend to collect at the bottom of the cloud, leaving positive charges at the top. The electrical interaction at a distance creates a strong attraction between the negative cloud bottom and the ground. If the force becomes great enough, it causes a sudden motion of the charges and a transfer of electrical energy between the cloud and the earth.

Approximately a century after Benjamin Franklin first showed that lightning was the uncontrolled transfer of electrical energy (Figure 19-2), we began to control electricity. Electrical forces can be used to explain our present methods. However, quantities that can be derived from the concept of electrical force—voltage and current—offer us a more convenient way of describing electricity. We turn now to a study of these quantities.

Figure 19-1
When the concentration of negative charges gets large enough, a bolt of lightning transfers electric charge (with energy!) to the ground below.

The Bettman Archive, Inc.

Figure 19-2
Benjamin Franklin demonstrated that lightning was the uncontrolled transfer of electrical energy.

ELECTRICAL CIRCUITS

Walk into a dark room and flip the switch. Energy in the form of light floods the room. Turn on the electric stove to cook supper. Thermal energy cooks the food. Turn on the television to catch the evening news. Energy—sound and light—gets the message across. Need to saw a board? Flip a switch; kinetic energy is instantly available. The dentist's office is on the fourth floor. Push the elevator button; your gravitational potential energy changes as you move upward. Countless devices convert electrical energy into other forms—light, heat, sound, kinetic energy, and gravitational potential energy.

We can describe the process that supplies these devices with energy in terms of our *energy-source–energy-receiver* model. A source, typically a battery or a wall socket connected to the local power plant, supplies the energy. Wires transfer this energy from the source to the receiver, like the television, which converts the energy into some other form. The energy source, the energy receiver, and the wires are collectively called an **electric circuit.** Let's examine the role of each component in delivering the energy we so readily consume.

Energy Sources: Voltages

Hook a flashlight bulb to a 1.5-volt (V) battery, and the light glows. Connect the same bulb to a 3-V battery, and it glows more brightly. Whatever voltage is, we see its effect in the brightness of the bulb. When connected to the same receiver, a 3-V battery delivers more energy than a 1.5-V battery.

A battery's voltage is related to its electrical potential energy. To understand how this energy arises, let's look at an analogy from gravitation. A hammer lies on the ground because gravitational forces draw it and the earth together. When we lift the hammer, we do work to overcome the forces of attraction between the hammer and the earth. The amount of work we do describes the amount of gravitational potential energy gained by the hammer. Much the same model can be applied to electrical charges. Positive and negative charges attract one another. Batteries and electrical generators pull negative charges away from positive ones, doing work to overcome the electrical

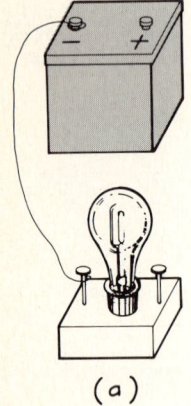

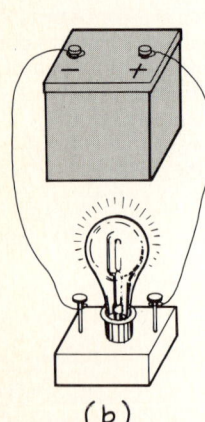

Figure 19-3

Energy can be transferred from the source to the receiver only in a closed circuit. **(a)** No conducting wire leads to the other side of the battery. **(b)** Conducting wires offer a complete path from source to receiver and back to source.

forces of attraction. The amount of work done depends on the number of charges and the distance they are separated. This work describes the electrical potential energy gained by the negative charges (electrons). Thus, work done by the battery or generator becomes available to the electric circuit as electrical potential energy.

When a battery separates two pairs of charges, it does twice as much work as when it separates one pair. However, the work done on each charge is the same in both cases and so is the change in the potential energy of each charge. Because each charge carries potential energy, the energy of each one is a convenient measure of the ability of the source to supply electrical energy.

The electrical potential energy which an energy source provides to each unit of charge is called the **voltage,** or **potential difference,** of the source. Thus

$$\text{Voltage} = \text{potential difference} = \frac{\text{potential energy}}{\text{electric charge}}$$

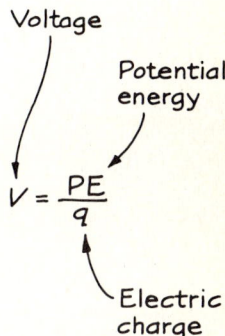

The units of voltage and potential difference are **volts** (V). A 1-V energy source supplies 1 joule (J) of energy to 1 coulomb (C) of charge.

As with other forms of potential energy, voltages are relative quantities. To state a voltage we must first choose a reference, then measure the voltage (potential difference) between the source and the reference. For many sources the reference is chosen to be the electrical potential energy of the charges on the surface of the earth, commonly called the *ground*. Most electrical power companies deliver a voltage of 120 V to your outlets. This statement means that the potential difference (voltage) between one of the prongs and the ground is 120 V. (The other prong is held at the same potential energy per charge as the ground. It is said to be *grounded*.) In some cases the reference is not the ground. For a battery either of its terminals is chosen as a reference, and the voltage of the other terminal is measured relative to it. To say your battery has a voltage of 1.5 V is shorthand for stating that the difference in voltage between the two terminals is 1.5 V.

Energy Transmission: Current

Attach a wire from a terminal on a battery to one side of a light socket (Figure 19-3(a)). Nothing happens. Add a second wire from the other side of the socket to the other terminal on the battery (Figure 19-3(b)). The light bulb glows. For electrical energy to move from the battery to the light bulb, a complete path must be available from one side of the battery to the other.

Lightning jumps from a cloud to the ground; current travels from one terminal to the other. In both cases, electrical energy moves from the point of higher voltage to that of lower voltage. You will recall that thermal energy always moves from regions of higher temperature to regions of lower temperature. In an analogous fashion, electrical energy always moves from a higher voltage to a lower voltage. When the voltage difference is enormous (as in a thunderstorm) and no established transmission path exists, the energy may be moved suddenly in the form of a spark or lightning. To be of any use,

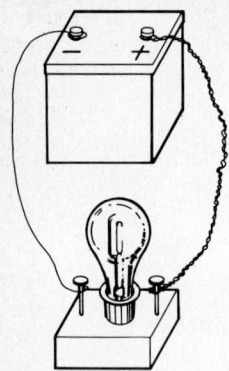

Figure 19-4
Metal wires work; strings do not.

however, the transfer of energy must proceed in a controlled fashion along an established path.

If you connect a wire between two places with different voltages, electrical energy moves from one to the other only until the voltages are equal. When lightning strikes, the voltages in the cloud and on the earth become the same for a brief moment. A similar result occurs when you connect a single wire to a light bulb. A tiny amount of charge moves from the battery to the lamp before the two attain the same voltage. However, this process occurs so quickly that you see no light. By connecting the bulb to the second terminal of the battery—completing the circuit—you create an entirely different situation. The battery maintains a constant voltage difference between its two terminals. Current continues to move from one terminal to the other until the battery runs down. An essential requirement for a current is a transmission path with its ends maintained at different voltages.

Consider an experiment in which one of the wires connecting the bulb and the battery is replaced with a string (Figure 19-4). The light goes off. Not just any material can provide a transmission path for current. Metal wires work; strings do not. Materials that allow the easy movement of charge are called **conductors,** while those that do not allow such easy motion are called **insulators.** Air is a good insulator—as you can see—because energy does not move from one terminal of a battery to the other through the air. Only tremendous voltages, such as those in a storm, can cause the transmission of electrical energy through air. Metals, on the other hand, are excellent conductors. Metal wires are the most commonly used transmission routes for electrical energy.

We can use these concepts to describe the two types of circuits shown in Figures 19-3 and 19-4. A **closed circuit** has a complete path of conductors from the one terminal of the source, through the energy receiver, and back to the other side of the source. Electrical energy will flow through these circuits, of which the circuit in Figure 19-3(b) is an example. Circuits through which electrical energy does not move are called **open circuits.** Figures 19-3(a) and 19-4 are examples. While a complete path may be present in Figure 19-4, at least part of the path includes an insulator. Thus no complete path of conductors is available to carry the energy. The controlled transfer of electrical energy can occur only in closed circuits.

The most common way to open and close a circuit is by using switches. In the off position (Figure 19-5(a)) the metal parts of a switch do not touch. The circuit is open. When the switch is turned to the on position, the metal parts touch (Figure 19-5(b)) and create a closed circuit. This opening and closing of circuits provides us some control over the transfer of electrical energy.

Figure 19-5
A switch can be used to **(a)** open and **(b)** close an electrical circuit.

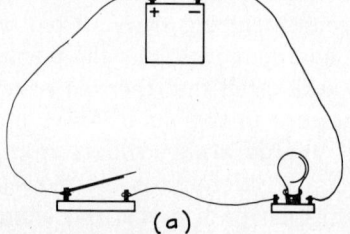
(a)

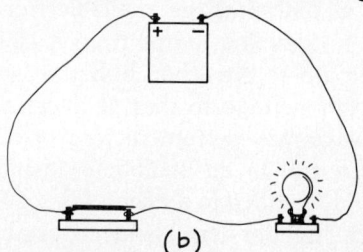
(b)

TAKING THE LIGHT WITH YOU

When you walk up the spiral staircase of your Victorian mansion, how can you be certain that the way is lighted and still minimize the cost of electricity? This question was on the mind of Armand Murat in 1895 when he invented the traveling stairway lamp. The lamp rides on two cables (a) and (b), which are electrical conductors. Thus when a switch (not shown) is closed, the conductors and the lamp (c) make a closed circuit. Because the conducting cables go from the ground floor to the top of the house, the circuit is closed at any point at which the lamp makes a connection between the two conductors. You will have light by which to see all along the stairs as you push the lamp with your cane (d). Just make sure that your cane is constructed from an insulator!

Once the circuit is closed, charges move from the energy source to the receiver. This motion of charges is an **electrical current,** defined as the amount of charge that moves past a point in a circuit divided by the time interval in which the measurement of charge is made:

$$\text{Electric current} = \frac{\text{amount of charge moving past a point}}{\text{time interval}}$$

When electric charge is measured in coulombs and time in seconds, the current is measured in **amperes** (A), often just called amps. One ampere is equal to 1 C of charge passing a point each second. **Ammeters** are used to measure current in a circuit.

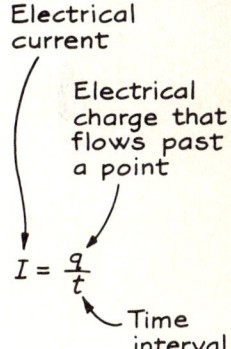

SELF-CHECK 19A

An electron has a charge of 1.6×10^{-19} C. What is the current when one electron passes a point per second? When 10^{19} pass in 1 second?

Energy Receivers: Resistance

Thousands of different energy receivers are used in electric circuits. All play the same role—converting the electrical energy supplied by the energy receiver into some other useful form of energy—heat, light, sound, kinetic energy, and so on. Yet each of the receivers affects the circuits in which they are placed differently. We can investigate these differences by measuring the current present when different energy receivers are connected to the same energy source.

Suppose you reach into a drawer full of old light bulbs and pull out three of them. Their markings are gone, but each seems a little different from the others. To see if they are really different, you hook each bulb into a circuit with an ammeter and connect each to an identical source. Although the voltage across each circuit is the same, each bulb burns with a different brightness, and the current in each circuit is different. The circuit containing the brightest bulb has the most current; that containing the dimmest bulb has the least current.

The characteristic of the energy receiver that determines the size of the current when the energy sources are identical is electrical resistance. As its name implies, **electrical resistance** describes how well a circuit component resists, or retards, the passage of electric current. The brightest bulb has the lowest resistance of the three and, therefore, allows the movement of the most current through it. Resistance is a characteristic of the entire circuit, but since the resistance of the transmitting wires is generally very small, we are usually concerned only with the resistance of the energy receiver.

Numerically, **electrical resistance** is defined as the ratio of the voltage of the energy source to the current moving through the energy receiver.

$$\text{Electrical resistance} = \frac{\text{voltage}}{\text{current}}$$

Electrical resistance → $R = \dfrac{V}{I}$ ← Voltage / Current

When voltage is measured in volts and current in amperes, resistance is determined in **ohms,** often represented by the symbol Ω. For a light bulb connected to a 120 V energy source in a circuit carrying a current of 1.4 A, the resistance is $(120 \text{ V})/(1.4 \text{ A}) = 85.7 \, \Omega$. If a different light bulb is placed in the same circuit, a current of 0.5 A moves through it. Then the resistance of this second light bulb is $(120 \text{ V})/(0.5 \text{ A}) = 240 \, \Omega$. As the resistance of an energy receiver increases, the current decreases.

Electrical resistance is a property of the material from which a device is made. If we compare the resistance of a piece of copper with an identically shaped piece of rubber (Table 19-1), we see the difference between electrical conductors and insulators. Copper, with its extremely low resistance, is used to make connecting wires in electrical circuits. Rubber, a substance with very high resistance, is used in a variety of electrical insulators. Between the extremes of insulators and conductors lie materials with resistances that range from a few ohms to a few million ohms.

Table 19-1	Electrical Resistance of Some Energy Receivers (in ohms)		
Copper wire*	1.7×10^{-8}	Light bulbs	100–3000
Aluminum wire*	2.7×10^{-8}	Distilled water*	5000
Salt water*	0.02	Glass*	10^{12}
Toaster	16	Rubber*	10^{15}
Vacuum sweeper	20		

*These resistances are for a cube of one meter on the side.

Ohm's Law

The equation defining resistance is often called **Ohm's law.** It describes the relationship between current, voltage, and resistance in an electrical circuit. The definition of resistance given above is one form of Ohm's law. When we know the voltage and the current, we can use it to calculate the resistance of an energy receiver. However, Ohm's law can be used to determine any one of the three variables if we know the other two. Ohm's law can be written as

$$\text{Voltage} = \text{current} \times \text{resistance} \qquad V = IR$$

and as

$$\text{Current} = \frac{\text{voltage}}{\text{resistance}} \qquad I = \frac{V}{R}$$

Together with the definition of resistance, these equations provide a way of calculating any one of the variables, given measurements of the other two.

Ohm's law tells us that we can control a current going to an energy receiver by changing either the voltage of the source or the resistance of the receiver. In many applications the voltage of the source is fixed, so we vary the resistance of the receiver. For example, a common three-way light bulb has resistances of 96 Ω, 144 Ω, and 288 Ω. With each change of the switch, we connect a different one of these resistances to the circuit and thus change the brightness of the light bulb.

SELF-CHECK 19B

Use Ohm's law to determine the current in the three-way lamp that has resistances of 96 Ω, 144 Ω, and 288 Ω. The lamp is connected to a 120-V energy source.

Circuits, Resistance, and Safety

Normally we do not think of the human body as part of an electrical circuit. However, any time we come into contact with an electrical energy source, we become part of a circuit whether we realize it or not. Knowledge of electrical resistance has played a large role in helping us protect ourselves in today's electrical world.

The damage done to the human body by an electrical shock is either tissue damage due to the conversion of electrical energy to heat or nerve damage due to disruption of normal nerve functions. Nerve impulses involve rather small currents. Consequently, it does not take much to disrupt the human body. However, the path of the current also determines its effect. Currents are most dangerous when they pass through the heart because they disrupt the normal electrical nerve pulses that pace the heart. Currents that would destroy the heart's operation can pass safely through one finger and out the next with little tissue damage. Table 19-2 describes the effects of different sizes of electrical currents on the human body.

Since currents in ordinary household circuits range up to several amperes, electrical circuits in the home pose a potential hazard. Fortunately, large currents frequently travel through two points close to one another on the body. For example, if you touch two adjacent fingers to bare wires, most of the current will go in one finger and out the other and probably cause little damage to the rest of the body. The most vulnerable organ, the heart, is somewhat protected. Additionally, your layer of skin provides a resistance of some 10,000 Ω to 100,000 Ω. If you accidentally come into contact with a 120-V outlet, the current moving through you would be (120 V)/(10,000 Ω) = 0.012 A to (120 V)/(100,000 Ω) = 0.0012 A. Even if it passes through your heart, you will probably survive, although you would certainly notice the effect.

The rather large resistance of your body helps assure that currents through your body remain low. However, we use other resistances to increase the total resistance and, thus, decrease

Table 19-2 Effect of Various Electric Currents on the Body

Current (A)	Effect
0.001	Can feel it
0.005	Painful
0.010	Involuntary muscle contractions (spasms)
0.015	Lose control of muscles (cannot let go)
0.070	If this current passes through the heart, serious disruption occurs. If a current of this size lasts for more than 1 s, it can be fatal.

the current between the user and the electrical device. A variety of common practices add this additional protection. Ordinary conducting wire is surrounded by plastic or rubber insulation that provides a resistance of millions of ohms. Electrical devices are covered by insulated cases that separate the user from the currents inside. Each practice increases the total resistance the system offers to the electric current, consequently decreasing the size of the current that might accidentally be established.

In spite of these safeguards, people are electrocuted accidentally while using ordinary household devices. The majority of these accidents occur because the devices are handled around water. Two effects occur when a hair dryer, for example, is used while taking a bath. Water has a fairly low resistance when it has anything, such as small amounts of salt, dissolved in it. Drops of water that collect around the on-off switch of a hair dryer can conduct current to the user. Touching it becomes similar to sticking your finger in a socket. Further, water on the surface of the skin dissolves the salt that is normally left from perspiration, leaving a layer of salt water, which lowers the skin's resistance to a few hundred ohms or less. If, while wet, you handle a hair dryer with a wet switch, substantial current can move through your body. Most electrical accidents that involve normal household voltages occur near water.

Other accidents occur because of voltage differences between the surfaces of two electrical appliances or between one appliance and the ground. In the kitchen, for example, the surface of a refrigerator may be at a different voltage than that of a stove. If an energy receiver, such as the cook, touches both simultaneously, current moves from one appliance to the other. While the current may flow for a very short time, it can deposit sufficient energy in the cook to do significant damage.

The difference in potential between two objects can occur in some older devices because of the way they were designed. Charges build up on their surfaces. In most modern appliances, such a voltage is most likely to occur because of a failure in the insulation around a wire. A bare or wet wire comes in contact with the metal exterior. Thus the exterior reaches the same voltage as the wall socket.

To prevent these kinds of problems, appliances have a built-in alternative to completing the circuit from the surface through the user. This alternate path, called the ground wire, is the round third prong of a three-wire electrical plug. The outside of each appliance is connected by a conductor to the ground plug. Then, all ground wires in all plugs are connected together through the house's wiring system. Finally, these connections are all attached to a long metal rod placed in the ground. (Hence, the name *ground wire*.) Because conducting paths always exist among the surfaces of appliances and between the appliances and the ground, no voltage differences can build up. Ground wires allow us to prevent unwanted currents, especially when we are likely to be the energy receivers.

The concepts of resistance and complete circuits tells us how to avoid shocks and more serious electrical hazards when using everyday voltages. Do not use electrical devices while wet or around water so you do not lower your resistance, and make sure all appliances are grounded so they cannot build up voltages.

ENERGY TRANSFER IN CIRCUITS

Ouch! The electric bill hit $60 this month and it is not even winter yet. How can I be using so much energy? Let's see—lights, hair dryer, toaster, oven (no, it's gas), stereo, dishwasher, microwave, clock, washer, and dryer. Hmm... wonder what uses so much electricity?

As energy costs soar, we have all had to reconsider the amount of electrical energy we consume. Some devices use little energy; others use lots. Some devices are used daily; others are used only a few times a month. Some jobs must be done electrically; others could be done by hand. Before we can list our priorities, we need to be able to describe the energy produced and consumed in electrical circuits.

Electrical Energy and Power

A light bulb converts electrical energy into light and, as we learn when we touch one, heat. How brightly the light bulb shines and how warm it gets provides us with a rough idea of how much electrical energy it has consumed. A light bulb burns more brightly when attached to a 3-V battery than when attached to a 1.5-V battery. Further, a light bulb burns more brightly in a circuit that has a large current than in one with a small current. Certainly, the longer the bulb stays on, the more energy it transforms into heat and light. Voltage, current, and time affect the amount of electrical energy transformed by the light bulb.

From these types of observations, we can conclude that voltage, current, and time each affect the amount of energy delivered to energy receivers such as light bulbs. The definitions of voltage and current are consistent with these observations. Voltage describes the amount of energy supplied to each unit of charge, while current describes the number of charges that pass through the receiver during each second. The product of voltage and current describes the rate (amount per second) at which energy is transferred to the receiver.

We use the concept of power to describe the rate of energy transfer. In a circuit the **electrical power** is the product of the voltage and the current.

$$\text{Electrical power} = \text{current} \times \text{voltage}$$

Electrical power
$P = IV$
Current
Voltage

When voltage is measured in volts and current is measured in amperes, the electrical power is given in watts. One **watt** (W) is equivalent to a rate of energy transfer of 1 joule per second (J/s).

The total energy transferred is the product of the power and the time. For a circuit the **electrical energy** becomes

$$\text{Electrical energy} = \text{current} \times \text{voltage} \times \text{time}$$

Electrical energy
$EE = IVt$
Current
Time
Voltage

Voltage is measured in volts; current, in amperes; time, in seconds; thus, electrical energy is given in **joules.**

Energy Transfer in Circuits   449

SELF-CHECK 19C

Use the current that you calculated in Self-Check 19B to determine the power of the three-way light bulb for each of the resistances.

Consuming Electrical Energy

All electrical appliances and tools carry information about the electrical power they consume. If you look on the bottom of an appliance, you will see information such as that shown in Figure 19-6. Suppose we have a toaster that uses electrical energy at a rate of 900 W and operates at a voltage of 120 V. The bottom of our vacuum sweeper shows similar information. This sweeper operates with a current of 6 A when connected to a 120-V circuit.

Which of the two devices, sweeper or toaster, costs more to operate? Comparing these costs involves two factors: the power required by each and the time during which each operates. We were told the power used by the toaster—900 W. We can calculate the power consumption of the vacuum sweeper from the voltage and current given on the tag. The sweeper uses (120 V) × (6 A) = 720 W. While the toaster consumes more power, we generally use it for shorter times than the vacuum sweeper. Suppose we spend 2 minutes per day toasting bread, for a total of 14 minutes per week. A fair estimate of a thorough vacuuming job might be 30 minutes per week. The total electrical energy consumed in 1 week would be:

Electrical energy = voltage × current × time

Toaster = (900 W) × (840 s)   Sweeper = (720 W) × (1800 s)
       = 756,000 J   = 1,296,000 J

Figure 19-6
Electrical devices carry information about the electrical power they consume.

The weekly cost of vacuuming exceeds that of toasting two slices of bread each morning.

We can analyze a variety of electrical devices in this manner. While the cost of electrical energy supplied by the power company depends on other factors, such as the time of day in which you are consuming and the total monthly consumption, a simple analysis like this one can enable you to get an idea of how much your electrical devices cost to operate. Such information is enormously valuable to consumers in selecting and operating electrical devices.

SELF-CHECK 19D

Before the late 1970s electric hair dryers required 750 W of power. The average person needed to use the dryers for 7 minutes (420 s) to have dry hair. A modern 1200-W hair dryer completes the same job in 3 minutes (180 s). Which dryer is less costly to operate?

MORE THAN ONE RECEIVER

Most circuits have more than one energy receiver. A string of Christmas tree lights may have 20 to 30 light bulbs. The circuit in your kitchen allows you to use the toaster, the blender, and the electric grill simultaneously. A car's electrical circuit includes the spark plugs, the ignition system, the starter motor, lights, radio, windshield wipers, and who knows what else. Using the shorthand notation provided by the circuit symbols listed in Figure 19-7, we can investigate ways in which additional energy receivers can be placed into a simple circuit.

If we construct a circuit with one energy source and two energy receivers, the two distinct configurations shown in Figure 19-8 are possible. These configurations, called series and parallel circuits, differ in terms of the routes that the electric current can take. In a **series circuit** the current must follow one route. It leaves the energy source, passes through both energy receivers, and then returns to the energy source. In a **parallel circuit** the current has more than one route available to it. The current leaves the energy source and separates into two routes at point A. Then, part of the current passes through each of the energy receivers. Finally, the two parts of the current recombine at point B and return to the energy source. We now examine each of these configurations and their effects on the electrical energy transferred in circuits.

Series Circuits

To examine what happens when energy receivers with identical resistances are added one at a time into a series circuit, consider the situation in Figure

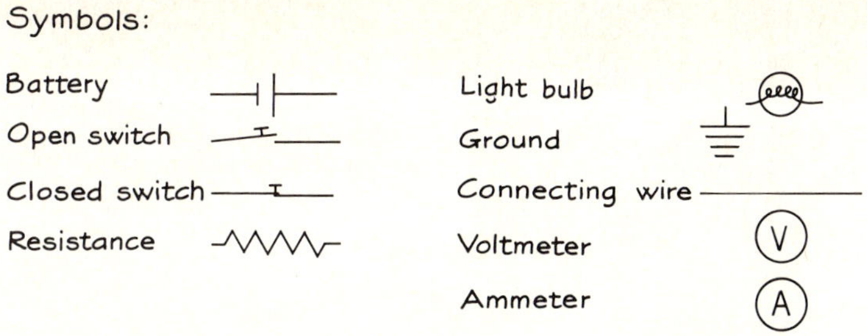

Symbols:

Battery —|⊢
Open switch —⊥—
Closed switch —⊥—
Resistance —/\/\/\—

Light bulb
Ground
Connecting wire ————
Voltmeter (V)
Ammeter (A)

Figure 19-7
Symbols are generally used to provide simpler pictures of electrical circuits.

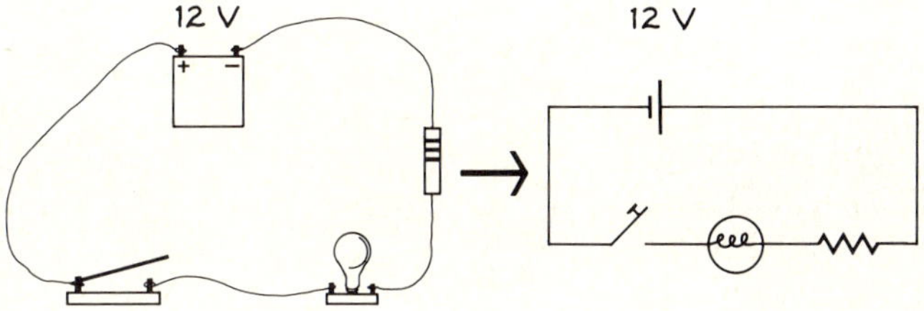

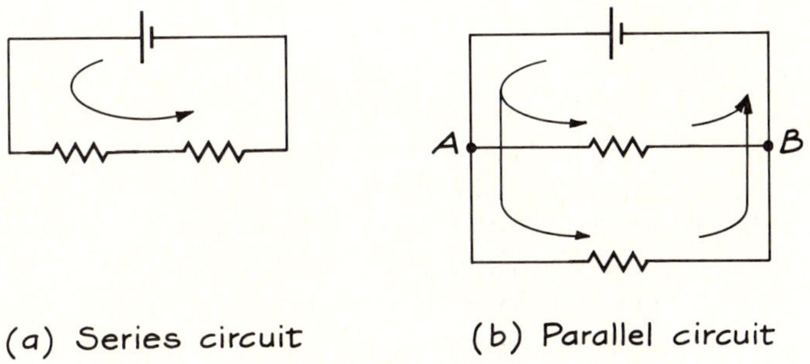

Figure 19-8
Two energy receivers can be placed in series or in parallel with one another. **(a)** In a series circuit, the current has only one route available to it. **(b)** In a parallel circuit, the current separates at point A and follows two separate routes before recombining at B.

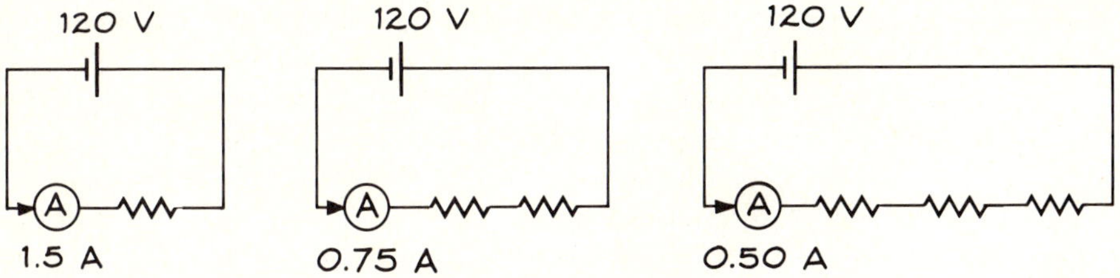

(a) Series circuit   (b) Parallel circuit

Figure 19-9
Adding energy receivers in series decreases the current in the circuit.

1.5 A   0.75 A   0.50 A

19-9. One resistance, then two, and finally three identical resistances are placed in series with one another. While the voltage provided by the source remains constant, the current in the circuit drops as the resistances are added. The current drops from 1.5 A with one resistance to one-half that value, 0.75 A, when a second identical resistance is added. A third identical resistance decreases the current to 0.5 A, one-third of the current in the original circuit. Adding energy receivers in series decreases the current in the circuit.

Figure 19-10

A rubber mat adds a second energy receiver in series with the machine operator. The two resistances add, decreasing any accidental currents to a safe level.

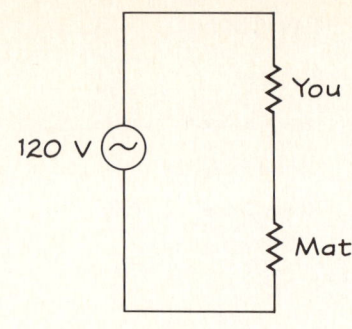

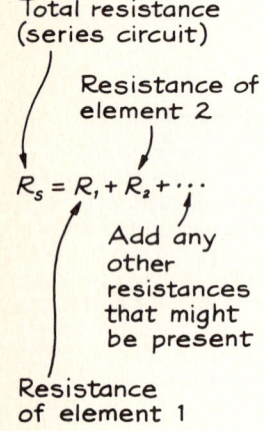

We can understand the decrease in current as energy receivers are added in series by looking at the way the electrical resistances of the receivers combine. Using Ohm's law, we find that the total resistance in circuit (a) is (120 V)/(1.5 A) = 80 Ω. In circuit (b) it is (120 V)/(0.75 A) = 160 Ω. The resistance in circuit (c) is (120 V)/(0.50 A) = 240 Ω. The pattern seems clear. Each energy receiver has an identical resistance of 80 Ω. If we place two of them in a series circuit, their resistances add to give us 160 Ω. Placing three of them in series gives us a resistance of 240 Ω. The total resistance in series is the sum of the resistances of the individual energy receivers.

Total resistance (series) = resistance 1 + resistance 2 + · · ·

We can apply the definition of electrical power to the circuits in Figure 19-9 and examine the way in which power is distributed as energy receivers are added in series. In circuit (a) the energy source provides (120 V) × (1.5 A) = 180 W. For circuit (b) the power drops to 90 W, while in circuit (c) it is only 60 W. As each additional resistance is added in series, the total power supplied by the energy source decreases.

Adding receivers in series also decreases the electrical power available to each receiver. Placed in a circuit by itself, each resistance uses 180 W of power. When placed in series with other resistances, each must now share the available energy. In Figure 19-9(b) the two receivers are identical, so each gets 45 W. The three receivers in 19-9(c) must share 60 W—each gets 20 W. As energy receivers are added in series, the power consumed by each decreases.

The most important characteristic of a series circuit is that additional energy receivers decrease the current in the circuit. Consequently, the power delivered to the whole circuit and to each individual receiver decreases. Depending on the situation, this decrease in current and power can be either beneficial or harmful.

Tackling the benefits first, we turn to electrical safety. People who routinely work with devices that operate at high voltages stand on rubber mats. We can understand the usefulness of this practice in terms of energy receivers connected in series. Suppose that a break in the insulation of a wire exposes a machine operator to 120 V. With an average body resistance of 10,000 Ω, the operator experiences a current of (120 V)/(10,000 Ω) = 0.012 A. As shown in Figure 19-10, a rubber mat adds a second energy receiver with a resistance of 1,000,000,000 Ω to the circuit. Because the mat and the person

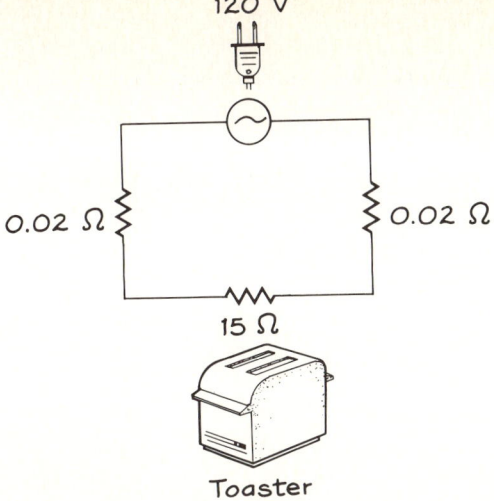

Figure 19-11
A one-meter-long extension cord adds two 0.02 ohm resistances in series with the toaster. The current in the circuit will drop slightly.

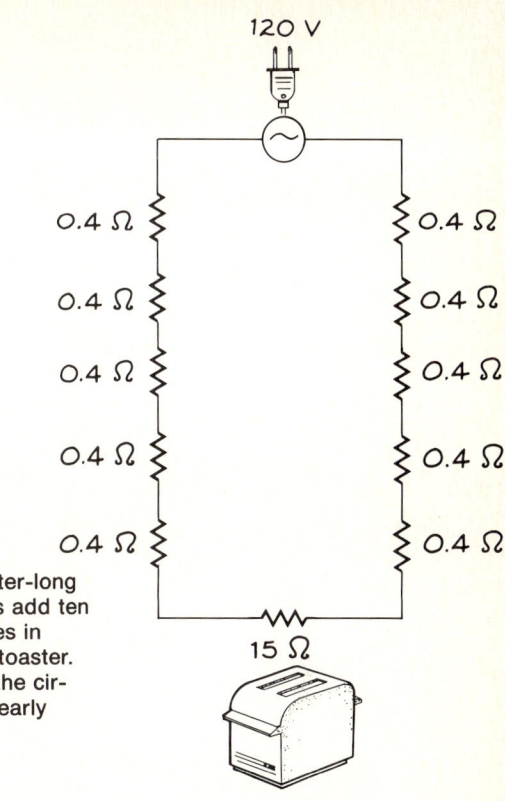

Figure 19-12
Five twenty-meter-long extension cords add ten 0.4 Ω resistances in series with the toaster. The current in the circuit drops by nearly 20%.

are in series, their resistances add. The current moving through each of them is (120 V)/(1,000,010,000 Ω), less than 1.2×10^{-7} A—not enough to cause any damage. Adding an energy receiver in series with a person can save a life!

The disadvantage of resistors in series is that this configuration places some limitations on the design of electrical circuits. A metal conductor, such as copper connecting wire, has such a small resistance that its effect on most electrical circuits is ignored. Nonetheless, each connecting wire does have a small resistance, which adds to the resistance of the circuit. For example, the resistance of a heavy-duty copper extension cord is about 0.02 ohms per meter (Ω/m). If we connect a toaster to a one-meter-long extension cord, we obtain the circuit in Figure 19-11. (The cord contains two 1-meter wires, one leading to the receiver and a second going back to the source.) The total resistance increases from 15 Ω for the toaster by itself to 15.04 Ω for the cord-toaster combination. When plugged into a 120-V outlet, the current drops from 8.0 A to 7.98 A.

For a toaster that operates at 8 A this may seem like no big deal. But suppose you were to decide to eat breakfast in your rose garden, 100 m from your house. You collect five 20-m extension cords, make sure the ground is dry, and string them together. The resistance of each single strand of a 20-m wire is (0.02 Ω/m) × (20 m) = 0.4 Ω. The new toaster-cord circuit, shown in Figure 19-12, has a resistance of 19 Ω. The current has been reduced to 6.3 A, and the power delivered to the toaster is nearly 20% less than when the toaster was sitting on the kitchen shelf. You will have to wait a lot longer for your toast!

SELF-CHECK 19E

One way to construct a circuit that allows you to dim a lamp is to use a variable resistance in series with the lamp. This circuit is shown in Figure 19-13, where the resistance of the lamp is 140 Ω and the arrow indicates that the resistance of the resistor can be changed. Suppose that the variable resistor (VR) has a value of 0 Ω. What are the total resistance, the current, and the power of the circuit? Calculate the same quantities when VR has a value of 60 Ω, and of 120 Ω. Which of these values will give the brightest light? The dimmest light?

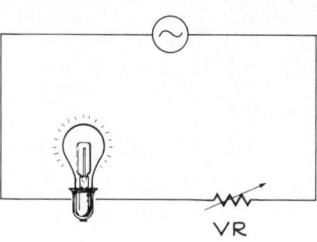

Figure 19-13

Parallel Circuits

Series circuits are useful in situations in which we would like to control the electric current supplied to an energy receiver—either for safety or for convenience. However, such arrangements would be very inconvenient in your home. If household devices were connected in series, the current supplied to each receiver would vary with the number of devices that were being used. If your brother dried his hair, your reading light would dim. Fortunately, there is an alternative.

Examine what happens to a circuit when identical bulbs are added one at a time in parallel. Figure 19-14 shows parallel circuits with (a) one light

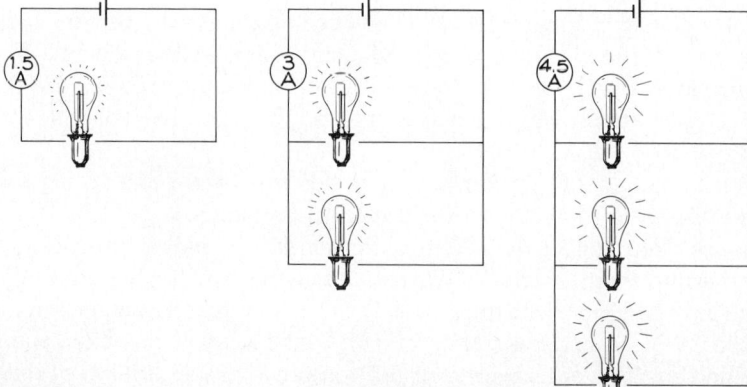

Figure 19-14
Adding light bulbs in parallel increases the current in the circuit.

Why Do My Christmas Tree Lights Burn Out So Fast?

On the boxes in which many Christmas tree lights are packaged is the admonition: Replace burned-out bulbs immediately to avoid additional burnouts. Strings to which this statement applies are connected in series, so to understand the warning we must learn a little more about series circuits.

A series circuit requires fewer wires than a parallel one, so manufacturers frequently use series Christmas light strings to lower the cost of their product. However, a series circuit of lights has a rather annoying property. When one bulb burns out, all the bulbs turn off. Several years ago, someone came up with a solution to this problem. Rather than becoming an open circuit, the burned-out bulb becomes a direct connection—its resistance becomes zero. This type of bulb allows current to move through it, so the rest of the lights stay on.

Even though the line of lights stays on, the package tells us that the other bulbs will burn out if we do not replace the one burned-out light. To see why, consider the whole circuit. In a series circuit every charge moves through every receiver. Thus the current is the same everywhere in the circuit. The voltage, however, is applied across the entire string, so each light gets only part of the total voltage. We can calculate the voltage across each lamp by using Ohm's law. Suppose you have a string of 20 lights, each of which has a resistance of 5 Ω. The total resistance of the circuit is 100 Ω. When connected to a 120-V source, the current is (120 V)/(100 Ω) = 1.2 A. This current moves through every light. Putting this value into Ohm's law, we can find the voltage across each individual light as voltage = (1.2 A) × (5 Ω) = 6 V. We can also determine the power of each light as power = voltage × current = (6 V) × (1.2 A) = 7.2 W. Repeated use of Ohm's law gives us the voltage across each light and the rate at which it transforms energy into heat and light.

Now, let us see what happens when one of these Christmas lights burns out. The resistance of that light is now zero, while the other 19 lights still have a resistance of 5 Ω each. The total resistance of the circuit becomes 19 × 5 Ω = 95 Ω, and the current through the lights is (120 V)/(95 Ω) = 1.26 A. So, the voltage across each lamp increases to (1.26 A) × 5 Ω = 6.30 V, and the new power is (6.3 V) × (1.2 A) = 7.94 W. Each light has increased the rate of energy transformation from 7.2 J/s to 7.94 J/s. This change does not seem like much, but it does cause the lights to heat up a little more. The heat will increase the probability of another burnout. When another light burns out, the power increases even more (to 8.89 W). Rather soon, you have a few bulbs burning very brightly . . . but not for long.

bulb, (b) two light bulbs, and (c) three light bulbs. As lamps are added, the voltage provided by the energy source remains constant. The current in the circuit, however, increases. The current in circuit (a) is 1.5 A. Adding a second light bulb increases the current to twice that—3 A—while the addition of the third light bulb increases the current to 4.5 A, three times the initial value. Adding energy receivers in parallel increases the current supplied by the energy source even though the voltage remains unchanged.

We can understand the increase in current as energy receivers are added in parallel by looking at the connections between the energy source and each of the receivers. For the circuit in Figure 19-14(c) we can trace a wire from each side of each light bulb to one side of the energy source. In effect, each light bulb is connected directly to the energy source. We conclude that, when connected in parallel, energy receivers act independently of one another. Each acts as if it were the only energy receiver in the circuit. Thus, in any parallel circuit the current through a receiver is the same as it would be when the receiver is the only device in the circuit.

The total current supplied by the energy source must be the sum of the currents required by each energy receiver. In Figure 19-14 each light bulb needs 1.5 A. Two light bulbs connected in parallel require 1.5 A + 1.5 A = 3 A. Three identical light bulbs require 1.5 A + 1.5 A + 1.5 A = 4.5 A. The total current supplied by the energy source must equal the sum of the currents in each branch of the parallel circuit.

Total current (parallel) = current 1 + current 2 + · · ·

$I_p = I_1 + I_2 + \cdots$

- Total current (parallel circuit)
- Current in branch 1
- Current in branch 2
- Current in other branches

As energy receivers are added in parallel, the total current supplied by the energy source must increase. If we compare the total resistances of the three circuits in Figure 19-14, this increase in current reflects a decrease in the total resistance of the circuit. Applying Ohm's law, we can calculate the total resistance of any circuit by dividing the energy-source voltage by the total current coming from the energy source. In the circuit of Figure 19-14(a), the total resistance is (120 V)/(1.5 A) = 80 Ω—the resistance of one light bulb. In the circuit in (b), the total resistance is (120 V)/(3.0 A) = 40 Ω—half the resistance of either one of the energy receivers. In the circuit of (c), the total resistance is (120 V)/(4.5 A) = 26.7 Ω—one-third the resistance of any one of the light bulbs. Adding energy receivers in parallel decreases the total resistance of the circuit.

If we apply the definition of electrical power to each circuit in Figure 19-14, we see that adding receivers in parallel increases the total power supplied by the energy source. The energy source supplies 180 W in the circuit in (a); twice that (360 W) in the circuit in (b); and three times (540 W) in the circuit in (c).

Adding energy receivers in parallel does not alter the electrical power available to each receiver. Placed in a circuit by itself, each light bulb would consume 180 W of electrical power. When two are placed in parallel, the energy source provides each one with 180 W. The light bulbs remain as bright in parallel with each other as they had been in a circuit by themselves.

SELF-CHECK 19F

A 10-Ω resistor and a 12-Ω resistor are connected in parallel to a 12-V energy source.
(a) What is the current through each resistor?
(b) What is the total current in the circuit?
(c) What is the total resistance of the circuit?
(d) Determine the electrical power of each resistor.
(e) Determine the total electrical power supplied by the energy source.

A STEP FURTHER—MATH

CALCULATING RESISTANCE IN PARALLEL CIRCUITS

To calculate the resistance in a parallel circuit, we may follow the steps described in Self-Check 19F. First, treat each energy receiver as if it were in the circuit by itself and use Ohm's law to calculate the current passing through it. Then, add all the currents to obtain the total current in the circuit. Finally, put the total current and the voltage in Ohm's law to determine the total resistance.

As an example, consider three receivers with individual resistances of 5 Ω, 10 Ω, and 20 Ω connected in parallel to a 60-V energy source. The current through the 5-Ω resistance will be (60 V)/(5 Ω) = 12 A; through the 10-Ω resistance, the current is (60 V)/(10 Ω) = 6 A; and through the 20-Ω resistance it is (60 V)/(20 Ω) = 3 A. Then, the total current in the circuit is 12 A + 6 A + 3 A = 21 A. Placing this value in Ohm's law we find a total resistance of (60 V)/(21 A) = 2.9 Ω.

A general equation for the total resistance of any number of devices in parallel with each other can be derived using the ideas described above. The equation is:

$$\frac{1}{\text{Total resistance}} = \frac{1}{\text{resistance 1}} + \frac{1}{\text{resistance 2}} + \cdots$$

So, for the 5-Ω, 10-Ω, and 20-Ω resistances, we have

$$\frac{1}{\text{Total resistance}} = \frac{1}{5\,\Omega} + \frac{1}{10\,\Omega} + \frac{1}{20\,\Omega} = \frac{7}{20\,\Omega}$$

$$\text{Total resistance} = \frac{20}{7}\,\Omega = 2.9\,\Omega$$

You can calculate the total resistance of a parallel circuit either by using this equation or by working through the three-step process described above. Both methods are equally valid and give the same answer.

$$\frac{1}{R_p} = \frac{1}{R_1} + \frac{1}{R_2} + \cdots$$

- Plus the resistances of other circuit elements
- Total resistance (parallel circuit)
- Resistance of Element 1
- Resistance of Element 2

Fuses, Brownouts, and Blackouts

Parallel circuits are useful in situations in which we would like to maintain the voltage supplied to each energy receiver at a constant value. Household wall sockets are wired in parallel with one another so that the current delivered to your study lamp does not change when your brother turns on his hair dryer. Groups of homes are wired in parallel so that the current delivered to your dishwasher will not change when your neighbor turns on a vacuum sweeper. The convenience provided by parallel circuits has a cost, however. Fuses blow, brownouts must be ordered, and, occasionally, blackouts occur.

Each household appliance that is plugged into a wall socket increases the total current moving through the wires. However, the current in these wires cannot grow without any bound. Electrical wires are rated in terms of the amount of electrical current they can carry before they begin to heat up appreciably. Overheating caused by excessive currents, frequently called **overloading the circuit,** can cause fires in the walls and insulation surrounding the wires. To prevent the homeowner from accidently overloading the circuit, fuses or circuit breakers are placed in series with the energy source. (A fuse is a wire that melts—or blows—when the current exceeds the value printed on it. A circuit breaker is an automatic switch that turns off—or trips—when the current exceeds its stated value.) When the current in the circuit begins to exceed the wire's current rating, the fuse blows or the circuit breaker trips, creating an open circuit.

For example, suppose that your household wiring is rated at 20 A. Twenty-ampere fuses have been placed in series with the wires entering your home. You get up in the morning and turn on the television. With a resistance of 60 Ω, your television has (120 V)/(60 Ω) = 2 A moving through it. A 15-Ω toaster adds more current—8 A. While watching television and waiting for the toast, you plug in a 12-Ω hair dryer. That is another 10 A. You are at the maximum allowed for the circuit—20 A. When you turn on the lamp... ZAP! There goes the fuse. The blown fuse is a signal that you have overloaded the circuit.

Power plants, like household circuits, can be overloaded. They are rated according to the maximum electrical power they can deliver. A 50-megawatt (MW; mega = million) power plant can deliver 50,000,000 W of electrical power. Any combination of energy receivers that requires more than this maximum available power will damage the electrical generators. Consequently, circuit breakers are used to open the circuit when the power required exceeds this maximum.

Most power plants have at least two options in responding to demand overloads. The plants are usually designed in networks so that excessive demands in one area can be supplied with extra power available in other areas. Power plants in northern cities often supply power to southern locations during periods of extreme hot weather. The second option is for the local power plant to reduce the voltage at which the electrical energy is supplied to consumers. Called a *brownout,* this procedure decreases the power consumed by each energy receiver without removing any energy receivers from the circuit.

Occasionally, accidents occur during periods of peak demands. Then, we learn how fragile our energy system is. In 1965 most of the northeastern section of the United States was blacked out when circuit breakers removed virtually all power plants from the energy distribution network. On a hot summer evening 12 years later, New York City experienced a blackout when lightning removed one major energy source at the same time that a second one had been shut down for repairs. Once the demand exceeds the power rating available from the plant, the circuit opens and we can be left in the dark.

CHAPTER SUMMARY

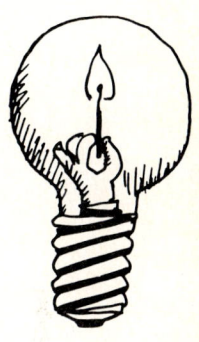

The process by which electrical energy is supplied to electrical devices can be analyzed in terms of our energy-source–energy-receiver model. An electrical energy source, like a battery, supplies electrical potential energy to a receiver, which converts the energy into another form such as light. The source, the receiver, and the transmission path are called an *electric circuit*.

We measure the energy of the source in terms of its *voltage*, the energy supplied per unit of electric charge. Energy can be transmitted from the source to the receiver only along paths (conductors) in which electrical charges can move from regions of higher voltage to lower voltage. *Electric current* describes the rate at which electric charge moves in a circuit. *Electrical resistance* is the characteristic of the energy receivers that determines the size of the current for an energy source of fixed voltage. It is defined as the ratio of the voltage to the current: resistance = voltage/current. This relationship allows us to determine one of the variables when we know the other two. *Ohm's law* can also be written as voltage = current × resistance and as current = voltage/resistance.

Voltage, current, and time affect the amount of energy delivered to the energy receiver. The product of voltage and current describes the rate at which electrical energy is transferred in the circuit; this rate is called the *electrical power*. The product of the electrical power and the time is the total *electrical energy* delivered to the energy receivers during some interval of time. These two concepts, electrical energy and electrical power, allow us to compare the energy consumed by common household devices.

Electrical circuits that have more than one energy receiver can be either series circuits or parallel circuits. In a *series circuit,* the electrical current has only one route available to it. Resistances add, so that the current in the circuit decreases as more energy receivers are added in series. Both the electrical power supplied to the circuit and the electrical power supplied to each individual receiver decrease as receivers are added in series. In a *parallel circuit* the electrical current has more than one route available to it. As energy receivers are added in parallel, the current in the circuit increases. Each receiver acts independently of the others, receiving the same current as though it were the only receiver in the circuit. As energy receivers are added in parallel, the electrical power supplied by the circuit increases, but the electrical power supplied to each individual receiver remains unchanged.

ANSWERS TO SELF-CHECKS

19A. Current = charge/time.

For one electron per second:
$$\text{Current} = \frac{(1.6 \times 10^{-19} \text{ C})}{(1 \text{ s})} = 1.6 \times 10^{-19} \text{ A};$$

For 10^{19} electrons per second:
$$\text{Current} = \frac{(1.6 \times 10^{-19} \text{ C} \times 10^{19})}{(1 \text{ s})} = 1.6 \text{ A}$$

19B. Current = voltage/resistance.

For 96 Ω: $I = \dfrac{(120 \text{ V})}{(96 \text{ Ω})} = 1.25 \text{ A}$

For 144 Ω: $I = \dfrac{(120 \text{ V})}{(144 \text{ Ω})} = 0.83 \text{ A}$

For 288 Ω: $I = \dfrac{(120 \text{ V})}{(288 \text{ Ω})} = 0.42 \text{ A}$

19C. Power = current × voltage.

For 96 Ω: P = (1.25 A) × (120 V) = 150 W

For 144 Ω: P = (0.83 A) × (120 V) = 100 W

For 288 Ω: P = (0.42 A) × (120 V) = 50 W

19D. Energy = power × time.

Old hair dryer: EE = (750 W) × (420 s) = 315,000 J

New hair dryer: EE = (1200 W) × (180 s) = 216,000 J

The new hair dryer uses less energy than the old one.

19E. The total power to the circuit decreases as the resistance of the variable resistance increases. When the variable resistance is zero, the total resistance is 140 Ω; the current is (120 V)/(140 Ω) = 0.86 A; and the power is voltage × current = (120 V) × (0.86 A) = 103 W.

When the variable resistance has a resistance of 60 Ω, the total resistance is 60 Ω + 140 Ω = 200 Ω. The current in the circuit is (120 V)/ (200 Ω) = 0.6 A; the power is (120 V) × (0.6 A) = 72 W.

With a variable resistance of 120 Ω, the total resistance becomes 120 Ω + 140 Ω = 260 Ω. The current is (120 V)/(260 Ω) = 0.46 A, and the power is (120 V) × (0.46 A) = 55 W.

The brightest light occurs when the power is the greatest (VR = 0 Ω). The dimmest light occurs when the power is the least (VR = 120 Ω).

19F. a. Each resistor acts as if it is in the circuit by itself, so the current through it is the same as if it were alone in the circuit. For the 10-Ω

resistance, the current is (12 V)/(10 Ω) = 1.2 A. For the 12-Ω resistance, the current is (12 V)/(12 Ω) = 1 A.
b. The total current is 1.2 A + 1 A = 2.2 A.
c. The total resistance of the circuit is the voltage divided by the total current: total resistance = voltage/current = (12 V)/(2.2 A) = 5.4 Ω.
d. power = voltage × current. For 10 Ω: P = (12 V) × (1.2 A) = 14.4 W; for 2 Ω: P = (12 V) × (1 A) = 12 W.
e. The total power is the sum of the individual powers: 14.4 W + 12 W = 26.4 W.

PROBLEMS AND QUESTIONS

A. Review of Chapter Material

A1. Define the following terms:
Conductor
Insulator
Voltage
Resistance
Electric circuit
Series circuit
Parallel circuit
Open circuit
Closed circuit
Electrical power
Ampere
Volt
Ohm

A2. Describe the process that causes lightning.

A3. Which term—resistance or voltage—is used to describe electrical energy sources? Which describes electrical energy receivers?

A4. Two lamps are connected to identical batteries. One is much brighter than the other. Which has a greater resistance?

A5. State Ohm's law. How is it useful?

A6. Describe how resistance is related to electrical safety.

A7. How are electrical power and energy related to current and voltage?

A8. As energy receivers are added in series, what happens to the total current? To the power delivered by the energy source? To the power received by each receiver?

A9. How does the resistance of a circuit change as energy receivers are added in series? In parallel?

A10. State the rule for calculating the total resistance of a series circuit.

A11. As energy receivers are added in parallel, what happens to the total current of the circuit? To the power delivered by the energy source? To the power received by each receiver?

A12. How would you determine the total resistance of a parallel circuit?

A13. Describe the process that causes a fuse to blow.

A14. Why can brownouts decrease electrical energy use?

B. Using the Chapter Material

B1. The mercury switch is used in thermostat and noiseless wall switches. It consists of two wires and a small amount of liquid mercury (a metal) in a glass capsule. Which of the positions shown below represents a closed mercury switch?

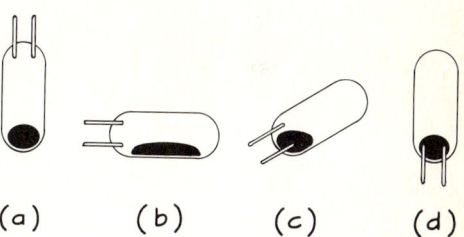

(a)   (b)   (c)   (d)

B2. Thirteen coulombs of charge pass a point in 26 s. What is the current at that point?

B3. A generator does 12 J of work on 2 C of electric charge. What voltage does it supply?

B4. A calculator operates on 8 V and 0.1 A. What is the power used by the calculator? How much energy does the calculator use in 200 s?

B5. What is the resistance of the calculator in problem B4? If the calculator were connected to 120 V, how much current would move through it? (High currents destroy electronic devices.)

B6. Some people determine if a 9-V transistor battery is functioning by placing its terminals against their wet lips. Why would you feel current in this situation and not when holding the terminals in your dry hands?

B7. How much current moves through your fingers (resistance = 1500 Ω) when you hold them against the terminals of a 1.5-V battery?

B8. Why can a wet electrical wire be a safety hazard even if it is covered with insulation?

B9. Three energy receivers with resistances of 5 Ω, 10 Ω, and 15 Ω are connected in series to a 60-V energy source. What is the total resistance of the circuit? What is the total current in the circuit?

B10. Automobiles have either two or four headlights. Are these headlights connected in series or parallel with each other? Explain how you obtained your answer.

B11. Three energy receivers with resistances of 3 Ω, 4 Ω, and 6 Ω are connected in parallel to a 12-V energy source. What is the current in each resistor? What is the total current in the circuit? What is the total power?

C. Extensions to New Situations

C1. The rear-window defrosters on automobiles consist of several strips of heater wire connected in parallel. Each wire typically has a resistance of 6 Ω. The car battery's voltage is 12 V.
   a. What is the total current through a defroster consisting of five wires connected in parallel?
   b. What is the total resistance of the five-wire defroster?
   c. What is the power consumed by the defroster?
   d. How much energy is consumed in 500 s?
   e. How much energy is needed to melt 0.1 kg of ice from the back window? (Refer to Chapter 10.)
   f. Can the defroster melt 0.1 kg of ice in 500 s?

C2. In normal household wiring, one side of the outlet is connected to the ground. This wire is said to be grounded. The other side, called the live wire, is connected to 120 V.
   a. Why does a health hazard exist for a person who holds only the live wire?
   b. Is a similar hazard present if the person holds only the grounded wire? Why or why not?
   c. If you were working on a live wire while standing on a ladder, would you prefer a wooden ladder or an aluminum one? Why?

C3. A three-way light can be constructed as shown below. The bulb has two separate connections, $B1$ and $B2$, at its base and one connection, S, on the side of the light bulb.

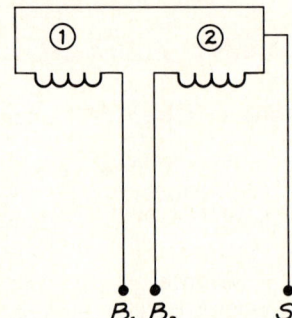

   a. To what points should the energy source be connected so that resistance 1 comes on by itself? So that resistance 2 comes on by itself? So that both resistances come on in parallel?
   b. When resistance 1 is 24 Ω and resistance 2 is 48 Ω, what is the current through each resistance when it is in the circuit alone? When both are in the circuit in parallel?
   c. What are the three different electrical powers this bulb can emit?

C4. For convenience we frequently desire switches at more than one location to control a single light. For example, rather than using the traveling stairway lamp described in the chapter, we may wish to turn on a light at the bottom of the stairs and turn it off at the top. To see how this works, we begin with two ordinary switches connected to a light, as shown on page 463.
   a. What happens to the light when either switch A or B is open?
   b. Using this circuit would you have complete control over the light at either switch? Why or why not?
   c. Now consider the switch shown in (b)

Problems and Questions   463

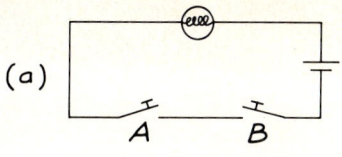

(a)

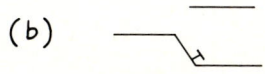

(b)

(c)

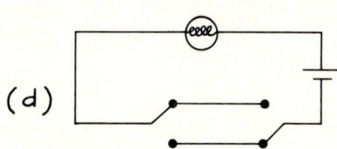

(d)

and (c). This switch is always connected to a wire; in (b) the lower wire is connected, while in (c) the upper wire is connected. As shown in (d), these switches can be placed in a lamp circuit. Using the idea of a closed circuit, show that you have complete control of the lamp at either switch.

C5. Some circuits can be combinations of series and parallel circuits. To learn one way to determine the total resistance of such a circuit, consider the circuit below.

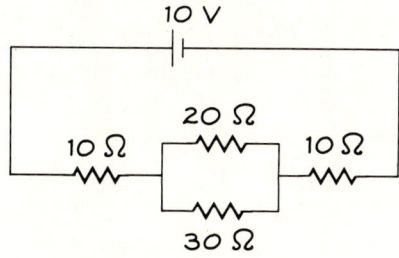

a. Identify the parallel part of this circuit. Draw a circuit consisting of the parallel part and a 10-V battery. Calculate the total resistance of this parallel circuit.
b. What single resistor could be used to replace the parallel resistor in part (a) and still have the same total current from the battery?
c. Replace the parallel resistors in the circuit drawing with the resistor from part (b). What is the total resistance of this circuit? (This is the total resistance for the combination circuit.)

C6. To see how brownouts can decrease power requirements, consider a home that has a television (resistance = 60 Ω), toaster (12 Ω), hair dryer (10 Ω), and lamp (120 Ω). All are connected in parallel.
a. What is the current through each component when they are connected to 120 V? What is the total current?
b. How much total power is used when all components are on?
c. The electric company decreases the voltage to 100 V. What are the new individual and total currents?
d. What is the new total power for all components?
e. Describe in words how the voltage decrease changes the power used.

C7. Here is a somewhat simplified version of a blackout. Suppose the energy sources are connected so that their energies add in the total circuit. Source A has a maximum power output of 5 kilowatts (kW); source B, 6 kW; and source C, 7 kW. If the power required by the receivers exceeds these values, a circuit breaker shuts off the source. The sources are connected to three homes. Each home has an air conditioner (2 kW) and a large freezer (1 kW).
a. What is the total power supplied by the sources?
b. What is the total power needed by all three homes?
c. Now, suppose a person in each home turns on a coffee pot (1 kW) and a microwave oven (1 kW). What is the total power requirement? Why should the electric company become concerned?
d. How would a brownout help alleviate the problem in part (c)?
e. Suppose that lightning destroys source A. What will happen to sources B and C? Why?

D. Activities

D1. Look inside a flashlight. Describe how the conductors change to open and close the circuit.

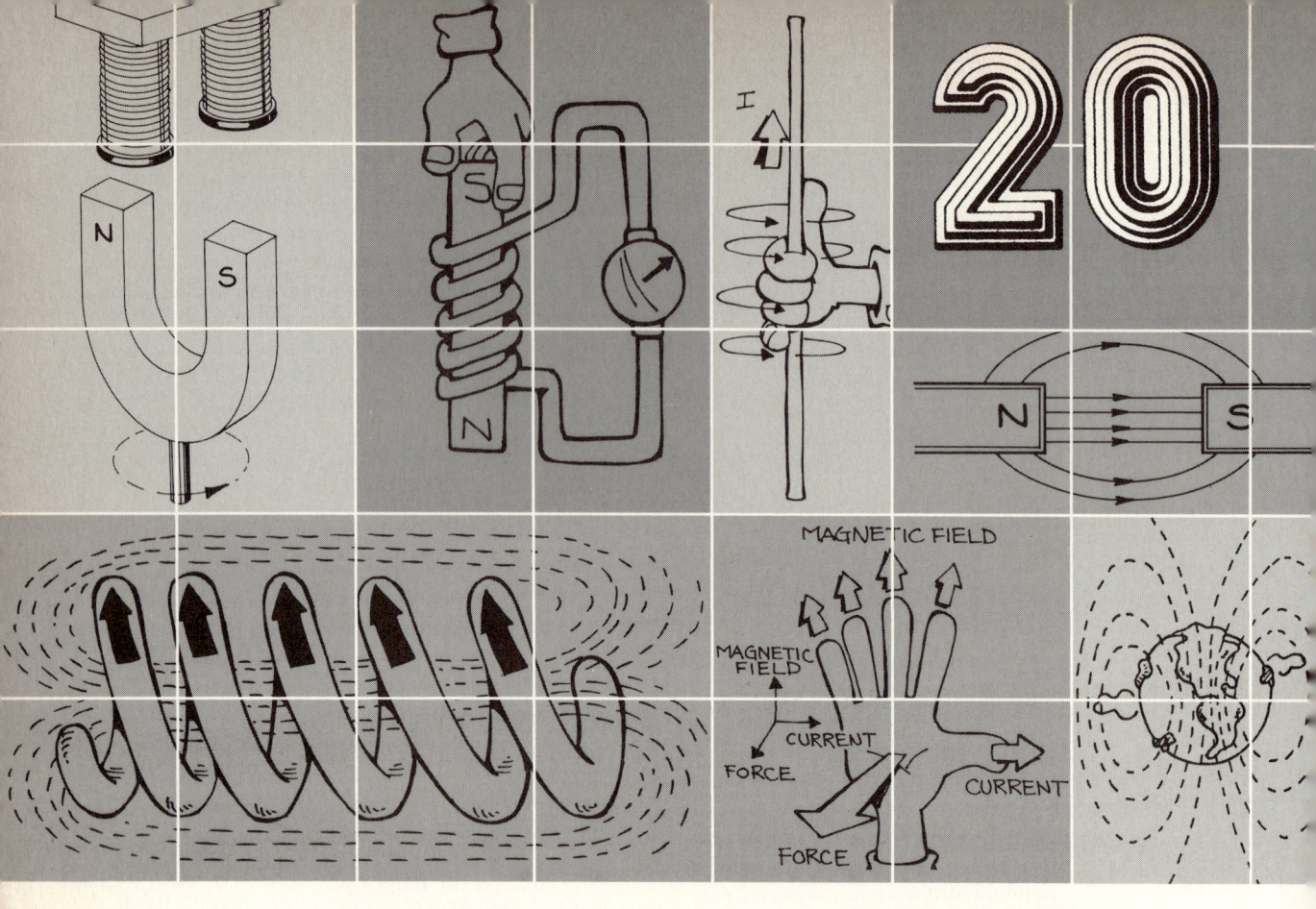

Electromagnetism

Man is a singular creature. He has a set of gifts which make him unique among the animals: so that, unlike them, he is not a figure in the landscape—he is a shaper of the landscape. In body and in mind he is the explorer of nature, the ubiquitous animal, who did not find but has made his home in every continent.

Jacob Bronowski, *The Ascent of Man*

Shaper of landscape... explorer of nature. These phrases from the opening paragraph of Jacob Bronowski's *The Ascent of Man* describe the inseparable relationship between technology and science. Nowhere is this relationship more obvious than in the development of electromagnetism during the nineteenth century. In 1820 Hans Christian Oersted discovered that electrical currents create magnetic fields. Four years later the first electromagnet was built. Michael Faraday discovered the reverse effect: Changing magnetic fields create a voltage. Less than a year later, in 1832, early models of the electric generator began to appear. James Maxwell published his complete theory of electromagnetism in 1873. Twenty-three years later the prototype of the electrical generating plant was delivering electrical energy to Buffalo, New York.

As explorers of nature, physicists during the nineteenth century forged one of the most awesome theories in the history of scientific thought. Rivaled in breadth only by Newton's mechanics, Maxwell's theory of electromagnetism unites phenomena ranging from the forces between electric charges to the movement of light waves. As shapers of the landscape, physicists and inventors revolutionized our way of life. Because of the ease with which electrical energy can be generated and distributed, it has completely altered our work, our leisure, and our landscape.

Beginning with an overview of *magnetism,* this chapter traces the discovery of the relationships between electricity and magnetism. Currents create *magnetic fields* and changing magnetic fields create voltages, a phenomenon called *electromagnetic induction*. These two principles provide the structure for the theory of *electromagnetism* and for the development of new technologies, including electric power, radio, and television.

MAGNETISM

Magnets come in a variety of shapes and sizes. Play around with a couple of magnets for a bit and . . . CLUNK, they stick together. Turn one of the magnets around, and you can chase the other one all over the table. Put a piece of paper over a magnet and sprinkle some iron filings about. The filings line up in a definite pattern. Put a magnet on a refrigerator door and it stays. Put it on an aluminum pan and it falls. These and a host of other observations are integrated into our present model of magnetism.

Magnetic Poles

Magnets exert forces on one another—forces that you can feel if you are holding the magnets or that you can see as one magnet pushes the other around the table. In many respects magnetic forces behave like electrical forces; they are exerted at a distance. Magnets do not need to touch each other in order to interact. Magnetic forces can be either attractive or repulsive, depending upon which ends of the magnet are held near one another. The strength of the magnetic force depends on the distance separating the two magnets. These and other similarities to the electrical interaction led physicists to develop a two-pole model of magnetism that is analogous to the two-charge model of electricity.

Magnets have at least two regions, called **magnetic poles,** which exert magnetic forces on other magnets. All magnets have two types of poles, called (by convention) **north** and **south** poles. For a bar magnet, these poles are located at the two ends. (A horseshoe magnet is just a bar magnet that has been bent, so its poles are at its two ends.) Others, such as ring magnets and ceramic magnets, have poles located on their faces. The two kinds of poles interact with one another, as do the two kinds of electric charge. Like poles (north-north, south-south) repel; unlike poles (north-south) attract.

While magnetic poles behave similar to electric charges in some ways, they exhibit one important difference. Isolated electric charges are commonplace, while isolated magnetic poles are not. If we break a bar magnet in half,

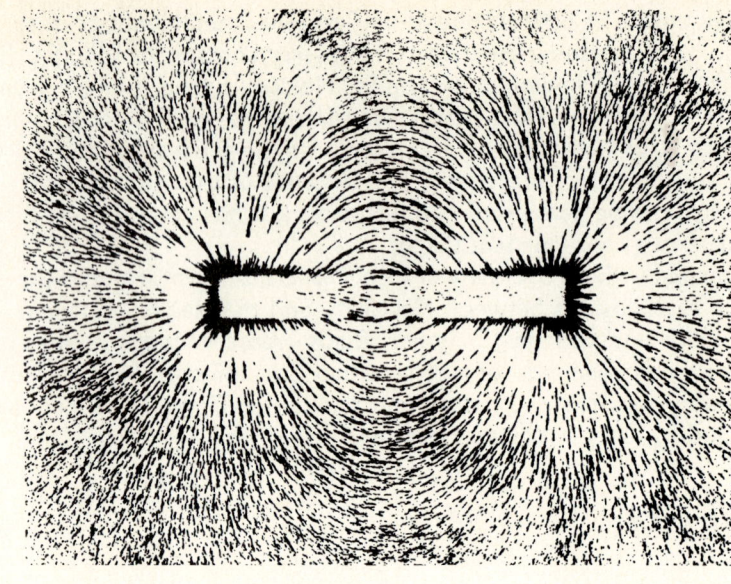

Figure 20-1
Iron filings trace out a pattern of lines called *magnetic field lines* in the space surrounding a magnet.

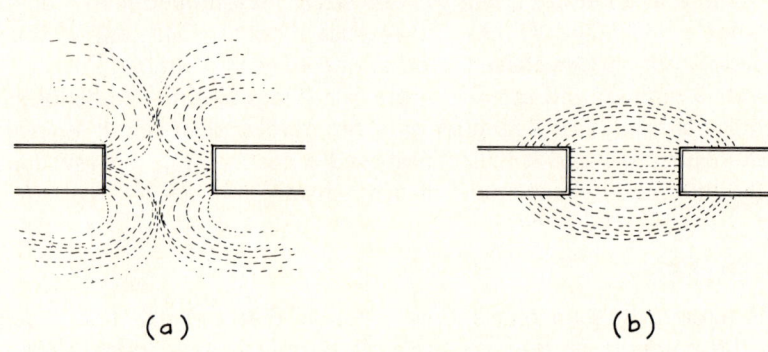

Figure 20-2
Magnetic field lines provide us a mental picture of the forces **(a)** between two like poles and **(b)** between two unlike poles.

each half still behaves as a complete magnet. Breaking each half gives us four fragments, each of which still behaves as a complete magnet. We could continue breaking off pieces of the magnet, but we would never isolate a single magnetic pole. Magnetic poles exist in pairs.

Magnetic Fields

We can demonstrate the interactions of magnetic poles using tiny iron filings. When sprinkled on a piece of paper, the filings do not form any special pattern. But, if we hold the paper over a bar magnet, the filings trace out a pattern of curved lines (Figure 20-1). Distinct lines spread out from one pole, circle one side of the magnet, and return to the other pole. Examine the patterns produced with two like poles (Figure 20-2(a)) and with two unlike poles (Figure 20-2(b)). The way the lines bend away or toward one another reflects our experience with magnets: Like poles repel and unlike poles attract.

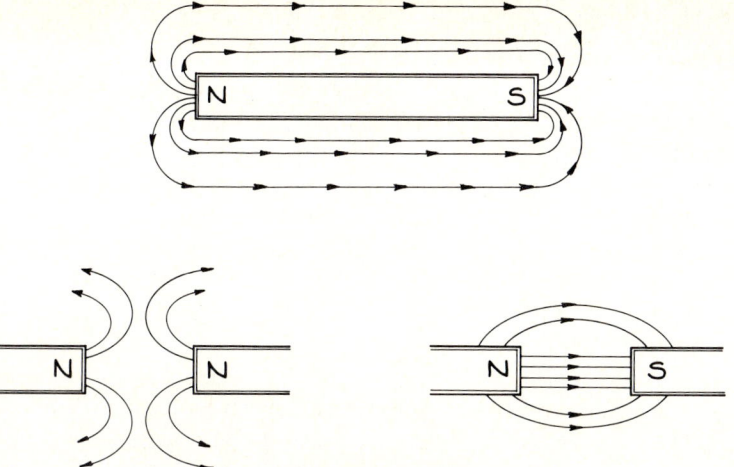

Figure 20-3
Magnetic fields have a direction that shows the force on a north pole at each point in the space surrounding the magnet.

The patterns produced by the iron filings represent the *magnetic fields* of the magnets. The field is a set of vectors that describes how a magnet affects other magnets. Another magnet, when placed in a magnetic field, will tend to line up its poles with the magnetic field. Thus the magnetic field provides us with a picture of the magnet's action at a distance.

The relative strength of the magnetic field can be determined by looking at the lines traced out by the magnetic field. Places where the lines are close together represent locations where the magnetic force is strong, while areas where the lines are far apart have weaker magnetic forces. The density of the lines in the drawing of a magnetic field shows the strength of the force.

Since a magnetic field represents the force due to a magnet, it must have a direction. By convention we define the direction of a magnetic field to be the direction of the force on a north magnetic pole. When a magnet is placed in the magnetic field of another magnet, the north poles of the two magnets will repel one another. Thus the direction of any magnet's field is away from its north pole and toward its south pole. The drawings in Figure 20-3 show these fields for a bar magnet and for the configurations shown in Figure 20-2.

The largest magnet on earth is the earth itself. The poles of the earth's magnet are located fairly close to the geographic north and south poles defined by the earth's rotation. As shown in Figure 20-4, the earth's magnetic field can be approximated by the field produced by a bar magnet tilted at about 11.5° relative to the earth's axis of rotation. The terms *north* and *south* as applied to magnetic poles are derived from the way in which magnets rotate in the earth's field. The north pole (more precisely called the *north-seeking pole*) of any magnet rotates so that it points north when the magnet is suspended in the earth's field. Since the north pole of a magnet is attracted to the earth's pole, which is located in northern regions, the earth's magnetic pole located up north is really the south pole of a magnet. Magnetic compasses are simply tiny magnets pivoted so that they are free to rotate. The tip of the needle is a north pole and will always point north—to the relief of many pilots, ship captains, and scouts.

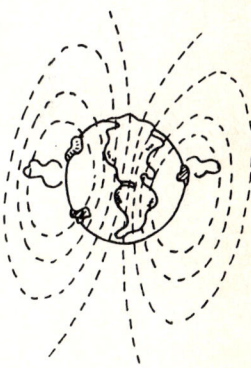

Figure 20-4
The earth's magnetic field can be approximated by the field produced by a bar magnet tilted 11½° relative to the earth's axis of rotation.

Making Magnets

Two paper clips sitting on your desk neither attract nor repel each other—they are not magnets. Yet, they are attracted to a magnet, just as though they were magnets themselves. In fact, the paper clips do acquire north and south poles when they are in a magnet's field. Paper clips can be magnetized.

When placed in a magnetic field, materials that can be magnetized form either permanent or temporary magnets. **Permanent magnets** continue to behave as magnets when the external magnetic field has been removed. They can attract or repel other magnets and attract materials made of iron. The various magnets we have discussed so far are permanent magnets. **Temporary magnets** behave as magnets only in the presence of an external magnetic field. Paper clips, for example, are strongly attracted to a magnet. Once the magnet is removed, however, they display no magnetic attraction or repulsion toward one another. Their magnetism is temporary.

Magnetism is a property of some, but not all, materials. A magnet will turn a refrigerator door—but not an aluminum pan—into a temporary magnet. Magnets attract safety pins and paper clips but not pennies. Bar magnets are made from alloys containing iron, nickel, or cobalt but not copper or zinc. Some materials can be magnetized; others cannot.

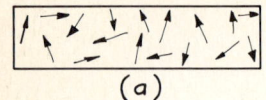
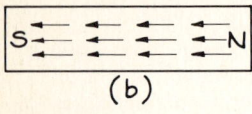

Figure 20-5
The magnetic behavior of different metals can be described in terms of the alignment of magnetic domains. **(a)** Oriented randomly, the magnetic domains produce magnetic fields that cancel one another. The metal demonstrates little, if any, magnetic properties. **(b)** Aligned with one another, the magnetic domains produce magnetic fields that add. The strength of the magnet depends on the number of domains that are aligned.

Magnetic Domains

The difference between magnetic and nonmagnetic materials is their ability to form magnetic domains. A **magnetic domain** is a small section of matter that behaves like a tiny magnet. When the domains are oriented in random fashion, the magnetic fields produced by each cancel the fields of others. When the domains line up with one another, their fields add (Figure 20-5). The strength of a magnet depends upon the number of magnetic domains that are aligned.

Magnetic domains enable us to understand the difference between permanent and temporary magnets. When some types of iron bars are placed in an external magnetic field, most of their magnetic domains are forced into alignment with the field. The alignment is so complete that the domains hold one another in place once the external field is removed. Thus, the iron bar becomes a permanent magnet. In a temporary magnet, fewer domains are forced into alignment. Once the external field is removed, molecular motion causes the aligned domains to return to a random arrangement. Pins and paper clips, for example, become temporary magnets as their domains align with an external magnetic field. When the field is removed, the domains return to randomness. Still other materials cannot have magnetic domains aligned at all. Aluminum, for example, does not become a magnet even temporarily.

SELF-CHECK 20A

If you drop a permanent magnet, it can lose some if its magnetic strength. Use the concept of magnetic domains to explain why.

Storing Information with Magnetic Domains

A spy in a hostile country, you arrange a secret code with an accomplice. You will place magnets along the inside of your house. Different orientations and strengths of the magnets will correspond to the different letters of the alphabet. Your accomplice need only walk by the outside of the house with a magnetic compass in hand to receive the message. While James Bond may never find this technique convenient, it illustrates the process by which magnetic domains can be used to store information.

Magnetic recording tapes and disks that are used in audio, video, and computer applications consist of a thin layer of iron or chromium oxide deposited on a plastic surface. Both iron and chromium oxide have magnetic domains that can be aligned by external magnetic fields. Information is stored on the tape or disk by aligning small sections of magnetic domains according to a coded format. The domains then remain aligned so that the tape or disk may be played back and the information retrieved. The player can collect the information by sensing the orientations of the magnetic domains. Consequently, information can be retrieved and still remain on the magnetic medium for future use. The information can be erased by moving the tape through a magnetic field so that all the domains are aligned in one direction rather than arranged according to a standard code. Thus, the information on a disk or tape can accidentally be removed if the medium is placed next to a magnet.

In a slightly different context, information about the earth's magnetic field is stored in the magnetic material trapped inside rocks. Many minerals contain bits of iron similar to the iron filings that we use to map magnetic fields. These rocks have been created from molten material of volcanic eruptions or from deposits at the bottom of oceans. At the time the rocks were being formed, iron was not trapped in a fixed position as it is now. Since they are magnetic, the small bits of iron rotated to align themselves with the earth's magnetic field. As the rocks hardened into their present form, the iron was trapped—recording the orientation of the earth's magnetic field. We can retrieve this information by looking at the bits of iron in rocks.

An examination of rocks from different periods of the earth's development provides us with a history of the earth's magnetic field. Investigations show startling changes. The earth's field has reversed itself fourteen times during the past 4.5 million years, changing the location of the magnetic north pole from one geographic pole to the other. Exactly how and why these changes occur is not well understood. The earth's field may die down to nothing and then build back up in the opposite direction, or it may tip over slowly somewhat like turning around a bar magnet. However, the time scale is well known. The earth reverses its field approximately once every half-million years and takes about 1000 years to complete the reversing process. Sometime in the next few hundred thousand years, the earth's field will reverse itself. Then, the magnetic north pole of the earth's field will be near the geographic north pole. That will make less confusion in describing the poles, but all our compasses will point in the wrong direction.

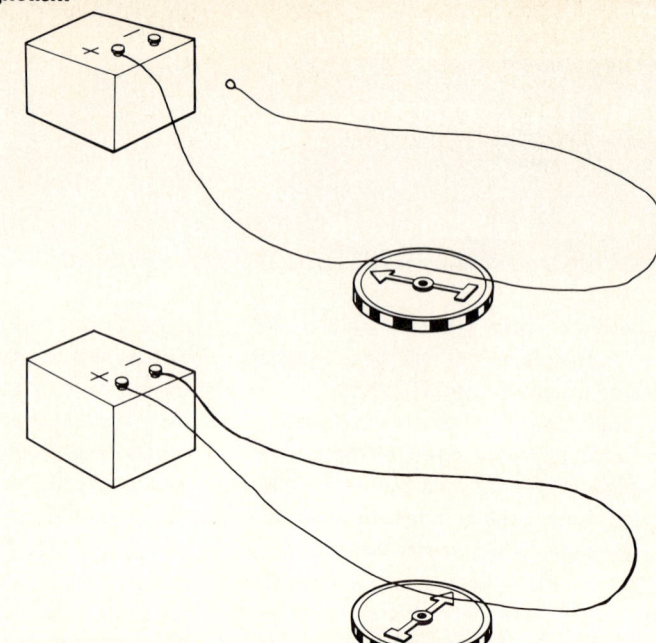

Figure 20-6
Oersted placed a long wire parallel to a magnetic compass. When an electric current moved through the wire, the compass rotated so that it was perpendicular to the wire.

ELECTRIC CURRENTS AND MAGNETIC FIELDS

Similarities between electricity and magnetism were so striking that by the beginning of the nineteenth century most physicists were convinced that the two phenomena were related. Reports circulated that lightning had magnetized sewing needles and tableware: yet physicists were unable to detect any such relationship. Finally, Oersted noticed that currents could affect a magnetic compass. By the end of the century, this single observation had grown into a comprehensive model of electromagnetism.

Moving Charges Produce Magnetic Fields

Oersted's experiment turned out to be remarkably simple. In fact, he first performed the experiment accidentally while giving a physics lecture to his students. A magnetic compass was lying on a table when he placed a current-carrying wire near it. Both the compass needle and the wire were oriented in a north-south direction (Figure 20-6). When an electric current moved through the wire, the compass needle rotated toward the east. When the direction of the current was reversed, the compass needle turned to the west. When the current was turned off, the needle again aligned itself with the earth's magnetic field. The behavior of the compass tells us that a current-carrying wire produces a magnetic field.

We can describe the shape and direction of this magnetic field in more detail with the help of iron filings and magnetic compasses. When the iron filings are sprinkled on a piece of paper that lies parallel to the current-carrying wire, no pattern is formed (Figure 20-7(a)). Perpendicular to the wire, however, the iron filings arrange themselves in a circular pattern (Figure 20-7(b)).

Electric Currents and Magnetic Fields 471

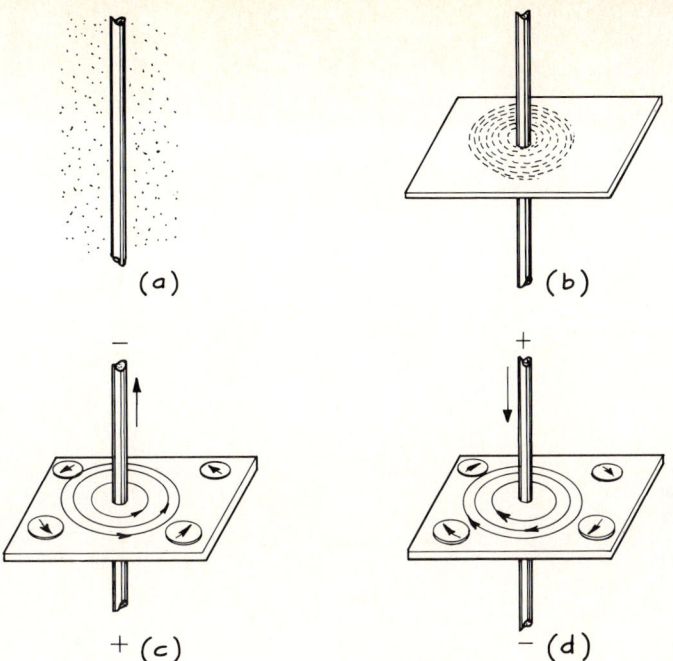

Figure 20-7
Iron filings and magnetic compasses trace out the magnetic field produced by the current-carrying wires. **(a)** No magnetic field exists in a plane parallel to the current-carrying wire. **(b)** Perpendicular to the wire, the iron filings arrange themselves in a circular pattern. **(c)** The magnetic field lines are counterclockwise when the current moves upward and **(d)** clockwise when the current moves downward.

The iron filings show the pattern but not the direction of the magnetic field. A series of small compasses can show both (Figure 20-7(c) and (d)). A circular magnetic field around a single wire is present whenever an electrical current exists in the wire.

This observation leads to the **right-hand rule** for determining the direction of a field created by a current in a wire. However, before we state the rule, we must define the direction of the electric current. Early researchers in electricity did not know that the moving charges were negative. So, they defined the direction of electric current to be from the positive side to the negative side of an energy source. When discussing magnetism, we still use this convention. Determine the direction of the current by assuming it moves from positive to negative. Then, hold your right hand so that four fingers form part of a circle and your thumb points away from the rest of your hand. If you place the thumb of your right hand in the direction of the electric current inside the wire, your fingers curl in the direction of the magnetic field (Figure 20-8). The right-hand rule reflects Oersted's discovery that the direction of the magnetic field changes in response to changes in the direction in which the electric current flows.

The strength of the magnetic field around a current-carrying wire is related to the size of the current in the wire. Measurements of the field strength and the current would show us that the magnetic field strength increases as the current increases. The relationship between the two is a direct one: Doubling the current causes the magnetic field strength to double.

Oersted's experiment provided the first reproducible observation of a relationship between electricity and magnetism. A current-carrying wire produces a magnetic field whose strength depends on the amount of the current flowing and whose direction depends on the direction that the current flows.

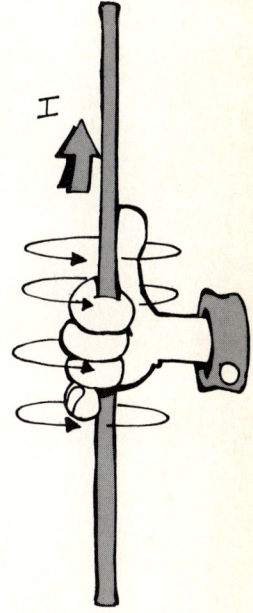

Figure 20-8
If you place the thumb of your right hand in the direction of the electric current, your fingers curl in the direction of the magnetic field lines.

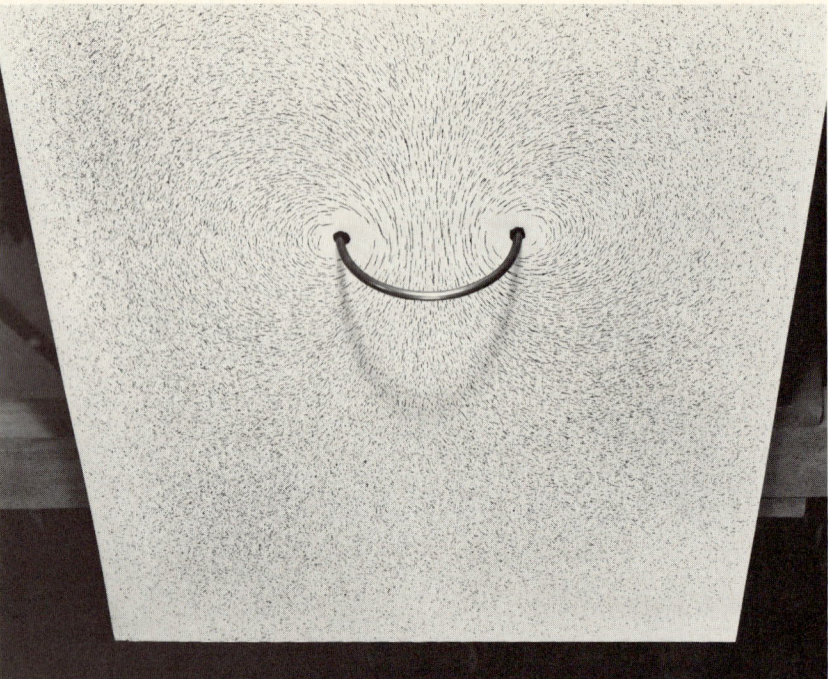

Figure 20-9
If we loop the wire, the magnetic fields produced by the different segments overlap. The fields add, producing a stronger magnetic field than that of the straight wire.

Clearly, the electric current itself gives rise to the magnetic field. Within a decade, this observation led to the development of powerful electromagnets and, ultimately, to an atomic explanation of magnetism.

Electromagnets

The magnetic field produced by an ordinary wire is not very large. A current of a few amperes produces a field that is barely large enough to rotate a compass needle. To be of practical value, fields need to be much larger. One way to create these larger fields is to arrange different segments of the same wire so that their fields add.

A single current-carrying wire produces a circular magnetic field along its entire length. If we loop the wire (Figure 20-9), the magnetic fields produced by different segments of the wire overlap. The fields add, producing a stronger magnetic field that looks somewhat like the field produced by a bar magnet. If we make a longer series of loops (a coil), the current-carrying wire produces a magnetic field that looks exactly like that produced by a bar magnet (Figure 20-10).

Placing a piece of iron inside the loops of a coil, we have an electromagnet. An **electromagnet** consists of a coil of current-carrying wire into which an iron bar has been placed. The magnetic field produced by the coil draws the magnetic domains in the iron into alignment, creating a magnetic field that can be a thousandfold stronger than the field produced by the coil itself. Even more useful is the control we have over the field. By turning off the current, we turn off the magnetic field. In allowing us control over the magnetic field, the electromagnet has been useful in a variety of applications.

MOVE, COW, MOVE

The application of circuits and magnetism to a typical farm problem resulted in an invention by Elmer Swensen in 1922. Mr. Swensen wanted to make sure that when something other than milk came out of his cows, it went into a place where he would not step in it. Electrical circuitry, magnetism, and a little knowledge of cow behavior provided a solution. Just before the dirty deed, the cow raises its tail, which is attached to a rope (29). The rope becomes slack and no longer holds open the switch (24). The closed switch completes the circuit between 17 and 30. The battery (15) then provides current to an electromagnet (19). In turn, the electromagnet attracts an iron switch, and a second circuit closes. This second circuit is attached by wire (33) to a pad (34) under the cow's feet and by a second wire (31) to the cow (32). Because these wires are connected to the battery, the cow now becomes part of a complete circuit. To stop the current the cow steps backwards and is in the correct position relative to the "toilet" (12).

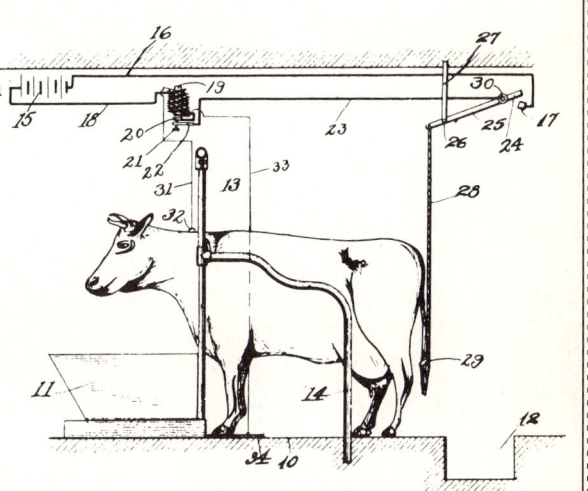

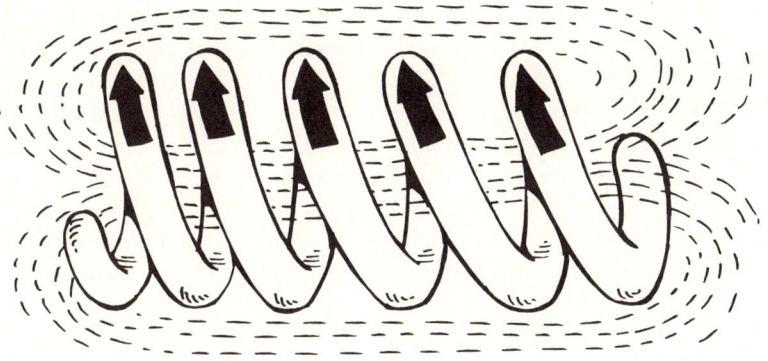

Figure 20-10
The magnetic field produced by a coil of wire looks very similar to the field created by a bar magnet.

SELF-CHECK 20B

The piece of iron in an electromagnet is removed and replaced by an identically shaped piece of aluminum. Will the strength of the electromagnet increase, decrease, or stay the same? Explain your answer using the field in the coil and the magnetic domains.

The Atomic Basis of Magnetism

The discovery that currents produce magnetic fields gave us more than just more powerful, controllable magnets. The realization that electric currents produce magnetic fields led many physicists to suspect that magnetic properties in general might arise from electric currents. Research into the internal structure of the atom revealed that they were right.

Atoms contain from one to more than a hundred electrons. Each electron moves constantly as it orbits the atomic nucleus and as it spins on its axis. These motions of electric charge are, in effect, small currents. Each produces a magnetic field. In most materials the electron spins and orbits are oriented so that the magnetic fields associated with them cancel each other, and each atom is left with no net magnetic field. However, a few materials, such as iron, nickel, and cobalt, have atoms that end up with a small net magnetic field. In these materials groups of atoms align themselves in each other's fields, forming magnetic domains. Once these domains are aligned, they produce the magnetic properties we observe.

Magnetic Fields Affect Moving Charges

If a wire is placed between the poles of a horseshoe magnet, the wire remains stationary only as long as no current flows in it. Once the current starts, the wire jumps away. Magnets exert forces on current-carrying wires.

We can explain this observation in terms of an interaction between two magnetic fields. Both the magnet and the current-carrying wire have magnetic fields. In the region where the two fields overlap, an interaction occurs, and the two fields repel one another. Since the wire is substantially less massive than the magnet, it jumps away.

We can describe both the direction and magnitude of the force exerted on the current-carrying wire without necessarily analyzing the two magnetic fields in detail. The direction of the force is perpendicular to both the current and the magnetic field; it can be determined from a *right-hand rule*. If you place the fingers of your right hand in the direction of the magnetic field produced by the magnet and your thumb in the direction of the electric current, your palm points in the direction in which the force acts (Figure 20-11). The magnitude of the force depends on the strength of the two magnetic fields. Increasing the strength of the magnet or the current in the wire increases the force on the wire.

These results can be generalized and applied to other situations. If we think of the magnet as supplying an external magnetic field and the current-

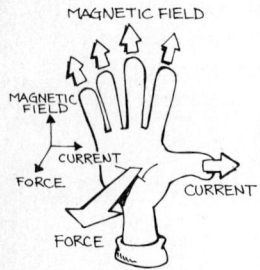

Figure 20-11

If you place the fingers of your right hand in the direction of the magnetic field and your thumb in the direction of the electric current, your palm points in the direction in which the force acts.

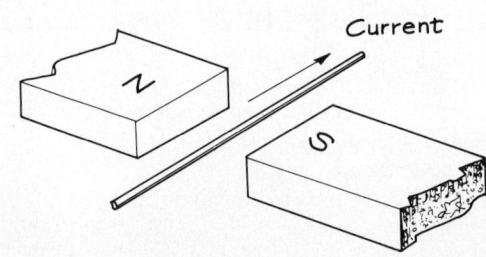

Figure 20-12

carrying wire as supplying a moving electric charge, then we can conclude that an external magnetic field exerts a force on a moving electric charge. The external magnetic field can be provided by a permanent magnet, an electromagnet, or even another current-carrying wire. The moving electric charge can be a stream of electrons, a stream of positively charged particles, or the conventional current in a wire.

SELF-CHECK 20C

A wire is placed between the poles of two magnets, as shown in Figure 20-12. In which direction will the wire move if a current is allowed to flow as shown?

Magnetic Forces Run Motors

The forces that magnetic fields exert on moving electric charges have been exploited in a number of applications, most notably the electric motor. An electric motor is a device that converts electrical energy into kinetic energy. In a simplified version (Figure 20-13), a motor consists of a wire coil placed in the magnetic field between two poles of a permanent magnet. When the current moves in the coil, the permanent magnetic field interacts with the coil's magnetic field and exerts a force that rotates the coil. A shaft attached so that it rotates with the coil then makes this motion available for performing useful work. The electrical energy supplied by the current is converted into kinetic energy of the shaft.

Though the idea is simple, it takes some ingenuity to make a motor actually work. Figure 20-14 shows a series of drawings as the coil rotates within the magnetic field. In (a) a current flowing from A to B to C to D results in the magnetic field applying a downward force on the coil segment AB and an upward force on segment CD. These forces combine to rotate the coil counter-

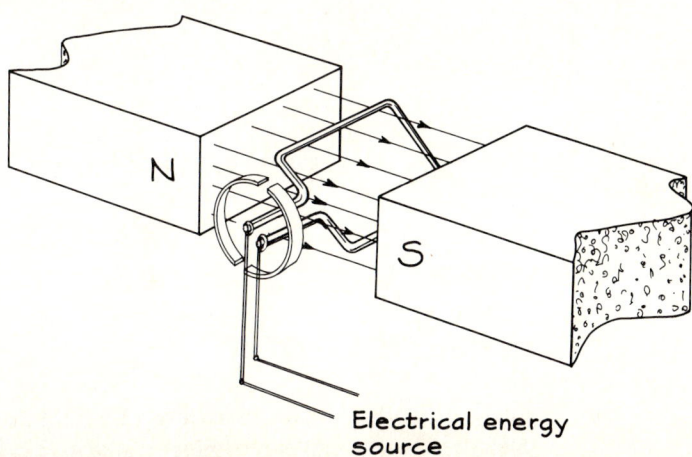

Figure 20-13

In simplified form, an electric motor consists of a wire coil placed in the magnetic field between two poles of a permanent magnet.

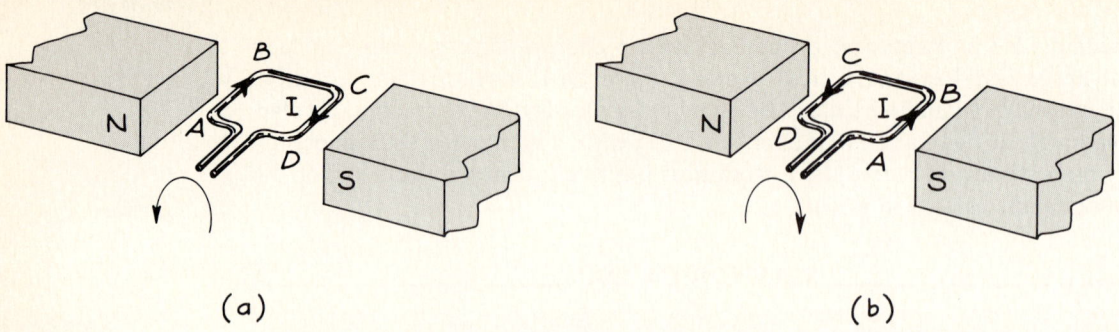

Figure 20-14
As current flows in the direction of ABCD, the permanent magnet exerts forces on segments AB and CD of the current-carrying coil.
(a) An upward force on CD and a downward force on AB rotate the coil counterclockwise.
(b) If the current continues in the same direction, a downward force on AB and an upward force on CD will rotate the coil back clockwise.

clockwise. In (b) the coil has rotated halfway around. If the current continues to flow in the same direction (A to B to C to D), a downward force will be exerted on the segment AB and an upward force on segment CD. These forces combine to rotate the coil clockwise—back in the direction from which it just came. To keep the coil rotating in the clockwise direction, we have to reverse the direction in which the current flows each time the coil completes half a revolution. The simplest way to accomplish this is to have the ends of the coil rotate about a cylindrically shaped pair of contacts so that each end touches first one contact then the other. Many motors use the arrangement shown in Figure 20-15. As the coil rotates, the end touching the positive and negative sides changes with each half revolution, and the current reverses direction.

In principle, all motors use the same design; an external magnetic field is used to rotate a current-carrying coil. In practice, motors vary depending upon the strength required in each application. The strength of a motor depends on the forces exerted on the coil which, in turn, depend on the magnetic fields. Motors used in small toys use permanent magnets to supply the external magnetic field. Larger motors use more powerful electromagnets powered by an external source of electricity. All motors increase the magnetic field produced by the coil by using many turns of wire, each of which carries the same current. Small motors use simple coils of a few hundred turns. Larger motors use thousands of turns wrapped around an iron core, forming an armature. In many cases a number of separate coils are arranged in different orientations to provide smoother rotation.

SELF-CHECK 20D

How will placing an iron core inside the coil increase the strength of the motor?

ELECTROMAGNETIC INDUCTION

Once Oersted had discovered that currents create magnetic fields, others set about looking for the inverse effect—magnetic fields that could create currents. At the beginning of the nineteenth century, the only current-producing

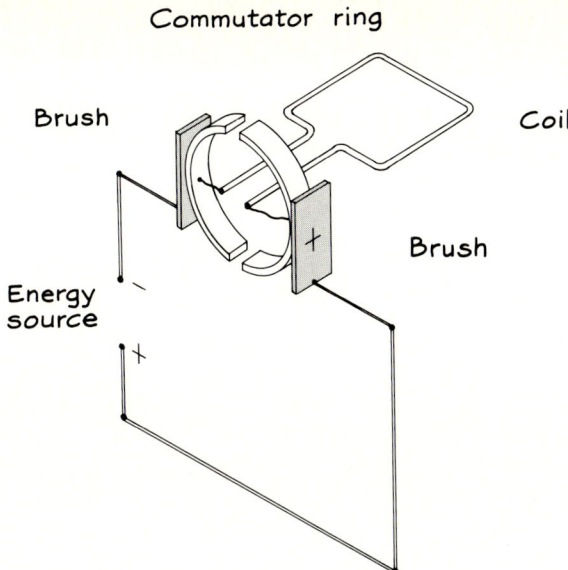

Figure 20-15
Many motors use splitring commutators. Each end of the coil is attached to half of the ring. An electric insulator separates the two halves. Two metal contacts, called brushes, are attached to an energy source. As the commutator rotates, each half alternates contact with the positive and negative brushes. This leads to a current in the coil that reverses direction every half revolution.

device was the voltaic cell—a forerunner of our present-day battery. Hardly cost-effective, these cells required large quantities of expensive metals to produce relatively small currents. Physicists and inventors alike saw the possibility of using magnetic fields to produce currents and hoped that the process would produce a cheaper and more substantial source of energy. You need only look about at the thousands of kilometers of electrical wires that span our country to realize how fruitful their search was to be.

Moving Magnets Produce Electric Currents

Michael Faraday was the first to observe the creation of electric currents by magnetic fields. He wrapped two coils around a circular piece of iron. One coil was attached to a battery so that it could be used like an electromagnet; the second was connected to a sensitive ammeter. Faraday had hoped that the magnetic field produced by the first coil would create an electric current in the second. His hopes were fulfilled, but differently than he had expected. A current was induced in the second coil, but only as the current to the first coil was being turned on or off. When the current from the battery remained constant, no current appeared at the ammeter. Only a changing current in one coil induces a current in the second.

A simpler demonstration of the effect is shown in Figure 20-16. A coil of wire is attached to a sensitive ammeter. We can place the coil and a permanent magnet near one another in a variety of orientations. As long as the coil and magnet remain stationary relative to each other, no current is induced in the coil. The relative motion between the coil and the magnetic field leads to an electric current.

In the experiment illustrated in Figure 20-16, a current was induced in the coil when relative motion between the coil and the magnet occurred. The coil experienced a changing magnetic field. Faraday's observations showed the same result. As the current in one coil was turned on or off, the magnetic

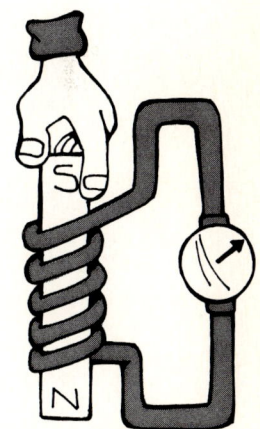

Figure 20-16
Only as the magnet moves in and out of the coil is a current induced in the coil.

Figure 20-17
If the magnet moves downward, the electrons in the wire move upward in the magnet's reference frame. An electron motion and a magnetic field toward you result in a force that pushes electrons along the wire. This force gives rise to an electric current in the wire.

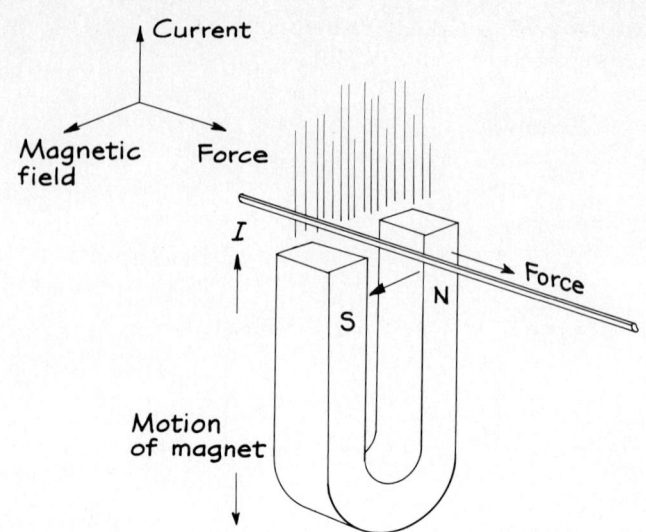

field that it produced changed. When the current remained steady, the magnetic field did not change. A current was induced in the second coil only while the magnetic field in the first coil was changing.

Faraday's Law

The process by which a changing magnetic field produces an electric current is called **electromagnetic induction.** Since a current arises from a voltage applied across the two ends of a coil, electromagnetic induction can also be described in terms of an induced voltage. Faraday found that identically changing magnetic fields induced the same voltages regardless of the composition of the wire. By contrast, the induced current depended on the resistance of the wire. Consequently, electromagnetic induction is more conveniently described in terms of induced voltages rather than currents. A voltage is induced across the two ends of a coil when the magnetic field inside the coil changes.

Electromagnetic induction arises from the forces that magnetic fields exert on moving charges. In an earlier section, we described how a current-carrying wire jumps away from the poles of a magnet. We showed that a right-hand rule describes the direction of the force applied to the wire in a magnetic field (Figure 20-11). If we look at this situation from the reference frame of the magnet, a current exists if the magnet moves past the electrons rather than the electrons moving past the magnet. For example, as the magnet in Figure 20-17 moves downward, the electrons in the wire move upward relative to the magnet. This motion constitutes a current. Placing the fingers of your right hand along the magnetic field and your thumb in the direction of the motion of the wire relative to the magnet, your palm points in the direction of the current we observe.

Faraday conducted a series of exhaustive experiments to determine the variables that influence the voltage induced by a changing magnetic field. He concluded that the voltage across the ends of a coil increases as the strength of the magnet increases or as the speed with which the magnet is moved increases. Both of these effects show that increasing the rate of change of the magnetic field increases the voltage. He also noted that as the number of turns of wire in the coil increases, so does the induced voltage. These results are summarized in **Faraday's law:** The voltage induced across the two ends of a coil is directly proportional to the number of loops in the coil and to the rate at which the magnetic field inside the coil changes.

SELF-CHECK 20E

Figure 20-18 shows a device designed in 1832. A permanent horseshoe magnet is rotated beneath two stationary coils. (a) Will this device produce an electric current? (b) What would you do to increase the induced voltage?

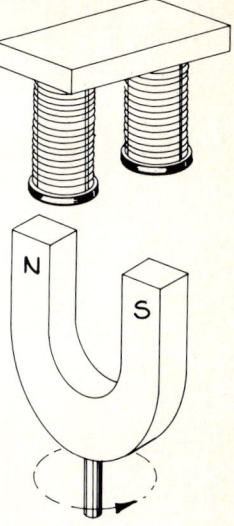

Figure 20-18

Generators and Alternating Current

Faraday's discovery ultimately led to the development of the electrical generator. A **generator** is a device that converts kinetic energy into electrical energy. It is essentially a motor running backward. Figure 20-19 shows a simple generator. A coil of wire that can be rotated by a crank is placed in a constant magnetic field. The number of magnetic field lines enclosed inside the coil depends on the orientation of the coil. In (a) the coil has the largest number of lines inside it. As the coil rotates, (b), it encircles fewer of the field lines until at (c) the coil lies along the field lines and encloses none of them. As

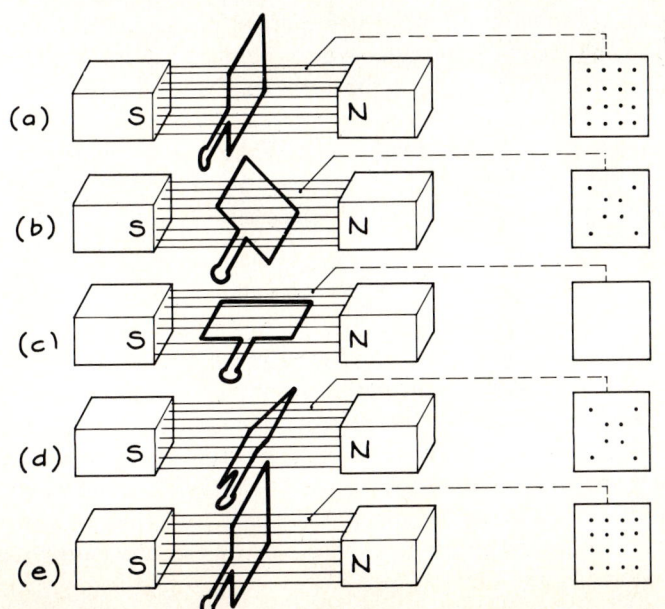

Figure 20-19

As the coil rotates, the number of magnetic field lines it encloses varies from a maximum at **(a)** to a minimum at **(c)** and back to a maximum again at (a).

Figure 20-20
As the coil rotates, both the magnitude and direction of the induced current change. One complete rotation of the coil leads to one complete cycle in current.

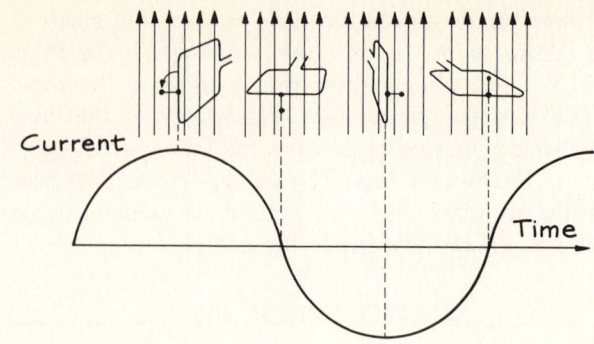

the coil continues to rotate, it includes more field lines (d) while it reaches a maximum at (e) once again. As the coil rotates, the magnetic field inside it changes, and a current is induced.

The current produced by a simple generator is an alternating current. In contrast to the direct current supplied by a battery, an **alternating current** changes magnitude and direction periodically (Figure 20-20). In a motor we had to reverse the current after each half revolution to keep the coil rotating in the same direction. It comes as no surprise that a coil that continues to rotate in the same direction produces a current that reverses direction after each half revolution. The alternating current we use in our homes is produced by generators standardized so that the current changes its magnitude and direction by going through 60 cycles each second.

Most power plants use more complex and more powerful variations of the simple generator. Powerful electromagnets supply the magnetic field. Huge coils of wire designed much like the armature of a motor are placed in the magnetic fields. The energy needed to rotate either the coils or the electromagnets is provided by a turbine. Energy from wind or falling water can cause the rotation, but in most commercial power plants the turbines are driven by moving steam. Steam generators require some external fuel to produce the steam needed to rotate the turbines. Natural gas, oil, and coal are the common fuels used. More recently, nuclear fuels have been introduced to replace dwindling supplies of fossil fuels. While the forms of energy used to create the kinetic energy vary, the process used to convert it to electrical energy is that illustrated by the simple generator.

Transformers

What made electrical energy so appealing in the nineteenth century was the ease with which it could be transmitted to other locations before being used. Steam engines required that the energy be used where it was produced. Electrical generating plants could transmit energy as far as the wires would reach, which has become as far as we want to string them. However, a wire is an energy receiver with a very small resistance, which increases as the wire gets longer. Some energy is changed into heat as electrical energy is transmit-

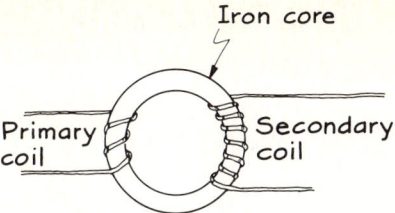

Figure 20-21
Faraday's induction coil is a simplified transformer. Alternating current to the primary coil induces an alternating current in the secondary coil.

ted along wires. These energy losses increase with the current in the wire. Thus, to make transportation of electrical energy over long distances practical, the current must be kept as small as possible. Because electrical power is equal to the product of voltage and current, we can transport energy with very low current by using high voltages. The transformer enables us to change high current and low voltage electricity to low current and high voltage electricity and, thus, is a vital link in the energy distribution network.

A **transformer** uses an alternating current and voltage in one coil to induce an alternating current and voltage in a second coil. In a simplified form, a transformer is identical to Faraday's induction coil (Figure 20-21). Alternating current supplied to the **primary coil** creates a changing magnetic field in the iron core. In turn, the constantly changing field induces an alternating voltage in the **secondary coil.** When the primary and secondary coils have different numbers of loops in their coils, the induced voltage in the secondary coil differs from the voltage in the primary.

Faraday's law provides us with a simple relationship between the two voltages and the number of loops in each coil.

$$\frac{\text{Primary voltage}}{\text{Number of loops in primary coil}} = \frac{\text{secondary voltage}}{\text{number of loops in the secondary coil}}$$

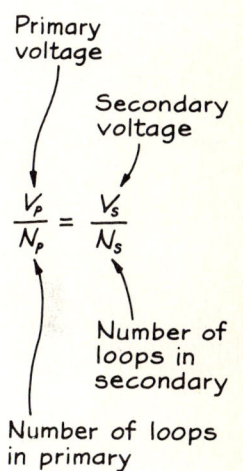

If we supply 120 volts (V) to a primary coil of 100 loops and the secondary coil has 200 loops, then the voltage in the secondary loop can be determined by (120 V)/(100 loops) = (secondary voltage)/(200 loops). Thus, the secondary voltage is 240 V.

This relationship allows us to increase or decrease the voltage of electrical energy at will. A **step-up transformer** is one that produces a larger voltage in the secondary than supplied to the primary. In this case the secondary coil has more loops than the primary. A **step-down transformer** is one that produces a smaller voltage in the secondary than that supplied by the primary. The secondary coil of a step-down transformer has fewer loops than the primary. Both types of transformers are important to the distribution of electrical power. The voltage induced by a typical electrical generating plant is 12,000 V. Before leaving the plant, this voltage is stepped up to 240,000 V. Near cities the voltage is stepped back down to 12,000 V for distribution throughout the city. Outside homes and businesses the voltage is stepped down again to provide the 120 V used in household circuits.

SELF-CHECK 20F

Transformers designed for use with typical electronic calculators transform the voltage supplied to the primary coil, 120 V, to the voltage needed by the calculator, typically 6 V. Is this a step-up or step-down transformer? How do the number of loops in the secondary coil compare to the number in the primary?

ELECTROMAGNETISM

By the middle of the nineteenth century the connections between electricity and magnetism were well established. At the same time, diffraction and interference had placed the wave model of light on very firm ground. However, the nature of light and its relation to other physical phenomena were still a puzzle. A major step toward the solution to that puzzle came from a better understanding of electricity and magnetism.

In the 1860s James Clerk Maxwell introduced a comprehensive theory of **electromagnetism,** a model that unites electricity and magnetism. In doing so, he used the idea of an electric field. This field surrounds electric charges in the same way that a magnetic field surrounds magnetic poles. Just as a magnetic field can be used to explain magnetic forces, an electric field is used to describe electric forces. With these two fields, he could express Oersted's and Faraday's results in broader terms:

A changing electric field produces a magnetic field.

A changing magnetic field produces an electric field.

A changing electric field could be produced by a current. Consequently, the first statement incorporates Oersted's results. An electric field in a wire would exert forces on the wire and create a current. Thus, the second statement includes Faraday's law. These general principles combine many of the ideas of electricity and magnetism.

However, Maxwell's theory of electromagnetism did more than just combine the results of earlier experiments. A vibrating electric charge creates a magnetic field that changes as the charge moves back and forth. This changing magnetic field creates a changing electric field, which, in turn, creates a changing magnetic field, and so forth. Maxwell was able to show that the motion of these changing fields as they move away from the source is a wave—an electromagnetic wave. Further, these waves travel in a vacuum with one speed—the speed of light. Light waves were a form of electromagnetic waves. In developing electromagnetism, Maxwell had described the wave nature of light. He also described a host of other wave phenomena, like X rays and radio waves, which were not discovered until after his death.

Human life has been radically transformed by the technology born of electromagnetism. Electromagnetic waves touch our lives daily. Electrical energy races continually across the landscape. As we use more electricity, our supply of traditional fossil fuels needed to drive our generators dwindles. Approximately 30 years after Maxwell completed his theory, a new energy source was discovered. Nuclear energy, the subject of the final two chapters of the text, stands as one of the more promising solutions to the dilemma of decreasing fossil fuels.

CHAPTER SUMMARY

All magnets have two *poles,* called *north* poles and *south* poles. Poles interact with one another: Like poles repel and unlike poles attract. *Magnetic fields,* illustrated by the patterns of iron filings sprinkled around magnets, describe the strength of magnetic forces surrounding a magnet. The magnetic properties of matter are described in terms of *magnetic domains,* regions that behave as tiny magnets. *Permanent* and *temporary* magnets differ in the number of magnetic domains that remain when an external magnetic field is removed. Magnetic domains arise when atoms with a net magnetic field become aligned. Electrons, which are charges moving within atoms, give rise to magnetic fields. In most atoms the fields from individual electrons cancel each other, but the atoms of a few elements have net magnetic fields. These atoms form magnetic domains.

All moving charges produce magnetic fields. A current-carrying wire produces a cylindrical magnetic field around itself. The strength of the field depends on the amount of current flowing, and the direction depends on the direction in which the current moves. The magnetic field of a current-carrying coil is similar to that produced by a bar magnet. Current-induced magnetic fields can be added to the field created when the domains of a magnetic material line up to create an *electromagnet.* A magnetic field exerts a force on a moving charge. A current-carrying wire placed in a magnetic field moves in a direction perpendicular to the current and the field. Motors use the interaction between a current-carrying wire and an external magnetic field to convert electrical energy into kinetic energy.

Electromagnetic induction is the process by which a changing magnetic field produces an electric current. *Faraday's law* summarizes the relationship between the changing magnetic field and the induced voltage: The voltage induced across the two ends of a coil is directly proportional to the number of loops in the coil and the rate at which the magnetic field inside the coil changes. Generators use the motion of a coil in an external magnetic field to convert kinetic energy into electrical energy. A simple *generator* produces an *alternating current,* which changes direction with each half-rotation of the coil. *Transformers* use the principles of electromagnetic induction to increase or decrease the voltage that is delivered to them.

Electromagnetism describes the single model which unites electricity and magnetism. A changing electric field produces a magnetic field. A changing magnetic field produces an electric field. When propagated outward at the speed of light, these two fields support one another. This leads to the creation of electromagnetic waves, which include light, radio waves, and X rays.

ANSWERS TO SELF-CHECKS

20A. Each time the magnet is dropped, some of its magnetic domains can be knocked out of alignment. As more domains are knocked out of alignment, the strength of the magnet decreases.

20B. The strength of the electromagnet decreases. Aluminum has no natural magnetic properties. Consequently, even when it is placed in the magnetic field of the coil, it does not become magnetized. The total magnetic field produced is simply that produced by the coil alone.

20C. We can apply the right-hand rule. If we place the fingers of our right hand in the direction of the magnetic field and our thumb in the direction of the electric current, our palm points downward. The wire jumps downward.

20D. The kinetic energy produced by a motor depends on the strength of the two magnetic fields. If we place an iron core inside the current-carrying coil, then the coil will produce an even stronger magnetic field. This increases the strength of the motor.

20E. a. Yes, because the horseshoe magnet rotates, the two coils are surrounded by a changing magnetic field.
b. If we had only the arrangement shown in Figure 20-18, we could increase the induced voltage by rotating the magnet faster. If we could alter the arrangement, then we might use a stronger magnet or replace the coils with ones that had more turns.

20F. Since the voltage supplied to the calculator is less than the voltage supplied to the transformer, this is a step-down transformer. The secondary coil has fewer loops than the primary coil.

PROBLEMS AND QUESTIONS

A. Review of Chapter Material

A1. Define each of the following:
Magnetic pole
Magnetic field
Magnetic domain
Permanent magnet
Temporary magnet
Electromagnet
Electric motor
Electromagnetic induction
Faraday's law
Generator
Alternating current
Transformer
Electromagnetism

A2. How is our model of magnetism similar to our model of electric charge? How are the two models different?

A3. Sketch the magnetic field produced by the two magnets shown below.

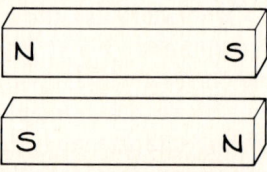

A4. Use the concept of magnetic domains to explain the different behaviors of permanent and temporary magnets.

A5. Describe Oersted's experiment. How did it demonstrate that a relationship existed between electricity and magnetism?

A6. Describe how Oersted's discovery eventually led to an atomic explanation of the nature of magnetism in matter.

A7. Use Oersted's discovery to explain how an external magnetic field can exert a force on a current-carrying wire.

A8. Describe Michael Faraday's experiment demonstrating that magnetic fields give rise to electric currents.

A9. What variables affect the strength of the induced voltage?

A10. How are electric motors and electric generators similar? Different?

A11. How does the number of coils in the secondary coil determine whether the transformer is a step-up or step-down transformer?

A12. Use the force exerted on a moving electric charge to explain electromagnetic induction.

A13. Describe how moving fields create electromagnetic waves.

B. Using the Chapter Material

B1. Describe the interaction that occurs when you hold:
   a. the north pole of a permanent magnet near the south pole of a different permanent magnet
   b. the north pole of a permanent magnet near the north pole of a different permanent magnet
   c. the north pole of a permanent magnet near an unmagnetized piece of iron
   d. the north pole of a permanent magnet near an unmagnetized piece of aluminum

B2. Describe how you could use a permanent magnet to locate the poles of a magnetized chunk of iron. Describe how you could use iron filings to locate these same poles. Which, permanent magnet or iron filings, would allow you to identify the north pole of the magnetized iron?

B3. If you heat a permanent magnet, some or even all of its magnetic strength can be lost. Use magnetic domains to explain why.

B4. Describe why the compass shown in the next column does not detect the presence of the magnetic field produced by the current-carrying wire.

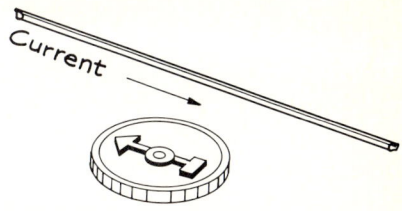

B5. Describe what you could do to the simplified motor shown in Figure 20-14 to build a more powerful motor.

B6. If you twirl an electrical wire through the air on earth, you can obtain a weak current in the wire. Explain why.

B7. Suppose an atom contains two electrons, one of which rotates clockwise about the nucleus while the other rotates counterclockwise. Why would you expect no magnetic field to be present as a result of these motions? Explain your answer.

B8. A magnet is dropped. As it travels downward it passes through two identical coils. The magnet passes through the first coil 1 second after it is dropped and the second 2 seconds later. Which coil shows a greater induced voltage? Why?

B9. A portable radio requires 12 V to operate correctly. If a 500-loop primary coil is connected to 120 V, how many loops must the radio's transformer have in the secondary loop?

B10. Some bicycles have electric generators that turn when the bike wheel turns. These generators provide energy for the bike's lights. When the bicycle is moving rapidly, the lights are bright, but a slowly moving bike produces a dim light. Explain why in terms of how a generator works.

C. Extensions to New Situations

C1. A telegraph system can be constructed as shown on page 486.
   a. Describe the magnetic fields when the switch is closed.
   b. Why is the L-shaped piece of iron attracted to the iron in the center of the coil?
   c. How could this system be used for communication?

C2. The motors and generators described in

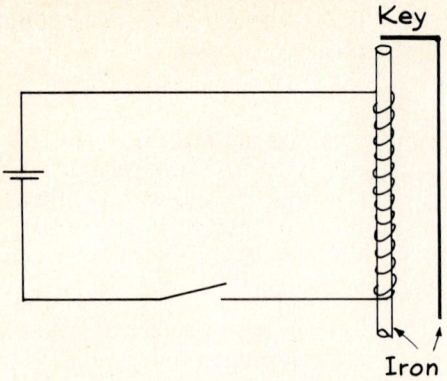

the chapter contained permanent magnets. Motors are frequently built without any permanent magnets.
  a. Draw the magnetic field of a motor with permanent magnets.
  b. Describe how this same field could be created by coils of wire. Include the position, orientation, and electrical connections of the coils.
  c. Use the results of part (b) to sketch a motor with no permanent magnets. Include all electrical connections.
  d. Could the motor you sketched in part (c) be used as a generator? Explain your answer.

C3. The relay uses electromagnetism for remote control of electric current. Use the figure below, which shows a simple relay, in answering the following questions.

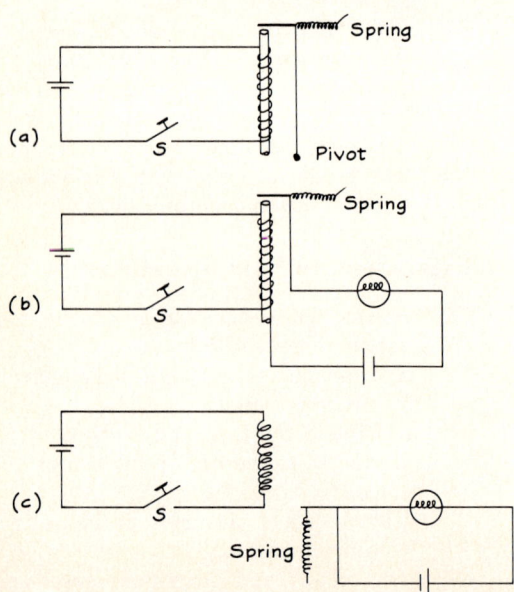

  a. Describe and explain what happens to the L-shaped iron attached to the spring when the switch in (a) is closed and when the switch is opened.
  b. In (b), will the light be on when the switch at S is opened? When S is closed? Explain your answers.
  c. How might you use these circuits to control a high current to an energy receiver with only a low current through the switch? (The starter system in an automobile works on this principle.)

C4. A relay can be built as described in Problem C3, or the iron inside the coil can actually move. In the latter situation, the position of the iron with no current in the coil is partway in the coil (see the figure below). Describe and explain the motion of the iron when current moves through the coil.

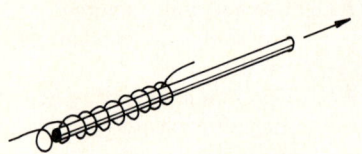

C5. The transformer enables energy to be transferred from one electrical circuit to another without a direct electrical connection.
  a. Two coils are placed next to each other, as shown below, part (a). Why will a current exist in circuit 2 only if the current in circuit 1 changes?
  b. An iron square is added to the circuit, as shown below, part (b). Then, when a current exists in circuit 2, it is much greater than that of part (a). Why?

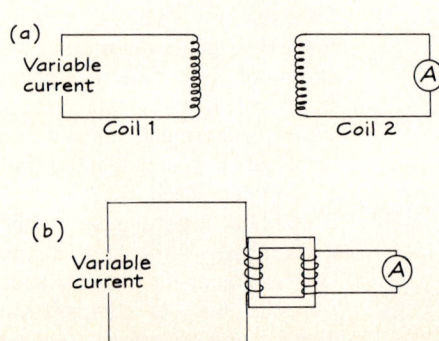

C6. A television picture is created by electrons, which cause light to be emitted when they hit the screen. In modern television sets, the motion of these electrons is controlled by magnetic fields. Describe how the coils of wire through which current flows can be used to control the location at which the electron strikes a television screen.

C7. Television pictures are created by moving electrons, which strike the screen. Color television screens are coated with a material that can very easily be magnetized permanently.
   a. If you bring a permanent magnet near a television, the picture is distorted. Why?
   b. When a magnet is brought near a color television, the distortion remains even after the magnet is removed. Why?
   c. Sometimes permanent distortion occurs on a color television after a vacuum sweeper has been operated very close to it. Why?

C8. When a steel can sits in one place for a long time, it becomes magnetized. Use an analogy with the discussion of information stored in rocks on earth to explain why.

D. Activities

D1. Tear apart an old motor, generator, or relay. Identify as many parts as you can.

D2. Use a magnetic compass to identify permanent magnets in your environment.

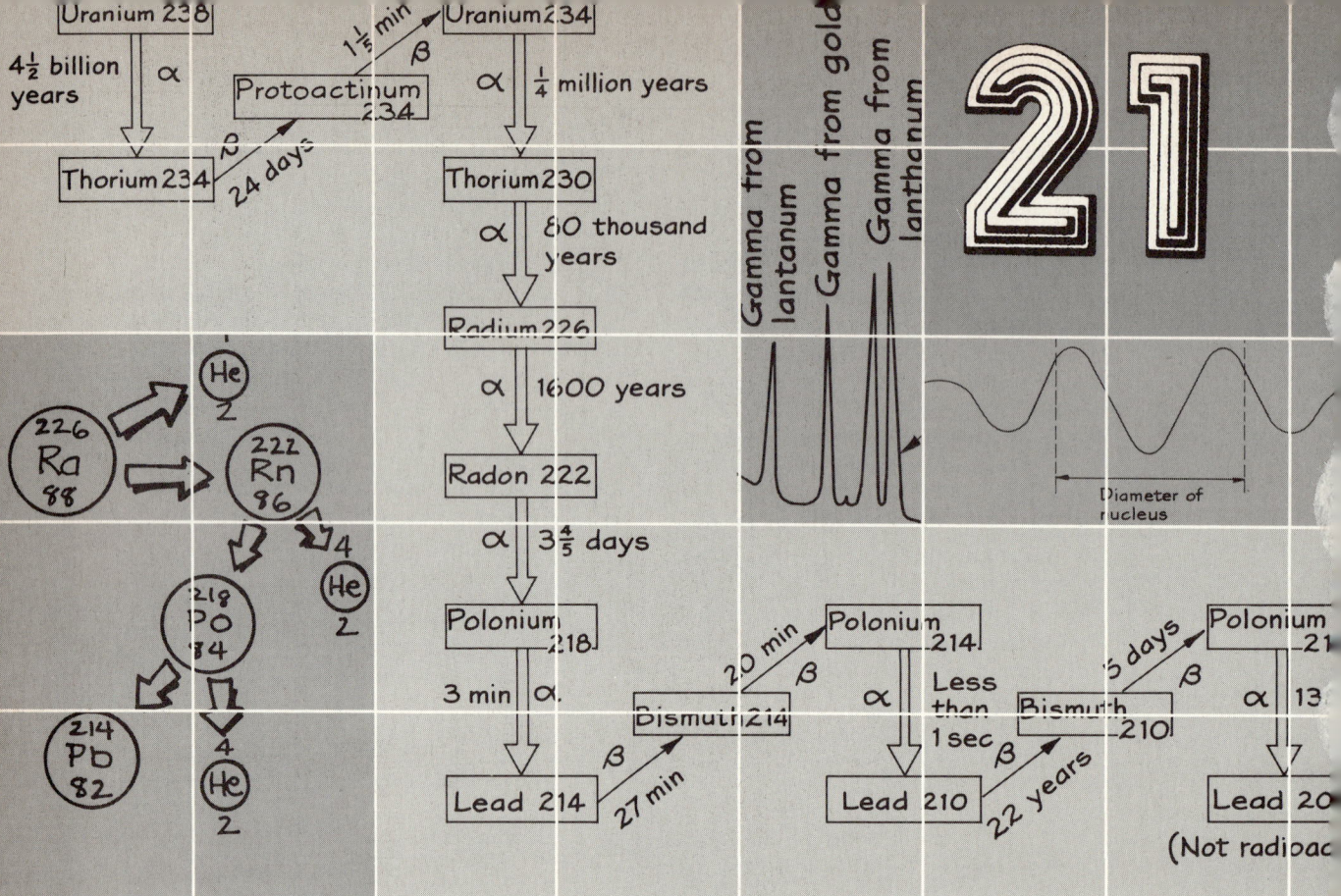

Radioactivity

Nuclear physics—a threat or a remedy? Does it reveal people's remarkable intellect or the stupidity with which people use their knowledge? Is it a means of vast destruction or a valuable diagnostic and research tool? Is it a vast source of energy or the most dangerous pollutant yet unleashed? The story of nuclear physics lies hopelessly entangled in the joys and sorrows of the twentieth century. It is with this topic that we close our book.

In 1896, Henri Becquerel noticed the radiation emitted by uranium salts. This radiation, which had much higher energy than any previously known, was named *radioactivity*. The emissions provided clues to the structure of the nucleus just as light spectra were providing clues to the structure of the atom. *Nuclear transformations* are the causes of the reactions that occur as the nuclear radiation is released. Nuclear radiation occurs in three major types, called *alpha*, *beta*, and *gamma* radiation. In alpha and beta radiation, one element is transformed into another. The *half-life* of a radioactive substance describes the rate at which these transformations occur.

THE NATURE OF THE NUCLEUS

Because the nucleus is so small (about 10^{-15} meters (m) in diameter), we cannot observe it directly. Like our knowledge of atoms, our knowledge of the nucleus comes from indirect evidence. Rutherford's scattering experiment, in which he fired alpha particles at thin gold foil, provided clues to both the size and the mass of the nucleus. Radioactivity, the energy and particles emitted by materials such as uranium, provides clues to the types of particles that would be found in the nucleus. The periodic table, Figure 21-1, organizes the various chemical elements according to several properties, some of which could be explained only in terms of nuclear structure. Before looking at the details of nuclear transformations, we briefly review the evidence that supports our present model of the nucleus.

Clues to the Structure of the Nucleus

In some respects understanding the nucleus means understanding the periodic table of elements. By the turn of the twentieth century, chemists had isolated and identified the physical and chemical properties of the seventy or so elements that are shaded in Figure 21-1. Elements with similar properties are grouped into families and listed vertically. For example, fluorine (F) and chlorine (Cl) have similar properties and are listed in the same vertical column. From upper left to lower right, the elements are placed in order of increasing

Figure 21-1

Modern version of the periodic table. The elements that are shaded were known at the turn of the twentieth century, when radioactivity was discovered.

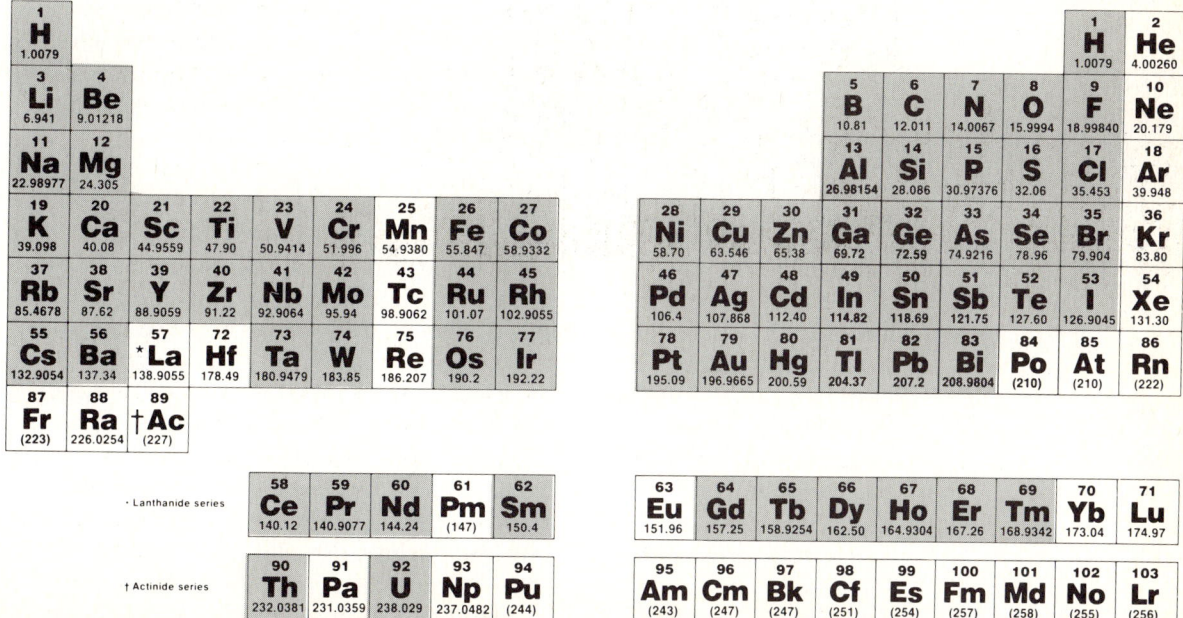

atomic mass. The resulting periodic table shows a relationship among the collection of known chemical elements. At first, however, no one understood why elements could be grouped in this way.

Rutherford's scattering experiment provided clues to the ordering of the elements. As you saw in Chapters 8 and 18, Rutherford showed that atoms consist of positively charged, massive nuclei that are surrounded by much less massive, negatively charged electrons. When the magnitude of the electrical charge of the nucleus is compared to the magnitude of the charge of a single electron, a pattern emerges. Carbon, the 6th element in the periodic table, has a nucleus with a charge 6 times the magnitude of the charge of an individual electron. Gold, the 79th element in the table, has a nucleus with a charge 79 times the magnitude of that of a single electron. The horizontal arrangement of elements in the periodic table is related to the amount of electric charge present in the nucleus. Since atoms have zero net charge, the number of electrons must be equal to the magnitude of the charge on the nucleus. Thus, the horizontal arrangement also describes the number of electrons present in each atom. Carbon atoms have 6 electrons; gold atoms have 79 electrons; and so forth.

The quantum mechanical model of the atom (Chapter 18) creates an image of a relatively stationary nucleus surrounded by electron clouds. While electrons change their average distances from the nucleus as atoms release or absorb energy, the nucleus never seems to change. Elements combine chemically as electrons of their atoms interact with those of neighboring nuclei. For example, the atoms of hydrogen gas can combine with those of oxygen to form molecules of water. The atoms that form water share electrons with each other. The nucleus, however, remains uninvolved in such interactions. An element's physical properties (how it is affected by conditions such as light and temperature) as well as its chemical properties (how it combines with other elements) depend on the number and energy levels of the electrons surrounding the nucleus. But the number of electrons in an atom depends on the charge of the nucleus. Ultimately, then, the structure of the nucleus determines the unique properties of every element.

Inside the Nucleus

The nucleus contains two particles, the proton and the neutron. Together these particles explain both the mass and electric charge associated with each nucleus. Protons have an electric charge of $+1.6 \times 10^{-19}$ coulombs (C), equal in magnitude but opposite in sign to the charge on the electron. They have a mass of 1.672×10^{-27} kilograms (kg). Neutrons have no electrical charge. Their mass, 1.675×10^{-27} kg, is approximately equal to the mass of the proton. As predicted by Rutherford's scattering experiment, the masses of the particles found in the nucleus are enormous compared to the mass of the electron. Protons and neutrons are each about 1850 times more massive than an electron. Because of their presence in the nucleus and their similar characteristics, protons and neutrons are collectively called **nucleons.**

The mass and electric charge of different atomic nuclei can be explained in terms of their nucleons. Protons establish the electric charge associated with the nucleus. The horizontal order of elements in the periodic table reflects the number of protons present in each nucleus. Hydrogen has one proton; helium has two; lithium has three; and so forth. Thus the characteristics of an element reflect the number of protons in its nucleus. Neutrons add mass to the nucleus without affecting its charge. Helium has only two protons, but its mass is about four times that of hydrogen. Lithium, with three protons, has a mass nearly seven times that of hydrogen. We explain these masses by assuming that the helium nucleus consists of two protons and two neutrons, while the lithium nucleus consists of three protons and four neutrons.

Since neutrons carry no electrical charge, they can be added to the nucleus without affecting its chemical identity. For example, a typical hydrogen atom has one proton and no neutrons in its nucleus. If we add a neutron, the nucleus is more massive but still has one proton. It is still a hydrogen nucleus. Careful measurements of atomic masses show that such variations do occur. Hydrogen nuclei commonly exist in three variations: nuclei with one proton and no neutrons, with one proton and one neutron, and with one proton and two neutrons. Nuclei of a single element that have different numbers of neutrons (and, therefore, different masses) are called **isotopes** of the element. The atomic masses listed in the periodic table are the average masses of the mixtures of isotopes commonly found for each element.

Investigators have now identified over 100 elements, each of which has several isotopes. This means that a lot of different kinds of nuclei exist. To identify a specific nucleus we use the shorthand notation shown in Figure 21-2. A one- or two-letter symbol (X in the figure) identifies the chemical element. For example, hydrogen is identified as H; helium, as He. A subscript (Z), called the **atomic number,** identifies the number of protons in the nucleus and is the number given to each element in the periodic table. A superscript (A), called the **mass number,** describes the total number of nucleons found in the nucleus. Thus the atomic number identifies the chemical element, while the mass number identifies the specific isotope of that element. The number of neutrons in a particular nucleus can always be determined by subtracting the atomic number from the mass number (number of neutrons = $A - Z$). Several examples are:

Hydrogen	one proton, no neutrons	$^{1}_{1}H$
Helium	two protons, two neutrons	$^{4}_{2}He$
Oxygen	eight protons, eight neutrons	$^{16}_{8}O$

We sometimes refer to an isotope by giving its chemical name and its mass number. For example, the isotopes listed above are called hydrogen 1, helium 4, and oxygen 16.

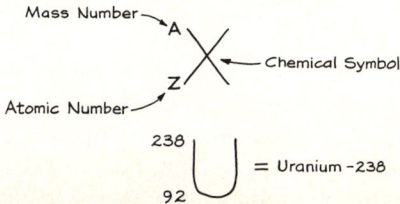

Figure 21-2

Shorthand notation used to describe nuclei.

We use a similar scheme for the electron. The symbol $_{-1}^{0}e$ identifies the electron as having a charge ("atomic number") of -1 and a mass of essentially zero when compared to the mass of the nucleons.

When expressed in kilograms, the masses of the nuclei are extremely small and somewhat cumbersome to use. So, we use a different unit of mass, the *atomic mass unit*. One **atomic mass unit** (amu) is 1.660565×10^{-27} kg, approximately equal to the mass of a nucleon. A proton has a mass of 1.006 amu, while the mass of a neutron is 1.0087 amu.

SELF-CHECK 21A

How many protons and neutrons are present in each of the following isotopes of uranium?

$$^{234}_{92}U \qquad ^{235}_{92}U \qquad ^{238}_{92}U$$

NUCLEAR TRANSFORMATIONS

The light emitted by matter provides us with a brilliant display of information for building a model of atomic structure. As electrons jump from one orbit to another, atoms emit or absorb photons of electromagnetic radiation, including visible light. But photons are not the only radiation emitted by matter. In 1896 Henri Becquerel conducted a series of experiments with uranium and uranium salts that led to the discovery of new radiations that were far more intense than any yet known. Unlike the radiation associated with light spectra, these emissions included particles with mass and charge. Such radiation could be understood only in terms of nuclear changes. Ultimately, they provided the information for building a model of the nucleus.

Radioactivity

Figure 21-3 describes Becquerel's initial observations. He had wrapped a photographic plate with several pieces of thick black paper to keep it from being exposed to ordinary light. He then placed a chunk of uranium salt on top of the paper and put the plate in a dark drawer for several days. Since the photographic plate was well protected from light, we would expect the developed plate to be uniformly gray. Instead, Becquerel found an intense silhouette beneath the location of the uranium salt. After a series of careful experiments he concluded that uranium emits some sort of radiation that can penetrate paper and expose photographic plates.

Marie and Pierre Curie undertook a systematic study of Becquerel's rays, which they named *radioactivity*. They demonstrated that certain known elements in addition to uranium are radioactive, and they isolated previously

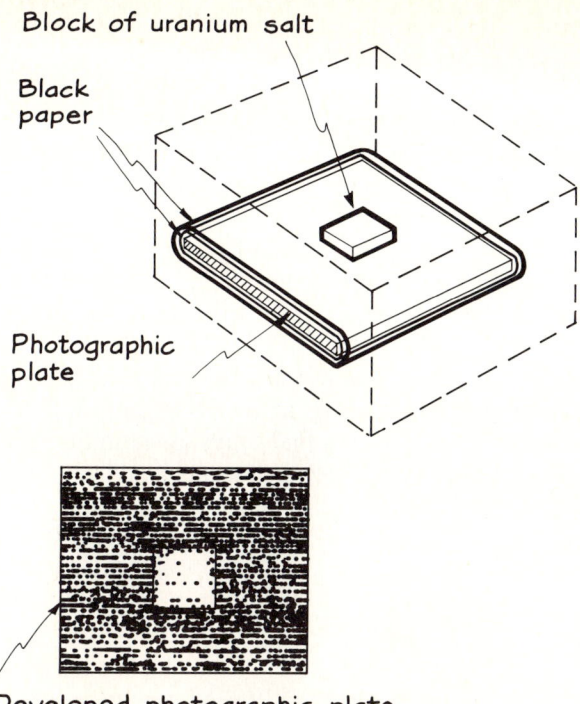

Figure 21-3
Uranium salts emitted radiation intense enough to expose a photographic plate that had been well protected from light.

unknown elements that are also radioactive. Rutherford and others concluded that the emissions are mixtures of three distinct types of radiations, which they called alpha, beta, and gamma.

The three kinds of radiation are distinguished by how hard they are to stop in matter and by their behavior when subjected to magnetic and electric forces. **Alpha radiation** is easily stopped by placing a few sheets of paper in its path. In the presence of a magnetic or electric force, alpha radiation is deflected. **Beta radiation** is not stopped by matter as easily as alpha, but it can be blocked by a rather thin sheet of aluminum. When magnetic or electric forces are present, beta radiation deflects in the opposite way from alpha. **Gamma radiation** is very difficult to stop; it can be stopped only by thick blocks of lead. It is not affected by either electric or magnetic forces. Subsequent experimentation revealed that all three radiations had already been known. Alpha radiation consists of helium nuclei (^4_2He) and beta of electrons ($^{\ \ 0}_{-1}e$). Gamma radiation (γ) is electromagnetic radiation but with a much shorter wavelength than any previously known.

The positively charged helium nuclei that form the alpha radiation interact electrically with the atoms in matter. Because of their relatively large mass and charge, they lose energy quickly during these interactions. So, alpha radiation can be stopped by a rather small amount of matter. The electrons that make up beta radiation have a smaller mass and electrical charge than alpha particles. Thus they lose energy at a slower rate and travel further before stopping. Because alpha and beta particles have opposite electric charges, they move in opposite directions in the presence of magnetic and electric forces. Because gamma radiation is electromagnetic energy, it has neither charge nor mass. Thus it interacts very little with matter and is not affected by magnetic or electric forces.

Of the three phenomena, gamma radiation seems at first glance the least surprising. After all, atoms emit a great variety of electromagnetic radiations as they change their energy states. But, when compared with the energy emitted by atoms, the energy of gamma radiation is astounding. Gamma photons carry, on the average, 6.6×10^{-13} joules (J)—more than a million times the maximum energy an electron in a hydrogen atom can emit. Clearly, the energy must come from an interaction different from the electron transitions in the atom. Gamma radiations must originate from interactions inside the nucleus.

Alpha and beta radiation are distinct from any other energy forms. In burning and other chemical reactions, atoms emit pure energy. Electromagnetic energy has no mass. Radiations that carry away mass as well as energy are clearly different. They raise a new question: If radium (88 protons, 138 neutrons) emits an alpha particle (2 protons, 2 neutrons), do we still have radium? Our intuitive sense of conservation tells us that the answer is no.

Nuclear Transformations and Conservation Logic

The emission of alpha or beta radiation causes a change, called a **nuclear transformation,** in the nucleus from which it was emitted. Both alpha and beta emissions result from nuclear transformations. A radium nucleus that emits alpha radiation, for example, is no longer radium. It has been transformed into the nucleus of a different chemical element and an alpha particle, which is separated from the nucleus. We write

$$\text{Old element} \rightarrow \text{new element} + \text{alpha particle } (^4_2\text{He})$$

where the arrow indicates *changed into.*

When radium emits an alpha particle, it is changed into radon gas. This transformation can be expressed as:

$$^{226}_{88}\text{Ra} \rightarrow ^{222}_{86}\text{Rn} + ^4_2\text{He}$$

We say that radium has decayed into radon.

Nuclear transformations occur within the boundaries of the conservation laws of physics. Conservation of electric charge can be verified directly from the notation describing the interaction. For example, the charge carried by the radium nucleus (+88) equals the sum of the electric charges of the radon nucleus (+86) and the alpha particle (+2). Electric charge is conserved in nuclear interactions as it is in all other interactions.

A new conservation principle involves the total **number of nucleons** in the interaction. This number does not change during a nuclear transformation. In the present example radium has 226 nucleons, which equals the sum of the nucleons in radon (222) and helium (4) nuclei. For an alpha decay, the nucleons do not change from one type to another; protons do not change into neutrons. However, beta transitions require that a neutron change to a proton. Even in this case, however, the total number of nucleons (protons + neutrons) is conserved.

In nuclear transformations, conservation of energy takes a slightly different form from most of our previous uses. Here, energy must include Albert Einstein's statement that mass is a form of energy and that the mass-energy equivalence is given by

$$\text{Energy} = \text{mass} \times (\text{speed of light})^2$$

(See Chapter 9). In the radium $\rightarrow$ radon + alpha decay, the mass of the radium (226.025406 amu) is greater than the sum of the masses of radon (222.017574 amu) and helium (4.002604 amu). Conservation of energy requires that this mass difference (0.005228 amu = 8.7×10^{-30} kg) result in 7.8×10^{-13} J of energy. If the radium nucleus was not moving initially, the radon and helium nuclei will move away from each other with kinetic energies that total 7.8×10^{-13} J. This energy is what we commonly refer to as *nuclear energy*.

Conservation of momentum places a restriction on the relative motions of the products in a nuclear transformation. If the initial nucleus was not moving, then the products must move away from each other in such a way that the vector sum of their momenta is zero. For the radium decay, this restriction means that the alpha particle, with a much smaller mass, must have a much higher speed than the more massive radon. Only then will their momenta add to zero.

Alpha, beta, and gamma emissions are the processes by which most naturally radioactive nuclei decay. As we shall see, alpha and beta transformations both lead to the formation of new nuclei. Gamma transformations involve a rearrangement of the nucleons within the same nucleus. While other radioactive processes have been discovered, alpha, beta, and gamma decay remain common to many nuclear reactions. We examine each process in more detail.

SELF-CHECK 21B

A nuclear transformation can occur only if no conservation laws are violated. Using conservation of nucleons and charge, determine whether the nuclear transformations listed below are possible. Explain your conclusions.

$$^{14}_{6}C \rightarrow {}^{14}_{7}N + {}^{0}_{-1}e$$

$$^{19}_{9}F \rightarrow {}^{3}_{2}He + {}^{16}_{8}O$$

ALPHA DECAY

As we discussed in Chapter 8, the struggle between the attraction of the strong nuclear interaction among all nucleons and the repulsion of the electrical interactions among protons can cause some nuclei to decay. Nuclei that decay are called *unstable,* while those that do not are *stable.* Alpha decay is a common nuclear transformation by which heavy, unstable nuclei decay to lighter, more stable nuclei. In alpha decay a nucleus splits into an alpha particle and a nucleus of an element two places to the left in the periodic table. Energy is released in the form of kinetic energy of the two products. Polonium 210, for example, changes into lead 206 by emitting an alpha particle. This transformation is written

$$^{210}_{84}\text{Po} \rightarrow {}^{206}_{82}\text{Pb} + {}^{4}_{2}\text{He}$$

Lead is two places to the left of polonium in the periodic table. Mass in the polonium nucleus is converted into the energy needed to force the lead nucleus and the alpha particle apart.

Among unstable nuclei, alpha decay is somewhat common. To understand the reasons for alpha decay, we must turn to the wave nature of matter and consider the matter waves of the alpha particle as it exists in the nucleus. Alpha decay occurs so frequently that physicists speculate that the alpha particle somehow retains an identity within the nucleus. While we are not certain that such an assumption is valid, it does provide us with a useful model to consider.

The matter wave associated with the alpha particle within the nucleus provides us with one way of understanding the conditions in which alpha decay occurs. As illustrated in Figure 21-4, a matter wave can be associated with the alpha particle in the nucleus. When compared to the size of the nucleus, the length of this wave describes the conditions under which alpha decay will occur. In all cases most of the wave exists within the boundary of the nucleus. Nuclei that decay by alpha particle emission, however, have significant fractions of the alpha particle matter wave extending beyond the nucleus. (Recall that the amplitude of the wave packet at any point is related to the probability that the particle will exist at that point.) Because of this wave extension, the alpha particle can exist outside and inside the nucleus.

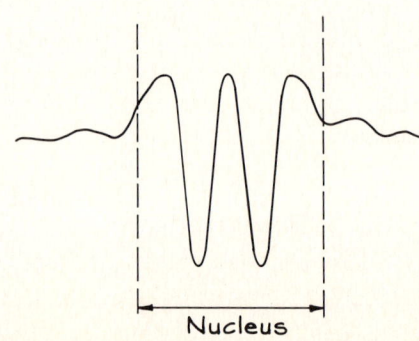

Figure 21-4
A wave packet can be associated with the alpha particle. If the wave packet spills out beyond the boundary of the nucleus, there is a finite probability that the alpha particle will move outside the nucleus briefly.

Inside the nucleus, the strong nuclear force of attraction exceeds the electric force of repulsion, and the alpha particle remains part of the nucleus. Once it wanders outside the nucleus, however, the strong nuclear force, which has a very short range, has less effect on the alpha particle. At some point, the electric repulsion takes over, and the alpha particle is permanently ejected from the nucleus. An alpha decay occurs, with a new nucleus being left behind.

BETA DECAY

Beta decay describes the process by which one element can be transformed into the next element to the right in the periodic table. For example, a nucleus of carbon, element 6, can be transformed into a nucleus of nitrogen, element 7. A beta particle (electron) and another particle called an antineutrino ($\bar{\nu}_e$) are released in this process.

$$^{14}_{6}C \rightarrow {}^{14}_{7}N + {}^{0}_{-1}e + \bar{\nu}_e$$

In some respects, beta decay is more of an enigma than alpha decay. While the assumption that the alpha particle somehow retains its identity within a nucleus is not well justified, it is not impossible to believe. A beta particle, on the other hand, is an electron. Electrons do not exist in the nucleus. Were it not for the observable transformation of a nucleus, as from carbon to nitrogen, we might suppose that the electron is released from the atom and not the nucleus. However, the nuclear transformation is unmistakable.

The Transformation of the Neutron

Observation of neutrons outside the nucleus provides some insight into the process of beta decay. Neutrons that are free from a nucleus are relatively unstable and decay fairly quickly into a proton, an electron, and an antineutrino.

$$^{1}_{0}n \rightarrow {}^{1}_{1}p + {}^{0}_{-1}e + \bar{\nu}_e$$

This decay led many people to speculate that a neutron consists of a proton, an electron, and an antineutrino bound together by some force. Through a series of careful experiments, physicists have tried to identify protons and electrons within the neutron. To date, these experiments have been unsuccessful. As far as we can tell, the proton, electron, and antineutrino do not exist inside the neutron. They appear to be created at the moment the neutron decays.

In general, beta decay shows many of the characteristics of neutron decay. As carbon decays into nitrogen, a neutron disappears. The number of protons increases by one and an electron and an antineutrino are ejected. However, neither the electron nor the antineutrino exist inside the nucleus before the time that the decay occurs.

Figure 21-5
Momentum is conserved in beta decay processes. **(a)** Measurements showed that the nitrogen nucleus and the electron did not move off in opposite directions. **(b)** Momentum would be conserved only if a third particle were released in beta decay.

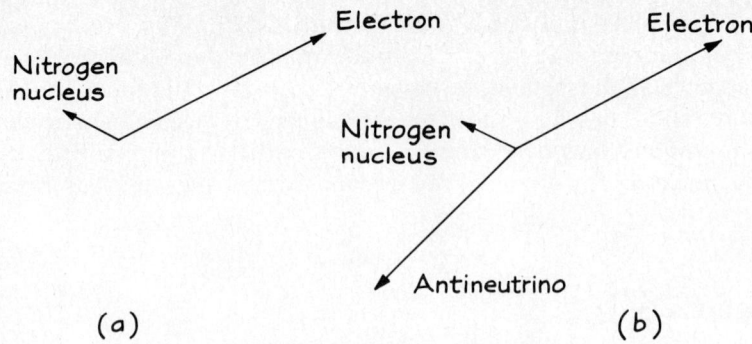

While alpha decay is characterized by the ejection of nucleons that are in the nucleus, beta decay involves the transformation of one type of nucleon into another by emitting particles that were not present until the transformation occurred. This difference, along with other more complex reasons, led physicists to conclude that a fundamental interaction different from the strong nuclear interaction was responsible for beta decay. It was clearly occurring in the nucleus. Further, careful measurements showed that the strength of the interaction was weaker than the strong nuclear force. Thus the interaction responsible for beta decay was named the *weak nuclear interaction*.

Antineutrinos and Conservation Logic

We have listed an antineutrino as a product of beta decay without explaining what the antineutrino is and why it is always included in these interactions. The story of its discovery illustrates the extent to which modern science relies on conservation logic.

When physicists first studied beta decay in detail, they detected the emitted electrons and the transformation of the nucleus. The carbon 14 reaction, for example, was first thought to be carbon transforming into a nitrogen nucleus and an electron only. Measurements of beta decay yielded two startling results. First, the beta particle had too little kinetic energy. The difference between the mass of the carbon 14 and the combined masses of the nitrogen 14 and the electron is 0.0002 amu. If all of this mass difference is converted into kinetic energy of the products, then the electron should have a kinetic energy of about 3×10^{-14} J. However, the kinetic energy of the electron ranged from zero to all the energy available, with the majority of the electrons being emitted with an energy of about 1.5×10^{-14} J. The second surprise was that when the carbon nucleus was initially not moving, the nitrogen nucleus and the electron did not move off in exactly opposite directions, as conservation of momentum requires (Figure 21-5(a)). Neither momentum nor energy appeared to be conserved in beta decay.

Rather than abandon the conservation principles, physicists suggested that yet another particle must be present in the products of beta decay. Conservation laws dictated that this new particle could have no electric charge (electric charge was already conserved), must have a very small mass, and must move off with a speed and a direction that would conserve momentum

Beta Decay in the Study of Art

Beta decay occurs in many naturally radioactive nuclei. In addition, stable nuclei can be turned into isotopes that emit electrons and antineutrinos by placing them in a beam of neutrons. The neutron is absorbed by a nucleus and a new, radioactive nucleus is created. A transformation that illustrates this process is

$$^1_0 n + ^{23}_{11}\text{Na} \rightarrow ^{24}_{11}\text{Na} \rightarrow ^{24}_{12}\text{Mg} + ^{\ 0}_{-1}e + \bar{\nu}_e$$

A neutron transforms the stable sodium 23 nucleus into an unstable isotope, sodium 24, which then emits an electron and antineutrino and becomes magnesium 24, another stable isotope. Hundreds of isotopes can be converted into beta emitters by this process.

The large number of nuclei that can be changed into beta emitters when they absorb a neutron has made beta decay an important tool for art historians. Typically, a painting is placed in front of a beam of neutrons. Nuclei of elements found in paint and charcoal are changed into unstable nuclei, which then transform into new stable nuclei by beta emission. A photographic plate placed over the painting reveals the location of these unstable nuclei as each emitted electron interacts with the photographic emulsion. The result is a "picture" of the locations of these unstable nuclei.

Moonlight by R.A. Blakelock (Figure A) was treated in this manner. One of the beta decay photographs showed the presence of a painting beneath the top painting (Figure B). Pigments used to paint the original figure of a woman included arsenic 76, antimony 12, and chromium 51. All these isotopes become beta emitters when they absorb a neutron.

Analyses such as these are important for two reasons. First of all, they provide information about the artist's style, techniques, and use of materials. Such information is valuable to art historians and collectors who want to identify forgeries. (Blakelock's work was often copied, but the forgers generally used slightly different pigment preparations.) Secondly, the analysis can be conducted without destroying the painting. Neutrons are absorbed by only a few of the many nuclei in the painting. So few nuclei change to different nuclei that the procedure produces no noticeable change in the original work.

Figure A

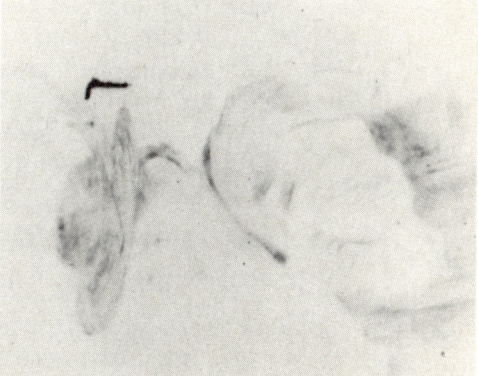

Figure B

"THE QUESTION NOW IS, 'HOW MANY NEUTRINOS CAN DANCE ON THE HEAD OF A PIN?'"

(Figure 21-5(b)). Italian physicist Enrico Fermi named the yet-to-be-discovered particle the *neutrino*—little neutral one. For more than 20 years the neutrino hypothesis remained unproved, though generally accepted. Finally, in the mid-1950s physicists detected a particle with properties identical to those predicted for the neutrino. Either a neutrino or its close relative, the antineutrino, is a part of every beta decay.

GAMMA DECAY

The third common form of nuclear radiation is gamma radiation. Having neither mass nor electric charge, this radiation is electromagnetic radiation of very high energy. Gamma decay releases energy from the nucleus but does not transform one isotope into another.

Gamma Decay Releases Electromagnetic Energy

Just as the electromagnetic energy emitted by atoms moves the atom from one energy state to another, gamma emission from nuclei moves the nucleus from one nuclear energy state to another. The nucleus just before a gamma transformation is in an excited, or high-energy, state. After the emission of

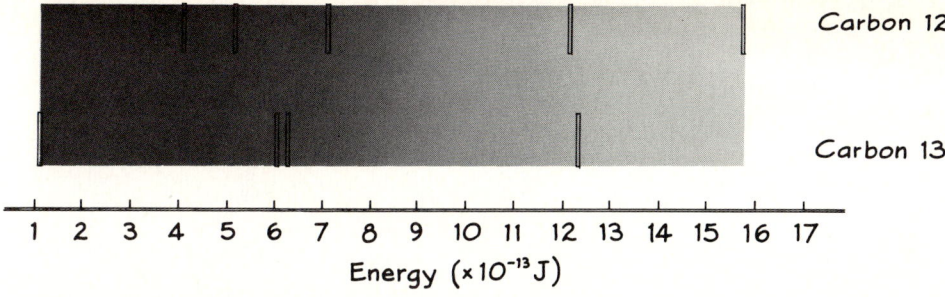

Figure 21-6
The energy spectrum of gamma radiation emitted by carbon 12 is distinctly different from that for carbon 13.

gamma radiation, the nucleus is in a lower-energy state. The difference in the nuclear energy states is the energy of the gamma radiation.

We represent the gamma transformation by a notation similar to that used for alpha and beta decay. However, we add an asterisk to identify a nucleus that is in an excited state. One gamma transformation is

$$^{12}_{6}C^* \rightarrow {}^{12}_{6}C + {}^{0}_{0}\gamma$$

The zeros as subscript and superscript on the gamma indicate that the gamma radiation carries no charge and no mass away from the transformation.

Gamma radiations that are emitted by the excited states of a nucleus have a unique spectrum associated with each isotope of an element. As illustrated in Figure 21-6, the gamma spectrum for carbon 12 is distinctly different from that for carbon 13. Just as we can identify atoms from their light spectrum, we can identify specific isotopes from the gamma radiation spectrum that they emit.

SELF-CHECK 21C

The difference between one of the excited states and the lowest energy state of nickel 60 is 2.1×10^{-13} J. What is the energy of the gamma radiation when nickel 60 undergoes a gamma transformation from the excited state to the ground state?

Gamma Spectra and Criminal Investigations

The uniqueness of the gamma spectrum emitted from a particular isotope has become a valuable tool to criminal investigators. One common application involves the use of gamma spectra to identify paint samples from automobiles.

Each time a batch of paint is mixed, the mixture is slightly different from other batches. While the colors may look identical, small differences in the mixing procedure or the cleanliness of the equipment can introduce slight impurities into the batch. These impurities, generally unique to each batch, can be used to link paint samples to a vehicle suspected in a hit-and-run accident.

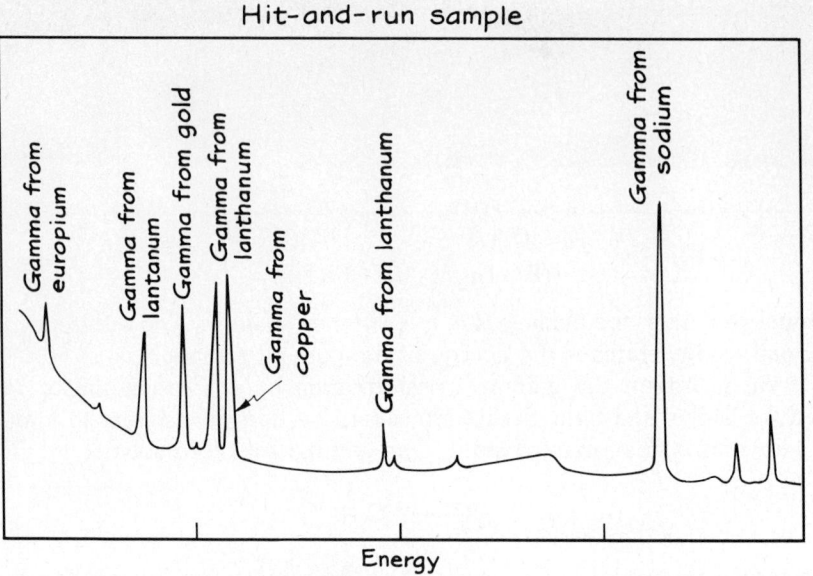

Figure 21-7
Differences in the gamma spectra of paint samples led detectives to conclude that the two could not be from the same vehicle.

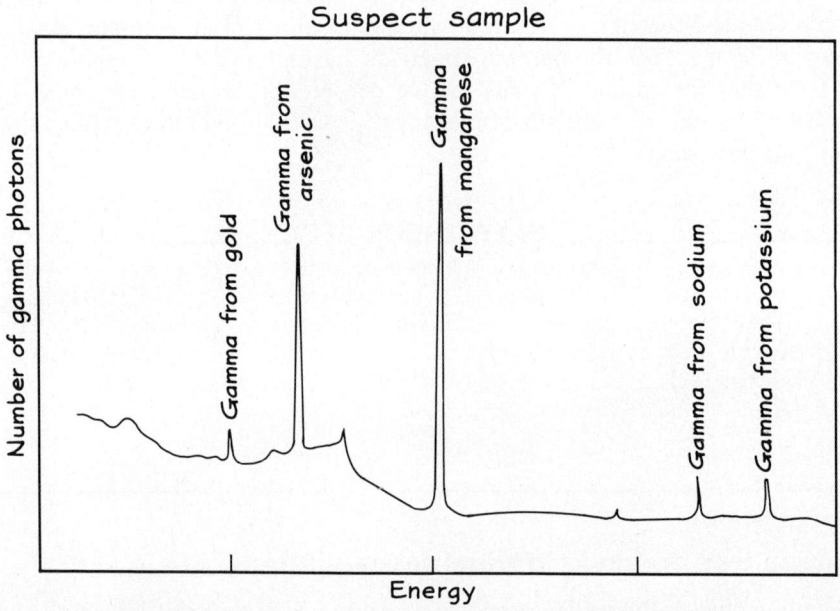

In a hit-and-run accident, paint from the car is usually transferred to the object struck. Using gamma spectra, detectives can compare a sample of this rubbed-off paint with a sample of paint from the suspect vehicle. Even if the two paint samples are the same color, they might not be from the same vehicle. The probability that they came from the same vehicle increases greatly if the detectives can show that the paint came from the same batch.

To determine if the paint samples were mixed in the same batch, scientists place the two samples in front of a beam of moving neutrons. These neutrons are absorbed by nuclei in the paint, changing them to excited states of different isotopes. These new isotopes move to their lowest energy state by the emission of gamma radiation. For example,

$$^1_0n + {}^{55}_{25}Mn \rightarrow {}^{56}_{25}Mn^* \rightarrow {}^{56}_{25}Mn + {}^0_0\gamma$$

By comparing the gamma spectra of the two samples, detectives can determine whether both samples came from the same batch of paint. If they did not, the suspected vehicle was not involved in the accident. Figure 21-7 illustrates two spectra obtained from samples suspected to be from the same car. The spectra clearly reveal that the two samples come from different batches. The suspected vehicle was not involved in the accident from which the other sample was taken even though both samples are the same color.

HALF-LIFE

In extending Becquerel's early work with uranium and uranium compounds, Marie Curie attempted to identify systematically other radioactive elements. Her measurements with uranium and thorium compounds allowed her to link the intensity of radioactive emissions to the amount of radioactive materials present. Curie found that the radioactivity emitted by pitchblende, an ore that is roughly 80% uranium oxide, was nearly five times that expected from the amount of uranium present. This result caused her to undertake an extensive chemical analysis of pitchblende, which turned out to contain three new radioactive elements—polonium, radium, and radon. A single ore, then, was found to have at least four radioactive elements.

As more radioactive isotopes were isolated, physicists realized that spontaneous transformations often lead to a sequence of reactions, as shown in Figure 21-8. Radium spontaneously emits an alpha particle and decays into radon. Radon decays into polonium; polonium decays into lead. Viewed in terms of such a sequence, Curie's results were hardly surprising. What was surprising was how relatively few radon and polonium atoms were present. Clearly, isotopes decay at different rates.

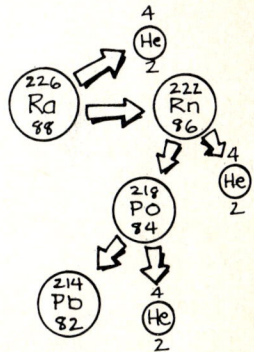

Figure 21-8
Spontaneous transformations often lead to a sequence of nuclear reactions.

Half-Life Describes the Rate of Decay

We can measure, theoretically, the rate at which a radioactive isotope decays by starting with a sample of a known number of nuclei and periodically counting the number of undecayed nuclei left. In practice, this turns out to be difficult. Instead, we count the number of radiations emitted. Each time one is detected, we know that one nucleus has decayed. Sample data for the decay of polonium are presented in Table 21-1. The same information is shown graphically in Figure 21-9.

The decay of an individual nucleus is random. We cannot know when any particular nucleus will decay, but we can look at collections of nuclei and describe them as an entity. Then, we describe the decay rate in terms of the time required for half of the remaining nuclei to decay. As you can see from Table 21-1, the half-life of polonium is a little more than 3 minutes (min).

Figure 21-9
Graph showing the number of polonium atoms left as a function of time. Most of the polonium nuclei decay during the first few minutes.

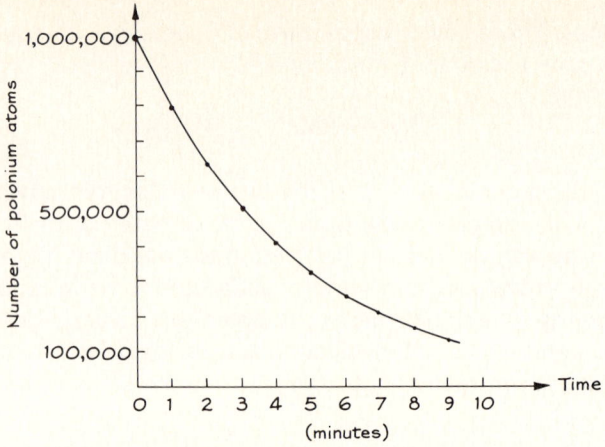

Table 21-1 Decay of Polonium into Lead

$$^{218}_{84}\text{Po} \rightarrow {}^{214}_{82}\text{Pb} + {}^{4}_{2}\text{He}$$

Time (min)	Number of Polonium Atoms	Number of Lead Atoms
0	1,000,000	0
1	796,751	203,249
2	634,812	365,188
3	505,787	494,213
4	402,986	597,014
5	321,080	678,920
6	255,821	744,179
7	203,825	796,175
8	162,398	837,602
9	129,391	870,609

Slightly more than half our initial sample (505,787) is left at the end of 3 min. After 3 min more, approximately half this number (255,281) remain, and so forth. During a time equal to one half-life, one-half of the remaining nuclei decay.

Each nucleus that undergoes a nuclear transformation has a well-defined half-life. Half-lives can range from a fraction of a second to thousands of years. The half-life of radium 226, for example, is 1620 years. Its decay product, radon 222, has a much shorter half-life of 3.82 days. As we just mentioned,

Half-Lives in Archeological Dating

The constancy of the half-life of a particular isotope has proven invaluable to archeologists. Because carbon is the primary building block of living tissue, many archeological studies concentrate on this element. Carbon exists in seven different isotopes, two of which (carbon 12 and carbon 13) are stable. The other isotopes are radioactive and have half-lives that range from 2.3 seconds to 5568 years. The 5568-year half-life of carbon 14 has made it useful to archeologists in dating artifacts.

Carbon 14 transforms to nitrogen 14 by beta decay,

$$^{14}_{6}C \rightarrow {}^{14}_{7}N + {}^{0}_{-1}e + \bar{\nu}_e$$

However, the percentage of carbon 14 found in the atomsphere has remained rather constant because it is continuously being replenished by neutron collisions in the upper atmosphere. Neutrons from space strike nitrogen 14 to produce carbon 14 and a proton. Consequently, decays of carbon 14 are balanced by its creation.

All living things continuously take up carbon from the environment. A tree, for example, takes in carbon dioxide in the process of photosynthesis and incorporates the carbon in its tissue. A known fraction of the carbon dioxide molecules contains carbon 14. Since the proportion of carbon 14 in the atmosphere remains constant, so should the proportion of carbon 14 in the tree. As long as the tree continues to take in carbon from the atmosphere, we should be able to detect about 15 beta emissions per minute for each gram of carbon in the tree.

When a tree is cut to provide wood for a house or a boat, for instance, it stops taking in carbon from the atmosphere. No more carbon 14 will enter the wood. However, the nuclei of carbon 14 already

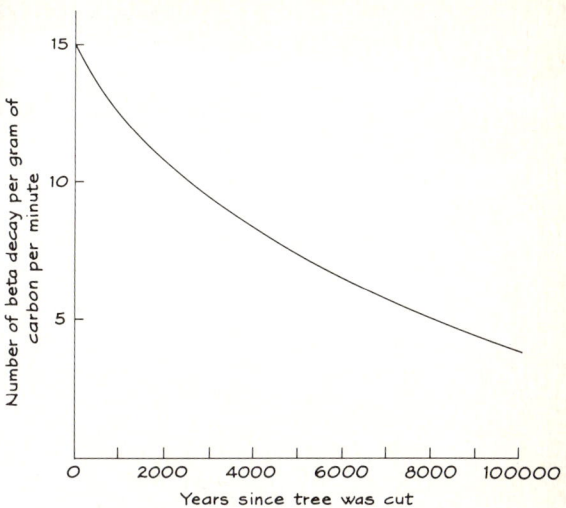

there will continue to decay with a half-life of 5568 years. If we measure 15 beta emissions per minute per gram of carbon at the time of cutting, then 5568 years later we should measure about 7 to 8 beta particles per minute per gram of carbon in the wood. By comparing the emissions of the cut sample with emissions of a living sample, we can infer the length of time that has elapsed since the tree was cut. The graph presents a rough estimate of the beta emissions found at various time intervals.

Archeologists use this technique extensively to date wooden artifacts and animal skeletons. The method does have limitations, however. First of all, the investigator must separate the beta particles emitted by carbon 14 from those emitted by other nuclei. This procedure is not simple. Secondly, we do not know the half-life of carbon 14 exactly. Finally, the number of beta particles emitted per gram of carbon is very small. The archeologist must use enough material to obtain a valid measurement and yet not destroy the artifact. These problems combine to give an uncertainty of about 15% in dates obtained by use of carbon 14 decays. A boat that is measured by carbon dating to be 500 years old may be as little as 425 years old or as old as 575 years. For some purposes this level of uncertainty is acceptable; for others, it is not.

polonium 218 has a half-life of only 3 min. The short half-lives of radon 222 and polonium 218 explain why Curie found such small amounts of these isotopes in her samples.

The half-life of a nucleus is remarkably constant for a given isotope. It is independent of any physical or chemical conditions in which the nucleus is found. The half-life of uranium 238 is 4.15×10^9 years regardless of whether it is combined chemically with other elements, whether it is found at the surface of the earth or 2 miles below, and whether the climate changes over the life of the isotope. This independence of the half-life from the surrounding conditions reflects the small range of the nuclear force. The very small nucleus is well isolated from its surroundings. Conditions affecting the atom, such as chemical combinations, primarily involve the electrons. Interactions involving the nuclear forces are not affected by these distant conditions.

SELF-CHECK 21D

Suppose you start with 256 nuclei of polonium 218, which has a half-life of 3 min. How many nuclei remain after 3 min? 6 min? 9 min?

Radioactive Decay Series and the Age of the Earth

The presence of so many radioactive nuclei in pitchblende led physicists to suggest that uranium, radium, radon, and polonium are linked by a series of radioactive transformations. A **radioactive-decay series** is a sequence of decay processes that leads from an unstable nucleus to a stable one through several nuclear transformations. Experiments identified the parent of this particular series as uranium 238. The stable nucleus that concluded the series was lead 206. As illustrated in Figure 21-10, the series transforms uranium 238 into lead 206 through eight alpha decays and six beta decays.

Many of the nuclei in the uranium 238 series have extremely short half-lives. Polonium 218 has a half-life of 3.05 min; that of lead 210 is 22 years. If we started with a small amount of either isotope, we could expect to have none of either after a few hundred years. However, the supply of polonium 218 and lead 210 is constantly being replenished by the decay of the parent element. Uranium 238 has such a long half-life (4.5 billion years) that we expect the decay series to be fed for billions of years to come.

The uranium 238 transformation series is one of four such series that have been identified. The parent of one series, neptunium 237, has a relatively short half-life (2.25 million years) and, thus, no longer occurs naturally. However, the series is produced artificially by creating neptunium 237 in the laboratory. All naturally occurring decay series end with an isotope of lead, while the neptunium series terminates with bismuth.

Because of the exceptionally long half-lives of the parent elements, radioactive-decay series provide a technique for estimating the ages of rocks

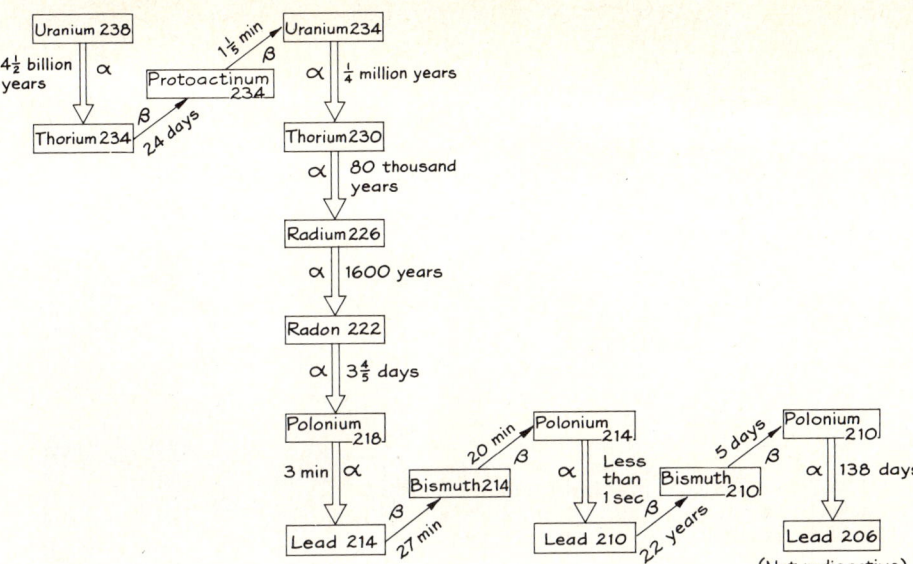

Figure 21-10
The uranium transformation shows a sequence of reactions leading from the uranium 238 nucleus to the lead 206 nucleus. Eight alpha decays and six beta decays are required to complete the process from unstable to stable nuclei.

on the earth and, to some extent, the age of the earth itself. The absence of neptunium 237 in natural rock implies that the earth is at least as old as the half-life of that element, 2.25×10^6 years. How much older? We can get some idea by comparing the ratio of the parent to the final product in each of the naturally occurring decay series. These and similar comparisons suggest that the earth is about 4.5 billion years old.

Tiny particles, bound in an uneasy truce within the incredibly small space of the nucleus, can in some cases fly apart, spontaneously releasing a small amount of energy. In the early years of this century, this energy was extremely useful for learning more about the structure of the nucleus. However, some of the early investigators saw no practical use for it (Figure 21-11). Only later did they discover and understand the enormous energies released and available for use in nuclear fusion and nuclear fission, the subjects of our final chapter.

CHAPTER SUMMARY

Chemical elements can be identified uniquely by the masses and electric charges carried by their nuclei. Both characteristics can be described by the presence of protons and neutrons in the nucleus. Protons have an electric charge equal in magnitude but opposite in sign to the charge on an electron. The proton mass is about 1850 times greater than the mass of the electron. A neutron carries no electric charge and has a mass approximately equal to that of the proton. Atoms with the same number of protons but different numbers of neutrons are called *isotopes*.

Some nuclei spontaneously emit radiations more intense than that seen in the spectra of atoms. Called *radioactivity,* this radiation was found to consist

Lord Rutherford Scoffs at Theory of Harnessing Energy in Laboratories

Atom-Powered World Absurd, Scientists Told

By the Associated Press

LEICESTER, England, Sept. 11—Lord Rutherford, at whose Cambridge laboratories atoms have been bombarded and split into fragments, told an audience of scientists today that the idea of releasing tremendous power from within the atom was absurd.

He addressed the British Association for the Advancement of Science in the same hall where the late Lord Kelvin asserted twenty-six years ago that the atom was indestructible.

Describing the shattering of atoms by use of 5,000,000 volts of electricity, Lord Rutherford discounted hopes advanced by some scientists that profitable power could be thus extracted.

"The energy produced by the breaking down of the atom is a very poor kind of thing," he said. "Anyone who expects a source of power from the transformation of these atoms is talking moonshine. ... We hope in the next few years to get some idea of what these atoms are, how they are made and the way they are worked."

Figure 21-11
Ernest Rutherford's crystal ball was slightly cloudy when he discussed practical applications of nuclear energy.

of alpha, beta, and gamma radiation. *Alpha radiation* consists of particles made up of two protons and two neutrons (that is, helium nuclei). The alpha decay of a nucleus can be explained in terms of a wave packet that extends beyond the nucleus. Once the alpha particle moves outside the nucleus, the electrical forces of repulsion exceed the nuclear forces of attraction and the alpha particle is ejected. *Beta particles* are electrons. Electrons and antineutrinos are ejected from the nucleus at the time of the transformation. *Gamma radiation* is electromagnetic radiation that is a million times more energetic than that emitted in atomic spectra. It occurs when a nucleus in an excited state moves to a lower energy state and emits energy.

Alpha and beta radiation arise because the nuclei undergo nuclear transformations in which a nucleus of one element changes into a nucleus of a different element. Such transformations proceed under the constraints of the conservation laws. The energy released in nuclear transformations comes from mass that has been transformed into energy in accordance with Einstein's principle of mass-energy equivalence.

Radioactive isotopes decay at a rate characteristic of each isotope. These rates are described by the *half-life*, the time required for half of the undecayed nuclei to decay. Half-lives can range from a fraction of a second to millions of years. The half-life is independent of any physical or chemical change the atom undergoes.

ANSWERS TO SELF-CHECKS

21A. $^{234}_{92}$U contains 92 protons and $(234 - 92) = 142$ neutrons.

$^{235}_{92}$U contains 92 protons and $(235 - 92) = 143$ neutrons.

$^{238}_{92}$U contains 92 protons and $(238 - 92) = 146$ neutrons.

21B. The first transformation is possible. Both nucleon number and electric charge are conserved. The second transformation is not possible. Electric charge is not conserved, although nucleon number is.

21C. Gamma radiation is emitted from the nucleus in much the same way that radiation is emitted by atoms undergoing electron transitions. The energy released is equal to the difference in energy of the excited and the ground state. The energy released is 2.1×10^{-13} J.

21D. Three minutes is one half-life. We will have $(\frac{1}{2}) \times 256 = 128$ polonium atoms left. Six minutes is two half-lives. We will have $(\frac{1}{2}) \times 128 = 64$ polonium nuclei left. Nine minutes is three half-lives. We will have $(\frac{1}{2}) \times 64 = 32$ polonium nuclei left.

PROBLEMS AND QUESTIONS

A. Review of Chapter Material

A1. Define each of the following terms:
Nucleon
Isotope
Atomic number
Mass number
Radioactivity
Nuclear transformation
Alpha decay
Beta decay
Gamma decay
Half-life
Radioactive-decay series
Atomic mass unit

A2. Describe the role of the proton and neutron in determining the mass and electric charge of a nucleus.

A3. How does the nucleus change if a proton is added? If a neutron is added? Which determines the chemical element?

A4. What observations allow us to conclude that the radiation emitted by uranium and radium consists of three distinct forms?

A5. What conservation laws govern nuclear transformations?

A6. From where does the energy that people refer to as *nuclear energy* come in a nuclear transformation?

A7. Under what conditions will alpha decay occur?

A8. In what ways are beta decay and the decay of a free neutron similar?

A9. Why is the third particle, the antineutrino, required to explain the results of beta decay?

A10. How is gamma decay different from alpha and beta decay?

A11. How do we describe the rate at which an isotope decays?

A12. With the help of Figure 21-10, describe the radioactive decay series in which uranium 238 is the parent nucleus and lead 206 is the stable end product.

B. Using the Chapter Material

B1. How many protons and neutrons are in each of the isotopes listed below?

$^{198}_{79}$Au $^{45}_{20}$Ca $^{36}_{17}$Cl $^{59}_{26}$Fe

In a neutral atom, how many electrons will surround each of these nuclei?

B2. Show that electric charge and nucleon number are conserved in the reactions shown below.

$^{90}_{39}$Y $\rightarrow$ $^{90}_{40}$Zr + $^{0}_{-1}$e + $\bar{\nu}_e$

$^{239}_{94}$Pu $\rightarrow$ $^{235}_{92}$U + $^{4}_{2}$He

B3. Can each of the reactions listed below occur? If the answer is no, explain how you reached your conclusion.

$$^{24}_{11}\text{Na} \rightarrow {}^{24}_{12}\text{Mg} + {}^{4}_{2}\text{He}$$

$$^{42}_{19}\text{K} \rightarrow {}^{42}_{20}\text{Ca} + {}^{0}_{-1}e + \bar{\nu}_e$$

B4. The figure below shows the wave packets for alpha particles in two different nuclei. The size of the nucleus has been superimposed over the wave packet. Which is more likely to undergo alpha decay? Explain how you reached your answer.

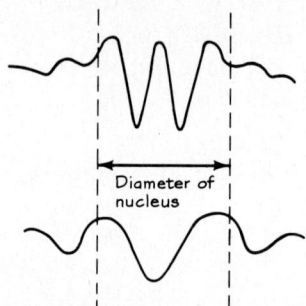

B5. When iodine 131 undergoes beta transformation, it emits electrons with energies ranging from 0 to 10×10^{-14} J. If an electron leaves the iodine with 4×10^{-14} J, how much energy does the accompanying antineutrino have? (Assume the antineutrino has zero mass.)

B6. Conservation of electric charge and conservation of nucleon number allow a free proton to undergo beta decay, releasing a neutron, a beta particle, and whatever other particles are required. Such a reaction never occurs, however. Use conservation of mass-energy to explain why.

B7. Oxygen 20 has an excited state that is 1.6×10^{-13} J above the ground state. What is the energy of the gamma radiation when this state transforms into the ground state?

B8. The half-life of strontium 90 is 29 years. Suppose you start with 1,000,000,000 nuclei of strontium 90. How many strontium 90 nuclei will remain after 29 years? After 58 years? After 116 years?

B9. Suppose you begin with a sample of 400,000 bismuth 210 nuclei. Fifteen days later you measure the number of bismuth 210 nuclei still present and find 50,000. What is the half-life of bismuth 210?

B10. A part of a tree is broken off. At the time of the break, the part had 2×10^5 nuclei of carbon 14 in it. Approximately 11,000 years later, how many carbon 14 nuclei are in the branch?

B11. For each of the reactions below, fill in the missing particles or nuclei.

$$^{35}_{16}\text{S} \rightarrow {}^{35}_{17}\text{Cl} + ?$$

$$^{210}_{84}\text{Po} \rightarrow ? + {}^{4}_{2}\text{He}$$

$$^{131}_{54}\text{Xe}^* \rightarrow {}^{131}_{54}\text{Xe} + ?$$

$$^{32}_{15}\text{P} \rightarrow {}^{32}_{16}\text{S} + ?$$

$$^{1}_{0}n + {}^{208}_{83}\text{Bi} \rightarrow ? + \gamma$$

$$^{19}_{9}\text{F} + ? \rightarrow {}^{22}_{11}\text{Na} + {}^{1}_{0}n$$

C. Extensions to New Situations

C1. Watch dials with radium in them will glow in the dark.
   a. Radium ($^{226}_{88}\text{Ra}$) transforms to radon ($^{222}_{86}\text{Rn}$). What particle is emitted in the transformation?
   b. Through what force can this particle interact with the electrons in the paint?
   c. During this interaction, what type of particle is exchanged between the alpha particle and the atomic electron? (You may wish to refer to the discussion of exchange particles in Chapter 8.)
   d. If an electron in an atom absorbs a photon, what will happen to it?
   e. How will the energy of the electron show up when it returns to the ground state?
   f. Why does radium in the dial paint make the dial glow in the dark?
   g. Some liquid crystal watches contain hydrogen 3, which undergoes beta transformation. Describe why they will glow in the dark.

C2. In the next chapter, we discuss the use of nuclear energy to generate electricity. One of the problems with nuclear energy is the radioactive waste products produced. An important component of this waste is cesium 137, which has a half-life of 30 years. Even though cesium has this relatively short half-life, it must be stored for thousands of years.

a. Suppose you have a sample with 10^{10} nuclei. After 10 half-lives, how many nuclei are left to transform later?
b. How many nuclei will decay between 300 years and 330 years?
c. If 10^5 radiations per year were considered dangerous, would you need to store the material longer than 330 years?
d. Why do you suppose that cesium 137 waste from reactors must be stored for such a long time?

(Note: The numbers in this problem were used to illustrate the difficulty and do not represent real situations.)

C3. The material at the center of the earth is very hot. The present theory is that this heat is the result of radioactive transformations in the earth. How do you think that such transformations could generate heat?

C4. One type of smoke detector contains a very small sample of americium 243 ($^{243}_{95}$Am).
a. The $^{243}_{95}$Am transforms to neptunium 239 ($^{239}_{93}$Np). What particle is emitted in the process?
b. The particles are emitted into a space that is in an electric field. If a large number of charged particles are there, a current will flow and the alarm will sound. The transformation products by themselves are not sufficient to cause the necessary current. Why do the smoke particles, which are not radioactive, cause the current to exist?
c. The smoke particles interact with the charged alpha particles. How can this interaction increase the number of electrically charged smoke particles?
d. The neptunium undergoes beta transformation. Into what does it transform?

C5. A sample of material containing nuclei that transform by all three processes is placed between two electrically charged plates. Plate A has a positive charge; plate B is negative. Describe the motion of each form of radiation in this situation.

C6. Persons who work around radioactivity or X rays must be sure that they do not receive too much radiation. To monitor the amount that reaches their bodies, they wear film badges. These contain small pieces of photographic film enclosed in paper wrappers. How can these devices determine the amount of radiation to which they are exposed?

C7. The uranium 235 decay series undergoes a series of transformations until it reaches the stable lead 207 isotope. The order in which the decay processes occur is listed below. Use this order and the periodic table in Figure 21-1 to construct a diagram similar to that shown for uranium 238 in Figure 21-10: alpha-beta-alpha-beta-alpha-alpha-alpha-alpha-beta-alpha-beta.

C8. Listed are the stable isotopes of lead, their relative abundances, and their masses. We could determine the average mass of lead by assuming that we had 100 atoms distributed as shown by their relative abundances. Calculate this value by:
a. multiplying the relative abundance of each isotope by its mass
b. adding the four values
c. dividing by the number of atoms in our sample, 100.

How does this value compare with the atomic mass given in the periodic table in Figure 21-1?

Isotope	Relative Abundance (%)	Mass (amu)
$^{204}_{82}$Pb	1.50	203.9731
$^{206}_{82}$Pb	23.60	205.9745
$^{207}_{82}$Pb	22.60	206.9759
$^{208}_{82}$Pb	52.30	207.9766

C9. Paintings can be dated using the analysis of the level of radioactivity emitted by lead-white paint, a common material used by artists for several hundred years. Lead-white, as the name implies, is made with lead removed from lead ore. Since most lead ores contain small quantities of uranium 238, isotopes within the uranium series are present in the ore. While the chemical separation of lead from the ore eliminates most of these radioactive isotopes, a small amount of radium remains. Consequently, the paint sample has some radioactivity present.
a. What radioactive isotopes can be present in lead-white paint?

b. When the paint is first made, some lead 210 is present naturally. What happens to this lead 210 over a few hundred years?
c. After the lead 210 mentioned in part (b) has decayed, some lead 210 will still be in the paint. From where will it come?
d. When will paint have greater amounts of lead 210, when it is first created or a few hundred years later?
e. How can the answers to parts (a)-(d) be used to detect art forgeries?

C10. As stated in the chapter, alpha particles can be stopped by paper and beta particles by a few centimeters of metal, but gamma particles can travel through many centimeters of material. Suppose you had a sample that emitted some type of nuclear radiation. How could you use the interactions with matter, stated above, to determine which type of radiation is present?

D. Activities

D1. If you can borrow a Geiger counter, see if you can find sources of radioactivity.
D2. Many science fiction stories contain references to radioactive transformations. Read such a story and see if the discussion contains correct physics.

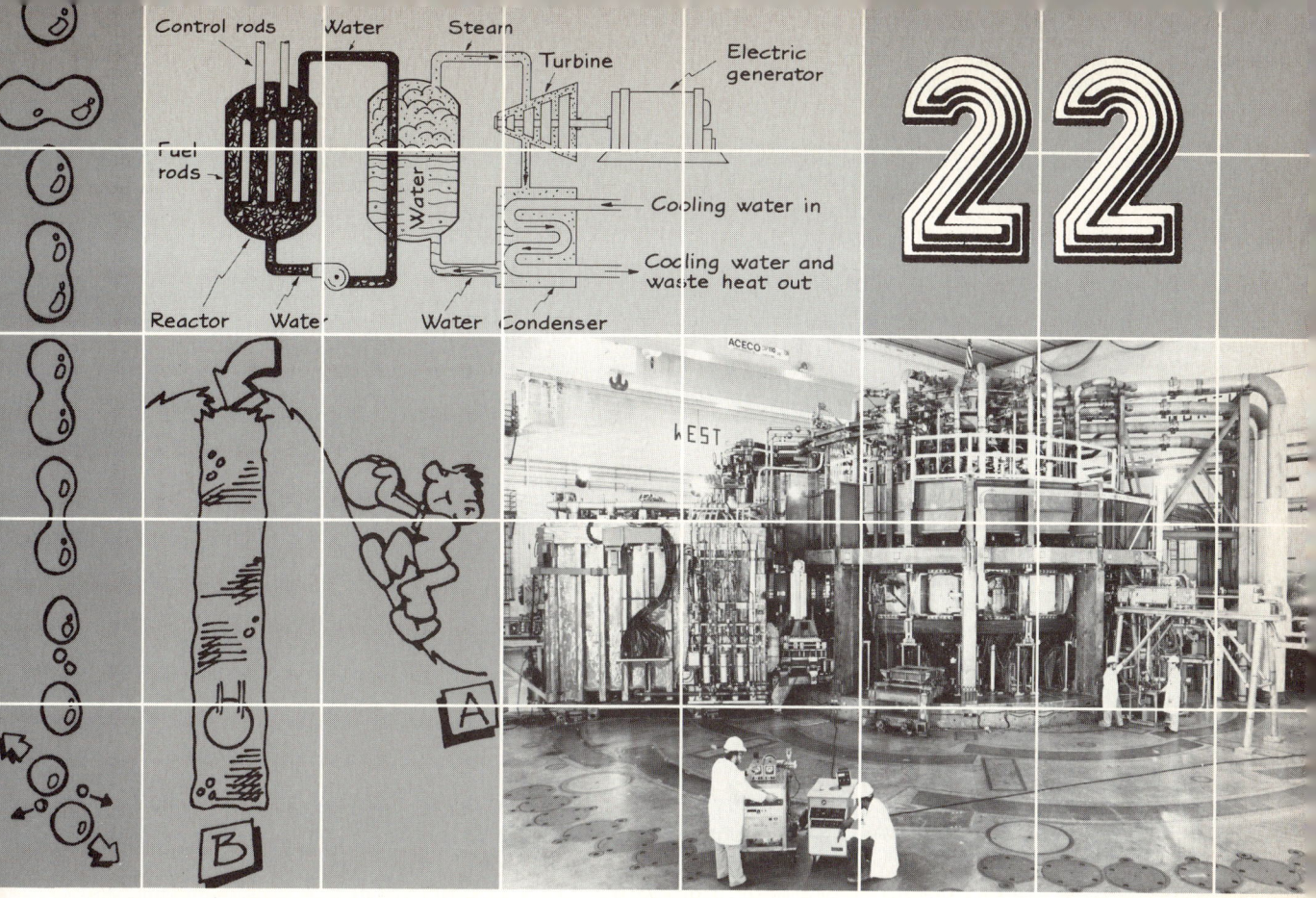

Nuclear Energy

The discovery of radioactivity in 1896 showed that energy stored in the nucleus could be released as the kinetic energy of radioactive particles. Thirty years later people eyed these reactions as a source of useful energy—energy to heat homes, to power tools, and to drive assembly lines. Scoffed at by some eminent physicists, nuclear power seemed at first an idle dream. The radiation emitted in most reactions carries little energy. The reactions themselves are not easily controlled and, in many cases, require more energy to induce than they release. Within a decade, however, fission and fusion reactions were discovered. Both produce enormous amounts of energy, more energy per kilogram of matter than had ever been imagined. The awesome release of this energy in atomic and hydrogen bombs converted the idle dreams into reality. Shortly after the first bombs, nuclear power plants were producing electrical energy. This final chapter turns to nuclear reactions and the risks and benefits we create as we reshape the landscape once more.

The energy released or absorbed in nuclear reactions arises from the different *binding energies* associated with each nucleus. Reactions that transform a loosely bound nucleus into a more tightly bound one release energy into the environment. Two processes, *nuclear fission* and *nuclear fusion,* accomplish this and release significant amounts of energy. Both processes are considered

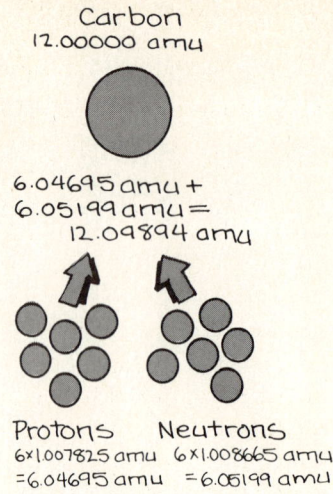

Figure 22-1
When nucleons combine to form a nucleus, the mass of the individual nucleons is greater than the final nucleus.

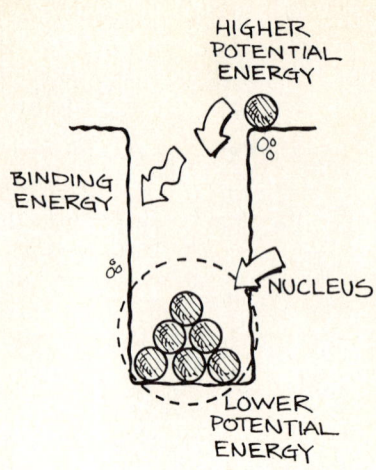

Figure 22-2
When a nucleon enters a nucleus, it gives up energy in a way that is similar to a rock falling in a well.

important alternative energy sources as the world's supply of fossil fuels dwindles. As the use of nuclear processes increases, the biological effects of alpha, beta, and gamma radiation becomes the subject of increasing interest.

BINDING ENERGY

If we compare the mass of a nucleus with the sum of the masses of its individual particles, we find that the two values do not agree. Figure 22-1 shows such a comparison for carbon. The carbon nucleus has a mass that is 0.098940 amu less than the sum of the masses of its six protons and six neutrons. There is a similar loss of mass, called **mass defect,** in all nuclei except hydrogen 1. We know that mass can disappear only by being changed into energy. When protons and neutrons combine to form a nucleus, some of their mass must be converted to energy and released. We now consider why this must happen.

Mass Decrease and Binding Energy

Since nucleons are attracted to one another by the strong nuclear interaction, energy must be supplied to pull them apart. If we were to pull a nucleon out of the nucleus, we would have to give it energy. A nucleon in a nucleus can be compared to a rock lying at the bottom of a well (Figure 22-2). The rock is attracted to the bottom by its gravitational interaction with the earth. To pull it out, we must supply it with some gravitational potential energy—the same amount it lost when it fell to the bottom of the well. A nucleus is a sort of "nuclear energy well." When nucleons bind to one another to form a nucleus, they fall into a nuclear well and lose some energy. To pull them out, we must supply the same amount of energy that they lost. This energy is called the *binding energy* of the nucleons.

The energy that a rock loses when it drops to the bottom of a well is determined by the strength of the gravitational interaction and by the depth of the well. Similarly, the energy that a nucleon loses when it binds to other nucleons is determined by the strength of the strong nuclear interaction and the distances between nucleons in the nucleus which is formed. Each kind of nucleus has a specific **binding energy**—the amount of energy lost when its nucleons came together. This is the same amount that would be needed to pull them apart.

When nucleons release binding energy, they experience a detectable loss of mass. This loss of mass, which is converted into energy, is the source of the mass defect of nuclei. The binding energy of a nucleus is related to its mass defect by the equation for mass-energy equivalence. For example, the mass defect of the carbon 12 nucleus is 0.098940 atomic mass units (amu), or 1.64×10^{-28} kilograms (kg). Its total binding energy is (mass defect) $\times$ (speed of light)^2 = 1.48×10^{-11} joules (J). Since carbon has 12 nucleons, the binding energy of each nucleon is $(1.48 \times 10^{-11} \text{ J})/12 = 1.2 \times 10^{-12}$ J. Each nucleon lost 1.2×10^{-12} J of energy when it united with the others to form carbon. We would have to supply that much energy to pull one nucleon free from a nucleus of carbon.

The Binding-Energy Curve

The total mass defect of the nucleus increases as we progress from lighter to heavier nuclei. We might guess that all nucleons lose about the same mass as they bind into a nucleus. Then, the mass defect would increase in proportion to the total number of nucleons. If this were correct, the carbon nucleus, with 12 nucleons, would have three times the mass defect as helium, with 4 nucleons. A quick check of the actual values for mass defects (Table 22-1) shows that this is not the case. The mass defect per nucleon for carbon is greater than the value for helium. We conclude that the binding energy of each nucleon varies from one nucleus to another.

Table 22-1 lists the mass defect per nucleon and the binding energy of each nucleon for a selection of nuclei. Both quantities increase rapidly from hydrogen 1 to nuclei with 40 to 50 nucleons and then remain fairly constant for nuclei of up to about 100 nucleons. From 100 nucleons on, the binding energy slowly decreases. This pattern, called the **binding-energy curve,** is illustrated graphically in Figure 22-3.

We can explain the binding-energy curve in terms of the attractive force within the nucleus. As the number of nucleons grows from 1 to about 50, so does the attractive force. With the increase in force the nucleons are pulled together more tightly. Thus the binding energy per nucleon increases. As the nuclei get very large, however, the repulsive electrical forces between protons become more important. This repulsive force adds as a vector to the attractive nuclear force, thus decreasing the net force on the nucleons. Thus the energy needed to pull a nucleon from a large nucleus is less than for smaller nuclei. The binding energy per nucleon decreases as the nuclei become very large.

Table 22-1 Mass Defect/Binding Energy Per Nucleon

Nucleus	Mass Defect (amu)	Number of Nucleons	Mass Defect per Nucleon (amu)	Binding Energy per Nucleon (J)
Hydrogen 1	0	1	0	0
Hydrogen 2	0.002388	2	0.001194	1.7796×10^{-13}
Hydrogen 3	0.008556	3	0.002854	4.2508×10^{-13}
Helium 3	0.007195	3	0.002398	3.5746×10^{-13}
Helium 4	0.030376	4	0.007594	11.318×10^{-13}
Lithium 7	0.042130	7	0.006019	8.9710×10^{-13}
Carbon 12	0.098940	12	0.008245	12.2888×10^{-13}
Nitrogen 14	0.112356	14	0.008025	11.9972×10^{-13}
Oxygen 16	0.137005	16	0.008563	12.8008×10^{-13}
Magnesium 24	0.206295	24	0.008596	12.8114×10^{-13}
Argon 40	0.359287	40	0.008982	13.3875×10^{-13}
Vanadium 51	0.466097	51	0.008147	13.0370×10^{-13}
Chromium 52	0.47683	52	0.009170	13.6674×10^{-13}
Iron 56	0.514291	56	0.0091838	13.6880×10^{-13}
Nickel 58	0.52844	58	0.009111	13.5780×10^{-13}
Germanium 72	0.65748	72	0.009132	13.6103×10^{-13}
Zirconium 90	0.81742	90	0.009108	13.5754×10^{-13}
Cadmium 114	1.01797	114	0.008929	13.3091×10^{-13}
Neodymium 142	1.2396	142	0.008730	13.0110×10^{-13}
Hafnium 180	1.519	180	0.008440	12.5794×10^{-13}
Lead 208	1.7121	208	0.008231	12.2683×10^{-13}
Radon 222	1.8351	222	0.008266	12.3571×10^{-13}
Radium 226	1.8590	226	0.008226	12.2963×10^{-13}
Uranium 238	1.884	238	0.007916	11.7988×10^{-13}
Fermium 253	1.9686	253	0.007781	11.5970×10^{-13}

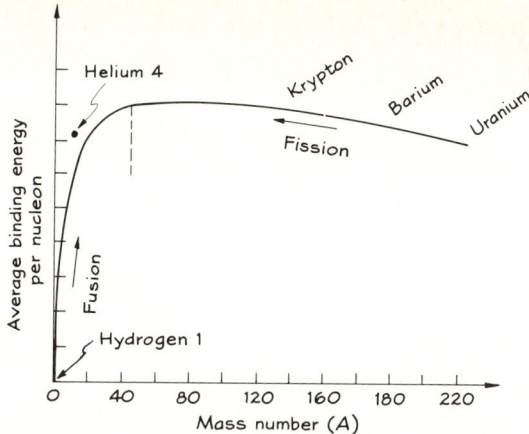

Figure 22-3
The binding energy per nucleon increases rapidly, reaches a peak and then gradually decreases. Fusion describes the process by which we combine lighter nuclei to form heavier nuclei. Fission describes the process by which we split heavier nuclei into light ones. Both processes move nuclei up the binding energy curve—from loosely bound nuclei to more tightly bound ones. Thus energy is released to the environment in both cases.

Binding Energy and Nuclear Reactions

The energy released when a nucleus is formed depends on the binding energy of the nucleus. A nucleus that is tightly bound releases more energy when formed than a nucleus that is less tightly bound. Thus a nuclear transformation that converts a loosely bound nucleus into a tightly bound one releases energy into the environment. We have already seen examples of this type of transformation in our study of radioactivity. When radium changes into radon and an alpha particle, it is changing from a nucleus with a total binding energy of $226 \times (12.2963 \times 10^{-13})$ J $= 2.779 \times 10^{-10}$ J to nuclei with binding energies of 2.743×10^{-10} J $+ 0.045 \times 10^{-10}$ J $= 2.788 \times 10^{-10}$ J. Because the products are bound more tightly, some energy is released in the transformation.

The energy released in radioactive decay is very small because it involves the rearrangement or removal of a small number of nucleons. The new nucleus has only a slightly different binding energy than the original one. But suppose a very heavy nucleus could be split into large fragments forming nuclei in the middle of the binding-energy curve. Or, suppose several very light nuclei combine to form a nucleus in the middle of the binding-energy curve. Both processes occur and release large amounts of energy. The former is called *fission;* the latter, *fusion.*

SELF-CHECK 22A

Can two hydrogen 2 nuclei (mass of each = 2.013553 amu) combine to create one helium 4 (mass = 4.001506 amu) and release energy?

NUCLEAR FISSION

One way to create a radioactive isotope is to place an element in front of a beam of neutrons. A nucleus absorbs a neutron and becomes an unstable isotope, which then decays. For example, when a neutron is absorbed by a nitrogen 15 nucleus, it forms nitrogen 16, which undergoes beta decay to become oxygen 16.

$$_0^1 n + {}_7^{15}N \rightarrow {}_7^{16}N \rightarrow {}_8^{16}O + {}_{-1}^0 e + \bar{\nu}_e$$

The net result of this and other transformations is that an element is changed into the next higher one in the periodic table.

In the 1930s, Enrico Fermi conducted an exhaustive study of neutron absorption of the various elements. At the time these experiments were conducted, uranium was the last known element in the periodic table. Intrigued with the idea of creating the element with an atomic number one greater than uranium—an entirely new chemical element—Fermi bombarded uranium nuclei with neutrons. In some cases the results were what Fermi expected, but in others nuclei with much smaller masses—nuclei of known elements—were the products of his experiments. Fermi had stumbled across the process of nuclear fission.

Nuclei Split into Smaller Fragments

The process by which massive nuclei split apart to form less massive nuclei is called **nuclear fission**. The most common fission reaction involves isotopes of uranium and plutonium. One example is uranium 236, an unstable nucleus that does not exist naturally. When we manufacture it, we find that it spontaneously transforms into isotopes of much less massive nuclei within 10^{-12} seconds (s). While the uranium 236 can split in many different ways, a common breakup is into barium, krypton, and three neutrons.

$$_{92}^{236}U \rightarrow {}_{56}^{142}Ba + {}_{36}^{91}Kr + {}_0^1 n + {}_0^1 n + {}_0^1 n$$

(236.045563 amu) = (141.912205 amu) + (90.91891 amu)
$\qquad$ + (3 × 1.008665 amu) + energy

This fission releases 2.8×10^{-11} J of energy because uranium 236 is much less tightly bound than the product nuclei, barium 142 and krypton 91.

The fission process involves two fundamental interactions—the strong nuclear and the electrical. Within the nucleus, the strong nuclear force dominates. In order for fission to occur, something must pull the nucleons far enough apart that the strong nuclear attractive force becomes less than the electrical repulsion of the protons. Then the nucleus can split. In some respects this situation is analogous to a ball sitting in a small rut at the top of a hill (Figure 22-4). The ball is quite stable; it will not move unless it receives a small push. Once pushed, the ball rolls all the way down the hill. By adding a small amount of energy to get it started, we enable the ball to give up much more energy when it reaches the bottom. Similarly, adding a little energy to the

Figure 22-4
Nuclear fission is in some ways like pushing a ball out of a rut at the top of a hill. A small amount of energy invested in the push allows the release of a much larger amount of energy.

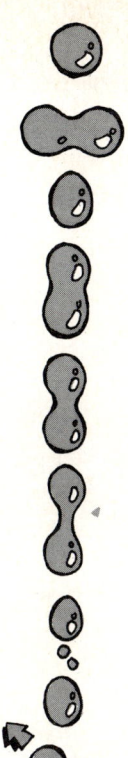

Figure 22-5
By supplying external energy, we can make a spherical drop of water vibrate so that it splits into two spherical fragments and several tiny droplets.

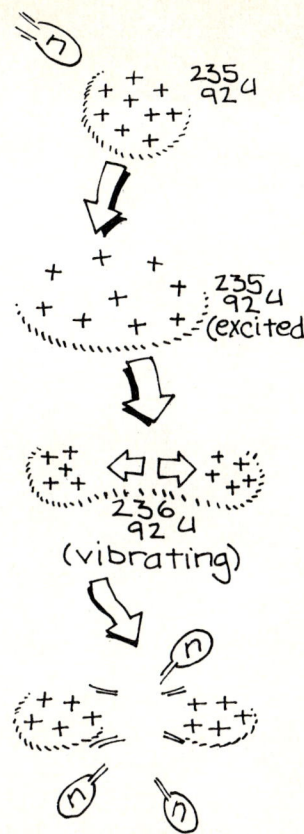

Figure 22-6
A neutron strikes the uranium 235 nucleus, transforming it to uranium 236 in an excited state. After a sequence of vibrations, the uranium 236 nucleus fissions to produce two smaller nuclei and three free neutrons.

nucleus—"pushing" it with the addition of a neutron—enables it to release much more energy in the process of fission.

We can describe the push that initiates a fission reaction by comparing a nucleus with a drop of liquid. The uranium 236 nucleus in its lowest energy state is much like a spherical drop of water. Strong nuclear forces bind the nucleus together in as small a space as possible. If we provide a drop of water with energy by flicking it with a finger, we can make it vibrate (Figure 22-5). If the vibrations match a natural frequency of the liquid drop, they become so large that the drop splits into two spherical fragments and a few tiny droplets. We imagine that much the same process occurs with the nucleus. If the massive nucleus is supplied with a small push, it begins to vibrate. As the vibrations increase, the fragments pull farther and farther apart. Finally, the repulsive electrical force exceeds the attractive nuclear force, and the nucleus splits into two fragments and several neutrons.

How do we push a uranium 236 nucleus? By producing it. If we force a neutron into a nucleus of uranium 235, we can convert uranium 235 into uranium 236. As the neutron is absorbed, it carries enough energy to excite the newly created uranium 236. Once excited, this nucleus can undergo fission. The entire process by which uranium 235 leads to fission is illustrated in Figure 22-6.

The energy released by one fission reaction seems quite small. However, it is about 10^8 times the energy released in a single chemical reaction, such as one carbon atom combining with two oxygen atoms in the burning of fossil fuels. In comparison with other energy-releasing methods, fission provides an enormous amount of energy per reaction.

SELF-CHECK 22B

Another fission reaction involving uranium 236 is:

$$_0^1n + \,_{92}^{235}U \rightarrow \,_{92}^{236}U \rightarrow \,_{38}^{90}Sr + \,_{54}^{135}Xe + 10\,_0^1n$$

$$(1.0087 \text{ amu}) + (235.0439 \text{ amu}) = (89.9073 \text{ amu})$$
$$+ (135.9072 \text{ amu}) + (10.0867 \text{ amu}) + \text{energy}$$

How much mass is converted to energy? How does the energy released in this transformation compare with that released in the transformation that produces barium 142 and krypton 91?

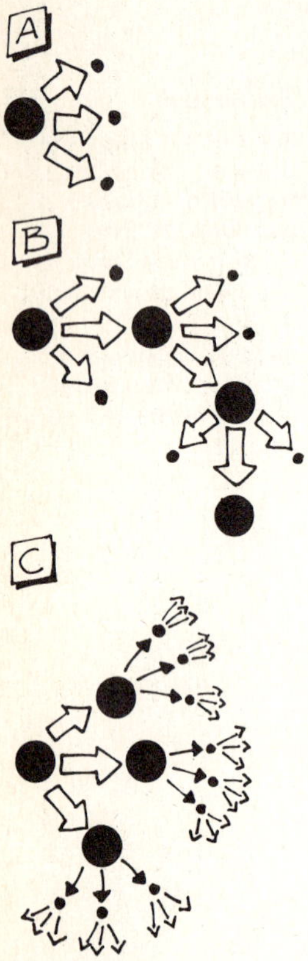

Chain Reactions

On the average three neutrons are released by each fission reaction. The destiny of these three neutrons determines whether the result is an isolated burst of energy, sustained energy production, or an explosion. The three possibilities are illustrated in Figure 22-7.

In (a) the neutrons are simply lost or absorbed by nuclei other than uranium 235. No further fission occurs. In (b) one of the neutrons released by each fission reaction is absorbed by a uranium 235 nucleus. This absorption creates another fission, which produces three neutrons, one of which is absorbed by another uranium 235 nucleus, and so forth. Fission continues at a constant rate. Finally, in (c) two or three of the neutrons released by each fission reaction is absorbed by uranium 235 nuclei. Each fission reaction creates more than one fission reaction and the process grows exponentially. An isolated fission event, such as that shown in (a), provides us with valuable insights into the structure of the nucleus, but it does not give us a useful quantity of energy. **Chain reactions** provide us with enormous amounts of energy, controlled in one case (b) and uncontrolled in the other (c).

The absorption of one neutron per fission event assures us of a sustained chain reaction in which each fission event leads to only one other fission event. Fission proceeds at a constant rate until all the uranium 235 is used up. We can use the energy released to drive electric generators.

The absorption of more than one neutron per fission results in an uncontrolled chain reaction. One neutron initiates one fission event, which produces three neutrons that initiate three fission events, which produces nine neutrons that initiate nine events, and so forth. The progression is 1; 3; 9; 27; 81; 243; . . . . The number of fission events very quickly becomes enormous, and a devastating quantity of energy is released, as documented by the atomic bombs dropped on Nagasaki and Hiroshima, Japan, in 1945.

Figure 22-7
Three possibilities exist for the neutrons released from a fission event. **(a)** When no neutrons are absorbed by other uranium 235 nuclei, the reaction stops. **(b)** When one neutron is absorbed by another uranium 235 nucleus, a sustained chain reaction occurs. **(c)** When two or three neutrons are absorbed by other uranium 235 nuclei, an uncontrolled chain reaction occurs.

Which of these processes occurs depends primarily on the number of uranium 235 nuclei near the fission event. Natural uranium consists of 0.7% uranium 235 and more than 99% uranium 238. This concentration of uranium 235 is not sufficient to produce a chain reaction. But, if we artificially increase the fraction of the uranium 235 to 3% of the total, an average of one neutron released by each fission will be absorbed by another uranium 235 nucleus. A sustained reaction will occur. If we increase the fraction of uranium 235 to 97%, almost all the released neutrons will be absorbed by uranium 235 nuclei. The number of fission reactions will grow exponentially, and an uncontrolled chain reaction will occur. Thus the ratio of material that can undergo fission to material that cannot determines the outcome of a fission event.

Even if the material contains 97% uranium 235, it may not explode. If it contains a very large surface area, the material may allow many of its neutrons to escape from it without encountering another uranium 235 nucleus. Designers of fission weapons use this factor. Masses of 97% uranium 235 are constructed to have a shape and density that allow most of their neutrons to escape. These masses are placed in close proximity. To cause an explosion, they are driven together by ordinary explosives such as TNT. Then, the shape, mass, and density of the uranium 235 reaches the place where an explosion occurs. The uranium 235 is said to have reached a **critical mass.**

SELF-CHECK 22C

Describe the progression of fission events if only two neutrons from each fission are absorbed by uranium 235 nuclei.

Fission Reactors

Today, more than 20% of our electrical energy is generated by nuclear **fission reactors.** As shown in Figure 22-8, the design and operation of these

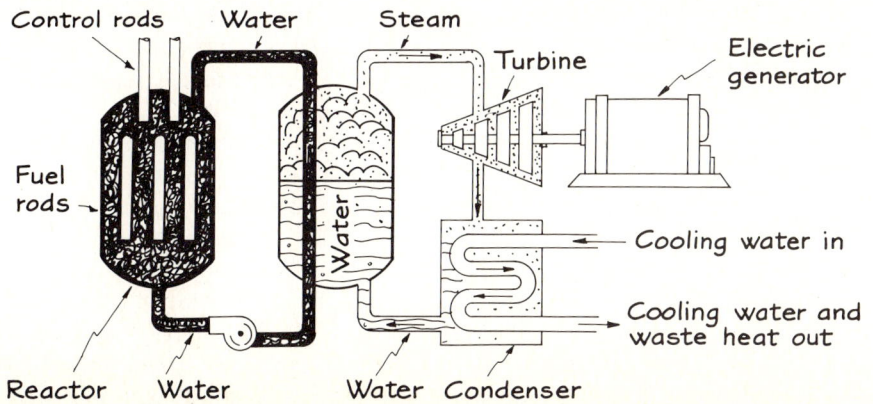

Figure 22-8

A schematic view of a typical fission reactor.

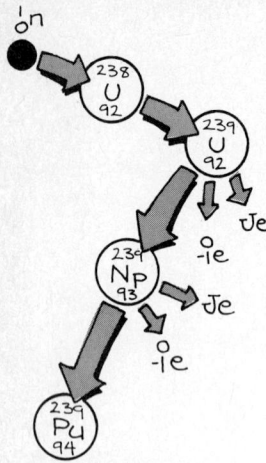

Figure 22-9
In breeder reactors, uranium 238 goes through a series of decays to produce plutonium 239, a fissionable nucleus.

generating plants are very similar to that of fossil fuel plants. Both use thermal energy to produce steam; the difference is the process used to produce the thermal energy. Fossil fuels are replaced by fission reactors. Chemical potential energy is replaced by nuclear potential energy.

The reactor itself contains three components: the nuclear fuel, the control rods, and water used to transfer the thermal energy from the reactor to the generator. The nuclear fuel consists of uranium in which the fraction of uranium 235 has been artificially increased to 3%. The control rods are made of a material that readily absorbs neutrons and does not undergo fission. They are used to control the number of neutrons by absorbing and removing some of them from the fuel. Water, which surrounds the nuclear fuel, absorbs the thermal energy generated by the fission process. This heated water transfers its thermal energy to a second water system and then returns to the fuel elements. The second water system operates the electrical generator in a conventional fashion. (Two isolated water systems are required because the water circulated near the nuclear fuel can become radioactive.)

Because only 0.7% of natural uranium is uranium 235, this isotope must be considered a very limited resource, like gas, coal, and oil. If we were to rely strictly on uranium 235 reactors for production of electrical power, we would very quickly end up in the same position with respect to fission fuel as we are with fossil fuels. Therefore, another fission process, using the more abundant uranium 238 isotope, is being developed as a commercial energy source. Called a **breeder reaction,** this process transforms uranium 238 into plutonium 239, which undergoes fission when it absorbs a neutron (Figure 22-9).

The fission of plutonium provides both the nuclear energy for heating the water system and neutrons needed to turn uranium 238 into plutonium 239. This process breeds its own fuel. To create the plutonium 239, some of the uranium 238 must be placed near a source of a large number of neutrons. The most abundant source today is a nuclear reactor. When uranium 238 is placed in a nuclear reactor, it absorbs neutrons, and plutonium is produced. This is the basic idea of the breeder reactor.

The design of a breeder reactor involves two regions. In the first, fission of either uranium or plutonium is taking place. Energy released in these fission processes is used to produce electricity. Surrounding the energy-producing area is a supply of uranium 238. As neutrons leave the fission area, they are absorbed by uranium 238 nuclei. The plutonium that is produced by this process is taken to a special processing plant, where it is chemically separated and turned into fuel for the energy-producing part of a reactor.

Breeder reactors hold the promise of being a seemingly inexhaustible energy source. The nuclear fissions in these reactors produce 100 million times as much energy per atom as the chemical reactions involved in burning coal or oil. In addition, if they become commercially cost-effective, they would create more fuel than they use. In spite of this overwhelming step forward in energy technology, fission reactors of all kinds come under heavy criticism. To see why, we must consider the advantages and disadvantages of this energy process.

Benefits and Risks of Fission Reactors

Fission reactors have two major advantages: availability of fuel and absence of chemical pollutants. The energy source for fission reactors is not a fossil fuel and, at present, does not come from foreign sources. Uranium is mined and processed within the United States, and the breeder process makes use of the abundant isotope uranium 238. Because they use nuclear rather than chemical processes, fission reactors do not release the chemical pollutants that are so objectionable in burning fossil fuels. In this respect, nuclear fission reactors are much cleaner than conventional power plants.

The major disadvantage of fission reactors is the radioactive material they produce. The neutrons released from the fission reactions interact with all parts of the reactor, producing a variety of nuclear transformations. The radioactive byproducts of these reactions range from probably harmless to lethal. Reactors are built with thick concrete and steel walls to absorb the radioactive emissions. These designs are so effective that the environmental radiation surrounding a nuclear fission plant is, in fact, less than that surrounding a coal-burning plant. (Coal contains minute quantities of radioactive material.) Because the fuel contains only 3% uranium 235, it cannot become an atomic bomb, which could cause vast destruction and harmful radiation. The issue, then, is not proper shielding from the nuclear pollutants or nuclear explosions. The major concerns are long-term storage of the radioactive waste products and protection against accidents and sabotage.

Every few years the uranium 235 in the reactor fuel is depleted to the point where it will not sustain a chain reaction. The material left in the fuel elements includes cesium 137 and strontium 90 as well as isotopes of plutonium and uranium. These products are highly radioactive and dangerous to living things. The huge quantities involved, the dangers in their radioactive emissions, and their relatively long half-lives mean that we must find places to

© 1983 by Sidney Harris.

store the waste products for more than 10,000 years. Storage sites must be geologically stable (no earthquakes for at least 10,000 years) and must be isolated from any ground water.

Some studies have proposed salt mines in the southwestern United States, which seem to meet both these criteria. If these sites prove satisfactory, then highly radioactive material must be transported there from all over the country. Of course, traffic accidents must be avoided. Further, residents of the region must be convinced that no harm will come to them or their environment from storage of the radioactive wastes. Finally, we as a society must answer the question of who will watch over these materials for the next 10,000 years—a time much longer than any civilization has survived on this planet.

Storage is a long-range problem; a more immediate problem is accidents. Reactors are complex mechanical and electronic machines—things can and do go wrong. One of the most serious possible accidents is called *loss of coolant*. In this situation, the cooling water does not circulate and cool the fuel. Heat builds up rapidly and, if the situation is not corrected, the fuel becomes hot enough to melt metal. The floor of the building in which the fuel is contained could melt, and radioactive material could escape. In the worst situation the extremely hot material would melt through the rock below the reactor until it hit water. The water would turn to steam and move upward into the air, carrying radioactive materials with it. Large areas of land could be covered with a dangerous radioactive material.

The worst possible accident has not occurred, but some very serious ones have. The most well-known incident occurred at the Three-Mile Island Reactor in Pennsylvania, where malfunctioning equipment caused a loss of coolant. The reactor fuel became extremely hot. For several days a complete meltdown of the reactor was possible. Finally, the fuel was cooled sufficiently that it would not melt through the building floor. However, in the process the reactor was destroyed.

Breeder reactors are susceptible to the same dangers. The first commercial breeder reactor built in the United States experienced a loss-of-coolant accident shortly after it was opened in 1967. While no radiation escaped, the reactor was permanently damaged and has not operated since.

The abundant plutonium that would be created by large-scale use of breeder reactors is also a concern. The plutonium must be processed chemically before it can be used as a nuclear fuel. This processing would require large-scale plants. Such plants can be built; however, two problems exist. First, plutonium is very toxic and must be handled with extreme care to avoid accidental poisoning. Second, a small group of knowledgeable people could steal enough plutonium to build a nuclear weapon. If large amounts of plutonium were being shipped from breeder to processing plants and back, the possibility of an accident or theft would increase greatly.

Nuclear fission can supply energy for our future. It is doing so successfully now. Proponents of fission fuels point to the past safety record of nuclear power plants. Even at Three-Mile Island, no one was seriously injured. The risks, they say, are small when compared to the benefits. The critics, on the other hand, state that one serious accident or theft of fuel could be devastating. They feel that the risks are much too great for the benefits received. As with all technological advances, this one has both advantages and disadvantages. The disadvantages always lead us to look for new solutions.

NUCLEAR FUSION

The sun releases some 240×10^{26} J of energy each minute, more energy than can be explained in terms of the energy released in ordinary chemical reactions. Ninety-nine percent of the sun's mass is hydrogen and helium. A look at the binding-energy curve (Figure 22-3) tells us what process must be involved in this enormous output of energy.

The binding energy per nucleon increases from hydrogen until it reaches a peak around iron, nickel, and cobalt. It then levels off and gradually decreases. The largest change in binding energies occurs as we move from hydrogen and its isotopes to helium 4. The binding energy of helium 4 is more than six times that of hydrogen 2. Thus, when hydrogen nuclei combine to make helium, large amounts of energy can be released.

Nuclei Combine into Heavier Elements

The process by which lighter nuclei join together to form heavier nuclei and release energy is called **nuclear fusion.** A combination of experimental and theoretical work has identified several fusion reactions that occur with lighter elements.

The most common fusion process begins with four hydrogen 1 nuclei and eventually transforms them into a helium 4 nucleus. (Since helium has only two protons and four hydrogens have four, two of the protons transform into neutrons. Two positively charged particles, called positrons, are also created. The positron is identical to an electron except that it has a positive charge. Its symbol is $_{+1}^{0}e$.) The hydrogen-to-helium fusion reaction is:

$$_{1}^{1}H + _{1}^{1}H + _{1}^{1}H + _{1}^{1}H \rightarrow _{2}^{4}He + _{+1}^{0}e + _{+1}^{0}e$$

$$4.03130 \text{ amu} = 4.00260 \text{ amu} + 0.000549 \text{ amu} + 0.000549 \text{ amu} + \text{energy}$$

The total mass of the four hydrogen nuclei is 4.03130 amu. The total mass of the helium nucleus and the two positrons is 4.003698 amu. A total of 0.027602 amu has been converted into energy—4.12×10^{-12} J. This energy arises from the large difference in binding energies in the nuclei.

Fusion does not normally occur on earth because of the electrical repulsion between the positively charged hydrogen nuclei. At a separation distance of more than 10^{-12} meters (m), the attractive nuclear interaction is essentially zero and the repulsive electrical interaction is dominant. To overcome this force, the hydrogen nuclei must have sufficient kinetic energy to move within 10^{-13} m where the nuclear attraction overwhelms the electrical repulsion.

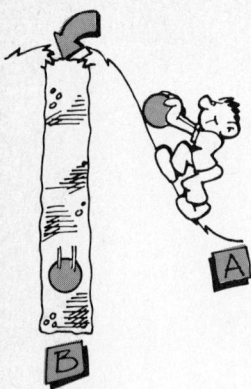

Figure 22-10

Nuclear fusion in some ways is like a ball that must roll uphill in order to fall into a well. Though the ball's gravitational potential energy at position A is greater than at B, the ball will not move until we give it a push.

This situation is somewhat analogous to rolling a ball into a deep well when the well is located at the top of a hill (Figure 22-10). Sitting at the bottom of the hill, the ball has a greater gravitational potential energy than it would have if it were at the bottom of the well. To reach the bottom of the well, however, the ball has to gain additional gravitational potential energy. Giving the ball a push can provide it with enough kinetic energy to reach the top of the hill and drop into the well. When it reaches the well bottom, it converts to other energy forms both the difference in gravitational potential energy between its starting point and the bottom of the well and the kinetic energy we gave it. Once the ball is located at the bottom of the well, it requires a lot of energy to get it out. The ball has given up energy and is more tightly bound than it was at the start.

When two hydrogen nuclei try to fuse, the hill they must overcome is the electrical repulsion between the two positive charges. To overcome this barrier, a hydrogen nucleus must be moving at a speed of about 10^5 m/s. When an environment produces hydrogen nuclei at these speeds, fusion will occur.

Fusion reactions have been produced on a small scale by accelerating hydrogen nuclei to the proper speed and then hurling them at targets of matter. While this process does produce fusion and provides knowledge about the process, it delivers much less energy than is required to accelerate the hydrogen. On a much larger scale, the huge temperatures and pressures in the sun sustain fusion reactions, which provide us with our solar energy.

On earth, large-scale fusion energy is produced in the hydrogen bomb. For a fusion bomb to explode, the hydrogen nuclei must move toward one another very rapidly. This motion is accomplished by surrounding the hydrogen with another bomb, usually a fission weapon. When the external bomb explodes on all sides at once, it drives inward (implodes) the hydrogen nuclei. The nuclei meet in the middle while traveling at tremendous speeds, and many fusion events occur so rapidly that a violent explosion occurs. Though the energy production is enormous in this case, it is so fast that it is uncontrollable and, except as a military threat, useless to us.

SELF-CHECK 22D

A second fusion reaction that has been observed involves deuterium (hydrogen 2):

$$^2_1H + ^2_1H \rightarrow ^3_1H + ^1_1H + 0.64 \times 10^{-13} \text{ J}$$

2.013553 amu + 2.013553 amu = 3.016050 amu
+ 1.007276 amu + energy

Explain why less energy is released in such a reaction than in the fusion of four hydrogen 1 nuclei.

THE BOMB

When Enrico Fermi first created a nuclear fission, he did not realize that he had done so. He knew that he had produced a new nucleus but did not know what it was. In 1938 two German chemists, Otto Hahn and Fritz Strassman, separated barium in the products of an experiment in which uranium had absorbed a neutron. They communicated their results to Lise Meitner, who had fled Germany and was working with Niels Bohr in Copenhagen. She analyzed the results and realized that the energy released in a single transformation was enormous compared to any other interaction. Soon afterward, Bohr came to the United States and discussed these conclusions with Fermi, Einstein, and others. At the urging of Leo Szilard, Einstein wrote a letter to President Franklin Roosevelt in which he stated "... it may become possible to set up a nuclear chain reaction in a large mass of uranium by which vast amounts of power ... would be generated." While Einstein was a pacifist, he was very concerned about what would happen if Hitler used this type of energy. Thus he described to Roosevelt the possibility that a fission bomb could be created.

This letter began one of the most concentrated efforts in the history of science and technology. When it was completed, two nuclear bombs had been dropped on Japan and had ended World War II. Since that time "the bomb" has become a part of our lives. We can only hope that, some day, people will learn better ways to discuss their differences than by threatening wholesale destruction. Maybe, then, the bomb will be as useless as some of the inventions we have described in this book.

Fusion Reactors

Nuclear fusion devices that produce energy in a controllable manner are called **fusion reactors.** Their design must deal with two related difficulties: temperature and containment. To drive the nuclei close enough to initiate fusion, we must heat the hydrogen nuclei to temperatures of about 10^8 K. Once at that temperature, the hydrogen nuclei must be contained. Unfortunately, all known materials melt at temperatures far below that required for fusion, so containment of the material must be accomplished by other than traditional methods. While physicists and engineers are studying several processes, they are currently devoting much attention to two of them: magnetic containment and laser-induced fusion.

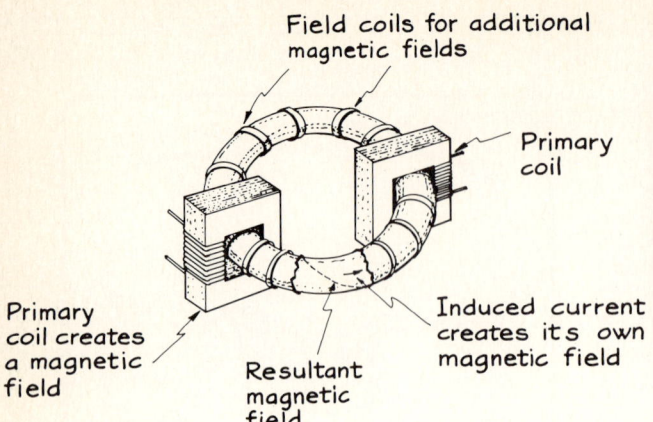

Figure 22-11
One magnetic confinement system, the tokamak, uses several magnetic fields. The field from the primary coil induces a current in the plasma. In turn, the plasma produces a magnetic field. All fields combine to produce a spiral field that contains the charged nuclei.

Magnetic containment takes advantage of the electrically charged nature of material at very high temperatures. Material raised far above its boiling point moves into a fourth state of matter, the plasma state. In a plasma all the electrons have been stripped from their nuclei, and the material is a swarm of positively charged nuclei and free electrons. Because a magnetic field deflects charged particles, it should be able to confine the electrically charged plasma.

Very complicated magnetic fields, which are created by a careful arrangement of current-carrying wires plus the magnetic field of the plasma itself (Figure 22-11), are arranged so that the charged particles always feel a force toward the inside of the plasma. Such a containment device is integrated into an electrical generating plant as illustrated in Figure 22-12. The intense heat generated by fusion is first absorbed by potassium and used to drive one generator, then transferred to water and used to drive a steam generator. This entire process is still under development.

Laser-induced fusion begins not with a very hot plasma, but with frozen hydrogen isotopes, hydrogen 2 and hydrogen 3. A pellet of the frozen material is placed in the path of a very intense laser beam. The laser light is directed toward the pellet equally from all directions. The amount of energy transferred to the pellet is enough to vaporize the surface layer of the hydrogen

Figure 22-12
A possible magnetic confinement electrical generating system.

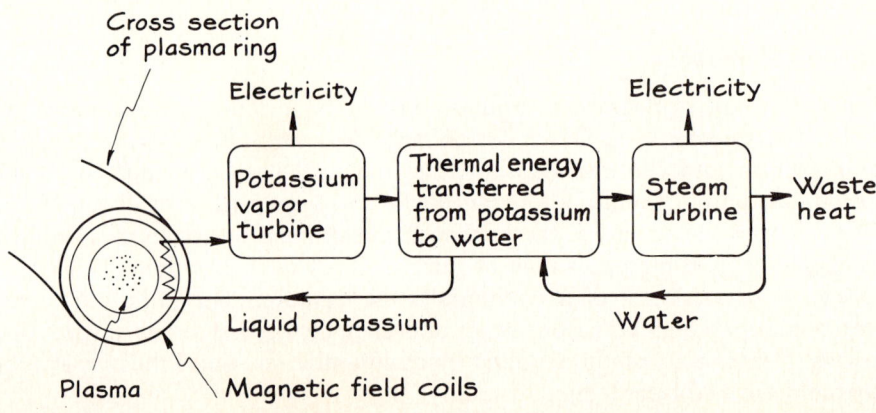

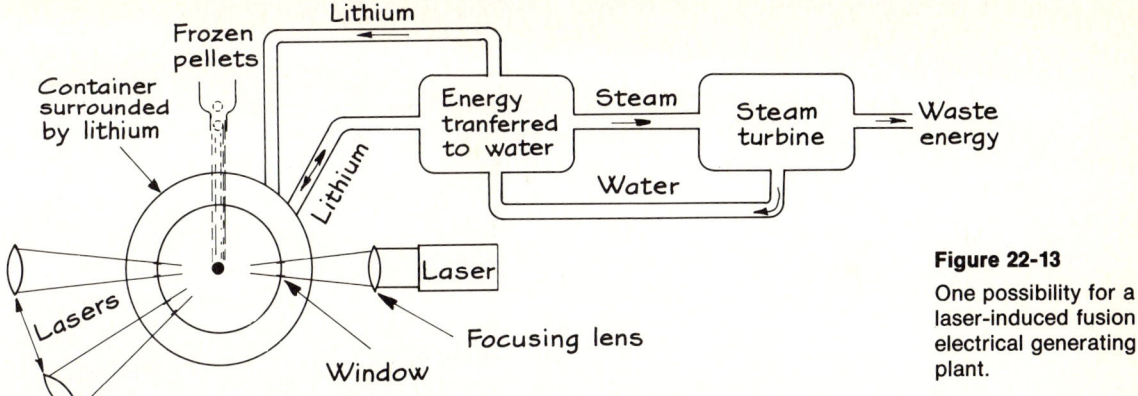

Figure 22-13
One possibility for a laser-induced fusion electrical generating plant.

isotopes. This sudden vaporization creates a shock wave that causes an implosion as it travels toward the center. As the pellet collapses in upon itself, the temperature of the isotopes rises to 10^8 K and fusion begins. A proposed power plant based on this idea is shown in Figure 22-13. Because of the range of temperatures involved, this process is sometimes called the "snowball-in-hell" approach to energy generation. This process, too, is still in the experimental stage.

Risks and Benefits of Controlled Fusion

In discussions of fusion reactors two concepts—scientific break-even and thermodynamic break-even—are frequently mentioned. These terms refer to breaking even on energy. A **scientific break-even** is reached when the

© 1979 by Sidney Harris.

energy released in the fusion reactions equals the energy consumed in creating them. Since the efficiency of converting the released nuclear energy to useful energy is necessarily less than 100%, scientific break-even is not good enough. We need **thermodynamic break-even,** in which the useful energy (for example, electrical energy) derived from the fusion reaction exceeds the energy consumed in creating it. Both magnetic containment and laser-induced fusion are approaching scientific break-even but are far from achieving thermodynamic break-even.

The magnetic fields needed to maintain magnetic containment must be very large. To keep them large, the electrical current to the primary coils must be enormous. At present, this electrical energy exceeds the energy released by the fusion. Additionally, the plasma leaks. Soon after the process begins, particles begin to leak out and the plasma current gradually collapses. New designs are needed to overcome this leakage.

The lasers required for laser-induced fusion are not ordinary grocery-store scanning lasers. They must produce large beams capable of reaching a power output of 10^{14} watts. One such laser, called SHIVA, has been built, but the energy consumed to produce its beam is far greater than the energy output of the beam.

Once these technical difficulties are overcome, then the advantages and disadvantages of fusion reactors are derived from comparisons with conventional power plants. Questions of fuel, efficiency, and safety become paramount.

One of the primary advantages of fusion reactors will be the cost and availability of fuel. The essential fuel in most fusion reactors, hydrogen 2 (deuterium), is found in ordinary water. The amount of deuterium in ocean water will produce one billion times the energy stored in the world's fossil fuel energy reserves. And, it is relatively inexpensive to separate.

Compared to fossil fuel power plants, fusion reactors should be much more efficient. The Carnot efficiency of a system increases as the difference between its high and low temperature increases (Chapter 12). Fusion reactors, because of their high operating temperatures, should be more efficient than the 35% efficiency of a typical conventional power plant. However, we have yet to learn how to handle large quantities of materials at these enormously high temperatures.

The most prevalent disadvantage of fusion reactors is the radiation that will accompany the reactions. The fusion of deuterium into helium releases neutrons, which interact with the walls and other reactor components to create other transformations. As components of the reactor need to be replaced, storage of large quantities of radioactive materials becomes a problem. In addition, the fusion process produces hydrogen 3 (tritium). Tritium is naturally radioactive, emitting low-energy electrons and antineutrinos. Outside the body, these radiations are not particularly harmful, but when they are generated inside our bodies, they can be dangerous. Since tritium can replace hydrogen in water and food, it must be handled extremely carefully. Any contamination of drinking water with tritium could be a very serious matter.

Fusion reactors are still decades away, but researchers are optimistic. Once the technology is mastered, a demonstration plant must be built and operated successfully. Only then will fusion reactors be built on any large scale. The most optimistic scientists predict that we will not see a commercial nuclear fusion power plant until the second or third decade of the 21st century.

NUCLEAR RADIATION AND THE HUMAN BODY

Until rather recently, post-World War II, our knowledge of the effects of radiation on the human body was meager at best. Early scientists had noticed the direct effects—severe burns on their hands as a result of handling radioactive materials. In most cases, however, the burns healed quite normally. Marie Curie and her daughter, Irene, both of whom performed many experiments with radioactive materials, died from leukemia. While we cannot be certain that their diseases were induced by their unusually large exposures to radiation, we know from experiments that there is a direct link between blood and bone disorders and unusually large radiation doses.

Victims of the Nagasaki and Hiroshima bombings have provided us with information on the effects of massive radiation on a large population. The people nearest the explosion suffered burns and tissue damage primarily related to the amount of energy released by the explosion. The same fatal effect would have resulted from an equivalent release of energy by a nonnuclear explosion. Survivors far from the blast had few immediate symptoms. Over many years, however, they showed patterns of illnesses that unveiled long-term internal damage done by nuclear radiation. In addition, they bore an unusually large number of deformed children, showing that their genetic material had been damaged by radiation.

Radioactive products interact with living tissue through two processes: direct collisions and electrical interactions. Neutrons and gamma rays carry no electrical charge and consequently interact with the atoms in our bodies via direct collisions. Through these collisions, they transfer energy to the electrons and often knock them free from their atoms. Because the electrons hold atoms together in molecules, these collisions can alter the molecular structure of millions of molecules in living cells. Alpha and beta particles both carry electric charge. They do not need to collide directly with atoms in order to interact with them. The electrical attractions and repulsions can cause molecules to be altered.

Interactions with both charged and uncharged particles lead to changes in the structure of matter. In living tissue, several things can happen. First, a cell can be damaged, die, and be replaced by a new cell. This is the least harmful effect of radiation. A more serious situation arises when a damaged cell survives but does not function properly. In some cases the cell repairs itself through natural processes, but in others the altered cell persists. The change in function is most serious when it is the result of damage to DNA, the genetic material. DNA molecules carry the codes for all the structure and activities of the cell and are passed from one cell generation to the next. Altering these molecules causes descendants of the original cell to be altered

Table 22-2 Typical Applications of Radioactive Isotopes

Isotope	Half-life	Important Uses
$^{3}_{1}H$	12 years	Traces water and organic substances.
$^{14}_{6}C$	5600 years	Studies of organic processes such as metabolism.
$^{24}_{11}Na$	15 hours	Studies of biochemical processes.
$^{32}_{15}P$	14 days	Studies of bone growth and treatment of blood diseases.
$^{60}_{27}Co$	5.3 years	Used in cancer therapy.
$^{131}_{53}I$	8 days	Treats and studies thyroid diseases.

in form and function. These altered cells may stop reproducing, or they may start to reproduce at an enormous rate, as is the case for cancer cells. Extensive cell damage is generally fatal to the entire organism. Finally, radiation can damage the genetic material of the reproductive cells in ways that do not appear until the next generation is born. These alterations can range from inconsequential to lethal.

Not all interactions between radioactivity and living cells are bad. As we gain more experience with the biological effects of radiation, we are able to direct radiation to therapeutic purposes. A summary of some important applications is included in Table 22-2. Controlled use of radiation allows us to kill cancer cells selectively. Radioactive isotopes that can circulate through the blood system are important in the diagnosis of heart and circulatory problems and thyroid, liver, and kidney ailments. The list of applications continues to grow as our understanding of biological processes increases.

Step by step we have explored our earth. We have examined the atoms and molecules from which it is constructed and the nuclei and electrons from which atoms and molecules are constructed. With each step we have gained knowledge—knowledge that has deepened our perceptions of ourselves and the altered landscape we share. As demonstrated by the concepts we build and the uses to which we put our knowledge, we have chosen to be active, not passive, observers. In so choosing, we accept both the risks and the benefits associated with change.

CHAPTER SUMMARY

The mass of a nucleus is always less than the sum of the masses of its constituents. This difference in mass, called the *mass defect,* is related to the binding energy that holds the nucleus together. The principle of mass-energy equivalence relates the mass defect and *binding energy* by the equation: binding energy = (mass defect) × (speed of light)². The binding energy per nucleon is not the same for all nuclei. Nuclei with atomic mass numbers that range from 50 to 100 have the largest binding energies per nucleon and are more tightly bound than either more massive or less massive nuclei. Nuclear reactions that transform loosely bound nuclei into more tightly bound nuclei release energy to the environment. The energy release can occur in two ways. A very massive nucleus can split to form smaller nuclei, and light nuclei can combine to form a more massive one.

The process by which massive nuclei split apart to form less massive nuclei is called *nuclear fission*. In a typical example a neutron penetrates into a uranium 235 nucleus, which then becomes an excited uranium 236 nucleus. This nucleus vibrates and splits into barium 142, krypton 91, and three neutrons. The uranium 236 can split into other products as well; however, on the average three neutrons are released per fission event. The neutrons released by the fission can be simply lost or they can be absorbed by other uranium nuclei to induce other fissions. The absorption of one neutron per fission assures us of a sustained *chain reaction*, which can be used to produce useful energy. The absorption of more than one neutron per fission results in an uncontrolled chain reaction, the process in the nuclear fission bomb. Fission reactors, which provide nuclear energy for generating electricity, use sustained chain reactions fueled by uranium 235 or by plutonium 239.

The process by which lighter nuclei join together to form heavier nuclei is called *nuclear fusion*. One possible reaction involves the fusion of four hydrogen 1 nuclei into one helium 4 nucleus and two positrons. To overcome the electrical repulsion between positively charged nuclei, the hydrogen nuclei need to be hurled at one another at speeds close to 10^5 m/s. While controlled fusion has been produced in the laboratory, the energy required to induce the controlled reaction has, thus far, exceeded the energy released. Fusion reactors are still decades away.

The products of radioactive decay can interact with living cells either by direct collision or through the electrical interaction. Damage to cells can range from inconsequential to fatal for the organism. Radioactive isotopes have also proven to be beneficial, particularly in diagnosing and treating illnesses.

ANSWERS TO SELF-CHECKS

22A. The mass of hydrogen 2 is 2.013553 amu. Thus the total starting mass is 2 × 2.013553 amu = 4.027106 amu. The final mass is that of one helium 4, 4.001506 amu. Because the final mass is less than the initial mass, energy can be released.

22B. The masses of the neutron and uranium 235 nucleus sum to give a total mass of 236.0526 amu before the fission. The strontium 90, xenon 135 nuclei and the ten neutrons sum to give a total mass of 235.9012 amu after the fission. The difference between the masses before and after fission is 236.0526 amu − 235.9012 amu = 0.1514 amu. This mass difference is less than that for the process that produces barium 142 and krypton 91. Thus, the energy released in this fission is smaller.

22C. One neutron initiates one fission event that produces two neutrons, which initiate two fission events that produce four neutrons, which initiate four fission events, and so on. The progression is 1, 2, 4, 8, 16, 32, 64, ... The reaction is uncontrolled.

22D. The mass before the fusion is 2×2.013553 amu $= 4.027106$ amu. After the fusion the total mass is 3.01650 amu $+ 1.007276$ amu $= 4.023326$ amu. The mass difference is 0.003780 amu, which is less than the mass difference for the four hydrogen 1 fusion.

PROBLEMS AND QUESTIONS

A. Review of Chapter Material

A1. Define each of the following:
Mass defect
Binding energy
Binding energy per nucleon
Nuclear fission
Sustained chain reaction
Uncontrolled chain reaction
Fission reactors
Nuclear fusion
Fusion reactors
Scientific break-even
Thermodynamic break-even

A2. Describe the process by which you could use the binding-energy-per-nucleon characteristic of each isotope to predict whether a reaction would proceed on its own.

A3. Use Figure 22-3 to explain why the process of nuclear fission releases energy.

A4. Describe the liquid-drop analogy used to understand the process by which a heavy nucleus splits into two smaller fragments.

A5. Distinguish between a sustained chain reaction and an uncontrolled chain reaction.

A6. Describe the operation of a nuclear fission reactor.

A7. What are two advantages to fission reactors?

A8. Why do the radioactive byproducts of nuclear fission pose a problem to us?

A9. Use Figure 22-3 to explain why the process of nuclear fusion releases energy.

A10. What two problems must be addressed in the design of fusion reactors?

*. Distinguish between scientific break-even and thermodynamic break-even.

A12. Describe the two processes by which radioactive products interact with living tissue.

B. Using the Chapter Material

B1. What is the total binding energy of iron 56?

B2. Could chromium 52 spontaneously transform into argon 40 and carbon 12? Explain your answer.

B3. If a hydrogen 1 nucleus fused with a vanadium 51 to produce chromium 52, how much energy would be released?

B4. Would the fusion of hydrogen 2 and iron 56 to produce nickel 58 release energy?

B5. Explain why nuclei which undergo fusion must be at very high temperatures.

B6. How many neutrons would be released if uranium 236 fissioned into barium 141 and krypton 91?

B7. Suppose a fission reaction which did *not* release any free neutrons existed. Would that process be useful for production of electrical energy? Why or why not?

B8. How much mass is converted to energy in the reaction:

$$^{235}_{92}\text{U} + ^{1}_{0}n \rightarrow ^{110}_{46}\text{Pd} + ^{110}_{46}\text{Pd} + 16^{1}_{0}n$$

The masses are: $^{235}_{92}\text{U} = 235.0439$ amu; $^{1}_{0}n = 1.008665$ amu; and $^{110}_{46}\text{Pd} = 109.9052$ amu.

B9. Will the reaction in Question B8 release more or less energy than the one given in Self-Check 22B?

B10. For the fission reactor and the fusion reactor, describe the changes in types of energy involved as the process goes from nuclear to electrical energy.

B11. Calculate the amount of energy released if 0.001 kg of mass were converted completely to energy.

C. Extensions to New Situations

C1. An unusual "fusion" reaction occurs when ordinary matter combines with material called *antimatter*. The antimatter of any particle is essentially identical to the particle, except that the electric charges of the matter and antimatter are opposite.
   a. Why can matter combine with antimatter when the particles are moving at very low speeds? (Recall the nature of the electric interaction between opposite charges.)
   b. What is the total electric charge of the products of a matter-antimatter collision?
   c. When an electron and an antielectron (positron) combine, the product is electromagnetic energy (which has zero mass). The electron and positron have masses of 9.11×10^{-31} kg each. How much mass is converted to energy in this process?
   d. What is the total energy released?

C2. Why would tritium create a much more serious problem if it were in drinking water than if it were in rocks?

C3. At some time in the future, the government research groups will need to choose between magnetic-containment or laser-fusion research. What type of information would you want if you had to make this decision? (Include both scientific and other information.)

C4. Suppose you had to make the decision to build a breeder reactor. What type of information would be useful to help you make that decision?

C5. If there were only two choices—fission or fusion—for future generation of electrical energy, which would you prefer? Why?

C6. One model suggested to explain the source of energy emitted by the sun is the carbon cycle, in which carbon undergoes nuclear transformations including fusion. Complete the steps of the cycle using the periodic table included in Chapter 21.

$$^{12}_{6}C + ^{1}_{1}H \rightarrow (\quad) + \gamma$$
$$(\quad) \rightarrow ^{13}_{6}C + ^{0}_{+1}e + \nu$$
$$^{13}_{6}C + ^{1}_{1}H \rightarrow (\quad) + \gamma$$
$$(\quad) + ^{1}_{1}H \rightarrow ^{15}_{8}O + \gamma$$
$$^{15}_{8}O \rightarrow (\quad) + ^{0}_{+1}e + \nu$$
$$(\quad) + ^{1}_{1}H \rightarrow ^{12}_{6}C + ^{4}_{2}He$$

C7. One fusion reaction that might occur in the sun is shown below. Use the information given in Table 22-1 to predict whether energy will be absorbed or released by the reaction.

$$^{4}_{2}He + ^{3}_{1}H \rightarrow ^{7}_{3}Li + \gamma$$

D. Activities

D1. Find out what types of nuclear reactors, if any, your local electrical utility has in operation. Learn about any future plans for nuclear power plants.

D2. Talk to someone who works around radiation (for example, an X-ray technician). Learn what they do to guard against biological damage.

Epilogue: Science, Technology, and Risk

Physicists generally investigate nature for the sheer enjoyment of learning. While it is rewarding when discoveries lead to tools for archeologists, medical diagnostic procedures, techniques to aid criminologists, and laser cash registers, these technologies have not generally been the motivations for their research. What motivates and guides basic research is a desire to build more comprehensive models of nature. Yet applications do evolve, regardless of the motivations of the discoverers. Our dual nature as explorers of nature and shapers of the landscape is undeniable. One activity is science; the other, technology.

Today, one of the most controversial technologies is the use of nuclear fission to generate electricity. What began as accidental discoveries, first by Becquerel and later by Fermi, has become an issue that spawns strong emotions. The arguments are not about scientific knowledge; they are about how science should be applied to our current energy dilemma. The question is no longer one of basic science, but rather a discussion of philosophies, of beliefs, and of social concerns.

The debate currently centers on the risks and benefits of nuclear power generation. Benefits include abundant energy, increased economic stability, anticipated growth in the quality of life and decreased chemical pollution. Risks include the introduction of significant amounts of radiation and radioactive waste products into our environment and the increased probability of severe nuclear accidents or sabotage. The unanswered—and perhaps unanswerable—question is whether the benefits outweigh the risks.

Every action we take involves weighing the benefits against the risks. Every time we cross a street, we run a risk of being struck by a car. By exercising proper caution, we can minimize that risk. The risk is relatively small; the benefit gained from freedom of movement is large. So we cross the street without thinking much about the risk involved.

Other situations involve more complex risk-benefit analyses. For example, our study of Newton's first law showed that the use of automobile seat belts is well founded in physical principles. The risk of not wearing one is a crippling or possibly fatal injury. The benefits of not using seat belts are the saving of a few seconds in buckling and unbuckling, a slight increase in the freedom of movement inside a car, and some psychological or emotional benefits that are difficult to define. The benefits and the belief that the probability of an accident is small convince most people to sit on top of their seat belts.

When individuals decide that the benefits of not using seat belts outweigh the risks, they are, for the most part, accepting a risk for themselves. Should an accident occur, they will receive the injuries. The rest of society is "injured" by having to share hospital and medical costs through increased insurance premiums and taxes. However, the societal injury is small compared to the individual's injury or loss of life.

As technology has advanced, the nature of the risks involved has changed. The pilots of commercial airliners realize that their actions can affect not only themselves, but hundreds of other people. The risk from an untrained airplane pilot is far greater than that from an untrained automobile driver. Risks from nuclear power plants are substantially greater than from coal-burning plants. An accident in a coal plant (or coal mine supplying it) can injure or kill hundreds. However, a nuclear power accident could affect millions and render huge areas of land uninhabitable.

Further, nuclear accidents introduce a new dimension—time—into the risk-benefit picture. The crash of an airplane produces problems that linger for only a few. The meltdown of a nuclear reactor could leave a radiation trail extending hundreds of years into the future. A single accident could affect our children, grandchildren, perhaps even our great-grandchildren, in addition to ourselves. Thus, the risk extends over time and space in a way that most other risks do not.

Who decides to assume the risks associated with a nuclear power plant? Is it the stockholders who decide to build it? Or should it be the regulatory agency charged with protecting society at large? Certainly the local residents who will shoulder a larger-than-average part of the risk need to be consulted. How about the coal miners and oil refinery workers who will be put out of work as nuclear fuels replace conventional fossil fuels? Cost-benefit analyses become increasingly complex as more people become involved, as technology becomes more complex, and as the decisions we make spread so far over space and time.

The risks from nuclear power are high, so the precautions must be great. If we can minimize the risks sufficiently, we can enjoy the benefits. Advocates of widespread implementation of nuclear power plants believe that the risks have already been minimized adequately—or can be soon. Opponents feel the risks are still too great. Some feel the risks will always be too great.

What is the role of science in all this? Scientific knowledge was used to develop the technology and, to a large extent, scientific knowledge can assist us in making the risk-benefit comparisons. Scientists can try to provide as much information as possible about the risks and benefits of nuclear power, and they can analyze this information statistically. As caring members of society, they share in the dilemma we face. But scientists cannot decide for society as a whole whether the risks are acceptable—that is a question of values. Knowledge brings us so far—we must take the last step ourselves.

APPENDIX A

Systems of Measurement

Twelve inches make one foot; three feet, a yard. Five and a half yards are one rod; 40 rods equal a furlong; eight furlongs give us a mile; and three miles equal one league. So when the submarine in Jules Verne's famous novel dove to 20,000 leagues beneath the sea, it was 60,000 miles; 480,000 furlongs; 16,200,000 rods; 89,100,000 yards; 267,300,000 feet; or 3,207,600,000 inches below the ocean surface. Unless, of course, the submarine's crew was measuring its depth in marine leagues, which equal three nautical miles. The nautical mile is defined as 10 cable lengths. Unfortunately, the cable length has three values, depending on whether you are using the U.S. Navy, British Navy, or ordinary definitions.

    Situations like this may seem somewhat absurd, and they are! However, they are also very real. Units of measurement vary from country to country and from one application to another. With many different systems of measurements communication is, at best, difficult. Thus, science (and most commerce in the world) relies on the metric system.

    In the metric system units have been carefully defined and agreed upon by international convention. For example, lengths are measured in meters. A meter is defined in terms of the wavelength of one color of light emitted by a cesium atom. Thus units of measure can be agreed upon by people at any location.

    More importantly, the units are related to one another by factors of ten: 1 m = 100 centimeters = 1000 millimeters = 0.001 kilometers. The conversion to different units of length is just a matter of moving the decimal. The same is true for other units in the metric system. We simply move the decimal point to change from one unit to another. As shown in the table, the prefix tells us the relation between the basic unit and other derived units.

Prefix	Relation to Basic Unit
Micro	0.000001
Milli	0.001
Centi	0.01
Kilo	1,000
Mega	1,000,000

Thus, a millimeter is one-thousandth of a meter; a kilogram is 1000 grams. The basic unit for the most common physical quantities are listed below.

Measurement	Metric Unit
Length	meter (m)
Mass	gram (g)
Force	newton (N)
Energy	joule (J)
Power	watt (W)
Electric charge	coulomb (C)
Electric current	ampere (A)
Potential difference	volt (V)
Electrical resistance	ohm (Ω)
Temperature	degrees Celsius or Kelvin (°C or K)

The one unit of measurement that has never been converted to one based on tens is the basic unit of time. The second, minute, and hour are used universally, even though conversion among them is somewhat cumbersome.

© 1979 by Sidney Harris.

"ALL RIGHT—NOW CONVERT THE WHOLE THING TO METRIC."

You have probably studied the metric system, but since you may not use it every day, some of the units may seem a little unfamiliar. To help you become more comfortable with them, we list the comparison between metric units and the traditional English units.

1 centimeter	= 0.3937 inches
1 meter	= 1.09 yards = 3.28 feet = 39.37 inches
1 kilometer	= 3,280 feet = 0.621 miles
1 kilogram	= 0.0685 slugs (See note below)
1 newton	= 0.22 pounds
1 joule	= 0.24 calories = 0.00024 (food) Calories = 0.00095 BTU
1 meter/second	= 3.28 feet/second = 2.2 miles/hour
1 kilometer/hour	= 0.621 miles/hour

Note: The differences between metric and U. S. measures become particularly muddled when one attempts to describe mass—the amount of matter present. In the metric system, mass is commonly described in kilograms. The corresponding English unit, the slug, is seldom mentioned. In the United States, we usually express mass in terms of weight (a force of gravity on a particular mass), given in pounds. On earth, a mass of 1 kilogram (0.0685 slugs) has a weight of 2.2 pounds.

APPENDIX B

Powers of Ten

The speed of light is 300,000,000 meters per second. The radius of the hydrogen atom is 0.000000000053 meters. Large and small numbers like these appear frequently in the study of physics. Dealing with all of those zeros can be confusing and always requires frequent checking to avoid errors.

To alleviate these problems, a shorthand notation has been developed for powers of ten. The basic idea is to express large numbers in terms of the number of times you need to multiply them by ten and small numbers in terms of the number of times you need to divide them by ten. For example, the speed of light can be written as:

$$3 \times 10 \times 10 \times 10 \times 10 \times 10 \times 10 \times 10 \times 10 \text{ meters/second}$$

That is, multiply 10 by itself 8 times then multiply that number by 3 and you have the speed of light in meters per second. Powers, $10^2 = 10 \times 10$, $10^3 = 10 \times 10 \times 10$, and so on, allow us to rewrite this long expression as 3×10^8 meters/second.

Likewise, we can obtain the radius of the hydrogen atom by dividing 5.3 by 10 a total of 11 times. Because negative exponents indicate division, the radius can be expressed as 5.3×10^{-11} meters.

The following are examples of numbers expressed as powers of ten:

$$10^{-8} = 0.00000001$$
$$10^{-6} = 0.000001$$
$$10^{-4} = 0.0001$$

$$10^{-2} = 0.01$$
$$10^{0} = 1$$
$$10^{2} = 100$$
$$10^{4} = 10{,}000$$
$$10^{6} = 1{,}000{,}000$$
$$10^{8} = 100{,}000{,}000$$

Multiplication and division of large and small numbers are relatively easy operations when we use powers of ten. To multiply, we add powers. Thus $(4 \times 10^{12}) \times (3 \times 10^{7}) = (4 \times 3) \times (10^{12+7}) = 12 \times 10^{19}$, which can also be written as 1.2×10^{20}. Division is accomplished by subtracting the powers: $(4 \times 10^{12})/(8 \times 10^{7}) = (4/8) \times (10^{12-7}) = 0.5 \times 10^{5}$, which may be written as 5×10^{4}. As you can see, powers of ten take much of the pain out of working with very large and very small numbers.

INDEX

Absolute zero, 212, 286
Absorption, wave, 318
Accelerated reference frames, 140-143
Acceleration, 30-35
  centripetal, 139
  direction, 31-32
  gravitational, 33, 35
  lunar, 35
  units, 31
Air bags, automobile, 113
Air resistance force, 137-138
Alpha decay, 496
Alpha radiation, 493
Alternating current, 479-480
Ammeters, 443
Ampere (unit), 443
Amplitude, 306
  loudness and, 341
Angle of incidence, 315, 317
Angle of reflection, 315
Angle of refraction, 317
Antineutrino, 498
Antinodes, 322
Aristotle, 175
Art and beta decay, 499
Atomic mass unit, 492
Atomic number, 491
Atomic structure, 162-164, 409-413
Atoms
  constituents, 164
  magnetism and, 474
  quantum mechanical model, 418-424
  structure, 162-164, 409-413
  thermal energy and, 281-294
Automobile, 260, 262
Avicenna, 175, 176

Background radiation, thermal, 357
Battery, 440
Becquerel, Henri, 488, 492
Beta decay, 497-499

Beta radiation, 493
Big bang model, 157
Binding energy, 514-517
  curve, 515
Blackouts, electrical, 458
Blakelock, R. A., 499
Bohr, Niels, 163, 401, 410, 527
Bohr's model of the atom, 163, 410-413
Boiling, 222
Born, Max, 435
Breakeven
  scientific, 529
  thermodynamic, 530
Breeder reactor, 522
Bronowski, Jacob (quote), 400, 464
Brownian motion, 296
Brownout, electrical, 458
Buoyant force, 235

Calorie (unit), 182
Carbon dating, 505
Carnot, Sadi, 272
Carnot efficiency, 272
Causality and the speed of light, 77, 83-84
Celsius temperature scale, 211
Centimeter (unit), 538
Centrifugal effect, 142
Centripetal force, 139
Chain reaction, 520-521
Change and interaction, 87
Change of state, 221-228
Charge, electrical, 160
Chemical potential energy, 197
Circuits
  closed, 442
  electrical, 440-457
  open, 442
  parallel, 454-457
  series, 450-454
Clock synchronization, 64-66
Closed system, 95

Coherent light, 414-415
Collisions, 92, 97, 100-102
Conduction, thermal, 238-246, 289
Conductors
  electrical, 442
  thermal, 239
Conservation
  of energy, 190-194
  logic, 94-96, 498
  of momentum, 96-102
  of nucleons, 494
  principles, 95
Convection, 234-238, 288
Coordinate system, 6-9
  rectangular, 9
Coulomb (unit), 160
Coulomb, Charles, 161
Coulomb's law, 161
Critical mass, 521
Curie, Irene, 531
Curie, Marie, 492, 501, 531
Curie, Pierre, 492
Current, electrical, 441-443, 456
  alternating, 479-480
  direction of, 471

Davy, Humphry, 232
De Broglie, Louis, 396
De Broglie waves, 396-397, 420
Diffraction, 376-381
  electron, 397-398
  light, 377-381
  resolution and, 379-380
  water, 376-377
Direction, reference, 4
Displacement, 12
Distance, 11
Doppler effect, 312
  light, 355-357
  sound, 346-348
Doppler shift, 313, 355-357
Doubling time, 298
Duality, particle-wave, 400-401

Echo, 342
Efficiency, 270-272, 275
   Carnot, 271
   energy, 271
   second law, 275
Einstein, Albert, 59, 60, 62, 392, 435, 527
Elastic potential energy, 197
Electrical
   charge, 160
   circuit, 440-457
   conductor, 442
   energy, 448
   forces, 159-164
   insulator, 442
   interaction, 159-164
   potential difference, 441
   potential energy, 159-164
   power, 448
   resistance, 444-447
Electric field, 482
Electricity and magnetism, 470-483
Electromagnet, 472
Electromagnetic
   induction, 476-479
   spectrum, 338
   waves, 348-357
Electromagnetism, 482
Electron, 163
Electron diffraction, 397-398
Electron microscope, 399-401
Elsewhere, 83-84
Emission spectra, 407
Energy, 180-277
   binding, 514-517
   chemical potential, 197
   conservation of, 190-194
   elastic potential, 197
   electrical, 448
   electrical potential, 197
   forms of, 195
   gravitational potential, 185-189
   kinetic, 189-191
   mass equivalence, 198, 495
   nonrenewable sources, 299
   nuclear potential, 197
   of position, 184-189
   receiver, 181
   relativistic, 198
   renewable sources, 299
   source, 181
   temperature and kinetic, 285
   thermal, 196-197, 209
   useful, 259
   wave, 198, 303-305
   work, 181-184
Energy-level diagram, 411-412
Engine
   heat, 257-259
   internal combustion, 262
Entropy, 264-270, 290-293
   and chance, 291
   and the future of the universe, 276
   and probability, 290
Equilibrium, 136
Escher, M. C., 269, 401
Evaporation, 226, 228
Exchange particle, 168
Excited state, 412
Exhaust, thermal, 259
Exponential growth, 298

Fahrenheit temperature scale, 212
Faraday, Michael, 469, 477
Faraday's law, 477
Fermi, Enrico, 500, 527
Fictitious force, 141-142
Field
   electric, 482
   magnetic, 466-467, 470-473
Fireplaces, 237-238
First law of thermodynamics, 257-263
Fission, nuclear, 518-525
Fission reactor, 521-525
   breeder, 522
Focal point, 342
Force, 107-152
   adding, 120-124
   air resistance, 137-138
   buoyant, 235
   centrifugal, 142
   centripetal, 139
   definition, 114
   electrical, 159-164
   equilibrium, 136
   exchange, 166-168
   fictitious, 141-142
   frictional, 118, 165
   gravitational, 116-117, 137
   identifying, 119
   magnetic, 467
   momentum change and, 110-117
   net, 122
   nuclear, 164
   reaction, 143-146
   restoring, 282
   strong nuclear, 164, 166
   weak nuclear, 166
Frame of reference. *See* reference frame
Franklin, Benjamin, 161, 247, 439
Frequency, 306
   fundamental, 324
   natural, 325
   pitch and, 341
Friction, 118, 165
   electrical charge and, 165
   kinetic, 118
   static, 118
Fundamental force, 153
   and exchange particles, 168
Fundamental frequency, 324
Fundamental interactions, 153-170
Fuse, 458
Fusion, nuclear, 525-531
Fusion reactor, 527-531
   break-even, 529-530
   laser-induced, 528
   magnetic confinement, 528

Galileo, 134, 176
Gamma decay, 500-503
Gamma radiation, 493
Gamow, George (quote), 58
Gases, 283
Generator, electrical, 479
Gravitational interaction, 154-158
Gravitational potential energy, 185-189
Ground, electrical, 441
Ground state, 412

Hahn, Otto, 527
Half-life, 503-507
Harmonic series, 324
Heat, 209-210
   capacity, specific, 213-221, 286
   engine, 257-259
   latent, fusion, 222-226, 284
   latent, vaporization, 222-226, 284
   pump, 274
Heisenberg, Werner, 426
Heisenberg uncertainty principle, 426-428
Hertz (unit), 307

Hertz, Heinrich, 336
Hologram, 374
Holography, 373-377
Hypotenuse, 14

Illusions of motion, 46-47
Inertia, 133-134, 176
Insulator
  electrical, 442
  thermal, 239
Interaction, 86-102, 118, 153-170
  at a distance, 118, 166-169
  change and, 87
  electrical, 159-164
  electromagnetic, 158-164
  frictional, 118, 165
  fundamental, 153-170
  gravitational, 154-158
  magnetic, 465-466
  momentum and, 86-102
  nuclear, 164-166
  strong nuclear, 164-165
  unified theory, 169-170
  weak nuclear, 165
Interference, 318-320, 364-377
  acoustics and, 376
  bands, 367
  colors, 371
  constructive, 320
  destructive, 320
  light, 364-375
  particles, 397
  sound, 376
  two-slit, 364
  water waves, 366, 368
  waves, 318-320
Internal combustion engine, 262
Isotope, 491

Joule (unit), 182
Joule, James, 232

Kelvin temperature scale, 212, 286
Kilogram (unit), 89-90
Kilometer (unit), 540
Kinetic energy, 189-191

Laser, 414-418
  helium-neon, 416-418
Latent heat
  of fusion, 222-226, 284
  of vaporization, 222-226, 284
Laws of motion, history, 175-177

Laws of thermodynamics
  first, 257-270
  second, 263-270, 290-293
Length contraction, 72-76
Lepton, 169
Light
  coherent, 414-415
  dispersion, 352
  interference, 364-377
  particle model of, 363, 381
  reflection, 350
  refraction, 351
  speed of, 59-61, 77, 83-84, 349
  wave model of, 363, 381
Lightning, 439
Liquids, 283
Loudness, 341

Magnetic domains, 468-469
Magnetic field, 466-467, 470-473
  earth's, 467
Magnetic poles, 465
Magnetism, 465-484
  atomic basis, 474
  and electricity, 470-476
Magnets, 468
Mass, 89-90
  atomic units, 492
  critical, 521
  defect, 514
  interaction and, 125
  number, 491
  relativistic, 125
  rest, 125
Mass-energy equivalence, 198-199, 495
Matter, states of, 207-208, 283
Matter wave, 398
Maxwell, James Clerk, 297, 464, 482
Maxwell's demon, 297
Maxwell's theory of electromagnetism, 482
Medium, 304
Meitner, Lisa, 527
Melting, 222
Mesons, 168
Meter (unit), 538
Metric system, 538
Michelson, Albert, 60
Michelson-Morley experiment, 60-61
Microscope, electron, 398-400
Microwave ovens, 354

Mirage, 353
Mirrors, 351
Molecules and thermal energy, 281-294
Momentum, 90-102
  conservation of, 96-102
  definition, 91
  and interaction, 90-92
Morley, Edward, 60
Motion, perpetual, 261, 271
Motor, electrical, 475
Muon, 66
Music, 344-346
Musical instruments, 345

Natural frequency, 323
Net force, 122
Neutrino, 498, 500
Neutron, 164
Newton (unit), 108
Newton, Isaac, 154, 176
Newton's law of universal gravitation, 155-158
Newton's laws of motion, 122, 133-148
  first, 133-136
  second, 122, 136-140
  third, 143-146
Nodes, 322
Nonrenewable energy sources, 299
Nonspontaneous processes, 264
Nuclear
  bomb, 521, 527
  decay, 496-508
  fission, 518-525
  fusion, 517, 520-522
  reaction, 517, 520-522
  reactor, 521-525, 527-531
  strong interaction, 164
  transformations, 494-495
  weak interaction, 166
Nucleon, 164, 490
Nucleus, 163, 489-492
  composition, 490-492
  isotopes, 491
  structure, 489

Object, reference, 4
Oersted, Hans Christian, 464, 470
Ohm (unit), 445
Ohm's law, 445
Origin, 7

Parallel circuit, 454-457
Particle-wave duality, 400-401
Perpetual motion, 261, 271
Philoponus, 175
Photoelectric
 cell, 393
 effect, 389-394
Photography, analyzing motion
   with, 35
Photon, 168, 391-393
Piaget, Jean, 94
Picasso, Pablo, 132
Pi-meson, 168
Pion, 168
Pitch, 341-342
Planck, Max, 392
Planck's constant, 392
Plasma, 528
Potential difference, 441
Potential energy, 184-188, 197-198
 chemical, 197
 elastic, 197
 electrical, 197
 gravitational, 184-188
 nuclear, 197
Power, electrical, 448
Powers of ten notation, 541-542
Pressure, 109-110
Priestley, Joseph, 161
Probability and entropy, 291-293
Probability cloud, 423
Proton, 164
Pythagorean theorem, 14

Quantization of energy, 393
Quantum, 393
Quantum mechanical model of the
   atom, 419-424
Quantum mechanics, 421-428
Quarks, 169

Radar, police, 356
Radiation, nuclear, 493
 biological effects, 531
Radiation, thermal background, 357
Radiation/absorption, thermal,
   246-249, 289
Radioactive
 dating, 505
 decay series, 506
 isotopes, 503

Radioactivity, 492
Reaction, nuclear
 binding energy and, 517
 breeder, 522
 chain, 520-521
Rectangular coordinate system, 9
Red shift, 355-357
Reference directions, 4
Reference frame, 5-6, 43-47
 accelerated, 140-143
 moving, 43-47
Reference objects, 3
Reflection, 314
Refraction, 316
Refrigerator, 224-225, 262-263
Relative
 motion, 43-52
 position, 3-6
 speed, 47-51, 59-60
 velocity, 47-51
Relativity
 principle of, 51-52
 special theory of, 58-84
Resistance
 electrical, 444-447, 452, 457
 thermal. See R-value
Resistivity, thermal, 239, 289
Resolution, 380
Resonance, 323-324
Right-hand rule, 471, 474
Rutherford, Ernest, 162, 410
Rutherford's atom, 162-164, 410
R-value, 241-242

Satellites, 154-155
Scalar, 13, 17
Scale, spring, 108
Schrödinger, Erwin, 435
Seat belts, 134-135
Second law efficiency, 275
Second law of thermodynamics,
   263-270, 290-293
Series circuit, 450-454
Simultaneity, 64-66
Societal values and science, 300
Solar collector, 249-250
Solar heating, 216
Solids, 283
Sound, 340-348
Space-time, 83-84
Special theory of relativity, 58-84
 postulates, 61-62

Specific heat capacity, 213-221, 286
 and molecular motion, 286
Spectrum
 absorption, 409
 atomic, 406-410
 continuous, 407
 discrete, 407
 electromagnetic, 338
 emission, 407
 line, 355
 mechanical wave, 335
 wave, 334
Speed, 24-25, 27-30, 59-60
 average, 27
 instantaneous, 27-30
 relative, high speed, 59-60, 62-64
 relative, low speed, 47-51
 wave, 308
Speed of light, 59-61, 77, 83-84,
   349
 and causality, 77, 83-84
 fastest speed, 77
Spontaneous process, 264
Spring scale, 108
Standing waves, 321-325, 344-345,
   354
 circular, 420
States, change of, 221-228, 283-284
States of matter, 207-208
Stimulated emission, 415
Strassman, Fritz, 527
Strong nuclear interaction, 164, 166
Superposition, 320
Synchronization, 64-66
System, 96
 closed, 96
 coordinate, 6-9
 rectangular coordinate, 9
Systems of measurement, 538-540
Szilard, Leo, 527

Tacoma Narrows Bridge, 323
Tail-to-tip method, 15
Temperature, 209-212
 kinetic energy and, 285
 measurement, 210-212
Thermal conduction, 238-246, 289
 equation, 243-244
Thermal energy, 196-197, 198,
   206-294
 atoms and, 281-299
 change of state and, 221-228

change of temperature and, 213–221
heat and, 210
molecules and, 281–291
transfer of, 216–219, 226, 227, 233–251
Thermal exhaust, 259
Thermodynamics, 256–276
first law, 257–263
second law, 263–270, 290–293
Thermometer, 210
Thought experiment, 62
Tides, 173
Time dilation, 66–71
Transformer, 480–481
Transverse waves, 305
Twin paradox, 72

Uncertainty principle, 426–428
Unified theory of interactions, 169–170
Universal law of gravitation, 155–156

Universe, expanding, 355–356

Vacuum, motion in, 175–177
Vector, 13–17
Velocity, 25–26
average, 27
instantaneous, 27–30
interactions and, 88
relative, high speed, 62–64
relative, low speed, 47–51
Vision, quantum effects, 394–395
Volt (unit), 441
Voltage, 440–441

Watt (unit), 448
Wave
packet, 422
receiver, 337–340
spectrum, 334
Wavelength, 306
Wave-particle duality, 400–401
Waves, 302–382

in the earth, 309–311
electromagnetic, 336–337, 348–357
longitudinal, 305
matter, 398
mechanical, 334–335
particle, 398
probability, 421
sound, 340–348
speed of, 308
standing, 321–322, 344–346, 354
transverse, 305
Weak nuclear interaction, 166
Weight, 118
Work, 181–184
W-particle, 168

X-ray diffraction, 398

Young, Thomas, 364
Young's double slit experiment, 364

PHOTO AND ILLUSTRATION CREDITS

Figures 1-4 and 6-C4: National Aeronautics and Space Administration (NASA).

Figures 1-6 (runner, tickets) and 11-9: Tom Ballard, EKM Nepenthe.

Figures 1-6 (thermometer, compass), 9-3a, b, c, 10-9, 15-5a, 17-5, and 19-6a, b: Photographs by Wayland Lee.

Box 1-1, p. 16: Reprinted from *Weird and Wacky Inventions* by Jim Murphy. Copyright © 1978 by Jim Murphy. Used by permission of Crown Publishers, Inc.

Figure 2-1a: © 1984 Emil Schultess from BLACK STAR.

Figure 2-1b: From the Gordon Parks Collection, Department of Art, Kansas State University.

Figures 2-1c and 2-9a, b: Dr. Harold Edgerton, MIT, Cambridge, MA.

Figure 2-1d: The Louise and Walter Arencsberg Collection, Philadelphia Museum of Art.

Figure 2-1e: Courtesy R. J. Ingebretson.

Figures 2-3, 7-C6, 9-C4, 14-25, 14-14, 15-21, 16-3, 16-4, 16-6, 16-7a, b, c, 16-11, 16-13a, b, c, 16-15, 17-7a, and 20-9: *PSSC Physics*, 2nd edition, 1965; D. C. Heath and Company with Education Development Center, Inc., Newton, MA.

Box 2-1 (p. 34): Wide World Photos.

Figure 2-B4: NFB Phototheque ONF ©.

Box 2-2 (p. 34), Box 10-1 (p. 215), and Box 19-1 (p. 443): American Heritage Division, McGraw-Hill.

Figure 3-C1: Whitin Observatory: Wellesley College Photograph.

Figures 4-1 and 19-2: Bettman Archive.

Figure 6-1a: © 1982 Mike Valeri, Freelance Photographers.

Figure 6-11: After an idea by Ted Geissert.

Figure 7-2: Auto Safety Insurance Institute.

Figure 9-4: © Jane Scherr, Jeroboam Inc.

Figure 10-1: Courtesy Ralph W. G. Wyckoff, and *PSSC Physics*, 2nd edition, D. C. Heath & Company with Education Development Center, Inc., Newton, MA.

Figure 10-6a, b: Tennessee Valley Authority Information Office.

Figure 10-11: Courtesy David Rayle, San Diego State University.

Figure 11-7: Courtesy of The Golden Flame, Redwood City, CA.

Figure 14-13: Battelle-Northwest Laboratories.

Figure 14-17: Fundamental Photographs.

Figure 14-23: Historical Photography Collection, University of Washington Libraries.

Figure 15-5b: The Arecibo Observatory is part of the National Astronomy and Ionosphere Center, which is operated by Cornell University under contract with the National Science Foundation.

Figure 15-3: The Time/Life Series in Sound and Hearing.

Figure 15-9b: © Zoological Society of San Diego.

Figure 15-9c: Courtesy Los Angeles Philharmonic Association.

Figure 15-9d: Courtesy Edmund Scientific Company.

Figure 15-11: Courtesy H. Kamimoto String Instruments, San Jose, California.

Figures 16-1 and 16-12: Cagnet et al., *Atlas of Optical Phenomena,* Springer-Verlag, 1971.

Figures 16-2 and 16-5: Holton and Roller: *Foundation of Modern Physical Sciences,* Addison-Wesley, Reading, Massachusetts, 1958.

Figure 16-16a: © Philip M. Rinard, Los Alamos, NM.

Figures 16-16b and 16-18: Sears, Zemansky, and Young, *University Physics,* Sixth Edition, © 1982, Addison-Wesley, Reading, Massachusetts. Pp. 788 and 799. Reprinted with permission.

Figure 16-C6: Courtesy Bruno Rossi, MIT.

Figure 17-8a: J. Vasek, *Introduction to Theoretical and Experimental Optics*; and *PSSC Physics*, 2nd edition, 1965; D. C. Heath and Company with Education Development Center, Inc., Newton, MA.

Figure 17-8b: From *Handbuch der Physik,* Volume 32 (Springer-Verlag, 1957). Courtesy H. Raether.

Figures 17-9a, b, c: Burton and Kohl: *Electronic Diffraction.* Reinhold Publishing Company.

Figure 17-C6: Courtesy Central Scientific Company.

Figure 18-13a: Courtesy Spectra-Physics.

Figure 18-13b: Courtesy Pioneer: Laserdisc Brand Videodisk Player LD-700.

Figure 18-13c: Hank Morgan, Rainbow.

Figure 18-19: © 1974, R. Eisberg and R. Resnick. *Quantum Physics of Atoms, Molecules, Solids, Nuclei, and Particles.* John Wiley and Sons.

Box 21-1a, b (p. 499): Maurice J. Cotter, Professor of Physics, Queens College of CUNY.

Box 21-2 (p. 508 photo): U.K. Atomic Energy Authority. Courtesy AIP Niels Bohr Library.

Box 21-2 (p. 508 text): Used by permission of The Associated Press.

Figure 22-11a, b: Courtesy of the Princeton University Plasma Physics Laboratory.

2795755
32
12185